Emerging Applications of Low Dimensional Magnets

Low-dimensional magnetic materials find their wide applications in many areas, including spintronics, memory devices, catalysis, biomedical, sensors, electromagnetic shielding, aerospace, and energy. This book provides a comprehensive discussion on magnetic nanomaterials for emerging applications.

Fundamentals along with applications of low-dimensional magnetic materials in spintronics, catalysis, memory, biomedicals, toxic waste removal, aerospace, telecommunications, batteries, supercapacitors, flexible electronics, and many more are covered in detail to provide a full spectrum of their advanced applications.

This book offers fresh aspects of nanomagnetic materials and innovative directions to scientists, researchers, and students. It will be of particular interest to materials scientists, engineers, physicists, chemists, and researchers in electronic and spintronic industries, and is suitable as a textbook for undergraduate and graduate studies.

Series in Materials Science and Engineering

The series publishes cutting edge monographs and foundational textbooks for interdisciplinary materials science and engineering. It is aimed at undergraduate and graduate level students, as well as practicing scientists and engineers. Its purpose is to address the connections between properties, structure, synthesis, processing, characterization, and performance of materials.

Automotive Engineering: Lightweight, Functional, and Novel Materials
Brian Cantor, P. Grant, C. Johnston

Multiferroic Materials: Properties, Techniques, and Applications
Junling Wang, Ed.

2D Materials for nanoelectronics
Michel Houssa, Athanasios Dimoulas, Alessandro Molle

Skyrmions: Topological Structures, Properties, and Applications
J. Ping Liu, Zhidong Zhang, Guoping Zhao, Eds.

Computational Modeling of Inorganic Nanomaterials
Stefan T. Bromley, Martijn A. Zwijnenburg, Eds.

Physical Methods for Materials Characterisation, Third Edition
Peter E. J. Flewitt, Robert K. Wild

Conductive Polymers: Electrical Interactions in Cell Biology and Medicine
Ze Zhang, Mahmoud Rouabhia, Simon E. Moulton, Eds.

Silicon Nanomaterials Sourcebook, Two-Volume Set
Klaus D. Sattler, Ed.

Advanced Thermoelectrics: Materials, Contacts, Devices, and Systems
Zhifeng Ren, Yucheng Lan, Qinyong Zhang

Fundamentals of Ceramics, Second Edition
Michel Barsoum

Flame Retardant Polymeric Materials, A Handbook
Xin Wang and Yuan Hu

2D Materials for Infrared and Terahertz Detectors
Antoni Rogalski

Fundamentals of Fibre Reinforced Composite Materials
A. R Bunsell. S. Joannes, A. Thionnet

Series Preface

The series publishes cutting edge monographs and foundational textbooks for interdisciplinary materials science and engineering.

Its purpose is to address the connections between properties, structure, synthesis, processing, characterization, and performance of materials. The subject matter of individual volumes spans fundamental theory, computational modeling, and experimental methods used for design, modeling, and practical applications. The series encompasses thin films, surfaces, and interfaces, and the full spectrum of material types, including biomaterials, energy materials, metals, semiconductors, optoelectronic materials, ceramics, magnetic materials, superconductors, nanomaterials, composites, and polymers.

It is aimed at undergraduate and graduate level students, as well as practicing scientists and engineers.

Proposals for new volumes in the series may be directed to Carolina Antunes, Commissioning Editor at CRC Press, Taylor & Francis Group (Carolina.Antunes@tandf.co.uk).

Emerging Applications of Low Dimensional Magnets

Edited by
Ram K. Gupta, Sanjay Mishra,
and Tuan Anh Nguyen

CRC Press is an imprint of the
Taylor & Francis Group, an **informa** business

First edition published 2023
by CRC Press
6000 Broken Sound Parkway NW, Suite 300, Boca Raton, FL 33487–2742

and by CRC Press
4 Park Square, Milton Park, Abingdon, Oxon, OX14 4RN

CRC Press is an imprint of Taylor & Francis Group, LLC

© 2023 selection and editorial matter, Ram K. Gupta, Sanjay Mishra, Tuan Anh Nguyen; individual chapters, the contributors

Reasonable efforts have been made to publish reliable data and information, but the author and publisher cannot assume responsibility for the validity of all materials or the consequences of their use. The authors and publishers have attempted to trace the copyright holders of all material reproduced in this publication and apologize to copyright holders if permission to publish in this form has not been obtained. If any copyright material has not been acknowledged please write and let us know so we may rectify in any future reprint.

Except as permitted under U.S. Copyright Law, no part of this book may be reprinted, reproduced, transmitted, or utilized in any form by any electronic, mechanical, or other means, now known or hereafter invented, including photocopying, microfilming, and recording, or in any information storage or retrieval system, without written permission from the publishers.

For permission to photocopy or use material electronically from this work, access www.copyright.com or contact the Copyright Clearance Center, Inc. (CCC), 222 Rosewood Drive, Danvers, MA 01923, 978–750–8400. For works that are not available on CCC please contact mpkbookspermissions@tandf.co.uk

Trademark notice: Product or corporate names may be trademarks or registered trademarks and are used only for identification and explanation without intent to infringe.

ISBN: 978-1-032-04874-1 (hbk)
ISBN: 978-1-032-05307-3 (pbk)
ISBN: 978-1-003-19695-2 (ebk)

DOI: 10.1201/9781003196952

Typeset in Times
by Apex CoVantage, LLC

Contents

Editor Biographies   ix
List of Contributors   xi

1. **Magnetic Nanomaterials in Catalysis**   1
 Bhagavathula S. Diwakar, B. Govindh, D. Chandra Sekhar, Venu Reddy,
 I.V. Kasi Viswanath, Ramam Koduri, V. Swaminatham

2. **Recent Advances in the Catalytic Applications of Magnetic Nanomaterials**   9
 B. Sehgal, G.B. Kunde

3. **Fabrication and Characterization of Two-Dimensional Transition Metal Dichalcogenides for Applications in Nano Devices and Spintronics**   33
 Geeta Sharma, Andrew J. Scott, Animesh Jha

4. **Spin Transistors: Different Geometries and Their Applications**   49
 Gul Faroz Ahmad Malik, Mubashir Ahmad Kharadi, Farooq Ahmad Khanday,
 Zaid Mohammad Shah, Sparsh Mittal

5. **Spin-Transfer Torque for Universal Memory Applications**   63
 Sameena Shah, Gul Faroz Ahmad Malik, Mubashir Ahmad Kharadi,
 Farooq Ahmad Khanday

6. **Nanowire Magnets**   77
 M. Boughrara, N. Zaim, H. Ahmoum, A. Zaim, M. Kerouad

7. **Spin Torque Devices**   93
 M. Shakil, Halima Sadia, M. Isa Khan, M. Zafar

8. **Nanosensors Based on Magnetic Materials**   115
 Kumar Navin, Rajnish Kurchania

9. **Role of Magnetic Nanomaterials in Biomedicine**   137
 Bhagavathula S. Diwakar, D. Chandra Sekhar, Venu Reddy, P. Bhavani,
 Ramam Koduri, S. Srinivasarao

10. **Recent Advances in Carbon-Based Nanomaterials for Spintronics**   147
 Trupti K. Gajaria, Narayan N. Som, Shweta D. Dabhi

11. **Rare Earth Manganites and Related Multiferroicity**   163
 Suresh Chandra Baral, P. Maneesha, Ananya T. J, Srishti Sen, Sagnika Sen,
 Somaditya Sen, E. G. Rini

12 Magnetic Nanofillers-PVDF Nanocomposite Laminated Structures for Broad-Band Electromagnetic Shielding Applications — 181
Soumyaditya Sutradhar

13 Iron-Based Materials to Remove Toxic Waste from the Environment — 197
Srimathi Krishnaswamy, Puspamitra Panigrahi, Ganapathi Subramaniam Nagarajan

14 Nanoferrite-Based Structural Materials for Aerospace Vehicle Radomes — 215
Manish Naagar, Sonia Chalia, Preeti Thakur, Atul Thakur

15 Miniaturization Techniques for Microstrip Patch Antenna for Telecommunication Applications — 245
Preeti Thakur, Shilpa Taneja, Atul Thakur

16 Applications of Magnetic Materials in Batteries — 263
Shiva Bhardwaj, Felipe M. de Souza, Ram K. Gupta

17 Recent Advancement in Magnetic Materials for Supercapacitor Applications — 283
Magdalene Asare, Felipe M. de Souza, Ram K. Gupta

18 Magnetic Nanomaterials for Flexible Spintronics — 303
Felipe M. de Souza, Ram K. Gupta

Index — 319

Editor Biographies

Dr. Ram K. Gupta is an associate professor at Pittsburg State University. Dr. Gupta's research focuses on nanomagnetism, nanomaterials, green energy production and storage using conducting polymers and composites, electrocatalysts for fuel cells, optoelectronics and photovoltaics devices, organic-inorganic hetero-junctions for sensors, bio-based polymers, bio-compatible nanofibers for tissue regeneration, scaffold and antibacterial applications, bio-degradable metallic implants. Dr. Gupta published over 250 peer-reviewed articles, made over 300 national/international/regional presentations, chaired many sessions at national/international meetings, edited/wrote several books/chapters for leading publishers. He has received over two and half million dollars for research and educational activities from external agencies. He is serving as associate editor, guest editor, and editorial board member for various journals.

Dr. Sanjay Mishra joined the Department of Physics at the University of Memphis in 1999. He has been consistently productive in research, instruction, and service to the University of Memphis (UoM). Dr. Mishra initiated an active multidisciplinary Materials Research program at the UoM since his inception at UoM. Before receiving postdoctoral experience from the Lawrence Berkeley National Laboratory (LBNL), the University of California-Berkeley at the Advanced Light Source Synchrotron Facility (ALS), he received his PhD in Physics from the Missouri University of Science and Technology-Rolla, MO, USA; MS from Pittsburg State University, Pittsburg, KS, USA; MSc from the South Gujarat University, Surat, India; and a postgraduate diploma in Space Sciences from Gujarat University, Ahmedabad, India. Dr. Mishra's research focuses on magnetic nanomaterials and nanocomposites of oxides and metals for energy applications, including energy-dense magnets, magnetocaloric materials, photocatalysts, and supercapacitors. Dr. Mishra has published over 300 peer-reviewed journal articles and has given numerous presentations at national and international conferences. He has secured over two and a half million federal grants over the years and received numerous awards from the UoM for his outstanding research accomplishments.

Tuan Anh Nguyen has completed his BSc in Physics from Hanoi University in 1992, and his Ph.D. in Chemistry from Paris Diderot University (France) in 2003. He was a visiting scientist at Seoul National University (South Korea, 2004) and the University of Wollongong (Australia, 2005). He then worked as a postdoctoral research associate and research scientist at Montana State University (USA), 2006–2009. In 2012, he was appointed as Head of Microanalysis Department at the Institute for Tropical Technology (Vietnam Academy of Science and Technology). He has managed four Ph.D. theses as thesis director and three are in progress. He is Editor-in-Chief of "Kenkyu Journal of Nanotechnology & Nanoscience" and Founding Co-Editor-in-Chief of "Current Nanotoxicity & Prevention". He is the author of four Vietnamese books and Editor of 32 Elsevier books in the Micro & Nano Technologies Series.

Contributors

A. Zaim
Physics of Materials and Systems Modeling Laboratory (PMSML), Faculty of Sciences, Moulay Ismail University
Zitoune, Meknes, Morocco

Ananya T. J
Delhi Public School Indore
Indore, India

Andrew J. Scott
Faculty of Engineering and Physical Sciences, School of Chemical and Process Engineering
University of Leeds
United Kingdom

Animesh Jha
Faculty of Engineering and Physical Sciences, School of Chemical and Process Engineering
University of Leeds
United Kingdom

Atul Thakur
Amity Institute of Nanotechnology, Amity University Haryana
Gurugram, Haryana, India

Atul Thakur
Centre for Nanotechnology
Amity University Haryana
Gurugram, India

B. Sehgal
Department of Applied Chemistry, Faculty of Technology and Engineering
The Maharaja Sayajirao University of Baroda
Vadodara, India

Bhagavathula S. Diwakar
Department of Engineering Chemistry, SRKR Engineering College
Bhimavaram, India

D. Chandra Sekhar
Department of Engineering Chemistry
SRKR Engineering College
Bhimavaram

E. G. Rini
Department of Physics
Indian Institute of Technology
Indore, India

Farooq Ahmad Khanday
Department of Electronics and Instrumentation Technology
University of Kashmir, Srinagar
Jammu and Kashmir, India

Felipe M. de Souza
Kansas Polymer Research Center
Pittsburg State University
Pittsburg, Kansas, USA

G. B. Kunde
Department of Chemistry
Indian Institute of Technology, Delhi

Ganapathi Subramaniam Nagarajan
Quantum Functional Semiconductor Research Centre (QSRC)
Nano Information Technology Academy (NITA), Dongguk University
Chung-gu, Seoul, Korea

Geeta Sharma
Faculty of Engineering and Physical Sciences, School of Chemical and Process Engineering
University of Leeds United Kingdom

Gul Faroz Ahmad Malik
Department of Electronics and Instrumentation Technology
University of Kashmir, Srinagar
Jammu and Kashmir, India

Contributors

H. Ahmoum
Physics of Materials and Systems Modeling Laboratory (PMSML), Faculty of Sciences
Moulay Ismail University
Zitoune Meknes, Morocco

Halima Sadia
Department of Physics
University of Gujrat
Gujrat, Pakistan

I. V. Kasi Viswanath
Department of H&S, NRI Institute of Technology
Vijayawada, India

Kumar Navin
Functional Nanomaterials Laboratory, Department of Physics, Maulana Azad National Institute of Technology
Bhopal, MP, India

M. Boughrara
Physics of Materials and Systems Modeling Laboratory (PMSML)
Faculty of Sciences, Moulay Ismail University
Zitoune, Meknes, Morocco

M. Isa Khan
Department of Physics, University of Gujrat
Gujrat, Pakistan

M. Kerouad
Physics of Materials and Systems Modeling Laboratory (PMSML)
Faculty of Sciences, Moulay Ismail University
Zitoune, Meknes, Morocco

M. Shakil
Department of Physics, University of Gujrat,
Gujrat, Pakistan

M. Zafar
Department of Physics, Govt. Rizvia Islamia post graduate college
Haroon Abad, Punjab, Pakistan

Magdalene Asare
Department of Chemistry
Pittsburg State University
Pittsburg, Kansas

Manish Kumar Bharti
Department of Aerospace Engineering
Amity University Haryana
Gurugram, Haryana, India

Mubashir Ahmad Kharadi
Department of Electronics and Communications Engineering
IIT Roorkee, Roorkee
Uttarakhand, India

Mubashir Ahmad Kharadi
Department of Electronics and Instrumentation Technology
University of Kashmir
Srinagar, Jammu and Kashmir, India

N. Zaim
Physics of Materials and Systems Modeling Laboratory (PMSML), Faculty of Sciences,
Moulay Ismail University
Zitoune, Meknes, Morocco

Narayan N. Som
Materials Design Division, Faculty of Materials Science and Engineering
Warsaw University of Technology
Warsaw, Poland

P. Bhavani
Department of Engineering Chemistry
SRKR Engineering College
Bhimavaram, India

P. Maneesha
Department of Physics, Indian Institute of Technology
Indore, India

Preeti Thakur
Department of Physics, Amity University Haryana
Gurugram, Haryana, India

Puspamitra Panigrahi
Centre for Clean Energy and Nano Convergence (CENCON), Hindustan Institute of Technology and Science
Padur, Kelambakkam, Chennai, India

Rajnish Kurchania
Functional Nanomaterials Laboratory, Department of Physics
Maulana Azad National Institute of Technology
Bhopal, MP, India

Ramam Koduri DIMAT
Universidad de Concepcion, Chile,
South America

Ram K. Gupta
Pittsburg State University, Pittsburg,
Kansas, USA

S. Srinivasarao
Department of Chemistry, V. R. Siddhartha Engineering College
Vijayawada, India

Sagnika Sen
Rankers International School,
Indore, 452016, India

Saikrishnan Madhavan
Department of Applied Physics, Delhi Technological University,
Delhi, India

Sameena Shah
Department of Electronics and Instrumentation Technology
University of Kashmir, Srinagar,
Jammu and Kashmir, India

Shilpa Taneja
Department of Physics, Amity University
Haryana, Gurugram, India

Shiva Bhardwaj
Department of Physics, Pittsburg State University,
Pittsburg, Kansas, USA

Shweta D. Dabhi
P. D. Patel Institute of Applied Sciences, Charotar University of Science and Technology
CHARUSAT Campus, Changa, Gujarat, India

Somaditya Sen
Department of Physics, Indian Institute of Technology Indore, India

Sonia Chalia
Department of Aerospace Engineering, Amity University Haryana, Gurugram
Haryana, India.

Soumyaditya Sutradhar
Department of Physics, Amity Institute of Applied Sciences
Amity University, Kolkata, India

Sparsh Mittal
Department of Electronics and Communications Engineering, IIT Roorkee
Roorkee, Uttarakhand, India

Srimathi Krishnaswamy
Centre for Clean Energy and Nano Convergence (CENCON), Hindustan Institute of Technology and Science
Padur, Kelambakkam, Chennai, India

Srishti Sen
School of Bioscience, Engg. and Technology, VIT Bhopal University
Bhopal, India

Suresh Chandra Baral
Department of Physics, Indian Institute of Technology
Indore, India

Trupti K. Gajaria
School of Science, GSFC University, Fertilizer Nagar, Vigyan Bhavan, Vadodara
Gujarat, India

V. Swaminatham
Department of H&S, Swarnandhra Engineering College
Narasapuram, India

Venu Reddy
Nanotechnology Research Center, SRKR Engineering College,
Bhimavaram

Zaid Mohammad Shah
Department of Electronics and Instrumentation Technology, University of Kashmir
Srinagar, Jammu and Kashmir, India

Magnetic Nanomaterials in Catalysis

Bhagavathula S. Diwakar, B. Govindh, D. Chandra Sekhar, Venu Reddy, I. V. Kasi Viswanath, Ramam Koduri, V. Swaminatham

Contents

1	Magnetic Nanoparticles as Catalysts	1
2	Classification of MNPs: Oxides & Ferrites	5
	2.1 Ferrites Using Shell	5
	2.2 Metallic	6
	2.3 Metallic Nanoparticles with a Shell	6
3	Concluding Remarks	6
References		6

1 MAGNETIC NANOPARTICLES AS CATALYSTS

Engineered nanomaterials as catalysts is an emerging area in material science. Many research groups around the globe put many efforts to develop strategies to prepare catalytically active nanomaterials. The advantages of catalytic nanoparticles include recycling and reuse, easy to separate, low quantity of material, less toxic, and easy to handle. Keeping in view, magnetic nanoparticles (MNPs) of iron core were focused to develop to address the requirements. Further, the MNPs were classified by the nature of magnetic core with reduced species such as oxides for example iron oxide, and its derivatives like cobalt ferrites, copper ferrites, etc., found numerous applications in oxidative reactions and coupling reactions in organic synthesis. The applications of MNPs in chemical synthesis as catalysts are widely studied in recent times for sustainable development in catalytic studies. Coated MNPs were also studied and explored as heterogeneous catalysts in organic transformations. Therefore, it was observed that the functionalization of MNPs sometimes acts as a bridge between the homogeneous phase and heterogeneous phase of the reaction medium to retain the qualities of both systems. This research path provides academicians and industries to innovate efficient and selective catalytic materials for diversified organic reactions in the liquid phase and multiphase conditions to minimize the cost of production and waste disposal. However, practical utilization of MNPs as catalysts in the liquid phase has some difficulties but many techniques were developed to bypass the complications.

DOI: 10.1201/9781003196952-1

2 Applications of Low Dimensional Magnets

Attractive nanoparticles are of expected use as catalysts and their supports [1–8]. In science, the catalyst is that material, typically strong with a high region, to which an impetus is joined [9–12]. The reactivity of heterogeneous catalysts happens at surface molecules. Subsequently, extraordinary exertion is framed to expand the space of a catalyst by circulating it over the help. The help could likewise be inactive or shared inside the synergist responses. Run of the mill upholds incorporate different types of alumina, carbon, and silica. Immobilizing reactant nanoparticles surface with surface-to-volume proportion resolves the issue. Inside the instance of smart nanoparticles, it adds the property of effortless detachment.

An initial model included rhodium-based catalysis appended attractive nanoparticles [13–16]. Its originated Candida rugosa lipase was arrested on attractive nanoparticles upheld ionic fluids taking distinctive cation chain lengths with C_1, C_4, and C_8 and anions like Cl^-, PF_6^-, and BF_4^-. Seductive nanoparticles upheld ionic fluids were gotten in covalent holding or ionic fluids silane on seductive silica nanoparticles. They are superparamagnetic with a size of around 55 nm. A huge amount of lipase was piled on the tool by ion adsorption. The effect of lipase immobilization was tested by an esterification reaction between monounsaturated fats and 1-butanol. The action of set lipase was 118.3% looked at thereto of the original lipase. Paralyzed lipase kept up with 60% of its beginning action in any event when the temperature was over to 800 °C. Also, paralyzed lipase held 60% of its underpinning movement after groups response, even though no action was linked after six cycles for the free chemical nanoparticles [17]. Turn to cover colloidal arrangements of monodisperse nanoparticles against level silica upholds facilitates simple and adaptable because of delivering model impetuses with clear cut impetus stacking and molecule size. By turn covering Fe_3O_4 (magnetite) nanoparticles by fluctuating molecule widths produced a model framework for Fe-based Fischer–Tropsch impetuses. [18]

Readiness of $CoFe_2O_4$ MNPs with normal sizes inside 40–50 nm had been accomplished through consolidated sonochemical and co-precipitation procedures in a fluid medium with no surfactant or natural covering specialist. The nanoparticles structure stable scatterings inside the dispersion medium. The uncovered nanocatalysts were used straightforwardly as a recyclable impetus for the Knoevenagel reaction [19]. Researchers reported an effective procedure for the manufacture of β,γ – unsaturated ketones by $CuFe_2O_4$ nanoparticles and were presented in Figure 1.1. The noteworthy benefits were cost-effective, heterogeneous reusable catalyst, smaller reaction time, without isomerization during the process throughout the reaction, and easy workup [20].

$CuFe_2O_4$ nanomaterial as a catalyst as a heterogeneous initiator of 1, 4-dihydropyridines synthesis was presented in Figure 1.2. In this reaction, aromatic aldehydes, ethyl acetoacetate, and ammonium

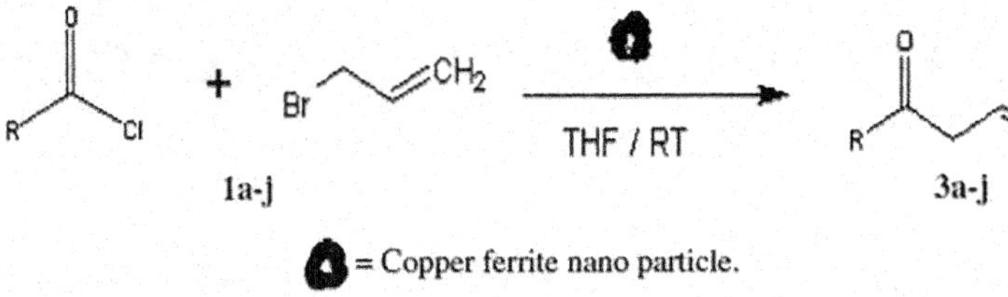

FIGURE 1.1 $CuFe_2O_4$ nanoparticles as a catalyst in c-c coupling reactions [20].
Source: © Indian Academy of Sciences.

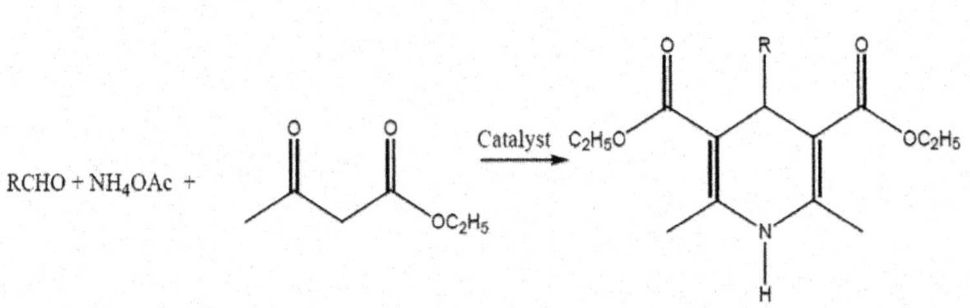

FIGURE 1.2 $CuFe_2O_4$ nanoparticles as a catalyst in the synthesis of 1, 4-dihydropyridines [21].

acetate were mixed with copper ferrite nanopowders at room temperature in solvent ethanol. The catalyst was recovered and reused [21].

Synergist movement of Fe$_3$O$_4$ nanoparticles (NPs) during comprising of sweet-smelling aldehydes, urea/thiourea, and a b-dicarbonyl under dissolvable conditions were explored. The response bears, the cost of the relating dihydropyrimidinones (thiones) in high yields. Therefore, the technique reliably gives a high return, simple attractive partition, less reaction time, and impetus reusable for contrasted and old-style Biginelli responses [22]. The catalytic action of nanosized Fe$_2$O$_3$ nanoparticles prepared by the sol-gel method was revealed in the synthesis of acridinedione was described in Figure 1.3. The results showed excellent yield and recyclability under solvent-free reactions deprived of using any additive/cocatalyst. In addition, the catalyst could be detached easily employing an external magnet and reused [23].

CoFe$_2$O$_4$ nanocatalysts with average sizes of 25 nm were used in the oxidation of alkenes along with tert–butyl hydroperoxide (t-BuOOH) in the reaction mixture. The impulses are constantly instantly disentangled by exercising an outside attraction and no conspicuous loss of exertion was seen when the motivation was reused for five nonstop runs. The results of certain boundaries, analogous to temperature, kinds of oxidants and solvents, on oxidation responses, were likewise examined. The issues presented by the use of CoFe$_2$O$_4$ with t-BuOOH as an oxidant in reaction with certain alkenes, superior issues are gotten varied with the maturity of the concentrated on similar ferrites [24].

Pd nanoparticles upheld on CoFe$_2$O$_4$ nanocatalysts were fulfilled by ultrasonicated co-precipitation inside the space of stabilizers or covering agents. The synergist prosecution of Pd consolidated CoFe$_2$O$_4$ NPs was audited in Suzuki coupling response in ethanol [25]. A simple and speedy strategy for the preparation of 5,5-disubstituted hydantoins attractive ferrite nanoparticles has been created. Multicomponent responses of carbonyl mixtures (aldehydes and ketones), cyanides, and carbonates were managed under free from solvent conditions to get different hydantoin subsidiaries. The attractive impetus may be promptly isolated by a magnet. This strategy enjoys many benefits, similar to the usage of a reusable attractive impetus, significant returns, short response times, straightforwardness, and truly ease with carrying out the technique [26].

Zinc ferrite (ZnFe$_2$O$_4$) nanoparticles and their photocatalytic color debasement capacity of Reactive Red 198 (RR198) and Reactive Red 120 (RR120) from shadowed wastewater were contemplated. Inorganic anions (nitrates and sulfates) were distinguished as color mineralization particulars. The issues further demonstrated that ZnFe$_2$O$_4$ may be employed as a seductive photocatalyst to debase colors from shadowed wastewater [27].

The organometallic section (MoI$_2$(CO)$_3$) was composed of Fe$_3$O$_4$ nanoparticles of different sizes (normal size of 11 & 30 nm) which are later covered with a silica shell and united with a pyridine attachment ligand was considered. Olefin epoxidation through organometallic nano-cross strains exercising TBHP as an oxidant was performed with brilliant issues. The synergist concentrates on showing that the impulses yield specifically the destined epoxides of a progression of olefins. Also, these impulses are planted to figure under a decent temperature range and further than many reactant cycles without ignominious prosecution mischance much of the time [28].

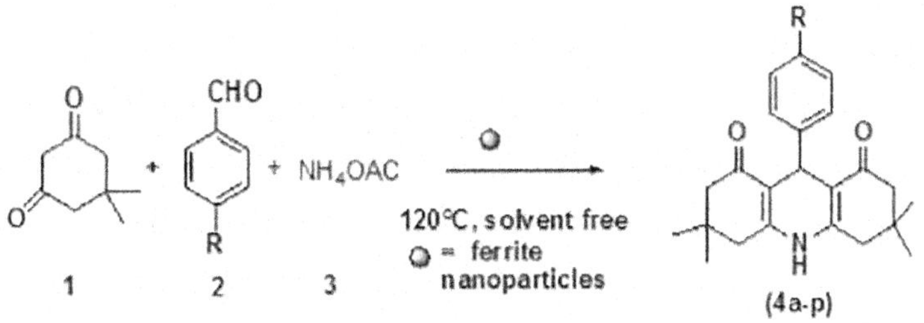

FIGURE 1.3 Fe$_2$O$_3$ nanoparticles catalyzed for acridinedione derivative [23].

Source: Copyright © 2021 The Chemical Society of Japan

Another attractively distinguishable impetus comprising of [Fe(HSO$_4$)$_3$] upheld on Si-covered NiFe$_2$O$_4$ NPs was ready. The new attractive impetus was demonstrated as a proficient heterogeneous impetus for a combination of 1,8-dioxodecahydroacridines under optimized conditions. The impetus was immediately recuperated via basic attractive decantation and reused a few times without huge loss of reactant movement [29]. The non-harmful seductive CuFe$_2$O$_4$ nanoparticles were incorporated, portrayed, and employed as an effective motivation for the admixture of the most recent inferiors for-tetrasubstituted imidazoles in brilliant yields. The intertwined fusions work out simple and sanitization of particulars is performed without chromatographic strategies. The motivation is constantly mended for the following responses and reused with no reliable mischance [30] (Figure 1.4).

A vigorous union for magnetic CuFe$_2$O$_4$ nanoparticles using an aqueous method was explored. The pre-arranged powder comprises small particles of nanometer size range with round shape and natural piece. Additionally, pre-arranged attractive CuFe$_2$O$_4$ nanoparticles are utilized as a proficient catalyzed for the union of tetrahydro pyridines and pyrrole subsidiaries in brilliant yields, simple workup, and cleaning of items by non-chromatographic strategies. The impetuses are frequently recuperated for the following responses and reused with no apparent misfortune [31]. Fe$_3$O$_4$@ polyhedral oligomeric silsesquioxanes with eight triethoxysilane arms (APTPOSS), as seductive nanocatalyst for high return combination of pyrans, has been achieved by the researchers with the help of magnetic ferrites [32].

Bacillus Subtilis was separated and estimated for protease chemical. Sanitized protease was effectively paralyzed on MNPs and employed for the admixture of a progression of glycinamides. The limiting of protease on MNPs were affirmed by FT-IR spectroscopy and Thermo Gravimetric Analysis examination. Running boundaries for the glycinamides union viz. temperature, pH, and reaction time were upgraded exercising response surface methodology (RSM) with Design Expert software. The topmost quantum of the yield of changed amides 2 butyramidoacetic sharp (AMD-1), 2-benzamidoacetic sharp (AMD-2) and ((carboxymethyl)amino-2-oxoethyl-2-hydroxysuccinyl-bis (azanediyl)) diacetic sharp (AMD-3) shaped were seen at pH-8, 30 min, and 50 °C. The integrated paralyzed protease held 70 of the underpinning movement indeed after 8 patterns of exercise [33].

A new and easy methodology for CoFe$_2$O$_4$ NPs were created by aqueous warming by tributylamine as a hydroxylating specialist and polyethylene glycol 4000 was employed as surfactant. The NPs were employed as attractively recoverable impulses inside the oxidation of alcohols to their relating aldehydes by oxyacid. This oxidative system is viewed as exceptionally effective bearing the cost of particulars in extremely high return and selectivity. The direct seductive partition of the motivation and effective reusability are crucial rudiments of this frame [34]. Vigorous union of seductive NiFe$_2$O$_4$ nanoparticles was delved through waterless strategy. The pre-arranged seductive NiFe$_2$O$_4$ nanoparticles were employed as a productive, modest, andeco-accommodating motivation for the Claisen–Schmidt buildup response between acetyl ferrocene and different aldehydes (fragrant or potentially heterocyclic) yielding acetyl ferrocene chalcones in phenomenal production, with simple workup and dropped response time movement. The motivation is incontinently mended by straightforward seductive decantation and might be reused

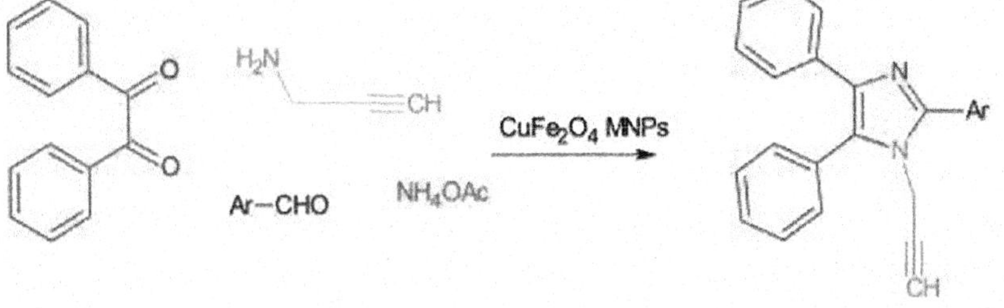

FIGURE 1.4 Magnetic CuFe$_2$O$_4$ nanoparticles in the synthesis of 1,2,4,5-tetrasubstituted imidazoles [30].
Source: Copyright © 2015 Elsevier Ltd

many times with no recognizable loss of reactant. Also, the pre-arranged chalcone inferiors were considered as their adversary of growth movement in discrepancy to three mortal cancer cell lines, specifically HCT116 (colon complaint), MCF7 (blood nasty growth), and HEPG2 (liver nasty growth), and showed great action against melanoma [35].

The admixture of mesoporous Si-covered Fe_3O_4 NPs (MMSNs) through sol-gel templated strategy exercising MNPs with tartrate as nucleation seeds were contemplated. The finagled boundaries are painstakingly controlled to the extent of the manufactured strategy to get biocompatible MMSNs per combination. MMSNs are employed to help to sort out titanium impulses for ring-opening of caprolactone [36]. A simple, minimum expenditure andeco-accommodating strategy for an admixture of Ni upheld Fe_3O_4 nanomaterials (Ni/Fe_3O_4 MNPs) exercising *Moringa oleifera* (MO) leaves liberate as a dwindling and covering specialist were created. The results displayed that these MNPs found a round shape with 16–20 nm size. The space temperature seductive estimations of the compound reveal ferromagnetic rates with absorption polarization (Ms) of 76.8 emu/g. The reactant action of MNPs towards the corruption of Malachite green (MG) color chosen by UV-Vis spectra. The outcomes show that catalysts are a complete motivation for the debasement of Malachite green when varied [37].

A straightforward and productive fashion to orchestrate attractively $NiFe_2O_4$@Cu nanocatalyst via co-precipitation conditions was monitored by spectroscopic examinations affirmed the enhancement of $NiFe_2O_4$@Cu nanoparticles. These nanoparticles displayed palatable reactant action for the drop of nitro to amines with exceptional returns. The motivation is constantly painlessly insulated by a magnet and again used with no conspicuous loss of action [38]. Marzouk et al. combined $ZnFe_2O_4$ seductive nanoparticles hydrothermally and portrayed by spectroscopic analysis [39]. In addition, it had been employed as a heterogeneous catalyst for a combination of benzyl, aldehydes, ammonium acetic acid derivate, amines, and 1-amino-2-propanol under dissolvable free conditions. The benefits of the reaction are more limited times, exceptionally high return, and a straightforward foundation. The thermally and instinctively steady, inoffensive, and practical motivation was easily mended exercising an external attraction and reused in at least five progressive runs.

2 CLASSIFICATION OF MNPS: OXIDES & FERRITES

Fe_3O_4 nanoparticles (iron oxides inside the gem construction of maghemite or magnetite) are the top delved seductive nanoparticles until this point. When the Fe_3O_4 come more modest than 128 nm they came superparamagnetic which forestalls agglomeration since they show seductive conduct just an outside seductive stir is applied. The shot of an attraction of ferrite nanoparticles is regularly tremendously expanded by controlled bunching of a multifariousness of individual superparamagnetic to superparamagnetic nanoparticles groups, specifically seductive nanobeads. With the outside seductive stir crackdown, the remanence falls back to nothing. The Fe_3O_4 nanoparticles are typically acclimated by surfactants, silica, silicones, or orthophosphoric sharp inferiors to expand their soundness in arrangement [40].

2.1 Ferrites Using Shell

Maghemite or magnetite seductive nanoparticles are fairly inactive and do not naturally permit solid covalent bonds with functionalization patches on their surface. Nevertheless, the catalytic activity of seductive nanoparticles is regularly bettered by covering a subcaste of silica on the superficial face. Silica shells are constantly easily acclimated with colorful face utilitarian gatherings using covalent connections between organo-silane and silica shells. Also, some fluorescein atoms are regularly covalently cleaved to the functionalized silica shell. Fe_3O_4 nanoparticle groups with thin size dispersion comprising of superparamagnetic material covered with silica shell enjoy many upper hands over metallic nanoparticles [41–43].

2.2 Metallic

Metallic nanoparticles could likewise be helpful for a couple of technical operations on account of their advanced shot of attraction through oxides being precious for biomedical uses. It also suggests that, for an identical alternate, metallic nanoparticles are constantly made more modest than oxides. On the contrary, metallic nanoparticles have maltreatment of being pyrophoric and responsive oxidizing compounds. It makes them to handle in difficult manner by enabling an undesirable side reaction which in turn is less suitable for biomedical operations. Colloidal development for metallic particles also undeniably a challenging task [44].

2.3 Metallic Nanoparticles with a Shell

The metallic center of nanoparticles could also be passivated by vigorous oxidations, the use of surfactants, polymers, and other precious essences. Co nanoparticles oxygen atmosphere can form antiferromagnetism CoO subcaste on the external subcaste of the Co nanoparticle. As of late, work has delved into the union and trade inclination impact in these Co center CoO nanoparticles with a gold external shell. Nanoparticles with a center made either out of Fe or Co with a nonreactive shell produced using graphene integration. The advantages varied with ferrite or essential nanoparticles are [45,46].

3 CONCLUDING REMARKS

Improving functionalization systems for attractive uncoated nanomaterials can combine numerous possibilities for immobilizing dynamic particles of reagents (organic catalysts, metal buildings, metal nanoparticles). On an attractive basis, it depends on the elimination of the impulse and the restoration of the forces at which the attractive partition operates in the group response. The attractive baffles are harmless to the ecosystem and can be used to isolate and rebuild impulses by limiting the use of solvents and auxiliary materials, shortening activity times, limiting impulse failures, and preventing large-scale setbacks and oxidation and energy.

REFERENCES

1. Schätz, A.A., Reiser, O., Stark, W.J., TEMPO supported on magnetic C/Conanoparticles: a highly active and recyclable organocatalyst. *Chem. Eur. J.* 16 (30), 2010, 8950–8967.
2. Yoon, T.J., Jong, T., Lee, W., Oh, Y.S., Lee, J.K., Magnetic nanoparticles as a catalyst vehicle for simple and easy recycling electronic supplementary information (ESI) available: XRD and FT-IR data, as well as the detailed experimental conditions for the catalytic hydroformylation reactions. *New J. Chem.* 27 (2), 2003, 227–229.
3. Rabias, I. Rapid magnetic heating treatment by highly charged maghemite nanoparticles on Wistar rats exocranial glioma tumors at microliter volume. *Biomicrofluidics* 4, 2010, 024111.
4. Kumar, C.S., Mohammad, F., Magnetic nanomaterials for hyperthermia-based therapy and controlled drug delivery. *Adv. Drug Deli. Rev.* 63, 2011, 789–808.
5. Kralj, S., Rojnik, M., Kos, J., Makovec, D., Targeting EGFR-overexpressed A431 cells with EGF-labeled silica-coated magnetic nanoparticles. *J. Nanopart. Res.* 15 (5), 2010, 1–11.
6. Willhelm, S., Analysis of nanoparticle delivery to tumors. *Nat. Rev. Mater.* 1, 2016, 16014.
7. Scarberry, K.E., Dickerson, E.B., McDonald, J.F., Zhang, Z.J., Magnetic nanoparticle-peptide conjugates for in vitro and in vivo targeting and extraction of cancer cells. *J. Am. Chem. Soc.* 130 (31), 2008, 10258–10262.

8. Parera, P.N., Kouki, A., Finne, J., Pieters, R.J., Detection of pathogenic Streptococcus suis bacteria using magnetic glycoparticles. *Org. Biomol. Chem.* 8 (10), 2010, 2425–2429.
9. Abdel-Rahman, L.H., Abu-Dief, A.M., Adam, M.S.S., Hamdan, S.K., Some new nano-sized mononuclear Cu(II) Schiff Base complexes: design, characterization, molecular modeling and catalytic potentials in benzyl alcohol oxidation. *Catal. Lett.* 146, 2016, 1373–1396.
10. Abdel-Rahman, L.H., Abu-Dief, Ahmed M., El-Khatib, Rafat M., Abdel-Fatah, S.M., Sonochemical synthesis, DNA binding, antimicrobial evaluation and in vitro anticancer activity of three new nano-sized Cu(II), Co(II) and Ni(II) chelates based on tri-dentate NOO imine ligands as precursors for metal oxides. *J. Photochem. Photobiol. B Biol.* 162, 2016, 298–308.
11. Abdel-Rahman, L.H., Abu-Dief, Ahmed M., El-Khatib, Rafat M., Abdel-Fatah, S.M., Some new nano-sized Fe(II), Cd(II) and Zn(II) Schiff base complexes as precursor for metal oxides: sonochemical synthesis, characterization, DNA interaction, in vitro antimicrobial and anticancer activities. *Bioorg. Chem.* 69, 2016, 140–152.
12. Abu-Dief, Ahmed M., Mohamed, W.S., a-Bi_2O_3 nanorods: synthesis, characterization and UV-photocatalytic activity. *Mater. Res. Express.* 4, 2017, 035039.
13. Koehler, F.M., Fabian, M., Rossier, M., Waelle, M., Athanassiou, E.K., Limbach, L.K., Grass, R.N., Günther, D., Stark, W.J., Magnetic EDTA: coupling heavy metal chelators to metal nanomagnets for rapid removal of cadmium, lead and copper from contaminated wate. *Chem. Commun.* 32 (32), 2009, 4862–4864.
14. Yang, H.H., Hao, H., Zhang, S.Q., Chen, X.L., Zhuang, Z.X., Xu, J.G., Wang, X.R., Magnetite-containing spherical silica nanoparticles for bio catalysis and bio separations. *Anal. Chem.* 76 (5), 2004, 1316–1321.
15. Norén, K.K., Kempe, M., Multilayered magnetic nanoparticles as a support in solid-phase peptide synthesis. *Int. J. Pept. Res. Ther.* 15 (4), 2009, 287–292.
16. Gupta, A.K., Gupta, M., Synthesis and surface engineering of iron oxide nanoparticles for biomedical applications. *Biomaterials* 26 (18), 2005, 3995–4021.
17. Jiang, Y., Guo, C., Xia, H., Mahmood, I., Liu, H., Magnetic nanoparticles supported ionic liquids for lipase immobilization: enzyme activity in catalyzing esterification. *J. Mol. Catal. B* 58, 2009, 103–109.
18. Moodley, F.J.E., Scheijen, J.W., Niemantsverdriet, P.C., Iron oxide nanoparticles on flat oxidic surfaces – introducing a new model catalyst for Fischer-Tropsch catalysis. *Catal. Today* 154, 2010, 142–148.
19. Senapati, K.K., Borgohain, C., Phukan, P., Synthesis of highly stable CoFe2O4 nanoparticles and their use as magnetically separable catalyst for Knoevenagel reaction in aqueous medium. *J. Mol. Catal. A* 339, 2011, 24–31.
20. Murthy, Y.L.N., Diwakar, B.S., Govindh, B., Nagalakshmi, K., KasiViswanath, I.V., Singh, R., Nano copper ferrite: a reusable catalyst for the synthesis of β, γ -unsaturated ketones. *J. Chem Sci.* 124 (3), 2012, 639–645.
21. Kasi Viswanath, I.V., Murthy, Y.L.N., One-pot, three-component synthesis of 1, 4-dihydropyridines by using nano crystalline copper ferrite. *Chem Sci. Trans.* 2 (1), 2013, 227–233.
22. Esfahani, M.N., Hoseini, S.J., Mohammadi, F., Fe3O4 nanoparticles as an efficient and magnetically recoverable catalyst for the synthesis of 3,4-dihydropyrimidin-2(1H)-ones under solvent-free conditions. *Chin. J. Catal.* 32, 2011, 1484–1489.
23. Mahesh, Palla, Guruswamy, Kolakaluri, Diwakar, B.S., Rama Devi, Bhoomireddy, Murthy, Y.L.N., Kollu, Pratap, and Pammi, S.V.N., Magnetically separable recyclable nano-ferrite catalyst for the synthesis of acridinediones and their derivatives under solvent-free conditions. *Chem. Lett.* 44 (10), 2015, 1386–1388.
24. Kooti, M., Afshari, M., Magnetic cobalt ferrite nanoparticles as an efficient catalyst for oxidation of alkenes. *Sci. Iran.* 19, 2012, 1991–1995.
25. Senapati, K.K., Roy, S., Borgohain, C., Phukan, P., Palladium nanoparticles supported on cobalt ferrite: an efficient magnetically separable catalyst for ligand free Suzuki coupling. *J. Mol. Catal. A* 352, 2012, 128–134.
26. Safari, J., Javadian, L., A one-pot synthesis of 5,5-disubstituted hydantoin derivatives using magnetic Fe3O4 nanoparticles as a reusable heterogeneous catalyst. *C. R. Chim.* 16, 2013, 1165–1171.
27. Mahmoodi, N.M., Zinc ferrite nanoparticle as a magnetic catalyst: synthesis and dye degradation. *Mater. Res. Bull.* 48, 2013, 4255–4260.
28. Fernandes, C.I., Maria, D.C., Liliana, P.F., Carla, D.N., Pedro, D.V., Organometallic Mo complex anchored to magnetic iron oxide nanoparticles as highly recyclable epoxidation catalyst. *J. Organomet. Chem.* 760, 2014, 2–10.
29. Khojastehnezhad, A., Rahimizadeh, M., Eshghi, H., Moeinpour, F., Bakavoli, M., Ferric hydrogen sulfate supported on silica-coated nickel ferrite nanoparticles as new and green magnetically separable catalyst for 1,8 dioxodecahydroacridine synthesis. *Chin. J. Catal.* 35, 2014, 376–382.
30. El-Remaily, M.A.A.A., Abu-Dief, A.M., CuFe2O4 nanoparticles: an efficient heterogeneous magnetically separable catalyst for synthesis of some novel propynyl-1H-imidazoles derivatives. *Tetrahedron* 71, 2015, 2579–2584.

31. El-Remaily, M.A.A., Abu-Dief, A.M., El-Khatib, R.M., A robust synthesis and characterization of super paramagnetic CoFe2O4 nanoparticles as an efficient and reusable catalyst for green synthesis of some heterocyclic rings. *Appl. Organomet. Chem.* 30, 2016, 1022–1029.
32. Ghomi, J.S., Nazemzadeh, S.H., Alavi, H.S., Novel magnetic nanoparticlessupported inorganic-organic hybrids based on POSS as an efficient nanomagnetic catalyst for the synthesis of pyran derivatives. *Catal. Commun.* 86, 2016, 14–18.
33. Sahu, A., Badhe, P.S., Adivarekar, R., Ladole, M.R., Pandit, A.B., Synthesis of glycinamides using protease immobilized magnetic nanoparticles. *Biotechnol. Rep.* 12, 2016, 13–25.
34. Paul, B., Purkayastha, D.D., Dhar, S.S., One-pot hydrothermal synthesis and characterization of CoFe2O4 nanoparticles and its application as magnetically recoverable catalyst in oxidation of alcohols by periodic acid. *Mater. Chem. Phys.* 181, 2016, 99–105.
35. Abu-Dief, Ahmed M., Hamdan, Samar K., Functionalization of magnetic nano particles: synthesis, characterization and their application in water purification. *Am. J. Nano Sci.* 2 (3), 2016, 26–40.
36. Cruz, P., Pérez, Y., Hierro, I., Titanium alkoxides immobilized on magnetic mesoporous silica nanoparticles and their characterization by solid state voltammetry techniques: application in ring opening polymerization. *Micropor. Mesopor. Mater.* 240, 2016, 227–235.
37. Prasad, C., Sreenivasulu, K., Gangadhara, S., Venkateswarlu, P., Bio inspired green synthesis of Ni/Fe3O4 magnetic nanoparticles using Moringa oleifera leaves extract: a magnetically recoverable catalyst for organic dye degradation in aqueous solution. *J. Alloys Compd.* 700, 2017, 252–258.
38. Zeynizadeh, B., Mohammadzadeh, I., Shokri, Z., Hosseini, S.A., Synthesis and characterization of NiFe2O4@Cu nanoparticles as a magnetically recoverable catalyst for reduction of nitroarenes to arylamines with NaBH4. *J. Colloid Interf. Sci.* 15, 2017, 285–293.
39. Marzouk, Adel A., Abu-Dief, Ahmed M., Abdelhamid, Antar A., Hydrothermal preparation and characterization of ZnFe2O4 magnetic nanoparticles as an efficient heterogeneous catalyst for the synthesis of multi substituted imidazoles and study of their anti-inflammatory activity. *Appl. Organomet. Chem.* 2017, e3794. http://doi.org/10.1002/aoc. 3794.
40. Kim, D.K., Mikhaylova, M., Anchoring of phosphonate and phosphinate coupling molecules on titania particles. *Chem. Mater.* 15 (8), 2003, 1617–1627.
41. Kralj, S., Drofenik, M., Makovec, D., Controlled surface functionalization of silica-coated magnetic nanoparticles with terminal amino and carboxyl groups. *J. Nanopart. Res.* 13 (7), 2017, 2829–2841.
42. Kralj, S., Makovec, D., Čampelj, S., Drofenik, M., Producing ultra-thin silica coatings on iron-oxide nanoparticles to improve their surface reactivity. *J. Magn. Magn. Mater.* 322 (13), 2010, 1847–1853.
43. Kralj, S., Rojnik, M., Romih, R., Jagodič, M., Kos, J., Makovec, D., Effect of surface charge on the cellular uptake of fluorescent magnetic nanoparticles. *J. Nanopart. Res.* 14 (10), 2010.
44. Grass, R.N., Robert, N., Athanassiou, E.K., Stark, W.J., Covalently functionalized cobalt nanoparticles as a platform for magnetic separations in organic synthesis. *Angew. Chem. Int. Ed.* 46 (26), 2007, 4909–4912.
45. Johnson, S.H., Johnson, C.L., May, S.J., Hirsch, S., Cole, M.W., Spanier, J.E., Co@CoO@Au core-multi-shell nanocrystals. *J. Mater. Chem.* 20 (3), 2010, 439–443.
46. Grass, R.N., Robert, N., Stark, J., 2006. Gas phase synthesis of fcc-cobalt nanoparticles. *J. Mater. Chem.* 16 (19), 2006, 1825.

Recent Advances in the Catalytic Applications of Magnetic Nanomaterials

2

B. Sehgal, G.B. Kunde

Contents

1	Introduction	10
2	Types of Magnetic Nanomaterials and Their Application in Catalysis	11
	2.1 Single Magnetic Nanoparticle Materials	12
	2.1.1 Oxides and Mixed Metal Oxide Magnetic Nanomaterial	12
	2.1.2 MNPs of Metal Ferrites and Rare Earth Orthoferrites	13
	2.1.3 Surface Modification of MNPs	14
	2.1.4 Ionic Liquid Coated MNPs	15
	2.1.5 Metal Carbide Magnetic Nanomaterials	15
	2.2 Magnetic Nanocomposite Materials	17
	2.2.1 Magnetic Frameworks Composites (MFCs)	17
	2.2.2 Magnetic Nanomaterial/Polymer Composites	17
	2.2.3 Carbon/Magnetic Nanomaterial-Based Nanocomposites	19
	2.2.4 Multicomponent MNP Based Composites	20
	2.3 Magnetic Bio-Nanocomposites	21
	2.3.1 Biochar Based Magnetic Nanocomposites	23
	2.3.2 Enzyme Immobilized Magnetic Nanomaterials	23
	2.3.3 Microbial/Magnetic Nanocomposites	24
3	Applications of MNPs in Catalysis	26
4	Conclusions	28
	References	28

1 INTRODUCTION

The phenomenon of catalysis is known for ages, but its understanding and mechanism were not clear at that time. In the early 19th century, the catalytic characteristics of several metals were extensively investigated and gained prominence due to their large-scale industrial applications. Catalysis has now become a popular approach in academia and industry since it allows the modification of the rate of a chemical reaction while utilizing sustainable resources and less polluting chemical processes. The approach of reducing both carbon emission and dependence on fossil fuels inevitably requires the reengineering of chemical pathways. The approach to achieve greener chemical routes necessitates the development of novel catalysts that fulfill the principles of Green Chemistry. In the chemical and energy industry, a major proportion of industrial catalysts remain heterogeneous due to their low operational cost and robustness. The stringent ecological and economical mandate for sustainable production has resulted in a paradigm shift from the old approach which focused on improving catalyst character and efficacy. Today, the recycling and reuse of catalysts with minimum contamination in reaction products and waste streams have become an important consideration during synthesis. With the benefit of being dissolved homogeneously with the reaction mass, homogeneous types of catalysts usually reveal greater catalytic character and can be selective, equally under moderate reaction conditions. Overcoming, the difficulty of removing these catalysts from the reaction mixture, they are reused in a solid-liquid and liquid-liquid medium. The liquid-liquid technique involves different non-miscible solvents for mixing the catalyst and end product separately followed by recovery via simple phase separation. The solid-liquid method depends upon immobilization of the metal nanoparticles on the solid support material. Another method of heterogenization of homogeneous catalytic systems is to modify homogeneous catalysts to "break" the homogeneity between the catalyst molecules and the products by separating them into separable phases. However, either the solubility of the reactants or the particle size limits these approaches. Furthermore, due to the difficulty of diffusion of reactants to the surface-anchored catalysts, a catalytic activity normally diminishes when homogeneous catalysts are immobilized. The use of porous and smaller-sized support materials has been employed to increase the surface area. In liquid phase catalysis, small particle size is used to minimize the mass transportation limitations. However, separation from the reaction medium is generally difficult when the particle size is exceedingly small. Moreover, on decreasing the dimension of the support to the nanometer level, the area of the surface will increase significantly and the support material will be homogeneously distributed in the solution, thereby maintaining the same problem of isolation and reuse of homogeneous kind of catalysts. In recent decades, the use of magnetic nanoparticles (MNPs) in catalysis gained a lot of consideration because of the development of bottom-up techniques. The rising utilization of magnetic nanomaterials in catalytic reactions is indicated by the number of publications in this area in the last decade (2010–2020). Figure 2.1(a) shows the tendency in publications related to MNPs, which show easy removal from the reaction mass utilizing the magnetic separation process. The quantity of publications has increased exponentially due to the higher stability of magnetic nanomaterials in both acidic pH and basic pH solutions and also in organic solvents. The bibliographic analysis is based on the search of "magnetic nanomaterials in catalysis" in Google Scholar including both articles and patents. Figure 2.1(b) shows the worldwide market share of magnetite, estimated at 43.9 million USD, in the year 2018 and projected compound annual growth (CAGR) estimated to be 10.3% during the period 2019 to 2025. The major consumers of magnetite are biomedical, energy, electronics, and wastewater industries [1].

Magnetic separation is a powerful technique for separating catalysts from reaction medium quickly and efficiently, and it provides an option for solvent, reaction time, and reaction energy-intensive separation processes. With large surface area, higher chemical, mechanical as well as thermal stability, these magnetic nanomaterials consist of a higher surface-to-volume ratio that allows for excellent loading of the homogeneous catalyst. The ease of recovery from the reaction mass and reusability many times by retaining effective catalytic character, is much easier with separations governed by magnetic characteristics,

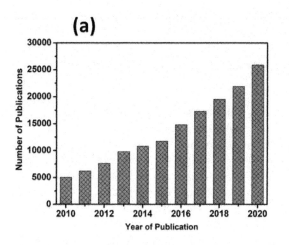 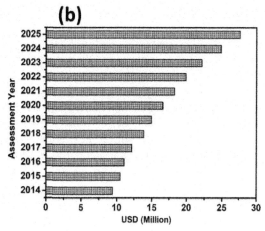

FIGURE 2.1 (a) Utilization of magnetic nanomaterials in catalytic applications during last decade (b) Market size of magnetite nanoparticles (USA) for 2014–2025 [1].

particularly when the catalyst dimensions can be in the sub-micrometer limit. Magnetic separation can thus be regarded as a catalytic tool for rapid and efficient separation with low energy consumption compared to catalyst isolation through filtration, liquid-liquid extraction, chromatography, or centrifugation.

MNPs can be used to close the gap between homogeneous and heterogeneous catalytic processes. MNPs act as catalytic supports, allowing homogeneous catalysts to be immobilized and magnetically recovered. Magnetic separation is now regarded as a more competitive technology than traditional procedures since it is quick, efficient, uses less energy with fewer solvents, and has minimum secondary wastes. Maintaining nanoparticle stability without precipitation or agglomeration is critical for most MNP applications. Due to the narrow space between the particles, higher amount of energy of the surface, and acting van der Waals attraction forces, MNPs frequently agglomerate into a large cluster. As a result, the magnetic nanomaterial catalyst disperses unevenly in reaction mass, resulting in a quick loss of catalytic and magnetic character. This problem can be worked out by coating the surfaces of nanomaterials by using particular protective materials, which can reduce interparticle contact thereby improving its chemical stability. MNPs can be stabilized either by adding monomers or by coating them with inorganic materials or organic matrices. Superparamagnetic NPs are prevented from aggregating into large ferromagnetic species by dispersing them in a matrix, such as a polymer, carbon, ceramic, ionic liquids, and other metals. The magnetic nanoparticles can be functionalized and modified for different functionality additionally to find their application in a variety of catalytic reactions of industrial importance. The combination of MNPs and catalytically active material offers the potential to provide a solution to several catalyst recyclability difficulties that no other filtration technology can easily address.

2 TYPES OF MAGNETIC NANOMATERIALS AND THEIR APPLICATION IN CATALYSIS

The early publications for the use of separation by the magnetic character in the area of catalysis concerned the employment of metal catalysts (Fe, Co, and Ni) with intrinsic magnetic properties. The concept of MNPs has rapidly evolved over the last decade to ease the recovery process in catalysis, as well as in biology and medicine. The addition of magnetic nanoparticles and nanocomposites mainly consisting of cobalt and iron oxides as solid supports has increased. Such solid supports are required to immobilize catalytic complexes of metals, biocatalysts, and organocatalysts. By application of external super magnet or magnetically supported cross-flow filtering and process of centrifugation, the catalytic NPs are isolated directly from reaction mass. For the sake of understanding, this chapter is divided into three divisions which include:

i) Single Magnetic Nanoparticle Materials: Those containing not more than one form of nanomaterial but may be improved through chemical modification. These consist of oxides, ferrites, rare earth orthoferrites, and metal carbide magnetic nanomaterials.

ii) Magnetic Nanocomposite Materials: Nanocomposites consisting of two or more distinct materials, one of which is magnetic, are included. Magnetic nanocomposites include nanoscale magnetic crystals embedded in an amorphous or crystalline matrix.

iii) Magnetic Bio-Nanocomposite Materials: These are composite materials in which MNPs are embedded in a matrix of biologically derived components such as enzymes, biochar, and microbes. The composite can play important role in combining biological processes with magnetic characteristics of the embedded MNPs.

2.1 Single Magnetic Nanoparticle Materials

2.1.1 Oxides and Mixed Metal Oxide Magnetic Nanomaterial

Iron oxide nanoparticles are the most basic type of magnetically retrievable catalysts. These nanoparticles are strong-featured with good stability in the air, easily functionalizable, and are easily suspendable in different solvents, such as water and other types of protic solvents. With tunable size, shape, and crystallinity, it is the most used magnetic nanomaterial. They are also known as ferrite nanoparticles (iron oxide exists in the crystalline structure of maghemite or magnetite). The ferrite particles which are lesser than 128 nm, can be converted to superparamagnetic, which prevents self-aggregation since it shows magnetic character when placed in the external magnetic field. Regulated clustering of many superparamagnetic types of nanoparticles into the superparamagnetic nanoparticle aggregates, which can be called magnetic nanobeads, greatly increase the magnetic moment of ferrite nanoparticles. The high magnetization of magnetite (Fe_3O_4) has made it one of the most applied nanoparticles of iron oxide for the magnetic type of separation. Fe_3O_4 is a mineral of a ferrimagnetic type developed in the form of spinel shape structure consisting of atoms such as Fe(II) and Fe(III). Although magnetite nanoparticles can be oxidized further to maghemite, they are not reactive under basic pH and utilized to make supported oxide/hydroxide catalysts. For instance, Aldol reactions are catalyzed by iron hydroxide supported on MNPs. Magnetite types of nanoparticles are well-known for their intrinsic catalytic characteristics in Fenton reactions, but they also have applications in the thiolysis of epoxides and oxidation. Maghemite (γ-Fe_2O_3) shows a lower scale of saturation magnetization compared to magnetite, but can be more unreactive and tolerant to acid and high temperatures than magnetite. Superparamagnetic type of pure Fe(0) nanoparticles can be utilized as magnetically retrievable catalysts, however, the synthesis can be performed in the inert atmosphere to avoid easy oxidation of Fe(0) in air. Zerovalent iron nanoparticles (ZVI) consist of pure iron and exhibit intriguing physicochemical characteristics that make them unique for the application as a material in the elimination of various contaminant species. Advanced oxidation processes (AOPs), which produce very reactive radicals, are very effective and important approaches in the removal of resistant organic contaminants. The effective oxidant in AOPs used for the elimination of organic contaminants from groundwater, wastewater and soil, is persulfate (PS). ZVI has been utilized to successfully activate PS that produces SO_4^{2-} for 4-chlorophenol, 2, 4-dinitrotoluene (DNT), polyvinyl alcohol (PVA), aniline, p-chloroaniline, and alachlor removal. ZVI is inexpensive, non-toxic in nature, and can be used as an alternate source of Fe^{2+} due to its ability to dissociate in the presence of both aerobic and anaerobic medium [2]. nZVI can generate Fe^{2+} by the below mentioned reactions given in eq. (I) to (V).

$$Fe^0 \rightarrow Fe^{2+} + 2e^- \tag{I}$$

$$Fe^0 + S_2O_8 \rightarrow Fe^{2+} + 2SO_4^{2-} \tag{II}$$

$$Fe^0 + H_2O + 0.5O_2 \rightarrow Fe^{2+} + 2OH^- \tag{III}$$

$$Fe^0 + 2H_2O \rightarrow Fe^{2+} + H_2 + 2OH^- \tag{IV}$$

$$S_2O_8^{2-} + Fe^{2+} \rightarrow Fe^{3+} + SO_4^{2-} + SO_4^- \tag{V}$$

The higher potential of SO_4^- makes it more efficient for the elimination of impurities. Besides, the higher surface area, nano scale particle size, and higher reactivity in the case of zero-valent iron (nZVI) increase the activation of PS. The magnetic characteristics of metallic Co and Ni are also notable. The intrinsic magnetic characteristics of Co and Ni nanoparticles could be working as a catalyst as well as a magnetic nanomaterial during separation by applying external magnetic field. Oxides of iron are less corrosive, less expensive, and easier to synthesis than Co and Ni, hence favored as catalyst support.

2.1.2 MNPs of Metal Ferrites and Rare Earth Orthoferrites

The catalytic scope of iron oxide nanoparticles could be expanded beyond coupling and oxidation type of reactions by the inclusion of a different metal in the spinel structures of Fe_3O_4. The excellent magnetic properties of spinel ferrite NPs are often accompanied by other functional properties like high chemical and thermal strength, in addition to the hardness of the material along with high surface area. Nanocatalysts of ferrite type are eco-friendly and obey Green Chemistry principles. Its recovery from the reaction mass and recyclability is appreciable. Furthermore, the magnetic character of the NPs can be modified to some extent by particle size and shape. Bimetallic ferrite spinel consisting of the common formula MFe_2O_4 (where M denotes divalent 3d transition metals such as Mn, Cu, Co, and Zn), has drawn much interest in wastewater treatment. For instance, $ZnFe_2O_4$ shows superb photocatalytic activity in the removal of dyes, $MnFe_2O_4$ and $CuFe_2O_4$ could be used as catalytic material in the removal of phenol and acetaminophen. Excellent catalytic activity is shown by $CoFe_2O_4$ nanoparticles in the catalytic initiation of peroxymonosulfate in the disintegration of organic pollutants. In some cases, the addition of a ligand that interacts with the NP surface can increase catalytic activity. For example, the addition of bipyridine made $CuFe_2O_4$ NPs catalytically active, whereas the application of a chiral type of ligand provides access during the asymmetric catalysis process. Mixed type spinel, doped spinel structures, and ternary spinel structures have been widely studied, and their catalytic characteristics for reduction of CO_2, electrochemical water splitting, water-gas shift kind of reactions, steam reforming of methane, decomposition of methanol, and biodiesel production has been evaluated.

The rare earth elements (REEs) comprise a group of transition elements, along with lanthanides, that have variable valence states and allow their use for a variety of catalytic applications besides being used in making permanent magnets. To influence the chemical and physical properties of different nanomaterials, rare earth elements are used as dopants. Because of their redox characteristics and good oxygen mobility, cerium-based materials are used as catalysts in Fischer–Tropsch synthesis, hydrogenation of CO_2, and methanation of CO_2. Elements of the type rare-earth orthoferrites with the formula $RFeO_3$ (R = Sc, Y, La-Lu) are a new type of acid-base catalyst with high activity and the ability to be recovered magnetically. They're mostly used in redox and photoinduced reactions. For example, the catalytic activity of holmium orthoferrite can be related to the existence of acid-base sites on the surface, which can be assessed for conversion of n-hexane at a temperature of 500 °C and 1 atmospheric pressure. The spent catalyst was removed from the reaction mass by using an applied magnetic field with 97% efficiency [3].

2.1.3 Surface Modification of MNPs

The numerous notable advances in the manufacturing of magnetic nanoparticles, useful for their stability over time without becoming agglomerated or sedimented persist as a challenge. In the application of magnetic nanoparticles, stability is essential. The transition group elements, for instance Fe, Co, and Ni, along with its alloys, are most reactive to air. Thus, the fundamental challenge in the application of transition metallic elements and their alloys proneness to oxidation, becomes more due to reduction in the size of particles. Every type of protection planning ends in the shape of a core-shell in the case of magnetic nanomaterials, in which the basic magnetic nanoparticle serves as a core and is protected from the environment by a shell. Organic shell coating, such as surfactants and polymers, or inorganic materials coatings, including silica, carbon, noble metals, or oxides, which can be formed by moderate oxidation to form nanoparticles' outer shells, constitute two major classes of coating techniques. The magnetization due to surface effects can cause nanomaterials, such as nanoparticle oxides, to decrease in comparison to the bulk value. The presence of a dead magnetic layer on the surface of the particle, the occurrence of canted spins, or a spin-glass like nature in the case of spins of the surface has all been linked to this reduction. Small metallic nanoparticles, such as cobalt, on the other hand, demonstrate an increase in the magnetic moment as their size decreases. In some circumstances, a direct link between magnetic characteristics and surface coating can be established. This is because the dipolar coupling is proportional to particles distances, which can rely on inert silica shell thickness. The coating of silica can be used to tailor the magnetic characteristics of nanomaterials. The magnetite or maghemite magnetic nanomaterial surface can be largely unreactive, hence, functionalization strategies involving strong covalent bonds are usually not possible. The reactivity of magnetic nanoparticles is however increased by covering them with a layer of silica. Superparamagnetic oxide nanoparticles covered with a shell of silica form ferrite nanoparticle aggregates having narrow particle size distribution and several benefits compared to metallic nanoparticles. Superior chemical inertness results in higher colloidal stability without agglomeration where silica surface enables covalent functionalization. A carbon or silica coating can also be used to avoid oxidation of metals and the functional groups are added via sp^2 carbon. Different catalytic groups can be added on the surface of metal nanoparticles through the functional groups through bonds with covalent or electrostatic nature. Similarly, metallic Ni and Co get easily oxidized, hence, it is required to coat them with strong coordinating ligands, inorganic oxides, or carbon to maintain stability. Because they have lesser magnetism than their corresponding metals, nickel and cobalt oxide nanoparticles are less desirable as supports. Although silica-coated magnetite NPs have received considerable attention, other coatings have also been investigated. To protect the cores of magnetic nanoparticles from oxidation, precious metals might be deposited on them. The magnetic characteristics get influenced by a precious metal coating surrounding the magnetic nanoparticles. Because of its low reactivity, gold appears to be a perfect coating. Cobalt nanoparticles coated with gold, for example, consist of lesser magnetic anisotropy as compared to untreated nanoparticles. On the other hand, coating of gold on iron nanoparticles increases anisotropy. Magnetic nanoparticles with gold coatings are particularly fascinating because the gold surface can be further activated with thiol groups. This approach lets functional ligands be linked together, potentially making the materials ideal for catalytic and optical applications. The magnetic properties of organic ligands, which are utilized to stabilize magnetic nanoparticles, are likewise affected. The anisotropy, as well as magnetic moments at the surface of the nanoparticles, can be changed by ligand. Organic ligands can stabilize cobalt nanoparticles, display a decrease of magnetic moment and anisotropy considerably. The surface magnetism of nickel nanoparticles cannot be changed by donor ligands but stimulate the development of rods, on the other hand, trioctylphosphine oxide reduces the magnetization of the particles. Carbon-protected magnetic nanoparticles have recently received a lot of attention since carbon-based materials have a series of benefits over polymer or silica in terms of thermal and chemical resilience and biocompatibility. The well-organized layers of graphitic carbon act as a barrier to chemical oxidation. These findings suggest that carbon-coated magnetic nanoparticles are thermally stable and resistant to oxidation and acid leaching. Furthermore, because carbon-covered nanoparticles are normally in the metallic state, they have a greater magnetic moment than their oxide counterparts. Surfactants or polymers can also be used during the synthesis of nanoparticles for the prevention of aggregation. Electrostatic repulsion is utilized for this purpose, and the surface charge is the most essential particle characteristic in this regard. The polyelectrolytes with high charge densities show significant characteristics in the coating of nanoparticles by stabilizing dispersed particles formed due to repulsive electrostatic charge. Polyzwitterions are a kind of polyelectrolyte that is utilized as a coating for MNPs. The important concept of chemisorption or physisorption of polyelectrolytes on the surface of nanomaterials depends on the electrostatic type of interaction between nanomaterial and polymer or interaction of hydrophobic type. Polymers or surfactants (by physical adsorption) form either monolayer or bilayer on the surface of nanoparticles. This layered structure provides enough repulsive forces to counteract the magnetic and van der Waals attractive forces working on the nanoparticles. The magnetic particles can be stabilized in suspension due to steric repulsion. Carboxylic acids, phosphates, and sulfates are examples of functional groups that can connect to the surface of magnetite. Poly(pyrrole), poly(aniline), poly(alkyl cyanoacrylates), poly(methylidene malonate), and polyesters like poly(lactic acid), poly(glycolic acid), poly(e-caprolactone), and their copolymers are some of the suitable polymers for coating. For instance, a series of magnetically retrievable catalysts for the conversion of syngas to methanol by thermal breakdown of metal acetylacetonates on magnetic silica support designed to investigate the efficacy of magnetic NPs [4]. When compared to $ZnO-SiO_2$, the activity of Zn-Cr-containing magnetic silica was found to be higher. The oxygen vacancies created by the intermixing of Zn and Cr species with Fe_3O_4 were responsible for this occurrence. As seen in Figure 2.2(a), magnetite NPs not only allowed for facile magnetic separation but also acted as a reservoir for the insertion of Zn and Cr ions.

2.1.4 Ionic Liquid Coated MNPs

Ionic liquid-coated MNPs have recently acquired a lot of attention because of their excellent catalytic activity of various industrial processes. Ionic liquids (ILs) are the combination of organic cation and inorganic anions or organic anions. It has melting point lower than water's boiling point. Many of these ILs have been studied as solvents and/or catalysts in different organic reactions. They have a broad liquid range with a wider potential window, excellent electrical conductivity and low vapor pressure. It can effectively functionalize the nanomaterials. The high viscosity, high cost, massive consumption, and homogeneity, however, limit the industrial applications of ionic liquids. Supported catalysts with ionic liquid can be an excellent alternative to conventional equivalents to alleviate these disadvantages. Ionic liquids have been loaded using a variety of support materials, such as zeolites, magnetic nanoparticles, mesoporous silica, polystyrene, and many more. Fe_3O_4@N-doped carbon NPs, for example, is an important oxygen-reduction electrocatalyst in lithium-ion batteries. A variety of materials can be used as N and C sources. Ionic liquids provided C and N in Fe_3O_4@void@N-doped carbon with a yolk-shell structure. In another synthesis strategy [5] polymer-coated magnetic nanomaterials can be used for the immobilization of ionic liquids e.g., ([Im][OH]/P(VBC-DVB)@MNPs). The three layers of the core-shell type of the catalysts consist of polymeric outer layer, intermediate shell and magnetic core. The magnetic core was made up of $CoFe_2O_4$ magnetic nanoparticles, to separate the catalyst with ease from the reaction media. To prevent magnetic nanoparticle corrosion in polar liquids, a protective layer was formed using a copolymer of divinylbenzene (DVB) and vinyl benzyl chloride (VBC). This increased the recyclability and stability of the product. The reaction took place in a catalytic layer made up of ionic liquid 1-propyl-3-alkyl imidazole hydroxide. Because of the huge amount of OH-ions placed on the surface, this catalyst demonstrated remarkable catalytic efficiency related to the Knoevenagel condensation reaction between ethyl cyanoacetate and benzaldehyde (Figure 2.2(b)). In addition, the catalyst had lower activation energy than the NaOH homogeneous catalyst. Dicationic ionic liquids are a new subclass of ionic liquids, which received a lot of interest recently. These molecules usually have a flexible spacer that binds two cationic groups and can also connect counter anions. Their chemical and physical characteristics can be tailored by defining the chain length and type of spacer or cation.

2.1.5 Metal Carbide Magnetic Nanomaterials

Metal carbide is a metal-carbon combination, specifically an intermediate transition metal carbide. Metal carbides have unique electrical, catalytic, and magnetic characteristics when compared to standard materials. Iron carbides (Fe_5C_2, Fe_3C, and Fe_2C) are infrequently researched despite their outstanding magnetic characteristics and stability. This is due to the difficulties encountered in their synthesis, which include regulating the size and shape. Transition metal carbides are metastable and have modest ferromagnetic characteristics. Nitrogen-doped core shell-structured porous $Fe/Fe_3C@C$ nano boxes supported on reduced graphene oxide sheets show enhanced electrical conductivity. Compared to commercial Pt/C catalyst, $Fe/Fe_3C@C$ exhibits good electrochemical catalytic characteristics, durability, and tolerance for methanol above the commercially available Pt/C type catalyst.

To build viable fuel cells, highly active and long-lasting catalysts are needed to promote the oxygen reduction process (ORR) at the cathode. Pt type materials can be the most well-known ORR catalysts. Pt materials have some major limitations that limit their widespread application, in addition to their expensive cost. When used as ORR catalysts, Fe_3C NPs with core-shell architectures show high activity and stability in alkaline media. Carbon nanotube/Fe_3C NMs, for example, have a greater ORR activity in KOH solution than a 20 wt% Pt/C catalyst. Fe_5C_2 NPs coupled with various metal ions showed significantly greater catalytic properties than pure phase Fe_5C_2 NPs. A Fe_5C_2 catalyst with Na and Zn modulation, for example, demonstrated excellent activity and good selectivity for alkenes (up to 79%) but low selectivity for CO_2 [6]. Metal carbides appear to be a promising catalyst and a viable alternative to ruthenium-based systems. Ruthenium catalysts have become increasingly important in the ammonia decomposition process for the production of CO-free hydrogen, which is useful in fuel cell applications. Catalysts other than the more expensive noble metal are needed to make the process feasible for large-scale production. In such instances, metal carbide could be a better solution.

Another major application involves the use of magnetic Mn-Fe oxycarbides (mMFC) as activators in peroxymonosulfate (PMS) oxidation system for butylparaben (BPB) oxidation via a radical reaction process. For BPB oxidation, the novel carbon-rich carbazides combine the properties of carbonaceous materials and metal oxides. Dual active sites are provided by the non-metal with $-C = O$ groups and the metal with Mn/Fe oxides to create $SO_4^{·-}$, $·OH$, $^1O_2$ and/or $O_2^{·-}$. When compared to nonmagnetic metal oxides, mMFC demonstrated not only greater stability but also desirable magnetism [7]. Table 2.1 shows examples of single MNPs and modified magnetic nanomaterials with catalytic applications [7–16].

TABLE 2.1 Examples of Single MNPs and Modified Magnetic Nanomaterials for Different Catalytic Reactions

MORPHOLOGY AND PARTICLE COMPOSITION OF MAGNETIC NANOMATERIAL	METHOD OF SYNTHESIS	SURFACE AREA $M^2 G^{-1}$	TYPE OF CATALYTIC REACTION	REF.
Rod like Mn/Fe oxycarbide	Solvothermal synthesis and calcination	179.6	PMS activation	[7]
Fe_3O_4 in nanoring form	Hydrothermal	109.3	Fenton	[8]
$CoFe_2O_4$	Microwave-assisted	N/A	Oxidation of styrene	[9]
$ZnFe_2O_4$	Sol-gel method	39.812	Claisen–Schmidt condensation of substituted benzaldehyde	[10]
Zero valent iron (nZVI)	N/A	N/A	Fenton oxidation of amoxicillin	[11]
Sulfur-modified nZVIs (S–nZVIs)	Liquid-phase reduction method	N/A	PMS activation	[12]
Carbon-modified nZVIs Flakes	Ball-milling	6.6	PMS activation	[13]
$CuFe_2O_4$	Sol-gel method	65.68	Hg^0 oxidation.	[14]
$ZnFe_2O_4$	citrate sol-gel method	61.7	Ozonation of aqueous oxalic acid	[15]
$CuFe_2O_4$	Sol-gel combustion method	27	disintegrating p-nitrophenol in aqueous solution (PS activation)	[16]

2.2 Magnetic Nanocomposite Materials

Various nanoparticles consist of different nanomaterials in a particular type of composite. This imparts the collective effect of properties of different nanomaterials on the performance of composite materials. Typically, the materials, known as nanocomposites, are made up of a variety of components, at least one of which has nanoscale dimensions. Nanocomposites are made up of two or more different materials, with a nanoparticle attached to or surrounded by another organic or inorganic substance, such as in a core-shell or heterostructure arrangement. The coating of a nanoscale substance with a polymer, SiO_2, or carbon is a common type of nanoparticle composite. Graphene or graphene oxide (GO) can also be used as supports for depositing nanoscale materials. Their use is largely due to the synergistic effects caused by the coexistence of two distinct materials and their interface, which result in characteristics that are often higher to those of their single-phase components.

2.2.1 Magnetic Frameworks Composites (MFCs)

Metal-organic frameworks (MOFs), which are made up of metal ions connected by organic bridging ligands, have become one of the most sought-after materials in recent decades. Due to their high porosity and well-defined channels, guest molecules can easily distribute within such materials. MOFs also provide size and shape selectivity for guest molecules due to their designable form and size. Magnetic framework composites (MFCs) with core-shells and non-core-shell forms are produced when functional MNPs and MOFs are combined. Specifically developed MFCs provide innovative characteristics which have a designable structure, higher surface area, ease of loading, and fast recovery. In different sectors such as optics, sensing, drug delivery, catalysis, and environmental applications, various nanoparticles (in the form of quantum dots, metal nanoparticles/nanorods, graphene, multiwall carbon nanotubes (MWCNT), and porous silica nano-sphere) can be merged with MOFs and their working characteristics are analyzed. MFCs can be thought of as one-of-a-kind host matrices that encase noble metal (e.g., Pd, Au, Ru, and Pt) or bimetallic alloy nanoparticles. The highly porous shell of MOF ensures easy transfer of mass among reactive sites of nanoparticles and reactants, increasing the MFC's endurance. The MOF shells' symmetrical cavities serve as docking sites and confinement nanoreactors, considerably improving their catalytic efficacy. The catalytic efficiency of Pd@Fe_3O_4@MOFs in the oxidative degradation of chlorophenols and phenols was outstanding. In Pd@Fe_3O_4 hybrids (core/shell Fe_3O_4 doped Pd nanoparticles (NPs)), many Fe(III) species are present in the Fe-MOF type of shell, Fe_3O_4 nanoparticles in the shell and inner Pd@Fe_3O_4 hybrids all worked as functional centers to activate H_2O_2 decomposition. The electrons moving from Pd NPs to Fe_3O_4 nanoparticles in the MOFs shell and internal Pd@Fe_3O_4 hybrids enhanced the Fe(III)/Fe(II) redox cycle, as seen in Figure 2.2(c). This ensured that H_2O_2 was continuously decomposed into OH• radicals, allowing for effective breakdown and decomposition of organic contaminants [17]. In another example, palladium doped amino-functionalized magnetic MOF-MIL-101 was efficiently utilized in the Heck coupling reaction in the presence of an inorganic base [18]. Figure 2.2(d) shows the activity of the hybrid catalyst with good reusability.

Multi-catalyst MFCs have recently gained a lot of attention for their ability to perform multistep reactions in a single pot. Because of their highly stimulated ligand-to-metal charge transfer, MOFs have inherent photocatalytic capabilities. As a result, MFCs can degrade organic dyes photocatalytically. Furthermore, the large hierarchical surface area of MOF combined with its exceptional porosity gives a high loading capacity and great affinity for enzymes, resulting in enzyme embedded MFCs.

2.2.2 Magnetic Nanomaterial/Polymer Composites

Nanocomposite development in future will certainly utilize polymeric nanomaterials. Polymers can connect a large number of nanoparticles in the form of aggregates. The chemical characteristics of polymers can help the nanocomposites perform the function for which they are designed. The loading with functional groups can be significantly raised by introducing a shell of dendrimers or polymers on the surface of the nanoparticles. Dendrimers are frequently added to magnetic nanobeads to optimize their dispersion in organic solvents and to increase the number of functional groups on the surface. While dendrimers are good at multiplying functional groups on the surface of magnetic nanoparticles, polymers have some advantages of their own: They usually require less effort to synthesize large molecular weight counterparts, which speeds up particle dispersion stability due to enhanced steric repulsion. The catalytic performance of biomolecule functionalized MNPs gained utmost priority due to efficient magnetic separation, large surface area, good biocompatibility, the presence of enormous surface hydroxyl functional groups, and high redox activity. For instance, in the complete reduction of 4-nitrophenol to 4-aminophenol, gold nanoparticles doped on sodium alginate (Alg) masked magnetite (Fe_3O_4@Alg-AuNPs) nanocomposite show good yield (99%) at a shorter time of reaction (1.5–4 min) in water medium at a normal temperature [19]. The shell of sodium alginate, which is a natural anionic polysaccharide, can act as a stabilizing species in the immobilization of gold nanoparticles on Fe_3O_4 microparticles core (Figure 2.2(e)). The integrated approach, which blends a magnetic type of separation with the surface complexation or reaction among

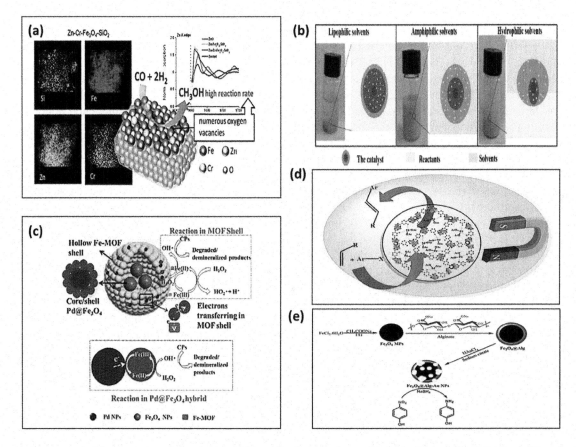

FIGURE 2.2 (a) Oxygen vacancies and metal ion distribution determines the methanol synthesis activity of magnetically recoverable catalysts, adapted with permission from [4]. Copyright (2017) ACS Publications; (b) Mechanism of Knoevenagel condensation catalyzed by [Im][OH]/P(VBCDVB)@MNPs in various solvents, adapted with permission from [5]. Copyright (2019) ACS Publications; (c) TCP degradation mediated by the yolk/shell Pd@Fe_3O_4@MOFs: a possible mechanism, adapted with permission from [17]. Copyright (2018) Elsevier; (d) Mizoroki–Heck Cross-Coupling Catalyst with Palladium Supported Amino Functionalized Magnetic MOF-MIL-101 as an efficient and Recoverable Catalyst [18]; (e) Synthesis and application of Fe_3O_4@Alg-AuNPs nanocomposite for nitrophenol reduction, adapted with permission from [19]. Copyright (2021) Elsevier.

metal nanoparticles and organic compounds, is critical for wastewater remediation because it produces no secondary waste and uses recyclable materials. During the electrostatic interface of reactive functional groups, some polymers can offer an effective host structure which is in the form of a coated layer on magnetic nanoparticles, resulting in a long-lasting stable structure with enhanced catalytic activity. PVP and TiO_2 polymer is commonly used as stabilizers in nanocomposites with improved photocatalytic activity. In the glucose oxidation and Suzuki coupling processes, resin ingrained gold nanoparticle/polyaniline and Pd doped polymer both work well as catalysts. Several research groups have shown polymer-metal nanoparticle composites materials with dynamic thermosensitive activity as a key benefit. Biopolymers are one of the most common materials used to make magnetic polymeric nanocomposites that have a wide range of uses. Cellulose, dextrin, alginic acid, polypyrrole gum, starch, and disaccharides such as lactose and maltose have all presently been employed to make magnetic nanocomposites. The abatement of contaminants of increasing concern (CECs), such as medicines and personal care items, in Fenton-like and photo-Fenton-like conditions can be carried out with Fe_3O_4 magnetic-based particles.

2.2.3 Carbon/Magnetic Nanomaterial-Based Nanocomposites

Carbon nanomaterials including activated carbon, carbon nanotubes, nanohorns, carbon fibers, and graphene have attracted increasing attention because of their distinctive properties. The ease of availability of starting material has made it an important material in the synthesis of new nanocomposite materials. Carbon-encapsulated MNPs are promising materials for removing heavy metal ions from aqueous solutions efficiently. With their large surface area, such nanocomposite catalysts show not only strong catalytic activity, but the carbon layer successfully prevents MNPs to dissolve in acidic environments. Besides, the outer carbon layer provides several functional groups (carboxylic, aldehyde, and hydroxyl groups) on the MNPs surface. In the reduction of 4-nitrophenol (4-NP) to 4-aminophenol (4-AP) in an aqueous solution using sodium borohydride as a hydrogen source, Au nanoparticles (Au NPs) coated Fe_3O_4@porous carbon core-shell NPs showed outstanding catalytic activity. Polydopamine (PDA) was used to encapsulate the magnetic microsphere, reducing the added chloroauric acid to Au NPs. Due to catechol groups acting as binding reagents, the Au NPs were subsequently loaded onto a magnetic microsphere. The inert carbon coating may not only stably bind Au NPs on magnetic microspheres but also protect the magnetic core from being oxidized or dissolved in harsh environments [20]. The schematic depiction for the preparation and application of Fe_3O_4@C-Au is shown in Figure 2.3(a). Other catalytically active species, such as enzymes and metal complexes, can also be immobilized on magnetic carbon composites in addition to metals and oxides. Magnetic mesoporous carbons, with incorporated magnetic nanoparticles (MNPs) into their matrix, have several desirable properties for immobilizing enzymes used for the magneto-bio-electrocatalysis process, which includes greater pore size and pore volume, the capacity to be placed with a magnet, as well as good conductivity of electrons. Carbon nanotubes are capable of binding iron and copper oxide nanoparticles via a liquid phase reduction process. A new catalyst (CNTs-Fe-Cu) was integrated with Al^0-CNTs to in situ generate H_2O_2 from activating O_2 for the degradation of the antibiotic sulfamerazine (SMR) in a Fenton-like system [21] (Figure 2.3(b)). A small quantity of rGO when combined with MNPs showed an important characteristic to be utilized as a chemical catalyst in the elimination of impurities from an aqueous medium. The catalytic system $ZnFe2O4/TiO2/RGO$ show a strong photocatalytic character and can be recycled easily. It can be used for the photocatalytic degradation of p-nitrophenol [22]. The primary active species in the photocatalytic degradation of p-Nitrophenol were superoxide anions (CO_2) and hydroxyl radicals (·OH), according to trapping experiments (Figure 2.3(c)). Less organized activated carbon was used as a surface for depositing MNPs with the potential of degrading common pollutants found leaching into the aquatic environment. Functional groups such as carboxylic acid or sulphonic acid can be used for the functionalization of magnetic tubular nanofibers. It can be further carbonized to obtain Fe_3O_4 nanoparticles. The magnetic tubular Fe_3O_4 nanoparticles can be effectively used in the adsorption of Uranium (VI) [23]. The polymeric nanofibers were produced by a self-polycondensation reaction, followed by carbonization and functionalization with SO_3H and COOH groups, as shown in Figure 2.3(d). The adsorption characteristics and magnetic nature of functionalized carbon nanomaterials for U(VI) adsorption from aqueous solution were improved after adding Fe_3O_4 nanoparticles using the decomposition method. Fenton's process for the removal of phenol from an aqueous medium can be catalyzed by using functionalized carbon nanomaterials. Persulfate activation can be aided by annealed nanodiamond as a carbon catalyst. The structure of annealed nanodiamond (900 °C) with sp^3 hybrid core of the nanodiamond and shell of graphitic carbon could be monitored by temperature for annealing and required time. The graphitic shell activated persulfate and oxidized water to form hydroxyl radicals. Persulfate was able to capture an electron from water, resulting in the formation of sulfate radicals. For selective removal of some pharmaceutical compounds, a catalyst in the form of nanodiamonds can be employed. Nanodiamonds facilitate the transfer of electrons from the phenolic group to persulfate. Ion implantation was used to synthesize a nanodiamond-based magnetic nanocomposite, while the crystal structure of the nanodiamonds does not change on annealing of the material. Even though it was not used as a catalyst in the case of persulfate activation, magnetic nanodiamond is thought to be a viable alternative. A variety of combinations are possible with iron nanocomposites as prospective materials. The development of novel magnetic nanocomposites may consist of a combination of iron oxide with copper oxide, gold, titanium oxide, vanadium oxide, etc.

2.2.4 Multicomponent MNP Based Composites

Due to the synergistic effects caused by interactions between different nanometer-scale components, multicomponent hybrid nanostructures that contain two or more nanometer-scale components have received much interest recently. Multicomponent structures have three distinct advantages. The first is to combine diverse functionalities

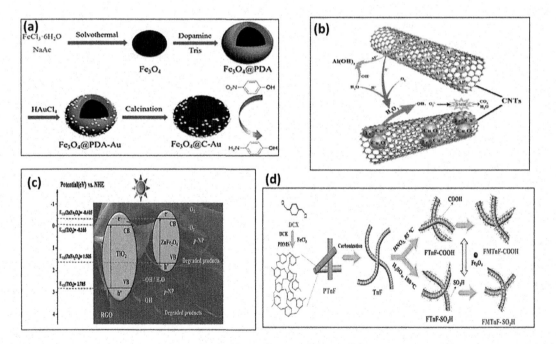

FIGURE 2.3 (a) Synthesis and application of nanostructured Fe_3O_4@C-Au, adapted with permission from [20]. Copyright (2020) Royal Society of Chemistry; (b) Al^0-CNTs/CNTs-Fe-Cu/O_2 system and its reaction mechanism, adapted with permission from [21]. Copyright (2020) Elsevier; (c) Schematic representation of a proposed photocatalytic process of ZTR 3 composite, adapted with permission from [22]. Copyright (2020) Royal Society of Chemistry; (d) Reaction mechanism for the preparation of sulphonic and carboxyl magnetic tubular nanofibers (FMTnF-COOH) and (FMTnF-SO_3H), adapted with permission from [23]. Copyright (2020) ACS Publications.

with individual component dimensions and material parameters optimized individually. The second benefit is that it allows for innovative functions that aren't possible with single-component materials or structures. The third advantage is the ability to achieve improved characteristics and overcome single-phase materials' inherent limitations. Covalent bonding links different nanoparticles together to form a hybrid structure.

The composite nanomaterials consist of oxides of metal, metal nanoparticles, and other nanomaterials in the form of core/shell nanostructures and other heterostructures of multicomponent content.

Because of their widespread use in catalysis, magnetic metal oxide, metallic type of core-shell nanoparticles can be considered as an important class of efficient nanomaterials. Ag nanoparticles have been studied extensively among the several noble-metal nanomaterials that can be utilized as the shell since they are substantially less expensive, non-toxic, and have specific optical, electrical, catalytic, and antibacterial capabilities. The iron oxide@Ag scheme has been extensively studied silver-based core-shell nanomaterials. The superparamagnetic property of iron oxide nanoparticles is useful in catalytic applications, while silver nanoparticles increase the strength and distribution of iron oxide. The nanoparticles work well in the catalytic reduction reaction and can be easily recycled by using an exterior magnet to separate them. Other core/shell nanoparticles with a variety of material combinations include metallic/magnetic $Fe_2O_3@SnO_2$ core-shell nanospindles structures to boost room-temperature HCHO oxidation [24].

The $Pt/Fe_2O_3@SnO_2$ demonstrated increased HCHO oxidation activity at room temperature when trace Pt nanoparticles were added. SEM and TEM images of pristine Fe_2O_3 and $Fe_2O_3@SnO_2$ nanospindles are shown in Figure 2.4(a). The molar ratios of Fe^{3+} and PO_4^{3-} have an impact on the morphology of α-Fe_2O_3, with lower and higher molar ratios resulting in the formation of α-Fe2O3 nanorods and nanospindles. The random growth and close packing of SnO_2 determine the surface roughness of $Fe_2O_3@SnO_2$, which promotes reactant adsorption. In another example, $ZnO/ZnFe_2O_4$ heterostructures [25] degrade Rhodamine B (RhB) and Methylene blue (MB) dyes in an aqueous solution with a relatively high catalytic performance. The flake-like nanosheets observed in Figure 2.4(b) are the result of a strong electrostatic repulsive force induced by Fe^{2+} ion additives. The strong coordination reaction in DMF solution causes Zn^{2+} ions and 2-methylimidazole to aggregates and amass into zeolitic imidazolate framework-8 (ZIF-8) MOF crystals. Dimer or oligomers heterostructures made up of specific components with diverse nanoscale properties may give rise to a blend of special properties. The non-symmetric structure may enable the insertion of an anisotropic distribution of various surface functional groups. In recent times, heterodimer (Pd@FexO) nanoparticles were used to synthesize a macroporous, hydrophobic, magnetically active, 3D template-free hybrid foam that could separate oil pollutants from water [26]. Pd domains in Pd@FexO heterodimers aided as nanocatalysts to bring about hydrosilylation of polyhydrosiloxane and tetravinylsilane, while the Fe_xO component imparted magnetic characteristics (Figure 2.4 (c)).

Photocatalytic dye degradation rates could be improved using hybrid nanostructures comprised of plasmonic noble metals known as PNMs and ZnO. This is because of the extraordinary properties of PNMs, such as the intense photons concentration, scattering of the photon, and the display of surface plasmon resonance (SPR). For catalytic processes, ZnO/Ag heterostructures are combined with Fe_3O_4 nanoparticles (NPs). For reusable type–surface plasmon resonance assisted photocatalysis, ZnO/Ag heterostructures can be connected to Fe_3O_4 nanoparticles [27]. SPR of Ag nanoparticles forms electrons in the Fermi level of energy of metal particles (E_{fm}) for illumination of visible photons. Later they plummet into the Fermi level of ZnO (E_{fs}) and are further caught by Fe^{3+} ions. The rate of methylene blue (MB) degradation is substantially accelerated by these electron separation processes as seen in Figure 2.4(d). The rate of degradation reaction of methylene blue (MB) in the presence of visible light illumination was used to assess the photocatalytic activities of the produced materials. The catalytic activity of multicomponent magnetic nanoparticle-based composites is shown in Table 2.2 [28–36].

2.3 Magnetic Bio-Nanocomposites

Bio-based composites are gaining popularity as means of reducing reliance on fossil fuels and transitioning to a sustainable material base. Bio-nanocomposites allow for the development of novel, elevated performance, light weight green nanocomposite materials that can be used to replace non-biodegradable

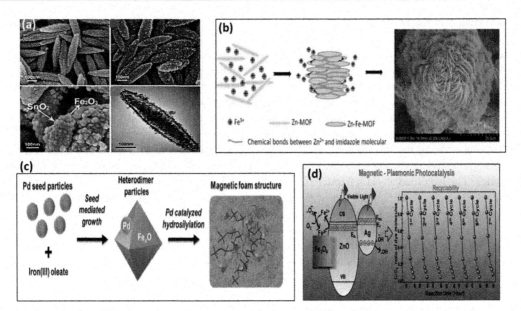

FIGURE 2.4 (a) Core-shell nanospindle structures of $Fe_2O_3@SnO_2$ effective for HCHO oxidation, adapted with permission from [24]. Copyright (2018) Elsevier; (b) Mechanism for the synthesis of nanosheet structures of Zn(Fe)-MOF, adapted with permission from [25]. Copyright (2017) Royal Society of Chemistry; (c) Pd@FexO heterodimer used in the preparation of magnetic foam for oil absorption and NPs for hydrosilylation reaction, adapted with permission from [26]. Copyright (2017) ACS Publications; (d) Surface plasmon resonance assisted photocatalysis with ZnO/Ag heterostructures associated with Fe_3O_4 NPs, adapted with permission from [27]. Copyright (2016) Elsevier.

TABLE 2.2 The Catalytic Activity of Multicomponent MNP-based Composites

CORE MNPS	OTHER COMPONENT	SURFACE AREA $M^2\ G^{-1}$	PREPARATION METHOD	TYPE OF CATALYTIC REACTION	REF
Fe_3O_4	Attapulgite/CuO/CeO_2	150.8	Mixing and calcination	Fenton	[28]
$CoFe_2O_4$	Diatomite	137.8	Citrate combustion method	PMS activation	[29]
Fe_3O_4	TiO_2/Ag–Au	N/A	Sol-gel combustion	catalytic reduction	[30]
Au	Fe_3O_4@$hTiO_2$	126	Stober's method and coating	Photodegradation and Catalytic reduction	[31]
Fe_3O_4	SiO_2/Ni	118.12	Solvothermal and sol-gel	Reduction and hydrogenation	[32]
$ZnFe_2O_4$	V_2O_5 (e-waste)	N/A	Solvothermal process	Fenton catalysis	[33]
Fe_3O_4	(MIL-100(Fe)	N/A	Solvothermal process (Fe_3O_4+ APTES+ $HAuCl_4$+ $FeCl_3 \cdot 6H_2O$)	In-situ catalytic oxidation of 3,3′,5,5′-tetramethylbenzidine (TMB) by $H2O2$ via a SERS technique	[34]
$CoFe_2O_4$	Mn-BDC	N/A	In-situ solvothermal ($CoFe_2O_4$ + DMF-MeOH + $MnCl_2$ + H_2BDC)	Multicomponent click coupling reaction to prepare pharmaceutically active 1,2,3-triazoles	[35]
Fe_3O_4	Polyvinyl alcohol(PVA)/ poly(acrylic acid)(PAA)/ MNPs with loaded AuNPs.	N/A	Electrospinning, Immobilization	Reduction of p-nitrophenol and 2-nitroaniline solutions	[36]

petroleum-based polymers. Coating MNPs with a biopolymer can help to prevent them from aggregating and enhance their application.

2.3.1 Biochar Based Magnetic Nanocomposites

Biochar is made up of fine-grained, carbon-rich, and porous material made by pyrolyzing biomass (it includes agricultural waste, the litter of animals, and sludge of municipal wastewater) at temperatures below 800 °C in an oxygen-deficient environment. It is widely used as a soil conditioner, carbon sequestration vehicle, and adsorbent in agriculture and the environment. Biochar could be utilized as an adsorbent to remove heavy metals and organic contaminants from the environment. Biochars made from agricultural waste are excellent dye absorbents when used for wastewater treatment. Many agricultural residues, including Azolla, corncobs, rice husk, coffee husk, fig leaves, sawdust, coconut shells, tea leaves, sugarcane, and cassava residues, are increasingly being reclaimed to generate biochar. Because of its porous properties, biochar plays a major role in catalytic remediation processes of environmental applications. Magnetic biochars (MBC) have recently received a lot of attention and have shown potential for the treatment of industrial effluents. The catalytic pathways of MBC have been studied in various systems including peroxydisulfate (PS), peroxymonosulfate (PMS), Fenton-like, photocatalysis, and NaBH4. For example, in the presence of peroxymonosulfate, $CoFe_2O_4$/HPC degraded bisphenol A significantly better than $CoFe_2O_4$ [37]. At the interface between biochar and $CoFe_2O_4$, biochar loaded with $CoFe_2O_4$ nanoparticles ($CoFe_2O_4$/HPC) produced through pyrolysis of maize straw, $KHCO_3$, and Fe and Co nitrates demonstrated excellent magnetic properties, ordered porous structure, and graphitized assemblies as shown in Figure 2.5(a). The activation performance of $CoFe_2O_4$/HPC for PMS to remove different organic contaminants such as tartrazine, BPA, phenol, sulfadiazine, and p-hydroxybenzoic acid was excellent. Increased PMS, the temperature of the reaction and a slightly lower pH were found to boost BPA removal efficiency. The free radical route, which was mostly provided by SO_4^{2-}, and the non-radical pathway, which was obtained by $^1O_2$ and transfer of an electron, were both involved in the degrading mechanism. Figure 2.5 (b) shows an insight into another mechanism of persulfate activation with zero-valent iron (nZVI) supported on biochar (BC) nanocomposite (nZVI/BC). The magnetic biochar nanocomposite was produced by dispersing zero-valent nano-iron (nZVI) on rice husk–derived biochar, which was utilized to activate persulfate for Nonylphenol degradation [2]. The presence of Fe^{2+} and –COOH, –OH groups on biochar or nZVI can activate persulfate molecules. PS could be activated indirectly by releasing Fe^{2+} from nZVI into aqueous systems, and then allowing the Fe^{2+} to directly activate persulfate to form free radicals. Because nZVI also functions as an electron donor, it can directly activate PS to create SO4·⁻ and·OH by electron transfer. Biochar was used as a carrier to disseminate nano-Fe^0 and prevent nanoparticle aggregation, which improved the catalytic activity of Fe^0. The graphitic biochar with high electron conductivity makes electron movement much easier. Furthermore, the magnetic biochar can adsorb the pollutants which reduce the distance of contact between pollutants and radicals, which accelerated the degradation process. Biochar is low cost as compared to other carbonaceous materials. Even though catalytic activity is not that appreciable in the case of pristine biochar, the reduced GO doping with change in processing conditions can improve it significantly. The catalytic activity of biochar can be improved either through doping with S, N, B, P, Fe, and Mn atoms, or by increasing the ratio of graphitic carbon by increasing the pyrolysis temperature, increasing porosity, and using the hydrothermal method of preparing biochar. Magnetic biochar composites are made by regulating the mass ratio of biochar and magnetic nanoparticles, optimizing the combining method, and selecting a suitable magnetic source that can generate magnetic biochar with high catalytic activity.

2.3.2 Enzyme Immobilized Magnetic Nanomaterials

The enzymes are known as naturally occurring types of biocatalysts that can be employed in a variety of products and processes, including pharmaceuticals, food processing, medicines, and biosensors, because of their high specificity, efficacy, and selectivity when compared to common catalysts used in chemical catalysis. Enzymes can catalyze a variety of reactions even under mild conditions. Soluble enzymes, on the other hand, do not remain stable in severe conditions during industrial operations, are expensive, and can be tough to recover from reaction mass for recycling. The recovery of the enzyme after a reaction for its recycling is, therefore, a tough task for industrial purposes. New ways for retaining enzyme activity and recycling without losing it are therefore in much demand. The carrier material and immobilization approach have a significant effect on the performance of immobilized enzymes. Various materials have been utilized for enzyme immobilization in the past, including organic or synthetic polymers, hydrogels, silica gels, and mesoporous materials. Nanoparticles, which have a large specific surface area, are a favorable material used for immobilizing and stabilizing enzymes. Because of their peculiar magnetic character, less toxicity, and chemically reactive surface, the core-shell type structured magnetic particles have recently acquired prominence. Cores of magnetic particles (Fe_3O_4) are usually applied in the enzyme immobilization because of several advantages, such as non-toxic nature, a large surface area, high mechanical characteristics, and the ability to modify the surface with different functional groups. Because bare magnetic nanoparticles are a non-porous kind of nanomaterial, mesoporous materials are frequently used to cover them. MCM-41 is one such type of mesoporous material is MCM-41. MCM-41 is employed as a support for immobilizing biomolecules due to the size of particles, kind of morphology, pore size, pore shape, low amount of toxicity, and thermal strength. Fe_3O_4@MCM-41 type core-shell nanoparticles, for example, were utilized to covalently immobilize various enzymes after being improved with functional molecules such as amino, epoxy, and thiol. Immobilization technology is one of the most effective ways to keep enzymes active. The carrier material and immobilization approach have a significant impact on the performance of immobilized enzymes. Nanoparticles, which have a large specific surface area, are a favorable material for immobilizing and stabilizing enzymes. It has a unique magnetic character, less toxicity, and a chemically changeable surface. The core shell-shaped magnetic particles have recently acquired prominence. For instance, core-shell CoFe2O4/Hydroxyapatite (HAP) nanoparticles showed high stability when functionalized with lipase enzyme through covalent bonding consisting of a molecular spacer [38]. The catalytic activity was evaluated and found to have high selectivity for the racemic mixtures of (R, S)-1-phenylethanol transesterification, where there is a full conversion of the (R)-1-phenylethanol enantiomer as shown in Figure 2.5(c). Another methodology of immobilization involves the use of 3-chloropropyl-trimethoxy-silane (CPTMS) an organic silylating compound. CPTMS can combine through covalent bonding with biological supermolecules such as proteins and enzymes. Covalent attachment occurs when chlorine groups and the enzyme's amino acid residues (–OH, –NH2, –SH) form a covalent bond. It is an effective immobilization approach for increasing enzyme immobilization yield and reusability. The covalent binding method of enzyme immobilization can firmly bind the enzyme to the carrier, preventing the shedding and leaking of the enzyme.

2.3.3 Microbial/Magnetic Nanocomposites

The high stiffness, good magnetic characteristics, and great stability of conventional magnetic nanoparticles have made them a commercially accepted material. However, these inorganic materials tend to agglomerate, resulting in poor mechanical and adsorption stability. To prevent aggregation, many chemical processes are available that can change the surface of inorganic materials. Microorganisms, particularly fungus, have recently been researched as a biological composite material in wastewater treatment due to their high surface-to-volume ratio. In comparison to inorganic materials, microbial nanoparticles have numerous advantages. To begin with, they are simple to obtain, can be artificially cultivated for a limited generation, and have a large number of chemical functional groups on their surface that can reduce secondary pollution by avoiding modification procedures. Furthermore, cells may direct and control the growth of inorganic materials. As a result, mycelium pellets can expand the length scale of nano-Fe_3O_4 to the macro-scale level as a living template. Finally, microbes have high mass transfer efficiency and are inexpensive. Microbial magnetic micro-materials have the potential to be used in wastewater treatment. The Saccharomyces cerevisiae fungus cell wall consists of chitin protein, lipid and glucan with functional groups like carbonyl, amino, hydroxyl, carboxylic acid, and phosphoryl group. Saccharomyces cerevisiae is also inexpensive, commonly available and eco-friendly. Because of such benefits, it is frequently utilized as a template for bio-micro composites. Derived from microorganism metabolites, the microbial extracellular polymeric substances also known as EPS is an intricate mixture of polysaccharides, proteins, and humic acids. Electrostatic interactions or complexations can link heavy metals to various functional groups related to EPS (e.g., phenolic, amino, hydroxyl, and carboxyl). As a result of their numerous functional groups, EPS could impact the subsistence of different heavy metals. For example, EPS can collect uranium (U) fallout from initial nuclear tests to avert it from dispersing. In recent years, microbial EPS has been shown to reduce radionuclides (U(VI) to U(IV), as well as heavy metals (Ag(I) to Ag(0) and Cr(VI) to Cr(III) and organic contaminants like nitrobenzene present in the environment. The reducing ability of EPS could be attributed to aniline, sulfhydryl, and phenols present in proteins, as well as hemiacetals found in polysaccharides. In one of the strategies, the concurrent reduction and adsorption of Sb(V) were achieved using microbial extracellular polymeric substances and a magnetic nanocomposite material. The functional groups of EPS could play an important role in the reduction of adsorption in microbial EPS-coated nano zero-valent iron, particularly under alkaline medium when nZVI is passivated [39]. Because Sb(III) is easier to take out than Sb(V), reduction adsorption is an efficient therapy for this pollutant of high priority. Inner Fe^0 might be protected from air by EPS, which would prevent nZVI from making a passivation layer on the surface and lower the concentration of hydroxyl radicals in the aqueous medium. This improved nZVI's reactivity while also improving its reducibility and adsorbability. Instead of inhibiting nZVI's electron transfer, the coated EPS can boost its reactivity, as seen in Figure 2.5 (d). As a result, microbial EPS may be useful in the decontamination of heavy metal polluted water. Nanocomposites have also been claimed to be produced by more evolved creatures. Live bacterial strains can be extracted from sludge laden with bacteria and utilized along with nano magnetite for the degradation of toxins. Examples of magnetic bio-nano composite materials containing biological component like biochar, microbes, and enzymes are included in Table 2.3 [40–48].

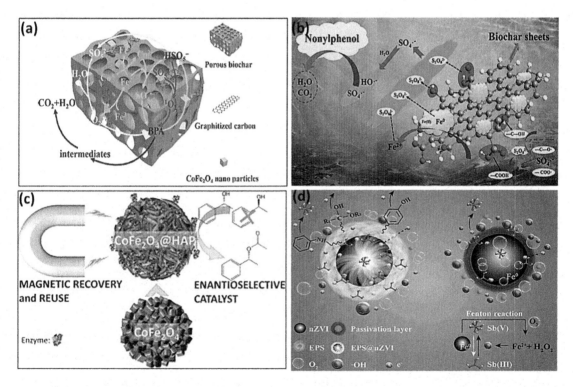

FIGURE 2.5 (a) General scheme for the activation of PMS for the degradation of bisphenol A using $CoFe_2O_4$ nanoparticles on graphene-like biochar. adapted with permission from [37]. Copyright (2020) Elsevier; (b) nZVI/BC structures effective in the mechanism of persulphate activation, adapted with permission from [2]. Copyright (2017) Elsevier; (c) Lipase immobilized $CoFe_2O_4$/HAP for enantioselective transesterification reactions, adapted with permission from [38]. Copyright (2020) Elsevier; (d) Enhanced reduction and absorption efficiency of EPS@nZVI/BC nanostructures for Sb(V), adapted with permission from [39]. Copyright (2020) Elsevier.

TABLE 2.3 Examples of Different Magnetic Bio-Nano Composite Materials with Catalytic Activity

MNPS	BIO-COMPONENT	SURFACE AREA $M^2 G^{-1}$	PREPARATION METHOD	TYPE OF CATALYTIC REACTION	REF
S–nZVI	EPS from sewage sludge	N/A	Liquid phase reduction	Anaerobic reduction	[40]
nFe_3O_4	Pine needles biochar	175.3	Pyrolysis and mixing	Fenton catalysis	[41]
$MnFe_2O_4$	Graphitized hierarchical porous corn stems biochar	389	Pyrolysis and graphitization	Activation of PMS	[42]
Fe_3O_4 MNPs	Lectin corona	N/A	Alkaline co-precipitation	Enzyme catalysis	[43]
Fe_3O_4@SiO_2	Immobilized lipase	N/A	Co-precipitation	Transesterification	[44]
nano-Fe_3O_4	Penicillium sp (fungus)	N/A	Self-assembly technique	Adsorbents for the removal of radionuclides	[45]
Hierarchical Fe_3O_4@MOF	Immobilized amidase	N/A	Solvothermal and modulator-induced defect-formation strategy	Synthesis of (S)-4-fluorophenylglycine	[46]
Fe_3O_4@SiO_2	Pd supported on silica – starch substrate (PNP-SSS)	N/A	Starch and chlorine-functionalized silica were used as support to immobilize Pd nanoparticles	Heck and Sonogashira coupling reaction	[47]
$\alpha Fe_2O_3/Fe_3O_4$	Bamboo biochar (HPA-Fe/C-B)	198.1	Coating α-Fe_2O_3/Fe_3O_4 onto bamboo biochar	Removal of phosphorus from aqueous solutions	[48]

3 APPLICATIONS OF MNPS IN CATALYSIS

Significant progress has been achieved for the advancement of novel catalytic processes that can be immobilized on magnetic nanomaterials in recent years. Generally, the non-magnetic inert material silica is used to create a magnetic core, which can protect it and maintain its magnetic properties. The silica shell is simple to functionalize and is effective in binding a wide range of catalytic species, including transition metal complexes. Inorganic phosphide complexes, for example, can easily be bonded between two nanoparticles by attaching the phosphide to the silica shell surrounding the nanoparticles. Nanoparticles of oxide of Iron can catalyze various coupling and oxidation types of reactions, adding a second type of metal dopant to the basic spinel structure of Fe_3O_4 broadening the scope of catalytic magnetic nanoparticles. The spinels with doping, mixed type spinel, and ternary spinels have all been widely studied, and their catalytic activity for reduction, hydrogenation, environmental contaminant degradation, coupling, condensation, oxidation has all been assessed [Table 2.4] [49–58].

The use of organic molecules as a catalyst is another form of catalyst that is relevant for organic synthesis. These compounds have a high degree of reaction specificity, which can permit an effective reaction compared to traditional processes in chemistry. MNPs have a variety of advantages when it comes to immobilizing organic catalyst molecules on heterogeneous substrates. They serve as a substrate for retrieving and recycling organic catalysts that would otherwise be impossible to extract from the reaction mixture. Organic catalysts are generally immobilized on magnetic core-shell nanostructures coated

TABLE 2.4 Examples of reactions catalyzed by magnetic nanomaterial

TYPE OF REACTION	MAGNETIC NANOMATERIAL	SYNTHESIS APPROACH	PERFORMANCE	METHOD OF SYNTHESIS	REF
Reduction	Au-Fe$_3$O$_4$@ MIL-100(Fe)	Layer-by-layer	Reduction of 4-nitrophenol	Fe$_3$O$_4$+ L-cysteine + chloroauric acid	[49]
	Cu-BTC@Fe3O4 composite	Secondary growth strategy	NaBH4 mediated reduction of nitroarenes to anilines	Carboxyl functionalized Fe$_3$O$_4$+Cu(OAc)$_2$•H$_2$O	[50]
Hydrogenation	Fe$_3$O$_4$@ZIF-8/Pd	Solvothermal	Semihydrogenation of phenylacetylene to styrene.	Core-Shell Pd-Based Catalyst (Fe$_3$O$_4$@M/Pd, M = SiO$_2$, AlOOH, TiO$_2$, Cu$_3$(BTC)$_2$ and ZIF-8)	[51]
	Pd@Hal@ Glu-Fe-C	Hydrothermal carbonation, surfactant-assisted sol-gel, coating	Hydrogenation of nitroarenes to corresponding anilines	Hydrothermally carbonized glucose (Glu) on halloysite nanoclay (Hal) + immobilization of (MNPs) + resorcinol-formaldehyde polymeric shell (RF) + carbonization + Pd NPs.	[52]
Coupling reaction	Fe$_3$O$_4$@PDA-Pd@[Cu$_3$(btc)$_2$]	layer-by-layer assembly	Suzuki-Miyaura coupling reaction.	core of polydopamine (PDA)-modified Fe$_3$O$_4$ NP + Pd NPs + Cu-based MOF outer shell	[53]
	EDTA-modified Fe$_3$O$_4$@SiO$_2$ nanospheres	Co-precipitation	Suzuki and Sonogashira cross-coupling reactions	Amino-functionalized Fe$_3$O$_4$@SiO$_2$+ cyanuric chloride group immobilized, +EDTA	[54]
Condensation reaction	ZIF-8@SiO$_2$@ Fe$_3$O$_4$	Encapsulation	Knoevenagel condensation between p-chlorobenzaldehyde and malonitrile	Encapsulation of SiO2@Fe$_3$O$_4$ into ZIF-8 via an in-situ approach	[55]
	Cu-BTC@SiO$_2$@ Fe$_3$O$_4$	ultrasonic assisted method	Pechmann reaction between 1-naphthol and ethyl acetoacetate	Magnetic SiO$_2$@Fe$_3$O$_4$ particles + thioglycolic acid (TA) + copper acetate solution	[56]
Oxidation	Core-shell Fe$_3$O$_4$@MIL-101(Fe)	PVP modified Solvothermal method, MOF-based catalysts for improved environmental remediation	Catalytic oxidation of Acid Orange 7	. Fe$_3$O$_4$ + FeCl$_3$.6H$_2$O + N, N-dimethylfor- mamide (DMF), terephthalic acid (H2BDC) + DMF	[57]
	Fe$_3$O$_4$@P4VP@ MIL-100(Fe)	Solvothermal method, encapsulation	aerobic oxidation of alcohols	Fe$_3$O$_4$@P4VP microspheres + FeCl$_3$.6H$_2$O + H$_3$BTC,	[58]

with silica or polymer. The silica shell can be employed as a functional substrate for attaching catalytic nanomaterials while stabilizing the magnetic core. However, there have been instances where the catalytic molecules can attach directly to the iron oxide nanomaterial surface. Considerable progress has been made in the synthesis of novel mesoporous MOF material coating on the surface of magnetic nanomaterials. It can act as a combined catalytic system with its magnetic and porous characteristics. The ligands of polydentate dendritic material can be immobilized on the magnetic nanomaterial. This helps in the removal of toxic catalytic substances during pharmaceutical drug synthesis. These processes are commonly used in different industries such as fragrances and flavors, pharmaceuticals, good industries etc. The development of novel magnetically immobilized enzymatic catalysts is another essential component. The sizes of biomolecules and nanoparticles are in nanoscale and can show similar properties of nanomaterials. Enzymes can also be called macromolecules and act as highly efficient target-oriented catalysts. The attachment of enzymes on the magnetic nanomaterial surface is through 1-Ethyl-3-(3-dimethyl-aminopropyl)carbodiimide bonding. These nanocomposites can then be employed in a variety of processes to produce valuable pharmaceuticals and organic compounds. Despite the numerous roadblocks and problems in this sector, the catalytic reaction carried by enzymes has significant scope in biopharmaceutical applications.

4 CONCLUSIONS

Over the last decade, magnetic nanoparticles have grown in popularity in the field of catalysis because they combine reactivity with a simple, cost-effective, and environmentally friendly form of recovery. Early recovery techniques centered on using nanoparticles in the form of the carrier for supporting additional nanomaterials or chemicals. Recent studies have revealed that bare magnetic nanoparticles can function both as a catalyst and magnetically retrievable entity in contrast to past strategies that focused on magnetic nanoparticles merely as a carrier for supporting other catalytic nanomaterials and chemicals. Simultaneously, a new concept of sustainability underlines the significance of the earth's plentiful and less hazardous minerals, particularly iron. Magnetic nanomaterials' catalytic character depends on the magnetic nanoparticle surface. The active species is either the magnetic material itself or another metal inserted within or connected to the MNP. Over the last several years, many design factors such as MNP size, crystallinity, shape, and composition, as well as the use of ligands or additions, have added to the improvement of MNPs' rich chemistry. MNP development for use as supports, as well as research into their uses in aqueous catalysis, is a key branch of green nanotechnology. MNPs have unlocked exciting new possibilities for the synthesis and design of catalytic processes using hybrid and multifunctional nanocatalysts. The use of a catalyst that is magnetically separable and reusable may make it possible to carry out reactions that are generally multistep synthesis in a single pot.

REFERENCES

[1] Magnetite NP Market survey report, Gd. View Res. (2021). www.grandviewresearch.com/industry-analysis/magnetite-nanoparticles-market%0A.
[2] I. Hussain, M. Li, Y. Zhang, Y. Li, S. Huang, X. Du, G. Liu, W. Hayat, N. Anwar, Insights into the mechanism of persulfate activation with nZVI/BC nanocomposite for the degradation of nonylphenol, *Chem. Eng. J.* 311 (2017) 163–172. doi:10.1016/j.cej.2016.11.085.
[3] K.D. Martinson, I.S. Kondrashkova, S.O. Omarov, D.A. Sladkovskiy, A.S. Kiselev, T.Y. Kiseleva, V.I. Popkov, Magnetically recoverable catalyst based on porous nanocrystalline HoFeO3 for processes of n-hexane conversion, *Adv. Powder Technol.* 31 (2020) 402–408. doi:10.1016/j.apt.2019.10.033.

[4] T. Oracko, R. Jaquish, Y.B. Losovyj, D.G. Morgan, M. Pink, B.D. Stein, V.Y. Doluda, O.P. Tkachenko, Z.B. Shifrina, M.E. Grigoriev, A.I. Sidorov, E.M. Sulman, L.M. Bronstein, Metal-ion distribution and oxygen vacancies that determine the activity of magnetically recoverable catalysts in methanol synthesis, *ACS Appl. Mater. Interfaces.* 9 (2017) 34005–34014. doi:10.1021/acsami.7b11643.

[5] Y. Ren, H. Li, W. Yang, D. Shi, Q. Wu, Y. Zhao, C. Feng, H. Liu, Q. Jiao, Alkaline ionic liquids immobilized on protective copolymers coated magnetic nanoparticles: An efficient and magnetically recyclable catalyst for knoevenagel condensation, *Ind. Eng. Chem. Res.* 58 (2019) 2824–2834. doi:10.1021/acs.iecr.8b05933.

[6] P. Zhai, C. Xu, R. Gao, X. Liu, M. Li, W. Li, X. Fu, C. Jia, J. Xie, M. Zhao, X. Wang, Y.W. Li, Q. Zhang, X.D. Wen, D. Ma, Highly tunable selectivity for syngas-derived alkenes over zinc and sodium-modulated Fe5C2Catalyst, *Angew. Chemie – Int. Ed.* 55 (2016) 9902–9907. doi:10.1002/anie.201603556.

[7] J.C.E. Yang, Y. Lin, H.H. Peng, B. Yuan, D.D. Dionysiou, X.D. Huang, D.D. Zhang, M.L. Fu, Novel magnetic rod-like Mn-Fe oxycarbide toward peroxymonosulfate activation for efficient oxidation of butyl paraben: Radical oxidation versus singlet oxygenation, *Appl. Catal. B Environ.* 268 (2020). doi:10.1016/j.apcatb.2019.118549.

[8] X. Li, J. Li, W. Shi, J. Bao, X. Yang, A fenton-like nanocatalyst based on easily separated magnetic nanorings for oxidation and degradation of dye pollutant, *Materials (Basel).* 13 (2020) 332. doi:10.3390/ma13020332.

[9] N.M.R. Martins, A.J.L. Pombeiro, L.M.D.R.S. Martins, A green methodology for the selective catalytic oxidation of styrene by magnetic metal-transition ferrite nanoparticles, *Catal. Commun.* 116 (2018) 10–15. doi:10.1016/j.catcom.2018.08.002.

[10] R.M. Borade, S.B. Somvanshi, S.B. Kale, R.P. Pawar, K.M. Jadhav, Spinel zinc ferrite nanoparticles: An active nanocatalyst for microwave irradiated solvent free synthesis of chalcones, *Mater. Res. Express.* 7 (2020). doi:10.1088/2053-1591/ab6c9c.

[11] S. Zha, Y. Cheng, Y. Gao, Z. Chen, M. Megharaj, R. Naidu, Nanoscale zero-valent iron as a catalyst for heterogeneous Fenton oxidation of amoxicillin, *Chem. Eng. J.* 255 (2014) 141–148. doi:10.1016/j.cej.2014.06.057.

[12] J. Qiao, Y. Guo, H. Dong, X. Guan, G. Zhou, Y. Sun, Activated peroxydisulfate by sulfidated zero-valent iron for enhanced organic micropollutants removal from water, *Chem. Eng. J.* 396 (2020). doi:10.1016/j.cej.2020.125301.

[13] X. Guan, X. Du, M. Liu, H. Qin, J. Qiao, Y. Sun, Enhanced trichloroethylene dechlorination by carbon-modified zero-valent iron: Revisiting the role of carbon additives, *J. Hazard. Mater.* 394 (2020). doi:10.1016/j.jhazmat.2020.122564.

[14] Y. Yang, J. Liu, Z. Wang, J. Ding, Y. Yu, Charge-distribution modulation of copper ferrite spinel-type catalysts for highly efficient Hg0 oxidation, *J. Hazard. Mater.* 402 (2021). doi:10.1016/j.jhazmat.2020.123576.

[15] F. Zhang, C. Wei, K. Wu, H. Zhou, Y. Hu, S. Preis, Mechanistic evaluation of ferrite AFe2O4 (A = Co, Ni, Cu, and Zn) catalytic performance in oxalic acid ozonation, *Appl. Catal. A Gen.* 547 (2017) 60–68. doi:10.1016/j.apcata.2017.08.025.

[16] J. Li, Y. Ren, F. Ji, B. Lai, Heterogeneous catalytic oxidation for the degradation of p-nitrophenol in aqueous solution by persulfate activated with CuFe2O4 magnetic nano-particles, *Chem. Eng. J.* 324 (2017) 63–73. doi:10.1016/j.cej.2017.04.104.

[17] H. Niu, Y. Zheng, S. Wang, L. Zhao, S. Yang, Y. Cai, Continuous generation of hydroxyl radicals for highly efficient elimination of chlorophenols and phenols catalyzed by heterogeneous Fenton-like catalysts yolk/shell Pd@Fe3O4@metal organic frameworks, *J. Hazard. Mater.* 346 (2018) 174–183. doi:10.1016/j.jhazmat.2017.12.027.

[18] A. Nuri, N. Vucetic, J.H. Smått, Y. Mansoori, J.P. Mikkola, D.Y. Murzin, Synthesis and characterization of palladium supported amino functionalized Magnetic-MOF-MIL-101 as an efficient and recoverable catalyst for mizoroki – Heck cross-coupling, *Catal. Letters.* 150 (2020) 2617–2629. doi:10.1007/s10562-020-03151-w.

[19] R. Ghorbani-Vaghei, H. Veisi, M.H. Aliani, P. Mohammadi, B. Karmakar, Alginate modified magnetic nanoparticles to immobilization of gold nanoparticles as an efficient magnetic nanocatalyst for reduction of 4-nitrophenol in water, *J. Mol. Liq.* 327 (2021). doi:10.1016/j.molliq.2020.114868.

[20] Q. Xu, T. Gao, S. Zhang, M. Zhang, X. Li, X. Liu, Synthesis of gold nanoparticle-loaded magnetic carbon microsphere based on reductive and binding properties of polydopamine for recyclable catalytic applications, *New J. Chem.* 44 (2020) 16227–16233. doi:10.1039/d0nj03216f.

[21] Y. Chen, Z. Yang, Y. Liu, Y. Liu, Fenton-like degradation of sulfamerazine at nearly neutral pH using Fe-Cu-CNTs and Al0-CNTs for in-situ generation of H2O2/[rad]OH/O2[rad]−, *Chem. Eng. J.* 396 (2020). doi:10.1016/j.cej.2020.125329.

[22] Y. Sun, J. Lei, Y. Wang, Q. Tang, C. Kang, Fabrication of a magnetic ternary ZnFe2O4/TiO2/RGO Z-scheme system with efficient photocatalytic activity and easy recyclability, *RSC Adv.* 10 (2020) 17293–17301. doi:10.1039/d0ra01880e.

[23] M. Ahmad, F. Wu, Y. Cui, Q. Zhang, B. Zhang, Preparation of novel bifunctional magnetic tubular nanofibers and their application in efficient and irreversible uranium trap from aqueous solution, *ACS Sustain. Chem. Eng.* 8 (2020) 7825–7838. doi:10.1021/acssuschemeng.0c00332.

[24] T. Lv, C. Peng, H. Zhu, W. Xiao, Heterostructured Fe$_2$O$_3$@SnO$_2$ core – shell nanospindles for enhanced Room-temperature HCHO oxidation, *Appl. Surf. Sci.* 457 (2018) 83–92. doi:10.1016/j.apsusc.2018.06.254.

[25] Y. Xu, S. Wu, X. Li, Y. Huang, Z. Wang, Y. Han, J. Wu, H. Meng, X. Zhang, Synthesis, characterization, and photocatalytic degradation properties of ZnO/ZnFe2O4 magnetic heterostructures, *New J. Chem.* 41 (2017) 15433–15438. doi:10.1039/c7nj03373g.

[26] M. Nawaz Tahir, M. Kluenker, F. Natalio, B. Barton, K. Korschelt, S.I. Shylin, M. Panthöfer, V. Ksenofontov, A. Möller, U. Kolb, W. Tremel, From single molecules to nanostructured functional materials: Formation of a magnetic foam catalyzed by Pd@FexO heterodimers, *ACS Appl. Nano Mater.* 1 (2018) 1050–1057. doi:10.1021/acsanm.7b00051.

[27] S. Chidambaram, B. Pari, N. Kasi, S. Muthusamy, ZnO/Ag heterostructures embedded in Fe3O4 nanoparticles for magnetically recoverable photocatalysis, *J. Alloys Compd.* 665 (2016) 404–410. doi:10.1016/j.jallcom.2015.11.011.

[28] T. Zhang, L. Dong, J. Du, C. Qian, Y. Wang, Cuo and ceo2 assisted fe2o3/attapulgite catalyst for heterogeneous fenton-like oxidation of methylene blue, *RSC Adv.* 10 (2020) 23431–23439. doi:10.1039/d0ra03754k.

[29] Y. Tan, C. Li, Z. Sun, R. Bian, X. Dong, X. Zhang, S. Zheng, Natural diatomite mediated spherically monodispersed CoFe2O4 nanoparticles for efficient catalytic oxidation of bisphenol a through activating peroxymonosulfate, *Chem. Eng. J.* 388 (2020). doi:10.1016/j.cej.2020.124386.

[30] J. Shen, Y. Zhou, J. Huang, Y. Zhu, J. Zhu, X. Yang, W. Chen, Y. Yao, S. Qian, H. Jiang, C. Li, In-situ SERS monitoring of reaction catalyzed by multifunctional Fe3O4@TiO2@Ag-Au microspheres, *Appl. Catal. B Environ.* 205 (2017) 11–18. doi:10.1016/j.apcatb.2016.12.010.

[31] J. Cheng, S. Zhao, W. Gao, P. Jiang, R. Li, Au/Fe3O4@TiO2 hollow nanospheres as efficient catalysts for the reduction of 4-nitrophenol and photocatalytic degradation of rhodamine B, *React. Kinet. Mech. Catal.* 121 (2017) 797–810. doi:10.1007/s11144-017-1185-z.

[32] J. Jiao, H. Wang, W. Guo, R. Li, K. Tian, X. Xu, Y. Jia, Y. Wu, L. Cao, In situ confined growth based on a self-templating reduction strategy of highly dispersed Ni nanoparticles in hierarchical Yolk – Shell Fe@SiO2 structures as efficient catalysts, *Chem. – An Asian J.* 11 (2016) 3534–3540. doi:10.1002/asia.201601196.

[33] H. Mohan, J.M. Lim, S.W. Lee, J.S. Jang, Y.J. Park, K.K. Seralathan, B.T. Oh, Enhanced visible light photocatalysis with E-waste-based V2O5/zinc – Ferrite: BTEX degradation and mechanism, *J. Chem. Technol. Biotechnol.* 95 (2020) 2842–2852. doi:10.1002/jctb.6442.

[34] X. Ma, S. Wen, X. Xue, Y. Guo, J. Jin, W. Song, B. Zhao, Controllable synthesis of SERS-active magnetic metal-organic framework-based nanocatalysts and their application in photoinduced enhanced catalytic oxidation, *ACS Appl. Mater. Interfaces.* 10 (2018) 25726–25736. doi:10.1021/acsami.8b03457.

[35] S. Yadav, S. Sharma, S. Dutta, A. Sharma, A. Adholeya, R.K. Sharma, harnessing the untapped catalytic potential of a CoFe2O4/Mn-BDC hybrid MOF composite for obtaining a multitude of 1,4-disubstituted 1,2,3-Triazole scaffolds, *Inorg. Chem.* 59 (2020) 8334–8344. doi:10.1021/acs.inorgchem.0c00752.

[36] R. Guo, T. Jiao, R. Xing, Y. Chen, W. Guo, J. Zhou, L. Zhang, Q. Peng, Hierarchical AuNPs-loaded Fe3O4/Polymers nanocomposites constructed by electrospinning with enhanced and magnetically recyclable catalytic capacities, *Nanomaterials.* 7 (2017). doi:10.3390/nano7100317.

[37] Y. Li, S. Ma, S. Xu, H. Fu, Z. Li, K. Li, K. Sheng, J. Du, X. Lu, X. Li, S. Liu, Novel magnetic biochar as an activator for peroxymonosulfate to degrade bisphenol A: Emphasizing the synergistic effect between graphitized structure and CoFe2O4, *Chem. Eng. J.* 387 (2020). doi:10.1016/j.cej.2020.124094.

[38] S. Saire-Saire, S. Garcia-Segura, C. Luyo, L.H. Andrade, H. Alarcon, Magnetic bio-nanocomposite catalysts of CoFe2O4/hydroxyapatite-lipase for enantioselective synthesis provide a framework for enzyme recovery and reuse, *Int. J. Biol. Macromol.* 148 (2020) 284–291. doi:10.1016/j.ijbiomac.2020.01.137.

[39] L. Zhou, A. Li, F. Ma, H. Zhao, F. Deng, S. Pi, A. Tang, J. Yang, Combining high electron transfer efficiency and oxidation resistance in nZVI with coatings of microbial extracellular polymeric substances to enhance Sb(V) reduction and adsorption, *Chem. Eng. J.* 395 (2020). doi:10.1016/j.cej.2020.125168.

[40] D. Zhang, Y. Li, A. Sun, S. Tong, X. Jiang, Y. Mu, J. Li, W. Han, X. Sun, L. Wang, J. Shen, Optimization of S/Fe ratio for enhanced nitrobenzene biological removal in anaerobic System amended with Sulfide-modified nanoscale zerovalent iron, *Chemosphere.* 247 (2020). doi:10.1016/j.chemosphere.2020.125832.

[41] J. Yan, L. Yang, L. Qian, L. Han, M. Chen, Nano-magnetite supported by biochar pyrolyzed at different temperatures as hydrogen peroxide activator: Synthesis mechanism and the effects on ethylbenzene removal, *Environ. Pollut.* 261 (2020). doi:10.1016/j.envpol.2020.114020.

[42] H. Fu, S. Ma, P. Zhao, S. Xu, S. Zhan, Activation of peroxymonosulfate by graphitized hierarchical porous biochar and MnFe2O4 magnetic nanoarchitecture for organic pollutants degradation: Structure dependence and mechanism, *Chem. Eng. J.* 360 (2019) 157–170. doi:10.1016/j.cej.2018.11.207.

[43] Y. Yong, R. Su, X. Liu, W. Xu, Y. Zhang, R. Wang, P. Ouyang, J. Wu, J. Ge, Z. Liu, Lectin corona enhances enzymatic catalysis on the surface of magnetic nanoparticles, *Biochem. Eng. J.* 129 (2018) 26–32. doi:10.1016/j.bej.2017.09.009.
[44] B. Thangaraj, Z. Jia, L. Dai, D. Liu, W. Du, Effect of silica coating on Fe3O4 magnetic nanoparticles for lipase immobilization and their application for biodiesel production, *Arab. J. Chem.* 12 (2019) 4694–4706. doi:10.1016/j.arabjc.2016.09.004.
[45] C. Ding, W. Cheng, Y. Sun, X. Wang, Novel fungus-Fe3O4 bio-nanocomposites as high performance adsorbents for the removal of radionuclides, *J. Hazard. Mater.* 295 (2015) 127–137. doi:10.1016/j.jhazmat.2015.04.032.
[46] C. Lin, K. Xu, R. Zheng, Y. Zheng, Immobilization of amidase into a magnetic hierarchically porous metal-organic framework for efficient biocatalysis, *Chem. Commun.* 55 (2019) 5697–5700. doi:10.1039/c9cc02038a.
[47] M. Tukhani, F. Panahi, A. Khalafi-Nezhad, Supported palladium on magnetic nanoparticles-starch substrate (Pd-MNPSS): highly efficient magnetic reusable catalyst for C-C coupling reactions in water, *ACS Sustain. Chem. Eng.* 6 (2018) 1456–1467. doi:10.1021/acssuschemeng.7b03923.
[48] Z. Zhu, C.P. Huang, Y. Zhu, W. Wei, H. Qin, A hierarchical porous adsorbent of nano-α-Fe2O3/Fe3O4 on bamboo biochar (HPA-Fe/C-B) for the removal of phosphate from water, *J. Water Process Eng.* 25 (2018) 96–104. doi:10.1016/j.jwpe.2018.05.010.
[49] F. Ke, L. Wang, J. Zhu, Multifunctional Au-Fe3O4@MOF core-shell nanocomposite catalysts with controllable reactivity and magnetic recyclability, *Nanoscale.* 7 (2015) 1201–1208. doi:10.1039/c4nr05421k.
[50] S. Yang, Z.H. Zhang, Q. Chen, M.Y. He, L. Wang, Magnetically recyclable metal – organic framework@Fe3O4 composite-catalyzed facile reduction of nitroarene compounds in aqueous medium, *Appl. Organomet. Chem.* 32 (2018). doi:10.1002/aoc.4132.
[51] L. Yang, Y. Jin, X. Fang, Z. Cheng, Z. Zhou, Magnetically recyclable core-shell structured pd-based catalysts for semihydrogenation of phenylacetylene, *Ind. Eng. Chem. Res.* 56 (2017) 14182–14191. doi:10.1021/acs.iecr.7b03016.
[52] S. Sadjadi, G. Lazzara, M.M. Heravi, G. Cavallaro, Pd supported on magnetic carbon coated halloysite as hydrogenation catalyst: Study of the contribution of carbon layer and magnetization to the catalytic activity, *Appl. Clay Sci.* 182 (2019). doi:10.1016/j.clay.2019.105299.
[53] R. Ma, P. Yang, Y. Ma, F. Bian, Facile synthesis of magnetic hierarchical core – shell structured Fe3O4@PDA-Pd@MOF nanocomposites: Highly integrated multifunctional catalysts, *ChemCatChem.* 10 (2018) 1446–1454. doi:10.1002/cctc.201701693.
[54] M. Esmaeilpour, S. Zahmatkesh, N. Fahimi, M. Nosratabadi, Palladium nanoparticles immobilized on EDTA-modified Fe3O4@SiO2 nanospheres as an efficient and magnetically separable catalyst for Suzuki and Sonogashira cross-coupling reactions, *Appl. Organomet. Chem.* 32 (2018). doi:10.1002/aoc.4302.
[55] Q. Li, S. Jiang, S. Ji, M. Ammar, Q. Zhang, J. Yan, Synthesis of magnetically recyclable ZIF-8@SiO2@Fe3O4 catalysts and their catalytic performance for Knoevenagel reaction, *J. Solid State Chem.* 223 (2015) 65–72. doi:10.1016/j.jssc.2014.06.017.
[56] Q. Li, S. Jiang, S. Ji, D. Shi, J. Yan, Y. Huo, Q. Zhang, Magnetically recyclable Cu-BTC@SiO2@Fe3O4 catalysts and their catalytic performance for the Pechmann reaction, *Ind. Eng. Chem. Res.* 53 (2014) 14948–14955. doi:10.1021/ie502489q.
[57] X. Yue, W. Guo, X. Li, H. Zhou, R. Wang, Core-shell Fe3O4@MIL-101(Fe) composites as heterogeneous catalysts of persulfate activation for the removal of Acid Orange 7, *Environ. Sci. Pollut. Res.* 23 (2016) 15218–15226. doi:10.1007/s11356-016-6702-5.
[58] Z. Miao, X. Shu, D. Ramella, Synthesis of a Fe3O4@P4VP@metal-organic framework core-shell structure and studies of its aerobic oxidation reactivity, *RSC Adv.* 7 (2017) 2773–2779. doi:10.1039/c6ra25820d.

Fabrication and Characterization of Two-Dimensional Transition Metal Dichalcogenides for Applications in Nano Devices and Spintronics

Geeta Sharma, Andrew J. Scott, Animesh Jha

Contents

1 Introduction	34
2 Structure	34
3 Properties	35
3.1 Optical Properties	35
3.1.1 Photoluminescence Spectroscopy	35
3.1.2 Raman Spectra	37
3.2 Magnetic Properties	38
4 Synthesis Methods	41
4.1 Vapor Phase Deposition	41
4.2 Exfoliations	41
5 Applications	42
5.1 Nanodevices	42
5.2 Spintronics	43
6 Conclusions	44
References	44

1 INTRODUCTION

With advanced fabrication methods, it is now possible to thin down many bulk layered materials down to a single monolayer. This has enabled the exploration of new low-dimensional physics such as anomalous quantum Hall effect and massless Dirac fermions in the case of graphene [1][2]. Transition metal dichalcogenides (TMDCs) form another class of layered materials, in which the interactions of d-electrons generate new physical phenomena [3][4]. TMDCs belong to the MX_2 class of semiconductors, where X is a chalcogen atom such as S, Se, or Te and M is a transition metal atom e.g., Mo and W [5,6]. Similar to graphite structure, single layers of the sandwich structure X–M–X can exist and multiple layers, stacked by weak van der Waals interactions, can form bulk solids [5,6]. The structure of TMDCs was first time determined by Linus Pauling in 1923 [7]. Later in 1963, ultrathin layers of MoS_2 were produced by Robert Frindt by exfoliation using adhesive tapes, and monolayer of MoS_2 suspensions was first produced in 1986 [8]. The discovery of graphene in 2004 fueled the development of techniques needed for working with layered materials leading to new studies related to 2D TMDCs both of fundamental and technological importance [9]. These materials find applications in optoelectronics, flexible electronics, spintronics, energy harvesting, sequencing of DNA, and customized medicines, etc. [8,10]. TMDCs are considered promising candidates to overcome the limitations belonging to zero-bandgap graphene, providing a possible solution for next-generation electronic applications. In particular, among the various TMDC materials that exhibit stable 2D crystalline structures (WSe_2, $MoSe_2$, WS_2, and MoS_2), MoS_2 has received considerable attention due to its unique optical, electronic, and other characteristics including its size-dependent bandgap [3–5][8,10].

In this chapter, taking MoS_2 as a benchmark material, a basic outlook of the large family of 2D TMDCs, highlighting their physical properties and introducing the recent preparation methods, are discussed. Finally, the emerging applications of MoS_2 in nanodevices and spintronics are presented.

2 STRUCTURE

TMDCs have a general chemical formula of MX_2, in which M represents transition metal and X stands for S, Se, and Te. In bulk form, these materials consist of X–M–X layers stacked together. These materials can exist in different structural phases (2H (thermodynamically stable), 1T (metastable form) [11,12]. MoS_2 is the most explored of all TMDCs for a variety of technological applications such as optoelectronics, nanoelectronics, and spintronics in different fields [8,10]. In a single layer of MoS_2, each Mo (+4) is surrounded by 6 S (−2) atoms and the Mo-S bond is mainly covalent. MoS_2 can exist in four structural phases: 1T, 1H, 2H and 3R [13]. The structural phases of MoS_2 depend upon the stacking of layers with respect to Mo coordination. For example, MoS_2 is mostly found in 2H phase in nature and this structure is formed due to layers stacked in an ABA type which leads to hexagonal symmetry because of two layers per unit cell. In the 1T and 1H phase, Mo atoms are octahedrally coordinated by six S atoms. 2H and 3R phases exhibit trigonal prismatic coordination around Mo atoms. The 3R structure with rhombohedral symmetry in MoS_2 is mostly obtained synthetically [12,13][14]. The unit cell consists of three layers. The 3R structure changes to the 2H type under heating due to its unstable nature. Mo hexagonal arrays are sandwiched between sulfur layers in both 2H and 3R forms. Whereas, in 1T crystal structure, the arrangement of layers follows ABC ordering. This is due to disorientation of one sulfur atom layer and the bond between Mo atoms (Mo-Mo bond) is symmetrical. 1T structure is metallic in nature.

The conduction band is formed by d orbitals of Mo whereas valence band is due to $2p$ orbitals of the sulfur atom. The hybridization of p_z orbital of S and d orbital of Mo leads to the formation of a bandgap between K and Γ. The bandgap changes as a function of a number of layers. The Γ point which represents valence band maximum falls below zero with a reduction in the number of layers

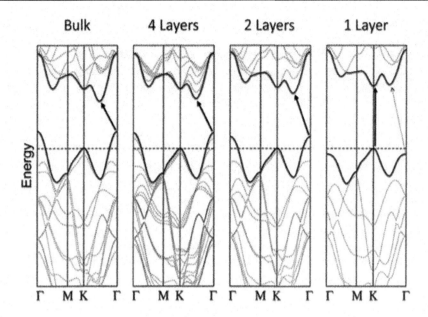

FIGURE 3.1 Variation in the energy dispersion curves in bulk and 2D materials. Adapted with permission from Reference [15].

Source: Copyright © 2010, American Chemical Society.

nevertheless the K point remains unchanged. Hence, in case of a single layer, the lowermost energy transition becomes vertical since energy required for indirect transition is more than the direct band edge transition. The band structure of MoS_2 varies as the function of a number of layers which in turn affects the chemical, physical and magnetic properties [11,12]. The bandgap changes from direct in monolayer MoS_2 (1.9 eV) to indirect (1.3 eV) in bulk. This change is due to long-range coulombic effects and quantum confinement in few-layer materials [28,29]. Figure 3.1 shows variation in the gap between valence band and conduction band with reduction the number of layers.

The semiconducting properties of MoS_2 in 2H phase are due to the presence of empty d_{xy} and $d_{x^2-y^2}$ and filled d_{z^2} orbitals. The 1T phase shows metallic properties as 4*d* orbitals split into e_g and t_{2g} states and two electrons are filled in the t_{2g} state. With reduction in number of layers, charge mobility drastically increases due to the direct bandgap and thinness of the material. These materials can be designed for different applications since their properties vary as a function of the number of layers. Further, these materials can be engineered by intercalation, reducing dimensions, and by forming a variety of heterostructures [12].

3 PROPERTIES

3.1 Optical Properties

3.1.1 Photoluminescence Spectroscopy

MoS_2 in the bulk form shows negligible photoluminescence and it has an indirect bandgap [16]. It becomes strongly photoluminescent when thinned to monolayer thickness. This control over emission characteristics at the atomic scale is due to the quantum confinement effect which provides freedom to engineer the matter at the nanoscale. The reduced dielectric screening in 2D semiconductors leads to very strong interactions between electrons, holes, and photons. This results in the formation of a bound quasi particle known as a neutral exciton (X) due to strong coulombic interactions between an electron and hole. The charged quasi particles consisting of either two electrons and one hole (known as negative trions and denoted as X– or T–) or one electron and two holes (known as positive trions, denoted as X+ or T+) are also formed. The large exciton binding energies in 2D TMDCs additionally leads to either neutral (XX) or charged (XT) biexcitons [17]. These excitonic species are elementary quasi-particles for 2D materials and control the optoelectronic properties of 2D materials.

The direct transition between valence and conduction band states around the K and K′ points in 2D materials occurs in the near-infrared and visible spectral region [16]. Monolayer TMDCs have theoretically very large exciton binding energies E_B (0.5–1 eV) for an exciton Bohr radius $a_B \approx 1$ nm [16,17]. The experimental absorption spectra of these materials show sharp resonance features due to strong excitonic effects [17]. Some recent experimental studies based on optical spectroscopy [16,18] have verified the large binding energies. Under pulsed laser excitation, trions and bi-excitons are formed in doped monolayer TMDCs [18,19]. Due to large binding energies of the excitonic particles present in 2D TMDCs their observation even at room temperature is quite possible. This makes trions very useful for electrical transport at room temperature. It is also possible to create high-density quantum coherent states of excitons [18].

Figure 3.2 shows optical reflection, Raman scattering, and photoluminescence spectroscopy [4] done by Andrea Splendiani et al. on few-layer MoS$_2$ structures. The A1 and B1 excitons show two prominent absorption peaks at 670 nm and 627 nm at the Brillouin zone K point in the spectrum. The spin-orbital splitting of the valence band causes the energy difference in the absorption peaks. Photoluminescence of single-layer MoS$_2$ is shown in Figure 3.2. It shows strong emissions due to A1 and B1 direct excitonic

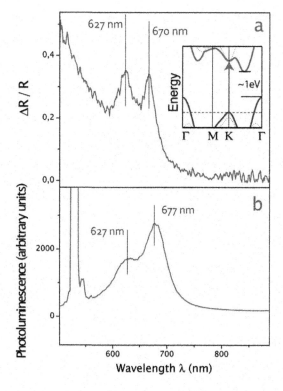

FIGURE 3.2 (a) Reflection difference due to ultrathin MoS$_2$ layers on a quartz substrate. The band structure of bulk MoS$_2$ is shown in the inset. (b) Photoluminescence spectra of ultrathin MoS$_2$ layers. The direct excitonic transitions are observed at 627 nm and 677 nm. Adapted with permission from Reference [4].

Source: Copyright © 2010, American Chemical Society.

transitions. These transitions are not present in bulk MoS$_2$. These transitions inherently belong to single-layer MoS$_2$ monolayer and external perturbations such as defect states might not affect them.

3.1.2 Raman Spectra

Raman spectroscopy is a popular technique to investigate number of layers, types of edges, controlling the quality, effect of an electric and magnetic field, attached chemical groups, strain, doping, etc. in 2D materials [20]. With rapid development in the Raman spectroscopy techniques and unique engineering of the materials, it is also used to probe the strength of interlayer coupling and interface coupling in the van der Waal heterostructures (vdWHs) [21]. MoS$_2$ has four Raman active modes, E_{1g}, E^1_{2g}, A_{1g}, and E^2_{2g}. The prominent peaks due to E^1_{2g} and A_{1g} are shown in Figure 3.3. E^1_{2g} and A_{1g} correspond to in-plane and out-of-plane vibrations of sulfur atoms and are generally used to understand the crystal structure of MoS$_2$ [12,20].

The typical changes in these Raman peaks while going from bulk to monolayer MoS$_2$ are listed below:

(i) E^1_{2g} shows a blue shift, whereas A_{1g} displays an opposite red shift. For single-layer MoS$_2$, E^1_{2g} is located at ~384 cm^{-1} whereas A_{1g} is at ~ 405 cm^{-1}.
(ii) The difference between E^1_{2g} and A_{1g} decreases as a function of the number of layers. The frequency difference changes from approximately 25 cm^{-1} for bulk to 19 cm^{-1} for monolayer MoS$_2$.
(iii) The peak intensities of these bands increase linearly up to four layers before decreasing for thicker MoS$_2$

Figure 3.3 shows variation of Raman peaks with increasing thickness. Raman characteristics of 2D materials in solution are quite different from solid-state 2D materials. Until now, Raman studies have focused on 2D material flakes in the solid-state only. Yanqing Zhao et. al. [20] have investigated 2D MoS$_2$ in solution using angle-resolved polarized, helicity-resolved, and resonant Raman spectroscopy. The ultrasound-assisted exfoliation of bulk MoS$_2$ in a solvent was done which resulted in the formation of a solution of 2D materials. This solution shows better and complete Raman spectra in comparison to 2D materials in the solid phase. Figure 3.4 reveals forbidden E_{1g} mode in the solution of 2D materials. The

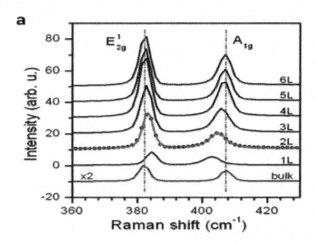

FIGURE 3.3 Variation in the Raman spectra of MoS$_2$ from thin layer (nL) to bulk form. Adapted with permission from Reference [22].

Source: Copyright © 2010, American Chemical Society.

38 Applications of Low Dimensional Magnets

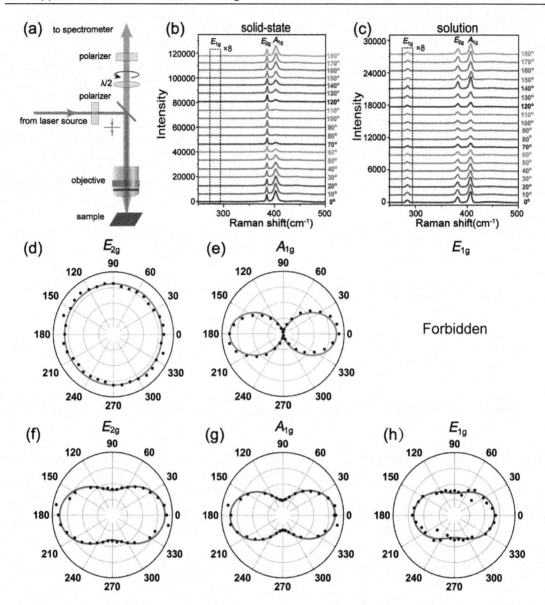

FIGURE 3.4 Angle-resolved polarized Raman spectroscopy of MoS$_2$ in solid-state and in solution. (a) Setup for the angle-resolved polarized Raman spectroscopy. (b) Raman spectra of solid-state MoS$_2$ flake with various angles (θ) between e$_i$ and e$_s$. (c) Raman spectra of MoS$_2$ solution. Adapted with permission from Reference [20].

Source: Copyright © 2020, American Chemical Society.

random orientation of dispersive nanosheets is found to be responsible for the unique Raman characteristic of 2D material in solution [20].

3.2 Magnetic Properties

Graphene and TMDCs are inherently non-magnetic in nature. Magnetic properties of these materials are modified using gating, doping, functionalization, etc. to obtain spin polarized states. In general, magnetism has two origins: the spin of elementary particles and moving charges. Both of these are hindered by a thermal disturbance in 2D materials. Magnetic properties of non-magnetic 2D materials are altered by suitable doping or by introducing vacancies. Due to these modifications, interactions between unpaired electrons give rise to local magnetic moments thus making it possible to realize long-range magnetism [23–26].

The optical, electrical, and mechanical properties of MoS$_2$ are well investigated both theoretically and experimentally. The magnetic response is mostly studied through theoretical calculations [27–31] but recently these materials have been studied experimentally [26,32–38]. According to theoretical predictions, different magnetic ground states exist due to different directions of termination of edges. For instance, armchair edges show stability in a non-magnetic state while the zig-zag edges possess net magnetic moment and are in a magnetic ground state. Hence, presence of zig-zag edges (providing the average grain size is small enough) leads to magnetism in MoS$_2$ nanoribbons, nanocrystalline thin films, and even in bulk structures. Sefaattin Tongay et al. investigated bulk MoS$_2$ in form of single-crystal experimentally, in the temperature range 300 K–10 K with an applied magnetic field ranging from 0 T-5 T [34]. It is found that magnetization of MoS$_2$ is possibly due to the existence of zig-zag edges with allied magnetism at grain boundaries and a temperature-dependent diamagnetic background. Figure 3.5 shows magnetic behavior of MoS$_2$ at different temperatures. From Figure 3.5 it is evident that magnetic response is subjugated by diamagnetic character however diamagnetic background is superimposed upon the ferromagnetic loop. With an increase in grain size (or ribbon width) it is expected that the net magnetic moment should decrease thereby diminishing ferromagnetic response in the bulk limit. However, there are a considerable number of grain boundaries present in these MoS$_2$ samples due to the small average grain size (75 nm). The ferromagnetic behavior of these samples is due to arbitrary distribution of zig-zag and armchair edges in a large number of grain boundaries. The estimated ferromagnetic signal from these zig-zag edges is 2.6×10^{-2} emu/g.

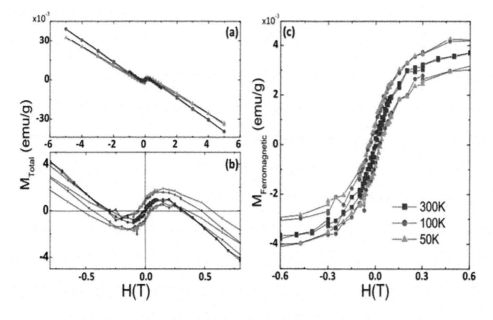

FIGURE 3.5 (a) Magnetization (M) vs applied field (H) curves of MoS$_2$ taken at different temperatures in the field parallel to the c-axis (b) M-H curves displayed at lower magnetic fields. (c) M vs H curves after subtracting out the diamagnetic background.

Source: Adapted with permission from Reference [34]. AIP Publishing.

In TMDC monolayers, electrical and optical properties are thickness-dependent but show non-magnetic behavior in their intrinsic form [32,39]. Atomically thin, dilute magnetic semiconductors (DMSs) are formed by doping with transition metal elements (V, Mn, and Fe) in TMDC due to magnetic coupling in these 2D structures. Doping of TMDC monolayers with transition metal ions is considered a promising way to realize DMSs having a Curie temperature above room temperature as predicted by first-principles studies [39]. Fe: MoS_2 monolayers reveal a well-defined M–H hysteresis loop both at low (5 K) and room temperature, suggesting that ferromagnetism in these materials can be observed even at 300 K [32][40]. Jieqiong Wang et. al. demonstrated robust ferromagnetism in Mn-doped MoS_2 nanostructures synthesized by a hydrothermal method [36]. Figure 3.6 shows the magnetic properties of MoS_2 and MoS_2 doped with Mn^{2+}. The observed ferromagnetism in Mn-doped MoS_2 shows strong temperature dependence and is quite different from defect-induced magnetism. Observed ferromagnetic phases with Curie temperatures of 80 K and 150 K are shown in Figure 3.6. It is found that all the Mn^{2+} dopant ions don't contribute equally to the observed magnetism. Mn^{2+} are paramagnetic if they are well isolated whereas the Mn^{2+} ions having nearest neighbors form clusters and are anti-ferromagnetic and do not contribute to the overall magnetism. Some Mn^{2+} ions order ferromagnetically via an indirect coupling mechanism. This is due to the tuning of spin polarization and magnetic ordering in TMDCs and can have applications in spintronic devices.

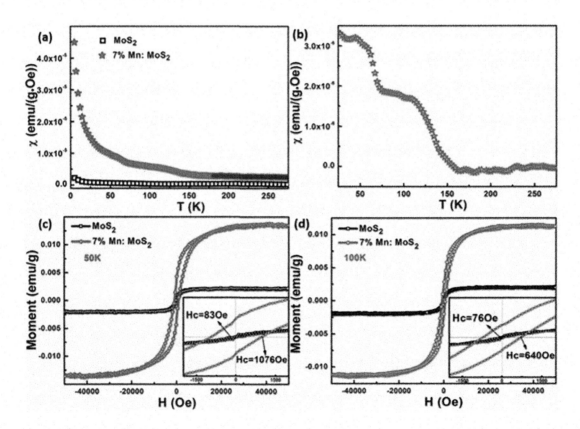

FIGURE 3.6 Magnetic properties of the MoS_2 and MoS_2 doped with Mn^{2+}. (a) Temperature dependence of magnetic susceptibility of the undoped and the 7% Mn^{2+} doped MoS_2 measured at a field of 500 O_e, and fitting of the high-temperature μ-T curve of the Mn-doped MoS_2 by the Curie-Weiss law (b) the μ-T curve of the 7% Mn^{2+} doped MoS_2 after paramagnetic background subtraction; two ferromagnetic transitions are observed. (c) Magnetic hysteresis loops of the undoped and the 7% Mn^{2+} doped MoS_2 measured at (c) 50 K and (d) 100 K, respectively.

Source: Adapted with permission from Reference [36]. AIP Publishing.

Peng Tao et al. have reported strain-induced magnetism in single-layer MoS$_2$ [31]. The delocalized electrons of Mo are believed to be responsible for this unique magnetism in the defect MoS$_2$. Due to these induced magnetic moments, MoS$_2$ can be used in the design of magnetic-switching or logic devices. The magnetic properties are also found to be structurally phase-dependent. S. Yan et al. investigated magnetic properties of 2H phase of MoS$_2$ (2H-MoS$_2$) and 1T phase of MoS$_2$ (1T-MoS$_2$) both experimentally and theoretically in single-layer MoS$_2$ sheets [33]. The pristine MoS$_2$ (2H-MoS$_2$) is found to be weakly diamagnetic. After exfoliation, significantly enhanced paramagnetism is observed due to transformation of the crystal structure from 2H to 1T.

4 SYNTHESIS METHODS

Synthesis techniques play an important role in obtaining material with desired properties. To make MoS$_2$ suitable for different applications, it is necessary to opt for a selection of an effective fabrication technique. The synthesis techniques for layered materials fall into two major categories: bottom-up and top-down methods. Exfoliations such as mechanical and in the liquid phase, chemical intercalation, etc. belong to the top-down class. The bottom-up process generally includes chemical vapor deposition and chemical synthesis etc. [41][11,12]. Some of these methods are discussed below:

4.1 Vapor Phase Deposition

This technique is used to prepare ultra-thin films by depositing vapor-phased compounds on the desired substrate (with or without chemical reaction) to form layers. The 2D materials formed have high crystallinity and uniformity and high layer controllability. Generally, this technique involves three main approaches: thermal decomposition of precursors, physical vapor deposition, and chemical vapor deposition.

- **Physical vapor deposition (PVD):** The PVD approach is to produce 2D materials by re-crystallization of materials through a vapor-solid process. Various 2D materials can be prepared by PVD. However, random nucleation of the crystals in PVD can make the layer thickness uneven; research is continuing to resolve this issue.
- **Chemical vapor deposition (CVD):** Since the size of MoS$_2$ films prepared by conventional methods such as exfoliation and vapor deposition are of the order of micrometers, considerable efforts have been put into preparing the large area, thin-layer MoS$_2$, though this is still a challenge. Chemical vapor deposition has gathered immense attention as 2D TMDCs have been successfully synthesized on a wafer-scale using this method. Another major advantage of this method is that interfacial contamination introduced during the layer-by-layer transfer process can be reduced and layered heterostructures can be grown using CVD.
- **Thermal decomposition:** Due to difficulties in controlling the thickness, uniformity, and polycrystalline nature of the films (such as deposited Mo film or (NH$_4$)$_2$MoS$_4$ thin film) this method is not widely used.

4.2 Exfoliations

- **Mechanical exfoliations:** To obtain one- or few-layer nano-sheets, mechanical exfoliation is a versatile and low-cost method in which the crystal structure and properties of a material are well maintained. The micromechanical exfoliation method was first used by Novoselov and Geim to obtain single-layer graphene from graphite.

- **Ultrasonic exfoliation**: This method employs the strategy of delaminating van der Waals solids into single-layer nanosheets. This method is more advantageous and effective compared to mechanical exfoliation. Various recent reports suggest solvents play an important role in stabilization of exfoliated nanosheets. Although this method has many advantages, high-purity single-layer 2D material is difficult to obtain using ultrasonic exfoliation. Recently, Chhowalla *et. al.* reported the synthesis of monolayer MoS_2 via lithium-intercalation and exfoliation [42].

5 APPLICATIONS

The various practical applications of MoS_2 are possible only due to its promising properties which are discussed above. These applications include biosensors, supercapacitors, solid lubricants, etc. These properties include a low coefficient of friction, high mechanical strength and high surface area, variable bandgap, excellent transport properties accounting for the large number applications [16,43,44] [25,27,28,32,36,37,40]. MoS_2, in particular, explored now being extensively studied for nanodevices and spintronics applications.

5.1 Nanodevices

The future semiconductor industry needs high-performance nanoelectronics, low-power, and multifunctional devices as it is not possible to further add new components into the existing silicon platform. Two-dimensional layered semiconductors possess ultrathin structures, atomic-scale smoothness, dangling bond-free surfaces, high carrier mobility, and a sizable bandgap. Flexible nanotechnology is desired for next-generation electronics and energy devices. 2D materials are suitable candidates for these applications such as sensors, thin-film transistors (TFTs), displays, solar cells, energy storage, etc. 2D TFTs based on MoS_2 operating at room temperature show a high on/off current ratio and current saturation. Electron mobility as high as 50 $cm^2V^{-1}s^{-1}$ and a current density of 250 mA/mm is observed from such TFTs [45]. Flexible monolayer MoS_2 TFTs offer robust electronic performance to 1000s of cycles of mechanical bending [45]. Because of these encouraging properties, there is great interest in their use for low-power RF TFTs for advanced flexible Internet of Things (IoT) and wearable connected nano-systems.

2D materials are considered as a reliable material for logic applications *such as* MOSFETs because these are compatible with the CMOS technology and can be easily scaled up and are quite cost-effective for large-scale synthesis. High mobility, band gap (~1 eV), and good ohmic contacts are basic requirements of material to be used in the aforementioned applications. The layered structure of these materials is bonded together by weak van der Waals forces [46]. The large relative effective mass of 2D TMDCs (~0.5 for electrons and ~0.66 for holes) as compared to Si (~0.29) leads to a reduction in the source-drain tunneling component in TFETs [47]. Intense research is being carried out to develop novel 2D materials with high performance and/or to improve performance of current 2D materials. Considerable progress in the field of 2D materials for state-of-the-art electronic nano-devices has been achieved in recent years However, there is still a paucity of 2D materials which can be potentially used for high-performance 2D-based electronic devices. MoS_2 has low electron mobility [48,49] and a relatively large bandgap [50–53] resulting in a high I_{on}/I_{off} ratio (~108) [48]. It has been reported by several researchers that the performance of electronic devices is affected by the number of layers, nature of substrate, and other 2D layered material coupled to them. Hence, to achieve the above-mentioned properties simultaneously, hetero-structures using different 2D materials, such as graphene and TMDCs can be used [11,54]. For the sake of thermal management and designing high-quality electronic devices, thermal conductivity and heat dissipation along with the above-mentioned properties also play an important role. Graphene and hexagonal-BN are still promising candidates for FET applications in comparison to 2D TMDCs despite these materials being widely investigated both experimentally and theoretically [55–57]. The difference in the thermal properties of monolayer MoS_2 (2D TMDC) compared with a one-atom thin graphene layer can be attributed to its sandwich structure due to the dominance of phonon contribution over the electron contribution in thermal conductivity [58]. Recent investigations revealed that the theoretical thermal conductivity of MoS_2 is 1.35 W m^{-1} K^{-1} [59][60] or 23.2 W m^{-1} K^{-1} [58]. Further, the values of thermal conductivity of MoS_2 with a few layers are found to be 1.59 W m^{-1} K^{-1} [61] and 52 W m^{-1} K^{-1} [62] experimentally at room temperature. For the realization of state of art electronic device applications, a deep insight into the thermal properties of MoS_2 (TMDCs), is of utmost importance.

The TMDCs are proving themselves very promising candidates for optoelectronic applications as single layers of these materials show excellent PL and electroluminescent properties and their flexibility allows novel innovative device designs. These materials show strong inter-band transitions due to heavy effective mass of "d" electrons and the van Hove singularity peaks in electronic density of states. An optical absorption as high as 107 m^{-1} is observed in a 300 nm thick TMD which leads to 95% absorption. The carrier mobility in 2D TMDCs is further enhanced by low density of traps. The carrier mobility of MoS_2 nanosheets is reported as 0.5 cm^2 V^{-1}s^{-1}–200 cm^2 V^{-1}s^{-1}[63]. This wide range is attributed to the sample quality, presence of adsorbates, suppression of impurities and grain boundaries, etc. The TMDC based photo-detectors have a limitation of low photo-response because active photo-response area is only limited to the regions at metal-TMDC interface as most of the incident light is not absorbed. Heterostructures of TMDCs with other 2D materials are particularly desired for practical optoelectronic application since there is better control of photocurrent separation and carrier transport. The TMDCs are also explored for photonic applications. It is found that MoS_2 shows greater saturation absorption near 800 nm but the lack of gain medium around this wavelength inhibits its usage. The saturation absorption is wavelength dependent, hence a high pump power is needed to saturate absorption at shorter wavelengths. Research related to the usage of MoS_2 in mode-locking laser applications is increasing these days [63] but there is limited understanding of the underlying physics of TMDCs for these applications hence more experimental work is needed.

5.2 Spintronics

The poor spin-orbit interaction on traditional spintronics materials permits operation of devices below room temperature. Further, ultra-pure materials are needed to avoid spin-flip scattering [64]. For next-generation nano-electronic and spintronic devices, new materials such as 2D materials are being investigated. High electron mobility and long-distance spin transport are possible in graphene at room temperature. However, high spin-orbit coupling and a direct bandgap are a primary need for a switching action in charge- or spin-based transistors. Graphene lacks a bandgap and spin-orbit coupling in its pristine state. The properties of graphene may be modified by doping, this being necessary for opening up the bandgap. Strong spin–orbit coupling induces spin splitting of up to 0.4 eV in 2D transition-metal dichalcogenides (TMDs). This enables the possibility of room-temperature operation of spintronics devices [65]. These properties make TMDCs suitable for novel spintronics applications for example magnetic sensors, ultra-thin high-density data storage devices, spin-field-effect switches, spin valves, magnetic logic gates, magnetic random-access memory, etc.

There are several challenges for developing two-dimensional spintronics: (i) increasing the magnetism in the system by spin injection and (ii) efficient manipulation of the spin [66,67]. Among various TMDCs reported, MoS_2 is special because of its interesting physical properties, which are a function of its thickness and structural phase transitions. Some of the properties which make MoS_2 a potential candidate for spintronic and nanodevice applications are absence of inversion symmetry, high atomic mass, very strong spin-orbit splitting due to confinement of electron motion in the plane, high on/off current ratios, and bandgap tunability (1.2–1.8 eV). Pristine 2H-MoS_2 has a non-magnetic ground state due to spin (S) = 0 Mo 4+ in a trigonal prismatic geometry, therefore a significant modification in the physical properties is essential before exploring it for spintronics applications. There are several reports of ferromagnetism in MoS$_2$ via transition-metal, non-metal, and lanthanide ion doping, adsorption of non-metals, structural phase transitions, and strain-induced defects.

For the practical realization of spin-based devices, electrical injection, transport, manipulation, and detection of spin-polarized carriers in MoS$_2$/ferromagnet (FM) heterostructures are primary requirements [65]. The electrons in two-dimensional systems possess valley degree of freedom which allows manipulation of physical properties and in turn leads to new device applications. At present there are few suitable two-dimensional valleytronic materials and search for such materials is very rigorous. Semiconducting behavior is exhibited by monolayer of MN$_2$X$_2$ and valence band maximum forms two degenerate valley. The strong spin-valley coupling is caused by large spin splitting (601 meV) [68]. The hole doping along with optical illumination in this system can lead to valley Hall effect. Information storage devices such as hard discs, solid state drivers, USB flash drivers etc. will use valleytronics. Hence it is very crucial to find out various experimental ways for the tuning of valley degree of freedom. Electrical, optical and magnetic methods are actively explored to achieve valley polarization.

6 CONCLUSIONS

Intensive research on MoS$_2$ is still going on, to exploit its favorable chemical, photonic, and electronic characteristics. These materials can be explored concerning the synthesis techniques and their compatibility with other materials. Some of the ongoing challenges are listed below:

(i) The separation of different 2D structures is still a challenge. Efforts are being made to develop synthesis techniques for the production of such materials. Additionally, the effect of moisture and environmental conditions on the stability of MoS$_2$ needs to be investigated.
(ii) MoS$_2$ layers under laser operation (when used as optical modulators) are prone to damage. A large amount of heat is produced while in operation due to their fast response. This can result in burning of the material.
(iii) Defects in MoS$_2$ structure such as point defects, dangling bonds on surface of MoS$_2$ layers grown by CVD, grain boundaries, etc can lead to low performance in device applications. These can be minimized using defect engineering methods.

REFERENCES

[1] J.W. McIver, B. Schulte, F.U. Stein, T. Matsuyama, G. Jotzu, G. Meier, A. Cavalleri, Light-induced anomalous Hall effect in graphene, *Nature Physics*. 16 (2020) 38–41. https://doi.org/10.1038/s41567-019-0698-y.
[2] K.S. Novoselov, A.K. Geim, S.V. Morozov, D. Jiang, M.I. Katsnelson, I.V. Grigorieva, S.V. Dubonos, A.A. Firsov, Two-dimensional gas of massless Dirac fermions in graphene, *Nature*. 438 (2005) 197–200. https://doi.org/10.1038/nature04233.
[3] W.H. Kim, J.Y. Son, Single-layer MoS2 field effect transistor with epitaxially grown SrTiO3 gate dielectric on Nb-doped SrTiO3 substrate, *Bulletin of the Korean Chemical Society*. 34 (2013) 2563–2564. https://doi.org/10.5012/bkcs.2013.34.9.2563.
[4] A. Splendiani, L. Sun, Y. Zhang, T. Li, J. Kim, C.Y. Chim, G. Galli, F. Wang, Emerging photoluminescence in monolayer MoS2, *Nano Letters*. 10 (2010) 1271–1275. https://doi.org/10.1021/nl903868w.
[5] H. Li, M.E. Pam, Y. Shi, H.Y. Yang, A review on the research progress of tailoring photoluminescence of monolayer transition metal dichalcogenides, *FlatChem*. 4 (2017) 48–53. https://doi.org/10.1016/j.flatc.2017.07.001.
[6] K. Khan, A.K. Tareen, M. Aslam, R. Wang, Y. Zhang, A. Mahmood, Z. Ouyang, H. Zhang, Z. Guo, Recent developments in emerging two-dimensional materials and their applications, Royal Society of Chemistry, *Journal of Materials Chemistry C*. 8 (2020) 387–440. https://doi.org/10.1039/c9tc04187g.

[7] K. Waltersson, The crystal structure of C$_s$[VOF$_3$]·12H$_2$O, *Journal of Solid State Chemistry*. 28 (1979) 121–131. https://doi.org/10.1016/0022-4596(79)90064-1.
[8] S. Manzeli, D. Ovchinnikov, D. Pasquier, O.V. Yazyev, A. Kis, 2D transition metal dichalcogenides, *Nature Reviews Materials*. 2 (2017) 17033. https://doi.org/10.1038/natrevmats.2017.33.
[9] M.J. Allen, V.C. Tung, R.B. Kaner, Honeycomb carbon: A review of graphene, *Chemical Reviews*. 110 (2010) 132–145. https://doi.org/10.1021/cr900070d.
[10] O. Samy, S. Zeng, M.D. Birowosuto, A. El Moutaouakil, A review on MoS2 properties, synthesis, sensing applications and challenges, *Crystals* 11 (2021) 355. https://doi.org/10.3390/cryst11040355.
[11] X. Li, H. Zhu, Two-dimensional MoS2: Properties, preparation, and applications, *Journal of Materiomics*. 1 (2015) 33–44. https://doi.org/10.1016/j.jmat.2015.03.003.
[12] N. Thomas, S. Mathew, K.M. Nair, K. O'Dowd, P. Forouzandeh, A. Goswami, G. McGranaghan, S.C. Pillai, 2D MoS$_2$: Structure, mechanisms, and photocatalytic applications, *Materials Today Sustainability*. 13 (2021). https://doi.org/10.1016/j.mtsust.2021.100073.
[13] H. Wang, C. Li, P. Fang, Z. Zhang, J.Z. Zhang, Synthesis, properties, and optoelectronic applications of two-dimensional MoS2 and MoS2-based heterostructures, *Chemical Society Reviews*. 47 (2018) 6101–6127. https://doi.org/10.1039/c8cs00314a.
[14] R.J. Toh, Z. Sofer, J. Luxa, D. Sedmidubský, M. Pumera, 3R phase of MoS2 and WS2 outperforms the corresponding 2H phase for hydrogen evolution, *Chemical Communications*. 53 (2017) 3054–3057. https://doi.org/10.1039/c6cc09952a.
[15] M. Chhowalla, H.S. Shin, G. Eda, L.J. Li, K.P. Loh, H. Zhang, The chemistry of two-dimensional layered transition metal dichalcogenide nanosheets, *Nature Chemistry*. 5 (2013) 263–275. https://doi.org/10.1038/nchem.1589.
[16] M. Tebyetekerwa, J. Zhang, Z. Xu, T.N. Truong, Z. Yin, Y. Lu, S. Ramakrishna, D. Macdonald, H.T. Nguyen, Mechanisms and applications of steady-state photoluminescence spectroscopy in two-dimensional transition-metal dichalcogenides, *ACS Nano*. 14 (2020) 14579–14604. https://doi.org/10.1021/acsnano.0c08668.
[17] K.F. Mak, J. Shan, Photonics and optoelectronics of 2D semiconductor transition metal dichalcogenides, *Nature Photonics*. 10 (2016) 216–226. https://doi.org/10.1038/nphoton.2015.282.
[18] T. Mueller, E. Malic, Exciton physics and device application of two-dimensional transition metal dichalcogenide semiconductors, *Npj 2D Materials and Applications*. 2 (2018) 1–12. https://doi.org/10.1038/s41699-018-0074-2.
[19] D. Van Tuan, A.M. Jones, M. Yang, X. Xu, H. Dery, Virtual trions in the photoluminescence of monolayer transition-metal dichalcogenides, *Physical Review Letters*. 122 (2019) 1–11. https://doi.org/10.1103/PhysRevLett.122.217401.
[20] Y. Zhao, Y. Sun, M. Bai, S. Xu, H. Wu, J. Han, H. Yin, C. Guo, Q. Chen, Y. Chai, Y. Guo, Raman spectroscopy of dispersive two-dimensional materials: A systematic study on MoS2Solution, *Journal of Physical Chemistry C*. 124 (2020) 11092–11099. https://doi.org/10.1021/acs.jpcc.0c01615.
[21] K.G. Zhou, F. Withers, Y. Cao, S. Hu, G. Yu, C. Casiraghi, Raman modes of MoS2 used as fingerprint of van der Waals interactions in 2-D crystal-based heterostructures, *ACS Nano*. 8 (2014) 9914–9924. https://doi.org/10.1021/nn5042703.
[22] C. Lee, H. Yan, L.E. Brus, T.F. Heinz, J. Hone, S. Ryu, Anomalous lattice vibrations of single- and few-layer MoS2, *ACS Nano*. 4 (2010) 2695–2700. https://doi.org/10.1021/nn1003937.
[23] Y. Liu, C. Zeng, J. Zhong, J. Ding, Z.M. Wang, Z. Liu, Spintronics in two-dimensional materials, *Nano-Micro Letters*. 12 (2020) 1–26. https://doi.org/10.1007/s40820-020-00424-2.
[24] K.F. Mak, J. Shan, Photonics and optoelectronics of 2D semiconductor transition metal dichalcogenides, *Nature Photonics*. 10 (2016) 216–226. https://doi.org/10.1038/nphoton.2015.282.
[25] I.S. Osborne, Nanoscale chiral valley-photon interface, *Science*. 359 (2018) 409B. https://doi.org/10.1126/science.359.6374.407-m.
[26] M.L.M. Lalieu, R. Lavrijsen, B. Koopmans, Integrating all-optical switching with spintronics, *Nature Communications*. 10 (2019) 1–6. https://doi.org/10.1038/s41467-018-08062-4.
[27] M.C. Wang, C.C. Huang, C.H. Cheung, C.Y. Chen, S.G. Tan, T.W. Huang, Y. Zhao, Y. Zhao, G. Wu, Y.P. Feng, H.C. Wu, C.R. Chang, Prospects and opportunities of 2D van der Waals magnetic systems, *Annalen Der Physik*. 532 (2020) 1–19. https://doi.org/10.1002/andp.201900452.
[28] X. Shi, Z. Huang, M. Huttula, T. Li, S. Li, X. Wang, Y. Luo, M. Zhang, W. Cao, Introducing magnetism into 2D nonmagnetic inorganic layered crystals: A brief review from first-principles aspects, *Crystals*. 8 (2018). https://doi.org/10.3390/cryst8010024.
[29] J. Pan, S. Lany, Y. Qi, Computationally driven two-dimensional materials design: What is next?, *ACS Nano*. 11 (2017) 7560–7564. https://doi.org/10.1021/acsnano.7b04327.
[30] B. Gao, C. Huang, F. Zhu, C.L. Ma, Y. Zhu, Magnetic properties of Mn-doped monolayer MoS2, *Physics Letters, Section A: General, Atomic and Solid State Physics*. 414 (2021). https://doi.org/10.1016/j.physleta.2021.127636.

[31] P. Tao, H. Guo, T. Yang, Z. Zhang, Strain-induced magnetism in MoS2 monolayer with defects, *Journal of Applied Physics*. 115 (2014). https://doi.org/10.1063/1.4864015.

[32] S. Fu, K. Kang, K. Shayan, A. Yoshimura, S. Dadras, X. Wang, L. Zhang, S. Chen, N. Liu, A. Jindal, X. Li, A.N. Pasupathy, A.N. Vamivakas, V. Meunier, S. Strauf, E.H. Yang, Enabling room temperature ferromagnetism in monolayer MoS2 via in situ iron-doping, *Nature Communications*. 11 (2020) 6–13. https://doi.org/10.1038/s41467-020-15877-7.

[33] S. Yan, W. Qiao, X. He, X. Guo, L. Xi, W. Zhong, Y. Du, Enhancement of magnetism by structural phase transition in MoS2, *Applied Physics Letters*. 106 (2015). https://doi.org/10.1063/1.4905656.

[34] S. Tongay, S.S. Varnoosfaderani, B.R. Appleton, J. Wu, and A.F. Hebard, Magnetic properties of MoS$_2$: Existence of ferromagnetism, *Appl. Phys. Lett.* 101, 123105 (2012). https://doi.org/10.1063/1.4753797.

[35] S. Mathew, K. Gopinadhan, T.K. Chan, X.J. Yu, D. Zhan, L. Cao, A. Rusydi, M.B.H. Breese, S. Dhar, Z.X. Shen, T. Venkatesan, J.T.L. Thong, Magnetism in MoS 2 induced by proton irradiation, *Applied Physics Letters*. 101 (2012). https://doi.org/10.1063/1.4750237.

[36] J. Wang, F. Sun, S. Yang, Y. Li, C. Zhao, M. Xu, Y. Zhang, H. Zeng, Robust ferromagnetism in Mn-doped MoS$_2$ nanostructures, *Appl. Phys. Lett.* 109, 092401 (2016). https://doi.org/10.1063/1.4961883.

[37] K. Kang, S. Fu, K. Shayan, Y. Anthony, The effects of substitutional Fe-doping on magnetism in MoS 2 and WS 2 monolayers, *Nanotechnology*. 32 (2020) 095708. https://doi.org/10.1088/1361-6528/abcd61

[38] H. Zheng, B. Yang, D. Wang, R. Han, X. Du, Y. Yan, Tuning magnetism of monolayer MoS2 by doping vacancy and applying strain, *Applied Physics Letters*. 104 (2014) 1–6. https://doi.org/10.1063/1.4870532.

[39] A. Ramasubramaniam, D. Naveh, Mn-doped monolayer MoS2: An atomically thin dilute magnetic semiconductor, *Physical Review B – Condensed Matter and Materials Physics*. 87 (2013). https://doi.org/10.1103/PhysRevB.87.195201.

[40] R. Pelosato, K. Kang, S. Fu, K. Shayan, Y. Anthony, The effects of substitutional Fe-doping on magnetism in MoS 2 and WS 2 monolayers, *Nanotechnology* 32 (2020) 095708. https://10.1088/1361-6528/abcd61.

[41] J. Kang, J.D. Wood, S.A. Wells, J.H. Lee, X. Liu, K.S. Chen, M.C. Hersam, Solvent exfoliation of electronic-grade, two-dimensional black phosphorus, *ACS Nano*. 9 (2015) 3596–3604. https://doi.org/10.1021/acsnano.5b01143.

[42] G. Eda, H. Yamaguchi, D. Voiry, T. Fujita, M. Chen, M. Chhowalla, Photoluminescence from chemically exfoliated MoS 2, *Nano Letters*. 11 (2011) 5111–5116. https://doi.org/10.1021/nl201874w.

[43] M. Terrones, A. Voshell, M.M. Rana, Review of optical properties of two-dimensional transition metal dichalcogenides, *Proc. SPIE 10754, Wide Bandgap Power and Energy Devices and Applications III*. 107540L (2018). https://doi.org/10.1117/12.2323132.

[44] M. Terrones, A review of defects in metal dichalcogenides: Doping, alloys, interfaces, vacancies and their effects in catalysis & optical emission, *Microscopy and Microanalysis*. 24 (2018) 1556–1557. https://doi.org/10.1017/s1431927618008267.

[45] J.S. Ponraj, Z.Q. Xu, S.C. Dhanabalan, H. Mu, Y. Wang, J. Yuan, P. Li, S. Thakur, M. Ashrafi, K. McCoubrey, Y. Zhang, S. Li, H. Zhang, Q. Bao, Photonics and optoelectronics of two-dimensional materials beyond graphene, *Nanotechnology*. 27 (2016). https://doi.org/10.1088/0957-4484/27/46/462001.

[46] K.S. Novoselov, V.I. Fal'Ko, L. Colombo, P.R. Gellert, M.G. Schwab, K. Kim, A roadmap for graphene, *Nature*. 490 (2012) 192–200. https://doi.org/10.1038/nature11458.

[47] Y. Naveh, K.K. Likharev, Modeling of 10-nm-scale ballistic MOSFET', *IEEE Electron Device Letters*. 21 (2000) 242–244. http://doi.org/10.1109/55.841309.

[48] B. Radisavljevic, A. Kis, Reply to "Measurement of mobility in dual-gated MoS2 transistors," *Nature Nanotechnology*. 8 (2013) 147–148. https://doi.org/10.1038/nnano.2013.31.

[49] M.S. Fuhrer, J. Hone, Measurement of mobility in dual-gated MoS2 transistors, *Nature Nanotechnology*. 8 (2013) 146–147. https://doi.org/10.1038/nnano.2013.30.

[50] K.F. Mak, C. Lee, J. Hone, J. Shan, T.F. Heinz, Atomically thin MoS2: A new direct-gap semiconductor, *Physical Review Letters*. 105 (2010) 2–5. https://doi.org/10.1103/PhysRevLett.105.136805.

[51] A.R. Klots, A.K.M. Newaz, B. Wang, D. Prasai, H. Krzyzanowska, J. Lin, D. Caudel, N.J. Ghimire, J. Yan, B.L. Ivanov, K.A. Velizhanin, A. Burger, D.G. Mandrus, N.H. Tolk, S.T. Pantelides, K.I. Bolotin, Probing excitonic states in suspended two-dimensional semiconductors by photocurrent spectroscopy, *Scientific Reports*. 4 (2014) 1–7. https://doi.org/10.1038/srep06608.

[52] F.A. Rasmussen, K.S. Thygesen, Computational 2D materials database: Electronic structure of transition-metal dichalcogenides and oxides, *Journal of Physical Chemistry C*. 119 (2015) 13169–13183. https://doi.org/10.1021/acs.jpcc.5b02950.

[53] K.K. Kam, B.A. Parkinson, Detailed photocurrent spectroscopy of the semiconducting group VI transition metal dichalcogenides, *Journal of Physical Chemistry*. 86 (1982) 463–467. https://doi.org/10.1021/j100393a010.

[54] A. Di Bartolomeo, Graphene Schottky diodes: An experimental review of the rectifying graphene/semiconductor heterojunction, *Physics Reports*. 606 (2016) 1–58. https://doi.org/10.1016/j.physrep.2015.10.003.
[55] A.A. Balandin, S. Ghosh, W. Bao, I. Calizo, D. Teweldebrhan, F. Miao, C.N. Lau, Superior thermal conductivity of single-layer graphene, *Nano Letters*. 8 (2008) 902–907. https://doi.org/10.1021/nl0731872.
[56] I. Jo, M.T. Pettes, J. Kim, K. Watanabe, T. Taniguchi, Z. Yao, L. Shi, Thermal conductivity and phonon transport in suspended few-layer hexagonal boron nitride, *Nano Letters*. 13 (2013) 550–554. https://doi.org/10.1021/nl304060g.
[57] S. Ghosh, I. Calizo, D. Teweldebrhan, E.P. Pokatilov, D.L. Nika, A.A. Balandin, W. Bao, F. Miao, C.N. Lau, Extremely high thermal conductivity of graphene: Prospects for thermal management applications in nanoelectronic circuits, *Applied Physics Letters*. 92 (2008) 1–4. https://doi.org/10.1063/1.2907977.
[58] Y. Cai, J. Lan, G. Zhang, Y.W. Zhang, Lattice vibrational modes and phonon thermal conductivity of monolayer MoS 2, *Physical Review B – Condensed Matter and Materials Physics*. 89 (2014) 1–8. https://doi.org/10.1103/PhysRevB.89.035438.
[59] L. Liu, S.B. Kumar, Y. Ouyang, J. Guo, Performance limits of monolayer transition metal dichalcogenide transistors, *IEEE Transactions on Electron Devices*. 58 (2011) 3042–3047. https://doi.org/10.1109/TED.2011.2159221.
[60] J.W. Jiang, H.S. Park, T. Rabczuk, Molecular dynamics simulations of single-layer molybdenum disulphide (MoS2): Stillinger-Weber parametrization, mechanical properties, and thermal conductivity, *Journal of Applied Physics*. 114 (2013). https://doi.org/10.1063/1.4818414.
[61] J.Y. Kim, S.M. Choi, W.S. Seo, W.S. Cho, Thermal and electronic properties of exfoliated metal chalcogenides, *Bulletin of the Korean Chemical Society*. 31 (2010) 3225–3227. https://doi.org/10.5012/bkcs.2010.31.11.3225.
[62] S. Sahoo, A.P.S. Gaur, M. Ahmadi, M.J.F. Guinel, R.S. Katiyar, Temperature-dependent Raman studies and thermal conductivity of few-layer MoS2, *Journal of Physical Chemistry C*. 117 (2013) 9042–9047. https://doi.org/10.1021/jp402509w.
[63] W. Choi, N. Choudhary, G.H. Han, J. Park, D. Akinwande, Y.H. Lee, Recent development of two-dimensional transition metal dichalcogenides and their applications, *Materials Today*. 20 (2017) 116–130. https://doi.org/10.1016/j.mattod.2016.10.002.
[64] R. Sanikop, C. Sudakar, Tailoring magnetically active defect sites in MoS2 nanosheets for spintronics applications, *ACS Applied Nano Materials*. 3 (2020) 576–587. https://doi.org/10.1021/acsanm.9b02121.
[65] A. Dankert, L. Langouche, M.V. Kamalakar, S.P. Dash, High-performance molybdenum disulfide field-effect transistors with spin tunnel contacts, *ACS Nano*. 8 (2014) 476–482. https://doi.org/10.1021/nn404961e.
[66] A.T. Neal, H. Liu, J. Gu, P.D. Ye, Magneto-transport in MoS2: Phase coherence, spin-orbit scattering, and the hall factor, *ACS Nano*. 7 (2013) 7077–7082. https://doi.org/10.1021/nn402377g.
[67] R. Sanikop, A.K. Budumuru, S. Gautam, K.H. Chae, C. Sudakar, Robust ferromagnetism in Li-intercalated and -Deintercalated MoS2Nanosheets: Implications for 2D spintronics, *ACS Applied Nano Materials*. 3 (2020) 11825–11837. https://doi.org/10.1021/acsanm.0c02349.
[68] K. Dou, Y. Ma, R. Peng, W. Du, B. Huang, Y. Dai, Promising valleytronic materials with strong spin-valley coupling in two-dimensional MN2X2(M = Mo, W; X = F, H), *Applied Physics Letters*. 117 (2020). https://doi.org/10.1063/5.0026033.

Spin Transistors
Different Geometries and Their Applications

4

Gul Faroz Ahmad Malik, Mubashir Ahmad Kharadi, Farooq Ahmad Khanday, Zaid Mohammad Shah, Sparsh Mittal

Contents

1 Introduction	49
2 Structural Development	52
2.1 Spin-Tunnel FET	52
2.2 Spin-MOSFET	52
2.3 Silicene and Graphene-Based spin-FET	53
2.4 Multi-Gate Spin-FET	54
2.5 Parallel and Anti-Parallel Spin-FETs	54
2.6 Spin-FETs for Reconfigurable Applications	55
2.7 Silicene Based Spin-TFET	55
2.8 Indium Phosphide Based Double Gate Spin-FET	56
3 Applications of Spin-Transistors	57
4 Issues Faced by Spin-Transistors	57
4.1 Low Spin Filtering Efficiency	58
4.2 Spin Precession	58
4.3 Low Current Drive	59
4.4 Less On-Off Ratio	59
4.5 Fabrication Platform Mismatch	59
4.6 Presence of Dresselhaus Field	59
5 Conclusion	59
References	60

1 INTRODUCTION

Device scaling has been one of the critical concerns of contemporary researchers. As the devices are scaled down to the nano dimensions, the number of devices on the integrated chip increases, enhancing

DOI: 10.1201/9781003196952-4

its functionality. Further, the power consumption of the devices has become a matter of grave concern. The need of the hour is devices with low power consumption [1]. Researchers have made various attempts to model the devices which consume low power and scale well to the nano dimensions. But as the devices are scaled-down, the short channel effects or second-order effects come into play. These effects deteriorate the performance of the transistor. Several alternative technologies and devices have been proposed in the open literature to address the aforementioned problems. One such technology is spintronics [2–6]. The field of spintronics uses the spin of electrons to store information, whereas conventional devices use charge to store data. It is easier to flip the spin of an electron than to move it from one terminal to another [7–10].

Various spintronic devices have been proposed to date and of them, magnetic tunnel junction (MTJ) and spin transistor or Datta-Das transistor (or spin-FET) are the most prominent devices [11–13]. Both these devices are three-terminal devices in which a non-magnetic material is surrounded by two magnetic materials [13]. In MTJ, the non-magnetic material can be a semiconductor, 2D-material, or insulator, while in a spin transistor, it cannot be an insulator. MTJ is a two-terminal device while a spin transistor is a three-terminal device. MTJ, being a two-probe device, shows ON and OFF state only on the change in the magnetization of its drain electrode. If the magnetization of the source and drain is parallel to each other, then all the spin-injected electrons from the source are accepted by the drain. As such, it shows minimum resistance and hence, is said to be in the ON-state. While if the magnetization of the source and drain electrode is anti-parallel to each other, the electrons whose spin is injected by the source electrode are rejected by the drain terminal. This reduces the current, and hence, the device acts in the OFF state.

The case of a spin transistor is different. Here, the transport of carriers through the channel is controlled by the potential applied at the gate terminal. The spin transistor, also called as Datta-Das transistor is, a normally-ON transistor [13]. The carriers which are spin-polarized by the ferromagnetic source are pumped into the channel region, where it is assumed that their spin does not relax (which usually is not the case), and the ferromagnetic drain collects them. The voltage applied at the gate changes the spin of electrons traveling in the channel region, due to the effect called the Rashba effect. Hence, the carriers are not collected by the drain terminal and as such, it turns OFF [14–18].

The spin polarizer polarizes the spin of electrons in a single direction and is illustrated by Figure 4.1(a) [19]. Any ferromagnetic material can act as a spin polarizer such as iron, nickel, or cobalt.

The spin-filter accepts all the electrons with a single spin orientation and discards all other electrons with a slightly different polarization direction; hence the name filter. It is shown in Figure 4.1(b) [19]. The spin filter is also made of ferromagnetic materials.

The spin polarizer and spin filter are the vital regions in a spin-FET and act as the two electrodes, similar to the source and drain (respectively) in conventional MOSFET. Figure 4.1(c) [19] shows the structure of a spin-FET, which consists of a spin polarizer (iron contact), a spin filter (again an iron contact). In between them is a semiconducting (non-magnetic) region acting as the channel for spin-polarized carriers.

Over the channel, a metal contact is placed, separated by an insulating layer (usually made of SiO_2). This metallic layer is called a gate terminal. The voltage applied at the gate terminal turns the spin-FET on and off. When no voltage is applied at the gate terminal, the carriers which are spin-polarized by the spin polarizer (source) travel to the spin filter via the semiconducting channel. The spin filter (drain) does not accept the carriers having a slight change in spin polarization. The change in spin polarization of electrons inside the channel is called spin precession. Less spin precession occurs when no voltage is applied at the gate terminal. Hence, maximum current flows through it and the device acts in the ON state. But when a voltage is applied at the gate terminal, the voltage-induced magnetic field called Rashba field is produced and changes the spin polarization of electrons. In other words, it causes spin precession inside the channel. When the spin polarization of electrons changes, the electrons are not accepted by the drain terminals. Hence, minimum current flows through it, and the device acts in the off-state. In this way, spin-FET acts as a normally-on device. But the working is true for parallel

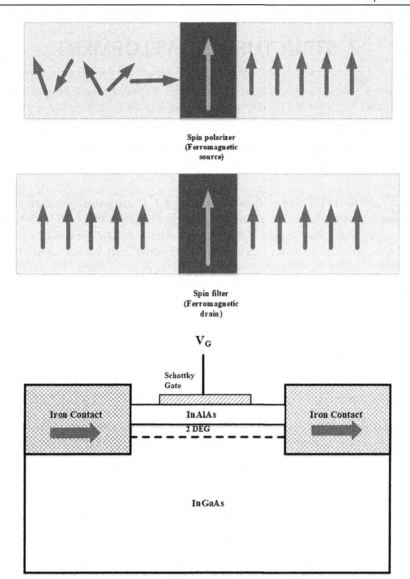

FIGURE 4.1 (a) Spin polarizer, (b) Spin filter, (c) Spin-FET.

configuration spin-FET, wherein the spin polarization of source and drain is in the same direction. The anti-parallel configuration works on the same principle, with the only difference being that it is a normally off device.

In this chapter, a comprehensive study of spin transistors and their different types has been presented. The structural developments, their applications, and the issues faced are also discussed in detail. Spin-FET-based digital design and reconfigurable design have also been discussed. Finally, a comparison based on the number of devices, transit time, and capacitance with conventional CMOS-based design is made to show the pre-eminence of the spin transistors in digital design applications. The rest of the chapter is organized as follows. Section 2 discusses the structural developments of spin transistors. Section 3 highlights various applications and Section 4 discusses the challenges faced by spin transistors. Finally, Section 5 concludes the chapter.

2 STRUCTURAL DEVELOPMENT

Spin transistors have seen rapid progress in structural design. Various structures highlighted in this chapter include spin-tunnel FET, spin-MOSFET, silicene-based spin-FET, graphene-based spin-FET, double and triple gate spin-FET, etc.

2.1 Spin-Tunnel FET

The structure of the spin-TFET [20] is shown in Figure 4.2. It has a dual-gated graphene/chromium triiodide/graphene tunnel junction. The ambipolar transport has been observed in this device, and the magnetic order of the chromium triiodide layer determines the conductance state of the device. The voltage applied gate terminal switches this chromium triiodide layer between antiferromagnetic and ferromagnetic states, switching between low and high conductance states. The ratio of high to low conductance achieved goes up to 400%. The device can be useful in the development of non-volatile memory.

2.2 Spin-MOSFET

Spin-MOSFET [21] is another developmental work in the case of spin transistors. Contrary to the spin-FET, the source and drain electrodes are half-metallic and not ferromagnets. Materials like CrO_2 are half-metallic and show ferromagnetic behavior, with an added advantage of high values (almost 100%) of spin polarization efficiency. Other benefits of the spin-MOSFET structure include reduced power-delay product, minute off-current, and ability of high current amplification. These devices are highly suitable

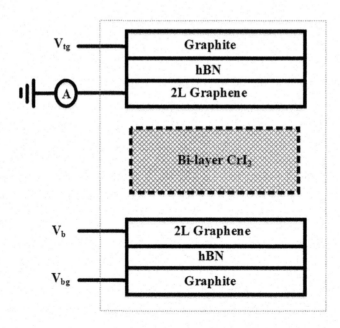

FIGURE 4.2 Spin-TFET.

for ULSI (ultra-high density integrated circuits) for switching and memory applications. Its structure is shown in Figure 4.3.

2.3 Silicene and Graphene-Based spin-FET

Graphene has been one of the highly researched materials for its miraculous electronic properties. Silicene, the silicon version of graphene, is another emerging two-dimensional material with excellent properties. Silicene is considered the best possible option to have compatibility with the already existing silicon-based platforms. These materials have been utilized to realize various spintronic devices, including magnetic tunnel junctions and spin-transistors. Figure 4.4 shows the structure of graphene/silicene-based spin-FET [22]. The use of these materials partially solves the problems of spin-FET such as the mismatch effect in spin injection and insufficient spin lifetime. They offer a high drive, high on-off ratio, and can be scaled to a maximum extent. The scalability of these devices is proven by the fact that various sub-10 nm spintronic devices have been realized using these materials.

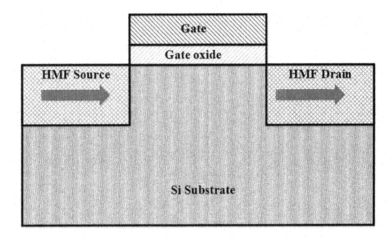

FIGURE 4.3 Spin-MOSFET.

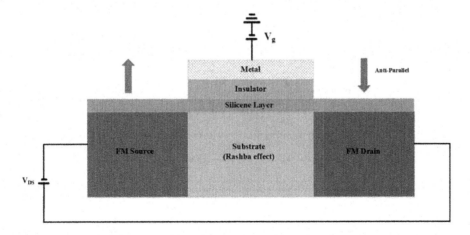

FIGURE 4.4 Silicene based spin-FET.

2.4 Multi-Gate Spin-FET

Like in conventional MOSFETs, the usage of multiple gates has also been implemented in spin transistors. Various efforts have been carried out to model multiple-gate spin-field effect transistors. Figure 4.5(a) shows the modeled spin-FET [23]. This approach has been further extended to the multiple-gate spin-FET, as demonstrated in Figure 4.5(b).

The use of multiple gates increases the control over the transport through the channel. It also enhances the current, hence, increasing the current drive. The devices with more drive are highly recommended for applications wherein there are multiple stages. When implemented for multi-stage applications, the device with less drive requires amplification stages in between, hence increasing the hardware, cost, and complexity.

2.5 Parallel and Anti-Parallel Spin-FETs

Another approach to utilize spin transistors in digital design applications is modeling of parallel and anti-parallel spin-FETs. Figure 4.6(a) shows parallel spin-FET while Figure 4.6(b) shows anti-parallel spin-FET [24]. This enhances the chances of spin transistors replacing the conventional CMOS design. The parallel and anti-parallel spin-FETs work in a circuit similar to how NMOS and PMOS work in CMOS design. Thus, spin-FET-based digital design can replace conventional FETs.

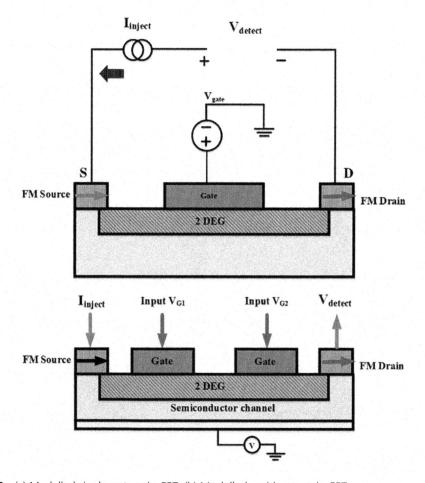

FIGURE 4.5 (a) Modelled single-gate spin-FET, (b) Modelled multi-gate spin-FET.

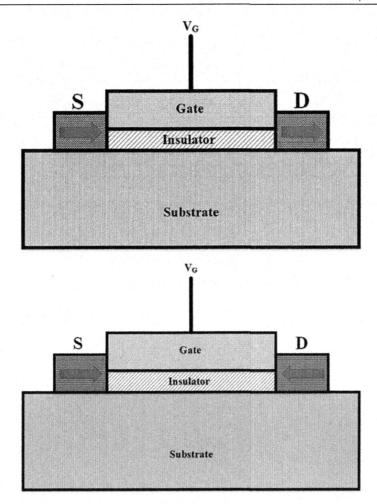

FIGURE 4.6 (a) Parallel spin-FET, (b) Anti-parallel spin-FET.

2.6 Spin-FETs for Reconfigurable Applications

Spin transistors have also been implemented for reconfigurable circuit applications [25]. The modeled multi-gate spin-FET is the basic building block of such a design. A triple-gate spin-FET can implement 2-input logic functions while the third terminal is used as the controlling terminal. The voltage applied at the controlling terminals changes the functionality of the gate/device. The most important advantage of this design, apart from electrical reconfigurability, is that this design is fast and less complex because only one device implements the whole logic design. Similarly, a quadruple-gate spin-FET is used to implement the 3-input logic functions wherein the fourth terminal is used as the controlling terminal. These devices can be cascaded to increase their functionality and widen the application spectrum. Figure 4.7 shows a triple-gate spin-FET implemented for various 2-input logic functions.

2.7 Silicene Based Spin-TFET

Silicene-based spin tunnel FET has been presented in which 98% spin polarization has been achieved [26]. It is a low-power transistor and operates at the voltage of 0.35 V. Spin-polarized edge states determine the

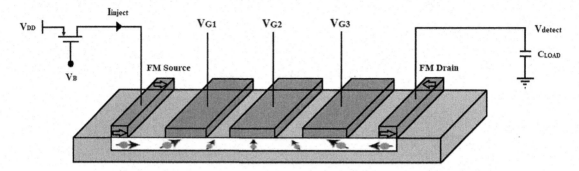

FIGURE 4.7 Triple-Gate Spin-FET Implemented for 2-Input Reconfigurable Logic Design.

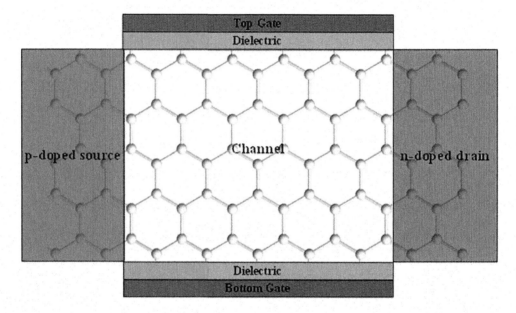

FIGURE 4.8 Silicene-based Spin-TFET.

operation of the device. Furthermore, the polarization can be tuned by modifying the bandgap, which is easy in a 2D material like silicene. The device also has the expected integration with the already existing silicon technology. The structure of the device is shown in Figure 4.8.

2.8 Indium Phosphide Based Double Gate Spin-FET

Indium phosphide channel and HfO_2 dielectric material-based double gate spin-field effect transistor [27] has been presented. Since the proposed device can be used to form parallel and anti-parallel configurations, it is well suited to replace CMOS technology. It offers a high drive current and high value of the on-off current ratio. This makes it feasible for use in futuristic digital-circuit applications. Its model is shown in Figure 4.9.

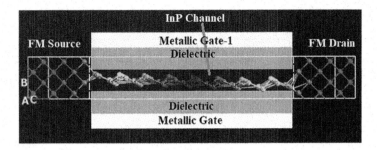

FIGURE 4.9 Indium Phosphide-based Double Gate Spin-FET.

3 APPLICATIONS OF SPIN-TRANSISTORS

Similar to conventional transistors, spin transistors also have a wide range of applications. The spin transistors can be a drop-in replacement of the traditional transistors. The modeling of parallel and antiparallel spin transistors makes them convenient for digital logic applications. Note that spin transistors consume less power than the conventional transistor because flipping of electron-spin requires 1/60th of the energy compared to that required for moving the electron. Power consumption is of utmost priority in the system design using devices. Apart from the power consumption, another performance parameter is the speed. The lesser the transit time of the device, the higher is the speed, and vice-versa. Table 4.1 [28] compares the properties of 3-input XOR Gate, 3-input XNOR Gate, 3-input Majority Gate, and 1-bit Full adder when designed with (1) CMOS (2) single gate spin-FET and (3) multiple gate spin-FET. It can be seen that all the parameters viz., no. of devices used, capacitance, and transit time improve in spin-FET-based design. The number of devices used determines the complexity, area, and cost. The higher the device count, the higher is the cost, area, and complexity of the design. A higher value of capacitance indicates higher power consumption, based on the formula $P = CfV_{DD}^2$. It can be seen from Table 4.1 [28] that spin transistor-based design offers lower capacitance and hence, has low power consumption. Finally, it is clear from the table that spin transistor-based design has less transit time, and thus it can be used for high-speed applications.

4 ISSUES FACED BY SPIN-TRANSISTORS

Although the various variants of spin transistors have shown significant promise, they face several challenges that need to be solved before they can be productized. The first and foremost issue is the prominence of silicon-based electronic devices. The spin-transistor devices have been theoretically simulated [29], [30], but achieving such performance in real-time applications is challenging. Further, the complete replacement of silicon-based CMOS devices by spin-based transistors presents several challenges. But since scaling is no more possible in silicon-based CMOS devices, the manufacturing of spin-based devices has started. Magnetic tunnel junction (MTJ), a spintronic device, has been implemented for the read heads in memories [31–40]. Spin transistors also face performance issues, such as low spin injection efficiency, low spin filtering efficiency, spin precession inside the channel of spin transistors, low current drive, less on-off ratio, fabrication platform mismatch, presence of Dresselhaus field, etc. In what follows, these issues are explained in detail [41–55]:4.1 Low Spin Injection Efficiency

TABLE 4.1 Comparison of CMOS-based Designs with Single Gate Spin-FET and Triple Gate Spin-FET Designs (TSMC 180nm) [28]

	NO. OF DEVICES USED		
FUNCTION/PARAMETER	CMOS	SINGLE GATE SPIN-FET	TRIPLE GATE SPIN-FET
3-input XOR Gate	12	2	1
3-input XNOR Gate	12	3	1
3-input Majority Gate	10	3	1
1-bit Full adder	28	5	2
Capacitance			
Function/Parameter	CMOS	Single gate Spin-FET	Triple gate Spin-FET
3-input XOR Gate	12CGATE +12CDRAIN +CLOAD	4CGATE +2CLOAD	3CGATE +CLOAD
3-input XNOR Gate	12CGATE +12CDRAIN+CLOAD	5CGATE +3CLOAD	3CGATE +CLOAD
3-input Majority Gate	10CGATE +10CDRAIN +CLOAD	6CGATE +3CLOAD	3CGATE +CLOAD
1-bit Full adder	28CGATE +28CDRAIN +CLOAD	10CGATE +5CLOAD	6CGATE +CLOAD
Transit time			
Function/Parameter	CMOS (ps)	Single gate Spin-FET (ps)	Triple gate Spin-FET (ps)
3-input XOR Gate	620	0.673	0.618
3-input XNOR Gate	590	1.03	0.832
3-input Majority Gate	1350	0.83	0.603
1-bit Full adder	1850	1.56	0.1

As shown in Figure 4.1(**a**), the source terminal injects the spin or simply polarizes the spin of electrons in a single direction. If the source terminal polarizes all the electrons, it is said to have 100% spin polarization efficiency. Theoretically, many materials like CrO_2 and other half-metallic material-based electrodes have been predicted to have 100% efficiency. However, it is still a distant dream to attain this efficiency value in reality. Low spin injection efficiency prohibits the ability of spin transistors to replace CMOS-based technology.

4.1 Low Spin Filtering Efficiency

Similar to spin polarization, spin filtering is another important performance parameter of spintronic devices in general and spin transistors in particular. The electrons coming from the channel are either accepted or rejected by the drain terminal depending upon the orientation of its spin polarization. If the spin-polarization direction of the incoming electron coincides with the polarization of the drain terminal, it is allowed to pass; otherwise, it is rejected. To attain 100% spin filtering is another challenge in the practical use of spin transistors.

4.2 Spin Precession

A non-magnetic (semiconducting) channel is sandwiched between the ferromagnetic source and drain terminal through which carriers' transport occurs. It is assumed that spin-polarized electrons while moving through this non-magnetic channel retain the polarization. The spin polarization flip takes place only when a voltage is applied at the gate terminal. But because of the presence of magnetic fields other than the Rashba field, the spin polarization of electrons changes and is called spin precession. This hampers the performance of the spin transistor and makes its behavior unexpected at times. Ideally, the spin precession because of other fields inside the channel must be zero.

4.3 Low Current Drive

Within the channel of the spin transistor, the transport of electrons occurs because of the presence of few electrons. Hence, they suffer from the problem of a low drive. The devices with low drive are not feasible for digital circuit design as the transistor in the first stage cannot drive the second stage transistor. This mandates the implementation of repeaters, which further increases the complexity, power consumption, and cost and decreases the speed. So spin transistors with a high current drive must be designed.

4.4 Less On-Off Ratio

The spin transistors also suffer the problem of a low on-off ratio of currents. The difference between on and off currents is small, and at times, the two are not distinguishable. This reduces the switching capability of these devices and limits their use in real-world applications. Thus, spin transistors with a high on-off ratio are needed.

4.5 Fabrication Platform Mismatch

Spin transistors also suffer from the platform mismatch as far as the fabrication processes are concerned. The existing platform is silicon-based, but several other types of semiconducting and magnetic materials are required in spin transistors. A drop-in replacement of silicon is possible only if spin transistors offer significantly superior properties.

4.6 Presence of Dresselhaus Field

The presence of the Dresselhaus field is another issue faced by spin transistors working based on the Rashba field. However, it has been used by several researchers to switch the transistor on and off. Their working principle is based on the Dresselhaus field. But in the devices where switching takes place because of the Rashba field, the Dresselhaus field is undesired but unavoidable. Hence, the Dresselhaus field deteriorates the performance of the spin transistor.

Apart from these effects, various assumptions are made for the proper working of the transistor, which includes:

(a) 100% efficiency of drain and source electrode
(b) At gate voltage equal to 0, no transverse magnetic field exists in the channel
(c) No magnetic field apart from the Rashba field exists in the channel
(d) No spin relaxation occurs in the channel, i.e., it is assumed that no DP (D'yakonov Perel), BAP (Bir-Aronov-Pikus), EY (Elliott-Yafet) relaxations, and hyperfine interaction exist in the channel

5 CONCLUSION

A study of spin transistors has been carried out, and the devices like Spin-FETs, Spin-MOSFETs, Spin-TFETs, 2D-material based spin transistors, etc., have been discussed in detail. A detailed description of the applications of spin transistors has also been incorporated in this chapter. Finally, it can be concluded that the spin transistors hold the promise to become the drop-in replacement of CMOS-based designs. The salient features of spin transistors include low power consumption, low transit time, and low device count required for the implementation of the device. Some spin transistors proposed in the literature have high on current (drive) and high value of on-off ratio. This chapter also discussed the issues faced by different variants of spin transistors. It is hoped that soon, spin transistors will be deployed in many real-world applications.

REFERENCES

[1] S. Bandyopadhyay and M. Cahay, *Introduction to Spintronics*. New York: CRC Press Taylor & Francis Group, Sept-2015.
[2] E. R. Weber, *Semiconductors and Semimetals*. Burlington: Academic Press, Elsevier, 1981.
[3] P. P. Freitas, F. Silva, N. J. Oliveira, L. V. Melo, L. Costa and N. Almeida, "Spin valve sensors", *Sensors and Actuators A*, vol. 81, pp. 2, Apr-2000.
[4] J. Wang, H. Meng and J. P. Wang, "Programmable spintronics logic device based on a magnetic tunnel junction element", *Journal of Applied Physics*, vol. 97, pp. 100509, May-2005.
[5] P. P. Freitas, L. Costa, N. Almeida, L. V. Melo, F. Silva, J. Bernardo and C. Santos, "Giant magnetoresistive sensors for rotational speed control", *Journal of Applied Physics*, vol. 85, pp. 5459, Apr-1999.
[6] W. J. Ku, P. P. Freitas, P. Compadrinho and J. Barata, "Precision X-Y robotic object handling using a dual GMR bridge sensor", *IEEE Transactions on Magnetics*, vol. 36, pp. 2782, Sept-2000.
[7] M. N. Baibich, J. M. Broto, A. Fert, F. Nguyen Van Dau, F. Petroff, P. Eitenne, G. Creuzet, A. Friederich and J. Chazelas, "Giant magnetoresistance of (001) Fe/(001)Cr magnetic superlattices", *Physics Review Letters*, vol. 61, pp. 2472, Nov-1988.
[8] J. S. Moodera, L. R. Kinder, T. M. Wong and R. Meservey, "Large Magnetoresistance at Room Temperature in ferromagnetic thin film tunnel junctions", *Physics Review Letters*, vol. 74, pp. 3273–3276, Apr-1995.
[9] M. Julliere, "Tunneling between ferromagnetic films", *Physics Letters A*, vol. 54, no. 3, pp. 225–226, Sept-1975.
[10] B. B. Aein, D. Datta, S. Salahuddin and S. Datta, "Proposal for an all-spin logic device with built-in memory", *Nature Nanotechnology*, vol. 5, no. 4, pp. 266–270, Feb-2010.
[11] M. Sharad, C. Augustine, G. Panagopoulos and K. Roy, "Ultra low energy analog image processing using spin based neurons", *2012 IEEE/ACM International Symposium on Nanoscale Architectures (NANOARCH)*, Amsterdam, Netherlands, Jul-2012.
[12] J. Kim, A. Paul, P. A. Crowell, S. J. Koester, S. S. Sapatnekar and J. P. Wang, "Spin-based computing: Device concepts, current status, and a case study on a high-performance microprocessor", *Proceedings of the IEEE*, vol. 103, no. 1, pp. 106–130, Jan-2015.
[13] S. Datta and B. Das, "Electronic analog of the electro-optic modulator", *Applied Physics Letters*, vol. 56, no. 7, pp. 665467, Feb-1990.
[14] Y. Sato, S. Gozu, T. Kita and S. Yamada, "Study for realization of spin-polarized field effect transistor in In0.75Ga0.25As/In0.75Al0.25As heterostructure", *Physica E*, vol. 12, nos. 1–4, pp. 399–402, Jan-2002.
[15] S. Meena and S. Choudhary, "Enhancing TMR and spin-filteration by using out-of-plane graphene insulating barrier in MTJs", *Physical Chemistry Chemical Physics*, vol. 19, pp. 17765–17772, Jun-2017.
[16] J. Nitta and T. Bergsten "Electrical manipulation of spin precession in an InGaAs-Based 2DEG due to the Rashba spin-orbit interaction", *IEEE Transactions on Electron Devices*, vol. 54, no. 5, pp. 955–960, May-2007.
[17] L. T. Chang, I. A. Fischer, J. Tang, C. Y. Wang, G. Yu, Y. Fan and K. L. Wang, "Electrical detection of spin transport in Si two-dimensional electron gas systems", *Nanotechnology*, vol. 27 no. 36, pp. 365701, Aug-2016.
[18] J. H. Kim, J. Bae, B. C. Min, H. Kim, J. Chang and H. C. Koo, "All-electric spin transistor using perpendicular spins", *Journal of Magnetism and Magnetic Materials*, vol. 403, no. 4, pp. 77–80, Apr-2016.
[19] G. F. A. Malik, M. A. Kharadi and F. A. Khanday, "Spin field effect transistors and their applications: A survey", *Microelectronics Journal*, vol. 106, no. 10424, Dec-2020.
[20] S. Jiang, L. Li, Z. Wang and J. S. K. F. Mak, "Spin tunnel field-effect transistors based on two dimensional van der Waals heterostructures", *Nature Electronics*, vol. 2, pp. 159–163, Apr-2019.
[21] Satoshi Sugahara and Masaaki Tanaka, "A spin metal – oxide – semiconductor field-effect transistor using half-metallic-ferromagnet contacts for the source and drain", *Applied Physics Letters*, vol. 84, no. 13, pp. 2307–2309, Jan-2004.

[22] N. Pournaghavi, M. Esmaeilzadeh, A. Abrishamifar and S. Ahmadi, "Extrinsic Rashba spin – orbit coupling effect on silicene spin polarized field effect transistors", *Journal of Physics: Condensed Matter*, vol. 29, no. 14, p. 145501, Feb-2017.
[23] S. M. Kang, Y. Leblebici and C. Kim, *CMOS Digital Integrated Circuits Analysis & Design*, 4th ed. New York: McGraw-Hill, Jan-2014.
[24] H. C. Koo, I. Jung and C. Kim, "Spin-based complementary logic device using datta – das transistors", *IEEE Transactions on Electron Devices*, vol. 62, no. 9, pp. 3056–3060, Sept-2015.
[25] G. F. A. Malik, M. A. Kharadi and F. A. Khanday, "Electrically reconfigurable logic design using multi-gate spin field effect transistors", *Microelectronics Journal*, vol. 90, pp. 278–284, Jul-2019.
[26] M. A. Kharadi, G. F. A. Malik, F. A. Khanday and S. Mittal, "Silicene-based spin filter with high spin-polarization", *IEEE Transactions on Electron Devices*, vol. 68, no. 10, pp. 5095–5100, Oct-2021.
[27] G. F. A. Malik, M. A. Kharadi, F. A. Khanday and K. A. Shah, "Performance analysis of indium phosphide channel based sub-10 nm double gate spin field effect transistor", *Physics Letters A*, vol. 384, no. 19, pp. 126498(1)–126498(7), Apr-2020.
[28] G. F. A. Malik, M. A. Kharadi, N. Parveen and F. A. Khanday, "Modelling for triple gate spin-FET and design of triple gate spin-FET-based binary adder", *IET Circuits Devices and Systems River Valley Technologies*, vol. 14 no. 4, pp. 464–470, Mar-2020.
[29] G. F. A. Malik, M. A. Kharadi, F. A. Khanday, K. A. Shah and N. Parveen, "Negative differential resistance in gate all-around spin field effect transistors", *Nano Systems: Physics, Chemistry, Mathematics*, vol. 11, no. 3, pp. 301–306, May-2020.
[30] M. A. Kharadi, G. F. A. Malik, F. A. Khanday and K. A. Shah, "Hydrogenated silicene based magnetic junction with improved tunneling magnetoresistance and spin filtering efficiency", *Physics Letters A*, vol. 384, no. 32, pp. 126826(1)–126826(10), Aug-2020.
[31] S. Shirotori, H. Yoda, Y. Ohsawa, N. Shimomura, T. Inokuchi, Y. Kato, Y. Kamiguchi, K. Koi, K. Ikegami, H. Sugiyama, M. Shimizu, B. Altansargai, S. Oikawa, M. Ishikawa, A. Tiwari, Y. Saito and A. Kurobe, "Voltage-control spintronics memory with a self-aligned heavy-metal electrode", *IEEE Transactions on Magnetics*, vol. 53 no. 11, pp. 27.6.1–27.6.4, Nov-2017.
[32] R. M. S. Andrade, H. Xu and M. Larsson, "Spin-FET based on InGaAs/InP heterostructure", Master Thesis at Lunds Universitet, Nov-2011.
[33] H. C. Koo, I. Jung and C. Kim, "Spin-based complemenary logic device using Datta-Das transistor", *IEEE Transactions on Electron Devices*, vol. 62 no. 9, pp. 3056–3060, Aug-2015.
[34] G. Wang, Z. Wang, J. O. Klein and W. Zhao, "Modelling for spin-FET and design of spin-FET-based logic gates", *IEEE Transactions on Magnetics*, vol. 53, no. 11, pp. 1–6, Nov-2017.
[35] G. Wang, Z. Wang, X. Lin, J. O. Klein and W. Zhao, "Proposal for multi-gate spin field-effect trasistor", *IEEE Transactions on Magnetics*, vol. 54, no. 11, pp. 1–5, Nov-2018.
[36] M. Kazemi, "An electrically reconfigurable logic gate intrinsically enabled by spin orbit materials", *Nature Scientific Reports*, vol. 17, pp. 15358, Nov-2017.
[37] P. N. Hai, S. Sughara and M. Tanaka, "Reconfigurable logic gates using single-electron spin transistors", *Japanese Journal of Applied Physics*, vol. 46, no. 10A, pp. 6579–6585, Apr-2007.
[38] S. Sugahara, T. Matsuno and M. Tanaka, "Novel reconfigurable logic gates using spin metal-oxide-semiconductor field-effect transistors", *Japanese Journal of Applied Physics*, vol. 43, no. 9A, pp. 6032–6037, Sept-2004.
[39] M. Tanaka and S. Sugahara, "MOS-based spin devices for reconfigurable logic", *IEEE Transactions on Electron Devices*, vol. 54, no. 5, pp. 967–976, May-2007.
[40] H. Dery, P. Dalal, L. Cywiński and L. J. Sham, "Spin-based logic in semiconductors for reconfigurable large-scale circuits", *Nature Letters*, vol. 447, pp. 573–576, May-2007.
[41] J. Kim, A. Paul, P. A. Crowell, S. J. Koester, S. S. Sapatnekar and J. P. Wang, "Spin-based computing: Device concepts, current status, and a case study on a high-performance microprocessor", *Proceedings of the IEEE*, vol. 103, no. 1, pp. 106–130, Jan-2015.
[42] W. H. Butler, X. G. Zhang, T. C. Schulthess and J. M. MacLaren, "Spin-dependent tunnelling conductance of Fe|MgO|Fe sandwiches", *Physical Review B*, vol. 63, no. 5, p. 054416, Jan-2001.
[43] J. Mathon and A. Umerski, "Theory of tunneling magnetoresistance of an epitaxial Fe/MgO/Fe (001) junction", *Physical Review B*, vol. 63, no. 22, p. 220403, May-2001.
[44] B. Dlubak, M.-B. Martin, C. Deranlot, B. Servet, S. Xavier, R. Mattana, M. Sprinkle, C. Berger, W. A. De Heer, F. Petroff, A. Anane, P. Seneor and A. Fert, "Highly efficient spin transport in epitaxial graphene on SiC", *Nature Physics*, vol. 8, pp. 557–561, Jun-2012.
[45] S. A. Crooker, M. Furis, X. Lou, C. Adelmann, D. L. Smith, C. J. Palmstrøm and P. A. Crowell, "Imaging spin transport in lateral ferromagnet/semiconductor structures", *Science*, vol. 309 no. 5744, pp. 2191–2195, Sept-2005.

[46] A. T. Hanbicki, B. T. Jonker, G. Itskos, G. Kioseoglou and A. Petrou, "Efficient electrical spin injection from a magnetic metal/tunnel barrier contact into a semiconductor", *Applied Physics Letters*, vol. 80, no. 7, p. 1240, Jan-2002.

[47] X. Jiang, R. Wang, R. M. Shelby, R. M. Macfarlane, S. R. Bank, J. S. Harris and S. S. P. Parkin, "Highly spin-polarized room-temperature tunnel injector for semiconductor spintronics using MgO(100)", *Physical Review Letters*, vol. 94, no. 5, p. 056601, Feb-2005.

[48] R. Wangb, X. Jiang, R. M. Shelby, S. S. P. Parkin, S. R. Bank and J. S. Harris, "Temperature independence of the spin-injection efficiency of a MgO-based tunnel spin injector", *Applied Physics Letters*, vol. 87, no. 26, p. 262503, Nov-2005.

[49] P. LeClair, J. K. Ha, H. J. M. Swagten, J. T. Kohlhepp, C. H. van de Vin and W. J. M. de Jonge, "Large magnetoresistance using hybrid spin filter devices", *Applied Physics Letters*, vol. 80, no. 4, pp. 625–627, Jan-2002.

[50] J. S. Moodera, X. Hao, G. A. Gibson and R. Meservey, "Electron-spin polarization in tunnel junctions in zero applied field with ferromagnetic EuS barriers", *Applied Physics Letters*, vol. 61, no. 5, p. 637, Aug-61.

[51] M. Gajek, M. Bibes, A. Barthélémy, K. Bouzehouane, S. Fusil, M. Varela, J. Fontcuberta and A. Fert, "Spin filtering through ferromagnetic BiMnO3 tunnel barriers", *Physical Review B*, vol. 72, no. 2, p. 020406, Jul-2005.

[52] U. Lüders, M. Bibes, K. Bouzehouane, E. Jacquet, J. P. Contour, S. Fusil, J. F. Bobo, J. Fontcuberta, A. Barthélémy and A. Fert, "Spin filtering through ferrimagnetic NiFe2O4 tunnel barriers", *Applied Physics Letters*, vol. 88, no. 8, p. 082505, Jan-2006.

[53] S. Matzen, J.-B. Moussy, P. Wei, C. Gatel, J. C. Cezar, M. A. Arrio, Ph. Sainctavit and J. S. Moodera, "Structure, magnetic ordering, and spin filtering efficiency of NiFe2O4(111) ultrathin Films", *Applied Physics Letters*, vol. 104, no. 18, p. 182404, May-2014.

[54] A. V. Ramos, M. J. Guittet, J. B. Moussy, R. Mattana, C. Deranlot, F. Petroff and C. Gatel, "Room temperature spin filtering in epitaxial cobalt-ferrite tunnel barriers", *Applied Physics Letters*, vol. 91, no. 12, p. 122107, Aug-2007.

[55] Y. K. Takahashi, S. Kasai, T. Furubayashi, S. Mitani, K. Inomata and K. Hono, "High spin-filter efficiency in a Co ferrite fabricated by a thermal oxidation", *Applied Physics Letters*, vol. 96, no. 7, p. 072512, Jan-2010.

Spin-Transfer Torque for Universal Memory Applications

5

Sameena Shah, Gul Faroz Ahmad Malik, Mubashir Ahmad Kharadi, Farooq Ahmad Khanday

Contents

1	Introduction	63
2	Magnetic Tunnel Junction (MTJ)	64
3	Magnetoresistive Random Access Memory (MRAM)	65
4	Spin-Transfer Torque MRAM (STT-MRAM)	67
	4.1 Issues with STT-MRAM	69
	4.2 Structural Improvements	70
	4.3 Novel Techniques for MRAM Switching	71
5	Conclusion	73
	References	73

1 INTRODUCTION

The scaling of Complementary Metal Oxide Semiconductor (CMOS) technology leads to an increase in leakage current and static power dissipation in memories and logic-based circuits. This limits the miniaturization of CMOS-based electronic devices. So, there is a need to replace these conventional charge-based electronic devices with some novel technologies. Spintronics is an emerging field that has the potential to replace mainstream electronics. Spintronics is a field where the spin of an electron is used to manipulate the data. Spintronics has developed into a key research area. The spin devices show better performance in many aspects. These devices use minimal power, have a fast response time, high density, scalability, and are non-volatile. Spin-based devices can be utilized in various memory and logic applications due to their potential advantages over conventional CMOS devices. Spintronics developed rapidly after the Giant Magnetoresistance (GMR) effect was discovered in 1988 [1],[2]. MTJ is one of the major developments in the field of spintronics which is based on the Tunnel Magnetoresistance (TMR) effect. One of the most significant applications of MTJ is the MRAM in which MTJ is employed as a fundamental memory element. MRAM combines the advantages of SRAM, DRAM, and flash memory i.e., high speed, high density, and non-volatility respectively. Therefore, MRAM has the potential to

DOI: 10.1201/9781003196952-5

replace conventional memory technologies and become universal memory. Two major breakthroughs i.e., the discovery of MTJs based on crystalline MgO barriers and spin-transfer torque (STT) magnetization switching enhanced the research & development in MRAM. MgO-based MTJs exhibit larger TMR ratios at room temperature (up to 600%) [3],[4] which provides faster read operation and improved read margin. STT-MRAMs provide better scalability and are compatible with the CMOS fabrication process. STT-MRAM has been identified as one of the most promising choices for next-generation universal memory technologies by the International Technology Roadmap for Semiconductors (ITRS). However, there is a need to overcome some key design issues with STT-MRAM for it to replace the conventional semiconductor memories, for example, reducing the high write currents required for STT switching and improving the reliability. One of the effective ways to lower the write currents in STT-MRAM is to utilize the MTJ with perpendicular magnetization (p-MTJ). Perpendicular Magnetic Anisotropy (PMA) based STT-MRAMs are being focused on research and development. These MRAMs are scalable to higher densities. Some other MRAMs have been developed to overcome the issues with STT-MRAM. These are Spin-Orbit Torque MRAM (SOT-MRAM) and Voltage Controlled Magnetic Anisotropy MRAM (VCMA-MRAM).

This chapter gives a brief description of MTJ, followed by the MRAM. Various types of MRAMs are mentioned. STT-MRAM is being studied in detail in the next section, its design issues and methods to overcome these issues are also being discussed.

2 MAGNETIC TUNNEL JUNCTION (MTJ)

MTJ is a magnetic device that is made up of two ferromagnetic (FM) layers with an insulating layer (ultrathin) placed in between them as shown in Figure 5.1. The insulating layer is called a tunnel barrier (TB). The orientation of the magnetization of one of the FM layers is fixed and is called a reference layer or pinned layer (PL). The other FM layer is called a free layer (FL) since its magnetization orientation is not fixed. The phenomenon behind the working of MTJ is the TMR effect which was actually discovered by Jullier in 1975 [5]. Spin-dependent tunneling is the principle behind the TMR phenomenon. By altering the direction of magnetization of the FL, also called as a storage layer, the MTJ configuration can be switched. This switching process can be achieved by any of the switching mechanisms available. The resistance of the MTJ device will be small if both FM electrode layers have their magnetizations in the same direction (i.e., parallel) and it will be high if the magnetization directions of electrodes are different (i.e., anti-parallel). Therefore, the parallel and antiparallel MTJ configurations have different resistance values. High resistance or antiparallel state is denoted as logic "1", and a low resistance or parallel state is denoted as logic "0". The TMR is observed by the tunneling current being dependent on the relative orientation of the magnetization of free and fixed layers. TMR value is chosen based on the device's intended use.

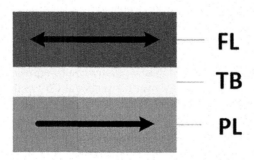

FIGURE 5.1 Magnetic Tunnel Junction (MTJ).

3 MAGNETORESISTIVE RANDOM ACCESS MEMORY (MRAM)

MRAM is a non-volatile memory in which magnetic anisotropy is used for retaining information and magnetoresistance is used to read information. Just like the DRAM consisting of a capacitor and a transistor, a typical MRAM cell also consists of a memory element (magnetoresistive element) and a transistor. MTJ is used as the memory element in MRAM making it a strong contender for universal non-volatile memory of the future. When it comes to DRAM, the charge stored on the capacitor determines the memory state while as in the case of MRAM, the resistance of the MTJ determines the memory state i.e., "1" and "0". A transistor is necessary for each MRAM cell as it helps during the operation of writing by providing the write current. Also, the voltage difference between the two states is insufficient, therefore, a transistor is required for its functioning. A memory element should satisfy some main requirements. It should be able to store data for longer periods (for non-volatile memory). Mechanisms should be available for reading data from and writing data into the memory device. In the case of MRAM, the resistance difference is sensed between the two states of a magnetoresistive device to carry out the read operation. To carry out the write operation, a variety of approaches can be used for changing the magnetization orientation of the FL. One such approach is inducing a magnetic field. Magnetic anisotropy of the storage layer is employed for the storage of data. The concept of storage in MRAM is based on the energy barrier necessary to transfer the magnetization from one direction to the other (Figure 5.2). If the barrier is sufficiently high so that it may overcome stray fields from outside, only then, the magnetization will be pinned in a specific direction. The magnetization orientation of the PL in an MRAM is fixed while the FL orientation is changed to store "1" or "0" state. The material used in the PL should have a significant energy barrier since its direction of magnetization must remain fixed and never change. The material employed in the FL must have a high magnetic anisotropy for storing the magnetization for a few years. MRAMs can be grouped into two categories based on the storage mechanism; (i) MRAMs using in-plane MTJ where the magnetization of the FM layers is in the plane of film and (ii) MRAMs using perpendicular MTJs where the magnetization of FM layers is perpendicular to the film plane.

The magnetoresistance effect (MR) is essential for accurate data reading. The potential difference between the two resistance states i.e., low and high states has to be more than 0.2 V for reliable data reading. Therefore, the MRAM is reliable only if the magnetoresistance is high enough. In GMR devices like spin valves, the magnetoresistance is not strong enough to provide a substantial voltage difference between low and high states of resistance. As a result, GMR devices are of no practical use in MRAM applications. TMR provided a boost to MRAM applications. MTJs with tunnel barriers of crystalline MgO enhanced the MRAM research and development. A TMR value of almost 600 percent has been

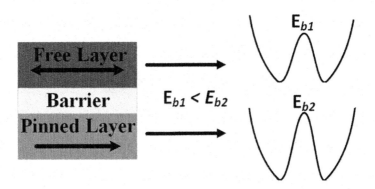

FIGURE 5.2 The relative energy barrier of the pinned layer and free layer in an MTJ.

observed in such devices in laboratory demonstrations and MRAM devices, a TMR value of 100%–200% has been achieved which ensures that the voltage difference between "0" and "1" states is sufficient.

Method of writing information is another requirement of a memory device. The use of magnetic fields is one of the earliest ways of writing data in MRAM. Since then, the data writing in MRAM has progressed and various writing techniques or switching techniques have been employed in MRAMs. Based on switching mechanisms, there are various types of MRAMs such as field-induced magnetization switching MRAMs (FIMS-MRAM), Toggle MRAMs, Spin-Transfer Torque MRAMs (STT-MRAM), Thermally Assisted MRAMs (TA-MRAM), Spin-Orbit Torque MRAMs (SOT-MRAM), Voltage Controlled MRAMs (VCMA-MRAM). In FIMS-MRAM, switching of MTJ occurs due to magnetic fields as a result of two orthogonal currents flowing across a word line (WL) and a bit line (BL). In such applications, each MTJ is situated between the two current lines i.e., WL and BL as shown in Figure 5.3. During a write operation to a particular MTJ or a particular MRAM cell, currents flow through the corresponding WL and BL producing magnetic fields. If the magnitude of currents and magnetic fields is sufficient enough, the magnetization in the free layer of the selected MTJ is switched and thus changing the stored information [6], [7], [8]. This method of switching suffers from some significant issues. When the current passes through the current lines, the other MTJs along the WL or BL experience the magnetic field (half selected MTJs) that can switch their state. Also, large currents are required to produce significant magnetic fields resulting in excessive energy consumption. Scalability is also limited due to the electromigration effect. To overcome the half selectivity issue, toggle switching was proposed [9], [10], [11]. However, this approach is a bit complex and does not overcome many of the issues.

Another milestone in the area of spintronics was the discovery of a switching mechanism based on Spin-Transfer Torque. For switching the magnetization state, the STT technique does not need any external magnetic field. In STT devices, the writing operation is carried out by a polarized current due to which a torque known as spin-transfer torque is exerted on the storage layer switching its magnetization direction. The architecture of STT-MRAM is simpler than that of FIMS-MRAM. Also, switching due to current is scalable, simplifying the MRAM design and process of fabrication. Due to the STT-MRAM's non-volatility, zero leakage, outstanding integration density, adequate reading, and writing performance, and compatibility with CMOS, the STT-MRAM has gained much interest. However, various issues/obstacles need to be addressed, for example, consumption of energy due to high current, long latency for a write operation, reliability of tunnel barrier, etc. Several techniques have been created to address the problems in

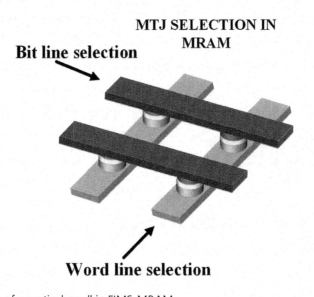

FIGURE 5.3 Selection of a particular cell in FIMS-MRAM.

STT mechanism, for example, utilization of perpendicular MTJ, Voltage Controlled Magnetic Anisotropy (VCMA), and Spin-Orbit Torque (SOT).

4 SPIN-TRANSFER TORQUE MRAM (STT-MRAM)

The most promising switching technique for MRAM applications is Spin-Transfer Torque switching. It was first observed in CoFeB/Al$_2$O$_3$/CoFeB MTJ in 2004 [12]. Diao et al. in 2005 demonstrated this effect in MgO-based MTJ where lower current density was required for switching [13]. An MTJ and an access transistor, typically an NMOS make up the conventional STT-MRAM cell structure as shown in Figure 5.4(**a**). This structure is known as the 1T-1MTJ STT-MRAM bit cell. In a memory array (Figure 5.4(**b**)), access to a particular MTJ is allowed by the access transistor. During the writing operation to a particular bit cell, the line connected to the gate terminal of the access transistor of the cell called WL (word line) is charged to Vdd. Due to this, the current can pass through the MTJ between BL(bit line) and SL(source line). To store "0" into the bit cell i.e., a low resistance state, the configuration of magnetization of the MTJ needs to be parallel. For that, the BL is charged to Vdd, and SL is discharged to the ground so that a current (I_{W0}) passes from FL to PL. Instead, if "1" (high resistance state) is to be written, the current (I_{W1}) direction needs to be reversed i.e., from PL to FL. In this case, BL is grounded and SL is charged to Vdd so that the MTJ configuration is antiparallel. The magnetization configuration changes only if the voltage Vdd is set such that the current through MTJ is greater or equal to the critical current (I_C), since the torque needs to be strong enough to change the state of MTJ. However, high currents lead to high power consumption. Also, the write current to store "1" i.e., I_{W1} should be a bit smaller than the write current to store "0" i.e., I_{W0} because of source degeneration of NMOS during the write "1" operation. Now, during the read process, the WL which is connected to the particular bit cell that needs to be read is charged to Vdd. A current called read current (I_R) is passed through the MRAM bit cell. Due to this, a voltage (V_R) gets developed across the BL and SL. This voltage is sensed by a voltage sensing amplifier which compares it to a reference voltage (V_{Ref}). Voltage sense amplifier senses "1" if $V_R > V_{Ref}$ and "0" if $V_R < V_{Ref}$. This is called a voltage sensing scheme. A current sensing scheme can also be employed where a current is sensed called read current and then it is to be compared to some reference current. During the read operation, the current passing through the MTJ needs to be small to avoid accidental overwriting into the cell.

The critical current density necessary to change the state of MTJ is calculated as

$$Jc = (\alpha/\eta)(2e/\hbar)M_s tH_k + 2\pi M_s \qquad (1)$$

where, "H_k" is anisotropy field, "t" is film thickness, "M_s" is saturation magnetization, "α" is Gilbert damping constant and "η" is the efficiency of STT related to spin polarization of current.

This expression is obtained from an equation called the Lifshitz-Gilbert (LLG) equation which models the STT switching mechanism describing the dynamics of magnetization of free layer [14], [15], [16]. The critical current density of an STT-MRAM is determined by the MTJ's material, structure, and size. If the writing current is higher than the critical current, the time required for switching will be less. Based on the pulse duration of the write current, there are two regimes for the switching process [17], [18], [19]. One is the precessional regime where the switching occurs due to the spin-transfer torque effect and another regime is the thermal activation one where the switching occurs because of thermal fluctuations. In the precessional regime, the current pulse duration is short (less than 10ns), and therefore the write current must be larger than the critical current to switch the magnetization of the FL. In the case of the thermal activation regime, the pulse duration of current is long (greater than 50ns) and the magnetization switching is possible for smaller currents (less than critical current). To reduce the write latency, a precessional regime is used in practice.

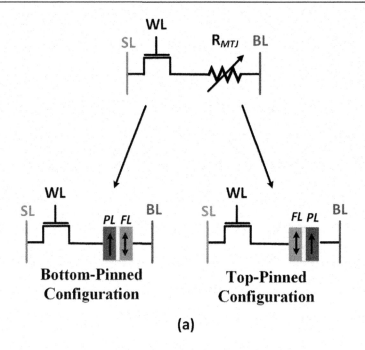

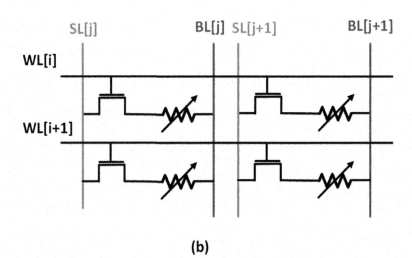

FIGURE 5.4 (a) STT-MRAM connected in two configurations: Bottom-Pinned and Top-Pinned. (b) 2x2 array of bit cells.

STT-MRAM must be thermally stable to function properly. The period for which the information can be retained by the memory depends on its thermal stability. A factor called thermal stability factor is defined for MRAMs which is given by [20].

$$\Delta = E_b/K_b T \qquad (2)$$
$$\text{and } t_{ret} = t_0 \exp(\Delta) \qquad (3)$$

where E_b is the energy barrier (between the two states), K_b is a constant called Boltzman constant, T is the temperature, t_{ret} is the retention time and t_0 is called the inverse attempt frequency. The higher the value of Δ, the more will be the data retention time. To retain information for longer periods (10 years and above), the factor Δ should be at least 60. For uniform magnetization rotation, E_b is equal to anisotropy energy given by

$$E_b = K_u V \tag{4}$$

where "K_u" is a constant called uniaxial anisotropy constant and "V" is volume and K_u is given by

$$K_u = H_K M_s / 2 \tag{5}$$

Therefore, the thermal stability factor can be calculated as

$$\Delta = H_K M_s V / 2 K_b T \tag{6}$$

According to the above expression, for the thermal stability factor to be equal to 60, the diameter of the memory cell has to be 60 nm with 5 nm thickness and $K_u = 8 \times 10^4$ ergs/cc for in-plane memory. Therefore, if a memory with a longer data retention period and smaller size (less than 60 nm) is required, perpendicular MRAMs need to be explored.

For an MTJ with Δ = 60, the read current I_R should be very small, typically less than $0.5 I_C$ to avoid accidental overwriting into the memory during a read operation. However, much small read current may result in the incorrect read operation. Generally, the current magnitudes should satisfy the following relation

$$I_{W0} > I_{W1} > I_C > I_R$$

4.1 Issues with STT-MRAM

Despite the fact that STT-MRAM is one of the most promising technologies for non-volatile memory, it suffers from some limitations. Since the same path is taken by both read and write operations across the MTJ, the read current can cause accidental overwriting into the device called read disturb. Another major drawback is that the switching process necessitates large currents with pulse widths in the nanosecond range which increases power consumption. Since the MgO barrier is ultra-thin, these large currents result in its breakdown [21], [22], [23], [24]. For better performance and functionality of STT-MRAM, the reliability of MTJ is crucial. As already mentioned, the STT-MRAM is having some reliability concerns which arise due to dielectric breakdown and resistance drift. Both dielectric breakdown and resistance drift leads to a reduction in the resistance of MTJ.

Any undesirable internal or external magnetic field influences the stability of the magnetic states in the ferromagnetic layers of MTJ [25]. Since MTJ consists of many ferromagnetic layers, the stray fields produced by these layers will definitely affect the stability of magnetization states of FL. The FM layers of neighboring cells also generate stray fields. The overall effect of these stray fields is known as magnetic coupling. Magnetic coupling may affect the thermal stability Δ of the MTJ. This effect also impacts the critical current density. The process variations in the access transistor and MTJ can also lead to some serious issues degrading the performance of STT-MRAM. Due to process variations, some MTJ parameters like magnetic anisotropy, saturation magnetization, TMR, barrier thickness, cross-sectional area can be affected. These variations are introduced at every stage of the manufacturing process of STT-MRAM. The change in dimensions of the free layer and thickness of the tunnel barrier can lead to a change in the behavior of MTJ (static and dynamic behavior) which in turn leads to failure in the operation of the STT-MRAM cell. Also, some key parameters of the access transistor-like threshold voltage and transistor size can be affected during the fabrication process due to which STT-MRAM performance degrades. The behavior of MTJ and CMOS transistor is temperature-dependent and if there are temperature variations, the performance and reliability of STT-MRAM is threatened [26]. Some MTJ parameters like TMR, Δ, and critical current get affected due to temperature fluctuations. Since the thermal stability factor Δ is inversely proportional to temperature according to equation (2), the data retention time decreases with temperature exponentially. According to Zhao et al., the resistance in antiparallel configuration decreases with an increase in temperature [27], [28], [29], due to which TMR value decreases. As a result, the read margin shrinks at high temperatures. One benefit at elevated temperatures is that the critical current decreases and this phenomenon are used in thermally-assisted MRAMs. The spin-transfer torque switching being stochastic in nature also creates some design challenges. Various efforts have been made by the researchers to tackle the above-mentioned design issues.

4.2 Structural Improvements

The design of the MTJ stack may be improved which solves many of the problems associated with the structure. To reduce the stray fields from the pinned layer, it can be replaced with a structure made of synthetic antiferromagnet (SAF). It is highly desirable to reduce the critical current in order to improve reliability, endurance, and power consumption. Reduced critical current lowers the stress on the MTJ. According to equation (1), the critical current can be minimized by lowering the value of α (damping constant) or by enhancing η (spin polarization efficiency). η can be increased by replacing the ferromagnetic layers of MTJ with half-metallic heusler alloys. The improvement in η also improves the magnetoresistance of MTJ. Decreased α also reduces the variations in critical current due to thermal fluctuations. The damping constant α depends on the thickness of the FL [30], [31]. However, optimizing FL thickness to reduce α while maintaining other magnetic properties is challenging.

Some alternative structures for MTJ have been explored by researchers to achieve improved STT-MRAM. Double barrier structures consisting of two reference layers and two tunnel barriers were investigated as shown in Figure 5.5. These structures showed an improvement in the spin filtering efficiency and reduction in critical current. It is to be noted that in double-barrier structures, the two PLs are antiparallel to each other and therefore one of the PL will be parallel to the FL. To distinguish between the two states of FL, the thickness of two tunnel barriers must be different. Another category of structures has been proposed where single or double barrier structures are introduced with synthetic antiferromagnetic

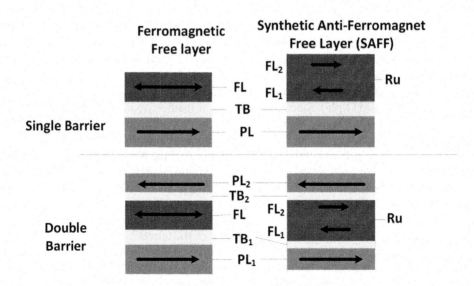

FIGURE 5.5 Single and Double barrier MTJ stack with and without Synthetic Antiferromagnetic FL (SAFF).

FL (SAFF). A pair of ferromagnetic layers that are magnetically coupled and sandwiching a thin layer of Ru (paramagnetic) make up the SAFF structure. This structure makes the FM layers more resistant to thermal fluctuations. Some other structures have also been developed to enhance the performance of STT based MRAM for example, MTJ with tilted magnetic anisotropy [32], which reduces the critical current. It was also proposed that an orthogonal MTJ stack be used to minimize critical current without losing its readability [33].

4.3 Novel Techniques for MRAM Switching

Various new switching mechanisms are being investigated to eliminate the problems with STT-MRAM. These include Spin-Orbit Torque (SOT), Voltage Controlled Magnetic Anisotropy (VCMA), Spin Hall Effect (SHE), etc. The MRAMs based on these switching mechanisms are considered as third-generation MRAMs, STT-MRAM being second-generation MRAM and FIMS-MRAM first-generation MRAM. These innovative switching strategies are designed to save energy while enhancing writing performance.

SOT magnetization switching has been developed to overcome the technical problems with STT switching such as long write latency, less reliability, high write current requirement i.e., large energy consumption. The magnetization of the free layer is reversed without passing the current across the tunnel in SOT switching scheme, separating the read and write routes and improving reliability. Also, the switching is faster in the case of the SOT mechanism. Spin-orbit torque is made up of two torques owing to the spin Hall effect and Rashba effect. These torques are orthogonal to each other. In SOT devices such as in-plane MTJ (iMTJ) consisting of a heavy metal/ferromagnetic structure (Figure 5.6(a-b)), the spin-polarized current can be produced in the heavy metal like Pt (with strong spin-orbit coupling), bypassing a charge current through it which causes the electrons with a different spin to accumulate laterally at the opposite sides on account of spin hall effect as shown in Figure 5.6(a). This spin current is perpendicular to both charging current and spin polarization. Due to this spin current, the spin-orbit torque (SOT) is generated in the ferromagnetic free layer of the device causing it to switch its magnetization [34], [35]. In the case of perpendicular MTJ (pMTJ), an externally applied magnetic field is necessary for switching since the anisotropy axis and spin direction are not collinear [36]. Much effort has been made to remove the requirement of field assistance for SOT switching in pMTJs. Many researchers have proposed field-free switching schemes based on different phenomena that have been used such as interaction between SOT and STT [37], [38], [39], interlayer exchange coupling [40], engineered tilted anisotropy [41], etc.

Both magnetic field and current-driven switching mechanisms suffer from some problems. It is not possible to reduce critical current without impacting the thermal stability in such mechanisms. The voltage-controlled approach employs voltage dependence of perpendicular magnetic anisotropy (PMA) to achieve voltage-controlled MTJ. This effect is called voltage-controlled magnetic anisotropy (VCMA). The states can be switched using low current voltage pulses across the MTJ terminals. Due to the bias voltage, an electric field will be generated which causes the charge to get accumulated at the metal barrier interface (Figure 5.6(c)) which ultimately reduces magnetic anisotropy. Reduced magnetic anisotropy makes switching possible with low power dissipation [42]. The energy barrier is lowered as a result of this phenomenon (between parallel and antiparallel states) during the writing process and therefore lower current will be required for switching. MgO-based MTJs with PMA are suitable for voltage-controlled switching, for example, in CoFe/MgO/Fe MTJ, a low switching current with sub-nanosecond voltage pulses was achieved [43]. This switching mechanism motivated researchers to develop non-volatile memories and hybrid circuits. A new generation of MRAM called VCMA-MRAM has been developed based on this effect. The VCMA effect has been used to assist STT switching to improve the write delay and dynamic power dissipation of MTJ. Precessional VCMA-assisted STT is one such switching mechanism that has proved to be very efficient. The development of voltage-controlled switching provides a new way to attain extremely low power consumption and very fast operation in future MTJ

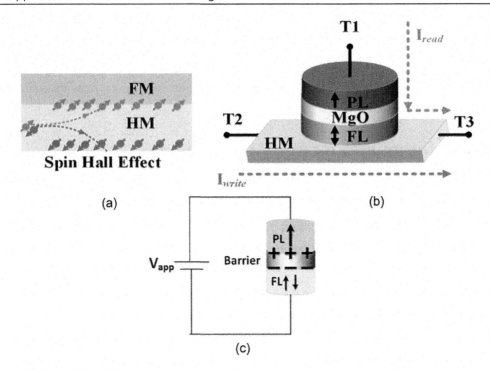

FIGURE 5.6 (a) Representation of spin hall effect (b) Spin-orbit torque memory device showing separate read and write paths (c) VCMA assisted pMTJ device.

based devices. Voltage-Gated Spin-Orbit Torque (VGSOT) has been proposed where VCMA is used to assist SOT switching to eliminate some of the issues associated with SOT such as switching reliability and energy dissipation [44]. However, the VCMA based devices suffer from reliability issues. Also, there is a need for an externally applied magnetic field to enable switching which makes it difficult to realize practical VCMA based memories. Deng et al. in 2017 proposed a perpendicular MTJ which is elliptical shaped to avoid the need for an external magnetic field. Further efforts need to be made to improve this switching mechanism.

The above switching mechanisms have been implemented in various spin devices, but their speed of operation is limited due to spin precession time. This time constraint needs to be eliminated to realize future fast spintronic applications. It has been seen that the optical switching mechanism has a great potential to be used for fast and energy-efficient memory applications. A mechanism called All-Optical Switching (AOS) that uses laser pulses with sub-picosecond pulse duration has recently been developed. A vast variety of material systems have been investigated for such effect which was discovered in materials like rare-earth and transition metal alloys such as TbFe, Gd(Fe, Co), TbCo [45], [46], [47], [48], [49], their ferrimagnetic multilayers [50] and also in multilayers of Co/Pt and Co/Ni which are ferromagnetic[51]. This effect was first observed in Gd(Fe, Co) alloy which attracted attention since its switching can be achieved using short single laser pulses [52]. Since the ferromagnetic multilayers are currently being utilized in spintronic devices, the research gained much attention when All-Optical Switching was observed in such multilayers. Chen et al. in 2017 demonstrated the AOS in GdFeCo based pMTJ (with GdFeCo as a free layer) using single sub-picosecond infrared laser pulses and it has been shown that the AOS switching is twice faster than other switching mechanisms. In 2020, Olivier et al. proposed that indium tin oxide (ITO) can be used as an optical access for MTJ in spintronic-photonic circuits [53]. A compact model was presented for AOS-MTJ by Pelloux-Prayer et al. in 2020 [54].

5 CONCLUSION

Since the discovery of TMR, three MRAM generations based on the mechanism of switching of MTJ have existed and are still under development. The first generation i.e., field-induced magnetization switching MRAM (FIMS-MRAM) faced many challenges, the most prominent being the scalability issue which obstructed its commercialization. MRAMs in their third technology generation are still in the early phases of development, requiring more advancements in manufacturing, materials, reliability enhancement, etc. On the other hand, STT-MRAM, which is regarded as the MRAM's second generation, is closest to widespread adoption for a variety of applications at the moment. In this chapter, a study of MRAMs with special emphasis on STT-MRAM has been carried out. The issues related to STT-MRAM and the methods for resolving these issues and improving the performance of STT-MRAM have been discussed. It can be concluded that MRAMs, especially STT-MRAMs have the potential to replace the conventional memory technologies and become the universal memory. STT-MRAMs can be used in numerous applications, for example, IoT (Internet of Things) and last-level caches. However, many difficulties must be dealt with before it can be mass-produced.

REFERENCES

1. M. N. Baibich, J. M. Broto, A. Fert, F. Nguyen Van Dau, F. Petroff, P. Etienne, G. Creuzet, A. Friederich and J. Chazelas. "Giant magnetoresistance of (001)Fe/(001)Cr magnetic superlattices," *Phys. Rev. Lett.*, vol. 61, pp. 2472, 1988.
2. G. Binasch, P. Grunberg, F. Saurenbach and W. Zinn, "Enhanced magnetoresistance in layered magnetic structures with antiferromagnetic interlayer exchange," *Phys. Rev. B*, vol. 39, pp. 4828–4830, 1989.
3. S. Yuasa, T. Nagahama, A. Fukushima, Y. Suzuki and K. Ando, "Giant room temperature magnetoresistance in single-crystal Fe/MgO/Fe magnetic tunnel junctions," *Nat. Mater.*, vol. 3, pp. 868–871, 2004. doi:10.1038/nmat1257.
4. S. S. P. Parkin, C. Kaiser, A. Panchula, P. M. Rice, B. Hughes, M. Samant and S.-H. Yang, "Giant tunnel magnetoresistance at room temperature with MgO(100)tunnel barriers," *Nat. Mater.*, vol. 3, pp. 862–867, 2004. doi:10.1038/nmat1256.
5. M. Julliere, "Tunneling between ferromagnetic films," *Phys. Lett.*, vol. 54, pp. 225–226, 1975.
6. R. H. Koch, J. G. Deak, D. W. Abraham, P. L. Trouilloud, R. A. Altman, Y. Lu and S. S. P. Parkin, "Magnetization reversal in micron-sized magnetic thin films," *Phys. Rev. Lett.*, vol. 81, pp. 4512, 1998.
7. S. Z. Peng, Y. Zhang, M. X. Wang, Y. G. Zhang and W. Zhao, "Magnetic tunnel junctions for spintronics: Principles and applications," in *Wiley Encyclopedia of Electrical and Electronics Engineering*, Hoboken, NJ: John Wiley & Sons, Inc., pp. 1–16, 1999.
8. V. K. Joshi, P. Barla, S. Bhat and B. K. Kaushik, "From MTJ device to hybrid CMOS/MTJ circuits: A review," *IEEE Access*, vol. 8, pp. 194105–194146, 2020. doi:10.1109/ACCESS.2020.3033023.
9. M. Durlam, D. Addie, J. Akerman, B. Butcher, P. Brown, J. Chan and S. Tehrani, "A 0.18/spl mu/m 4Mb toggling MRAM," *Electron Devices Meeting, 2003. IEDM'03 Technical Digest*. IEEE International, pp. 34–36, 2003.
10. C. K. Subramanian, T. W. Andre, J. J. Nahas, B. J. Garni, H. S. Lin, A. Omair and W. L. Martino Jr, "Design aspects of a 4 Mbit 0.18 mm 1T1MTJ toggle MRAM memory," *International Conference on IEEE Integrated Circuit Design and Technology, ICICDT'04*, pp 177–181, 2004. https://ieeexplore.ieee.org/document/1309940. doi:10.1109/ICICDT.2004.1309940.
11. B. N. Engel, J. Akerman, B. Butcher, R. W. Dave, M. DeHerrera, M. Durlam and S. Tehrani, "A 4-Mb toggle MRAM based on a novel bit and switching method," *IEEE Trans. Magn.*, vol. 41, pp 132–136, 2005.
12. Y. Huai, F. Albert, P. Nguyen, M. Pakala and T. Valet, "Observation of spin-transfer switching in deep submicron-sized and low resistance magnetic tunnel junctions," *Appl. Phys. Lett.*, vol. 84, no. 16, pp. 3118_3120, Apr. 2004.

13. Z. Diao, D. Apalkov, M. Pakala, Y. Ding, A. Panchula and Y. Huai, "Spin transfer switching and spin polarization in magnetic tunnel junctions with MgO and AlO x barriers," *Appl. Phys. Lett.*, vol. 87, pp. 232502–232502, 2005.
14. J. Slonczewski, "Current-driven excitation of magnetic multilayers," *J. Magn. Magn. Mater.*, vol. 159, pp. L1 – L7, 1996. doi:10.1016/0304-8853(96)00062-5.
15. J. Sun, "Spin-current interaction with a monodomain magnetic body: A model study," *Phys. Rev. B*, vol. 62, pp. 570–578, 2000. doi:10.1103/PhysRevB.62.570.
16. A. Aharoni, "Micromagnetics: Past, present and future," *Phys. B: Condens. Matter*, vol. 306, pp. 1–9, 2001. doi:10.1016/S0921-4526(01)00954-1.
17. W. H. Butler, Tim Mewes, Claudia K. A. Mewes, P. B. Visscher, William H. Rippard, Stephen E. Russek and Rank Heindl, "Switching distributions for perpendicular spin torque devices within the macrospin approximation," *Trans. Magn.*, vol. 48, pp. 4684–4700, Dec. 2012. doi:10.1109/TMAG.2012.2209122.
18. D. Bedau, H. Liu, J. Z. Sun, J. A. Katine, E. E. Fullerton, S. Mangin and A. D. Kent, "Spin-transfer pulse switching: From the dynamic to the thermally activated regime," *Appl. Phys. Lett.*, vol. 97, pp. 262502, 2010. doi:10.1063/1.3532960.
19. E. B. Myers, F. J. Albert, J. C. Sankey, E. Bonet, R. A. Buhrman and D. C. Ralph, "Thermally activated magnetic reversal induced by a spin-polarized current," *Phys. Rev. Lett.*, vol. 89, pp. 196801, Oct. 2002. doi:10.1103/PhysRevLett.89.196801.
20. A. V. Khvalkovskiy, D. Apalkov, S. Watts, R. Chepulskii, R. S. Beach, A. Ong, X. Tang, A. Driskill-Smith, W. H. Butler, P. B. Visscher, D. Lottis, E. Chen, V. Nikitin and M. Krounbi, "Basic principles of STT-MRAM cell operation in memory arrays," *J. Phys. D: Appl. Phys.*, vol. 46, pp. 074001, 2013. doi:10.1088/0022-3727/46/7/074001.
21. A. A. Khan, J. Schmalhorst, A. Thomas, O. Schebaum and G. Reiss, "Dielectric breakdown in Co-Fe-B/MgO/Co-Fe-B magnetic tunnel junction," *J. Appl. Phys.*, vol. 103, 2008. doi:10.1063/1.2939571.
22. C. Yoshida, M. Kurasawa, Y. M. Lee, K. Tsunoda, M. Aoki and Y. Sugiyama, "A study of dielectric breakdown mechanism in CoFeB/MgO/CoFeB magnetic tunnel junction," *IEEE International Reliability Physics Symposium*, pp. 139–142, 2009. https://ieeexplore.ieee.org/document/5173239. doi:10.1109/IRPS.2009.5173239.
23. M. Schafers, V. Drewello, G. Reiss, A. Thomas, K. Thiel, G. Eilers, M. Munzenberg, H. Schuhmann and M. Seibt, "Electric breakdown in ultrathin MgO tunnel barrier junctions for spin-transfer torque switching," *Appl. Phys. Lett.*, vol. 95, pp. 1–4, 2009. doi:10.1063/1.3272268.
24. A. A. Khan, "Analysis of dielectric breakdown in CoFeB/MgO/CoFeB magnetic tunnel junction," *Microelectron. Reliab.*, vol. 55, pp. 894–902, 2015. doi:10.1016/j.microrel.2015.02.018.
25. Y.-H. Wang, S.-H. Huang, D.-Y. Wang, K.-H. Shen, C.-W. Chien, K.-M. Kuo, S.-Y. Yang, D.-L. Deng, "Impact of stray field on the switching properties of perpendicular MTJ for scaled MRAM," *IEEE International Electron Devices Meeting*, Dec. 2012. https://ieeexplore.ieee.org/document/6479127. doi:10.1109/IEDM.2012.6479127.
26. L. Zhang, Y. Cheng, W. Kang, L. Torres, Y. Zhang, A. Todri-Sanial and W. Zhao, "Addressing the thermal issues of STT-MRAM from compact modeling to design techniques," *IEEE Trans. Nanotechnol.*, vol. 17, no. 2, pp. 345–352, March 2018. doi:10.1109/TNANO.2018.2803340.
27. X. Kou, J. Schmalhorst, A. Thomas and G. Reiss, "Temperature dependence of the resistance of magnetic tunnel junctions with MgO barrier," *Appl. Phys. Lett.*, vol. 88, pp. 1–4, 2006. doi:10.1063/1.2206680.
28. H. Zhao, P. K. Amiri, Y. Zhang, A. Lyle, J. A. Katine, J. Langer, H. Jiang, K. L. Wang, I. N. Krivorotov and J. P. Wang, "Spin-transfer torque switching above ambient temperature," *IEEE Magn. Lett.*, vol. 3, pp. 3 000 304–3 000 304, 2012. doi:10.1109/LMAG.2012.2195775.
29. L. Zhang, Y. Cheng, W. Kang, Y. Zhang, L. Torres, W. Zhao and A. Todri-Sanial, "Reliability and performance evaluation for stt-mram under temperature variation," *International Conference on Thermal, Mechanical and Multi-Physics Simulation and Experiments in Microelectronics and Microsystems*, Apr. 2016. https://ieeexplore.ieee.org/document/7463380. doi:10.1109/EuroSimE.2016.7463380.
30. F. Schreiber, J. Pflaum, Z. Frait, T. Mühge and J. Pelzl, "Gilbert damping and g-factor in FexCo1x alloy films," *Solid State Commun.*, vol. 93, no. 12, pp. 965–968, Mar. 1995.
31. T. Devolder, P.-H. Ducrot, J.-P. Adam, I. Barisic, N. Vernier, J.-V. Kim, B. Ockert and D. Ravelosona, "Damping of CoxFe80xB20 ultrathin films with perpendicular magnetic anisotropy," *Appl. Phys. Lett.*, vol. 102, no. 2, 2013, Art. no. 022407.
32. N. N. Mojumder and K. Roy, "Proposal for switching current reduction using reference layer with tilted magnetic anisotropy in magnetic tunnel junctions for spin-transfer torque (STT) MRAM," *IEEE Trans. Electron Dev.*, vol. 59, no. 11, pp. 3054–3060, Nov. 2012.
33. H. Liu, D. Bedau, D. Backes, J. A. Katine and A. D. Kent, "Precessional reversal in orthogonal spin transfer magnetic random access memory devices," *Appl. Phys. Lett.*, vol. 101, no. 3, 2012, Art. no. 032403.
34. L. Liu, C. F. Pai, Y. Li, H. W. Tseng, D. C. Ralph and R. A. Buhrman, "Spin-torque switching with the giant spin hall effect of tantalum," *Science*, vol. 336, pp 555–558, 2012.

35. C. F. Pai, L. Liu, Y. Li, H. W. Tseng, D. C. Ralph and R. A. Buhrman, "Spin transfer torque devices utilizing the giant spin Hall effect of tungsten," *Appl. Phys. Lett.*, vol. 101, no. 12, Sep. 2012, Art. no. 122404.
36. M. Cubukcu, O. Boulle and M. Drouard, "Spin-orbit torque magnetization switching of a three-terminal perpendicular magnetic tunnel junction," *Appl. Phys. Lett.*, vol. 104, no. 4, 2014, Art. no. 042406.
37. Z. Wang, W. Zhao, E. Deng, J.-O. Klein and C. Chappert, "Perpendicular-anisotropy magnetic tunnel junction switched by spin-Hall-assisted spin-transfer torque," *J. Phys. D: Appl. Phys.*, vol. 48, pp. 065001, 2015.
38. M. Wang, W. Cai, D. Zhu, Z. Wang, J. Kan, Z. Zhao, K. Cao, Z. Wang, Y. Zhang, T. Zhang, C. Park, J.-P. Wang, A. Fert and W. Zhao, "Field-free switching of a perpendicular magnetic tunnel junction through the interplay of spin – orbit and spin-transfer torques," *Nat. Electron.*, vol. 1, pp.582–588, 2018.
39. S. Pathak, C. Youm and J. Hong, "Impact of spin-orbit torque on spin-transfer torque switching in magnetic tunnel junctions," *Sci. Rep.*, vol. 10, pp. 2799, 2020.
40. Y. C. Lau, D. Betto, K. Rode, J. M. D. Coey and P. Stamenov, "Spin_orbit torque switching without an external _eld using interlayer exchange coupling," *Nat. Nanotechnol.*, vol. 11, no. 9, pp. 758–762, Sep. 2016.
41. L. You, O. Lee, D. Bhowmik, D. Labanowski, J. Hong, J. Bokor and S. Salahuddin, "Switching of perpendicularly polarized nanomagnets with spin orbit torque without an external magnetic field by engineering a tilted anisotropy," *Proc. Nat. Acad. Sci. USA*, vol. 112, no. 33, pp. 10310–10315, Aug. 2015.
42. J. Deng, G. Liang and G. Gupta, "Ultrafast and low-energy switching in voltage-controlled elliptical pMTJ," *Sci. Rep.*, vol. 7, no. 1, pp. 1–10, Dec. 2017.
43. Y. Shiota, S. Miwa, T. Nozaki, F. Bonell, N. Mizuochi, T. Shinjo and Y. Suzuki, "Pulse voltage-induced dynamic magnetization switching in magnetic tunneling junctions with high resistance-area product," *Appl. Phys. Lett.*, vol. 101, p 102406, 2012.
44. K. Zhang, D. Zhang, C. Wang, L. Zeng, Y. Wang and W. Zhao, "Compact modeling and analysis of voltage-gated spin-orbit torque magnetic tunnel junction," *IEEE Access*, vol. 8, pp. 50792–50800, 2020. doi:10.1109/ACCESS.2020.2980073.
45. C. D. Stanciu, F. Hansteen, A. V. Kimel, A. Kirilyuk, A. Tsukamoto, A. Itoh and T. Rasing, "All-optical magnetic recording with circularly polarized light," *Phys. Rev. Lett.*, vol. 99, pp. 047601, 2007.
46. S. Alebrand, M. Gottwald, M. Hehn, D. Steil, M. Cinchetti, D. Lacour, E. E. Fullerton, M. Aeschlimann and S. Mangin, "Light-induced magnetization reversal of high-anisotropy TbCo alloy films," *Appl. Phys. Lett.*, vol. 101, pp. 162408, 2012.
47. T. A. Ostler, J. Barker, R. F. L. Evans, R. W. Chantrell, U. Atxitia, O. Chubykalo-Fesenko, S. El Moussaoui, L. Le Guyader, E. Mengotti, L. J. Heyderman, F. Nolting, A. Tsukamoto, A. Itoh, D. Afanasiev, B. A. Ivanov, A. M. Kalashnikova, K. Vahaplar, J. Mentink, A. Kirilyuk, T. Rasing and A. V. Kimel, "Ultrafast heating as a sufficient stimulus for magnetization reversal in a ferrimagnet," *Nat. Commun.*, vol. 3, pp. 666, 2012.
48. A. Hassdenteufel, B. Hebler, C. Schubert, A. Liebig, M. Teich, M. Helm, M. Aeschlimann, M. Albrecht and R. Bratschitsch, "Thermally assisted all-optical helicity dependent magnetic switching in amorphous Fe100−xTbx alloy films," *Adv. Mater.*, vol. 25, pp. 3122, 2013.
49. M. S. E. Hadri, P. Pirro, C. H. Lambert, S. Petit-Watelot, Y. Quessab, M. Hehn, F. Montaigne, G. Malinowski and S. Mangin, "Two types of all-optical magnetization switching mechanisms using femtosecond laser pulses," *Phys. Rev. B*, vol. 94, pp. 064412, 2016.
50. S. Mangin, M. Gottwald, C. H. Lambert, D. Steil, V. Uhlir, L. Pang, M. Hehn, S. Alebrand, M. Cinchetti, G. Malinowski, Y. Fainman, M. Aeschlimann and E. E. Fullerton, "Engineered materials for all-optical helicity-dependent magnetic switching," *Nat. Mater.*, vol. 13, pp. 286, 2014.
51. C. H. Lambert, S. Mangin, B. S. D. C. S. Varaprasad, Y. K. Takahashi, M. Hehn, M. Cinchetti, G. Malinowski, K. Hono, Y. Fainman, M. Aeschlimann and E. E. Fullerton, "All-optical control of ferromagnetic thin films and nanostructures," *Science*, vol. 345, pp. 1337, 2014.
52. J.-Y. Chen, L. He, J.-P. Wang and M. Li, "All-optical switching of magnetic tunnel junctions with single subpicosecond laser pulses," *Phys. Rev. Appl.*, 7(2). doi:10.1103/physrevapplied.7.021001.
53. A. Olivier, L. Avilés-Félix, A. Chavent, L. Álvaro-Goémez, M. Rubio-Roy, S. Auffret, L. Vila, B. Dieny, R. C. Sousa and I. Prejbeanu, "Indium Tin Oxide optical access for magnetic tunnel junctions in hybrid spintronic – photonic circuits," *Nanotechnology*, vol. 31, no. 42, pp. 425302, 2020.
54. J. Pelloux-Prayer and F. Moradi, "Compact model of all-optical-switching magnetic elements," *IEEE Trans. Electron Dev.*, vol. 67, no. 7, pp. 2960–2965, July 2020. doi:10.1109/TED.2020.2991330.

Nanowire Magnets

6

M. Boughrara, N. Zaim, H. Ahmoum, A. Zaim, M. Kerouad

Contents

1	Introduction	77
2	Common Synthesis Techniques	78
	2.1 Electrodeposition Method	78
	2.2 Electroless Deposition	79
	2.3 Sol-Gel Method	79
	2.4 Chemical Vapor Deposition	79
	2.5 Electrospinning Method	80
3	Theoretical Investigations of Nanowires	80
4	Experimental Investigation of Nanowires	84
5	Applications of Magnetic Nanowires	88
	5.1 Nanowires for Nonvolatile Memory Application	88
	5.2 Nanowires for Biomedical Applications	89
References		89

1 INTRODUCTION

Condensed matter physics research on the nano-scale has gained particular attention, in as much as it reveals novel effects from fundamental relevance and paves the way for future technologies [1]. Subjects of research are so-called nanostructured materials, which can be classified as objects with a characteristic length scale in the order of a few nanometers in at least one dimension [2]. In fact, dimensionality plays a crucial role in the adaptation of materials properties. The theoretical and experimental researchers have demonstrated a great interest in nanostructures and low dimensional systems [3, 4]. These systems need realistic and efficient ways to describe them. Among the possible nanostructured materials investigated tremendously in literature, nanowires have been widely studied due to their unique properties such as a large proportion of surface atoms and high length to diameter ratio [5]. Also, nanowire systems exhibit unusual physical properties that make the fabrication of such systems a challenge.

Over the past 20 years, the progress of the chemical synthesis methods allows producing magnetic nano-objects of complex shapes, in particular very well-defined nanowires. This particular shape gives the objects remarkable magnetic behavior. In addition, these objects are based on 3d transition metals (cobalt, iron, or alloys) to have materials exhibiting both strong saturation magnetization, high coercive field linked to their morphological anisotropy, and good resistance to a temperature of their magnetic

DOI: 10.1201/9781003196952-6

properties. Many experimental studies of magnetic nanowire were realized by using many experimental techniques [6, 7].

To investigate theoretically the magnetic properties of nanowires, some theoretical methods are available including the Mean Field Theory (MFT) [8], Effective Field Theory (EFT) [9], Monte Carlo Simulation (MCS) [10], Density Functional Theory (DFT) [11]. MCS incorporates a broad class of methods that can be used to study magnetic nanowires. The common idea of all Monte Carlo methods is to use random rather than deterministic processes [12]. When the system size is bigger (i.e., the nanostructures exhibit properties like the bulk ones) a quantum mechanical treatment is too expensive (and thus, sometimes undesired) and a semi-classical model can give a good description of the system properties. Notably, Monte Carlo methods allow the calculation of non-equilibrium properties, and hence a description of the hysteretic properties of a nanostructure. On another hand, when the system being studied has few atoms a popular tool is the DFT. The DFT allows obtaining accurate ground-state magnetic and electronic properties. In this context, using the density functional theory the filling of nanowires with magnetic elements like Fe, Co, and Ni or other transition metals (TM) are extensively examined [13, 14].

This chapter intends to give a review of the most relevant results obtained by the researcher's groups on the properties of magnetic nanowires obtained in theoretical and experimental studies. The chapter is organized as follows: first, the common experimental techniques used in magnetic nanowires synthesis are briefly recalled in section 2. Next, a review of theoretical studies given by DFT and MCS are presented and discussed in section 3. In section 4, the hysteresis behavior, the magnetic saturation, and the critical phenomena are discussed throughout some experimental studies that treat the effect of synthesis parameters and the size of nanowire on the magnetic properties. Finally, the last section is devoted to some applications of magnetic nanowires.

2 COMMON SYNTHESIS TECHNIQUES

There are a wide variety of templates capable of synthesizing nanoparticles, the template-filling methods are used to synthesize nanowires. In this part, we discuss synthesis techniques that involve templates such as electrodeposition, electroless deposition. With this template-filling method, the dimensions and composition through choosing the proper templates and the experimental conditions can easily be controlled. We will also discuss the sol-gel deposition and the chemical vapor deposition.

2.1 Electrodeposition Method

Electrodeposition (or electrochemical deposition) is a general growth technique [15] that is based on the reduction of metallic ions from an electrolyte, usually an aqueous liquid, controlled by an external direct current controlled in two modes namely galvanostatic and potentiostatic modes. With the potentiostatic mode, the control is given over crystallography and mainly composition in segmented structures and alloys. The galvanostatic mode enables an easier tuning of the length of the nanowire. In fact, the best control of the growth rate is given when the quantity of matter is directly proportional to the passed electric charge. Other interesting characteristics of this technique are profitability, operation simplicity, and the possibility of depositing on substrates with complex geometry and very high vertical aspect ratios. Compared to other physical or chemical methods, electrodeposition offers several possibilities for local deposition in low dimensions which present a high technological potential.

Several techniques based on self-assembly in insulating supports are used to obtain arrays of magnetic nanowires with high aspect ratios, that are difficult to form through physical processes. The interesting templates for obtaining regular wire networks are porous Al_2O_3 aluminas. The use of templates offers the possibility of controlling the shape, size, density, and distribution in a hexagonal network of its nanopores with simple control of the various production parameters. Unlike other methods, electrodeposited wires tend to be continuous, dense, and highly organized.

2.2 Electroless Deposition

Electroless deposition is also one of the techniques used for the growth of nanowires [16]. It is an autocatalytic process in which the substrate (porous membranes) develops potential in a deposition bath containing metal ions to be reduced without using electrical energy. It is a chemical deposition process that implies the use of an appropriate reducing chemical agent such as a complexing agent, a stabilizer, or other components to cover, with the material to be deposited, the wall of the pores then continues towards the inside. One of the advantages of this technique is that it does not require electrically conductive substrates or deposited materials which leads to the absence of edge effects. The nanowires' length depends only on the deposition channels or pore's length. Varying the time of deposition would give different wall thickness of the nanotubes and prolonged deposition very rarely gives nanowires. However, this chemical deposition process is itself unstable, its stability depends entirely on the substrate, the pre-treatment process, the type of solution used, as well as the pH and temperature during deposition. In addition, the spontaneous nature of the deposit formation process imposes thermodynamic and kinetic conditions which must be observed to allow the deposit formation.

2.3 Sol-Gel Method

Sol-gel method [17] is also used to synthesize magnetic nanowires. During the sol-gel synthesis, a colloidal solution should be prepared with nanoparticles (sol) and converted to gel. Conventional precursors used to prepare the gel are alkoxides, these alkoxides are often obtained by deposition of an alkaline salt of metal and mixed with alcohol and additives. Usually, the heating assists the precipitation of colloidal particles and later removes solvent for gelation.

2.4 Chemical Vapor Deposition

The chemical vapor deposition [18] method refers to a whole family of techniques already well established on an industrial scale for the production of thin films. This family involves one or more dispersed compounds most in a gaseous form called "precursors" which are molecules containing the chemical elements necessary for the formation of the material (for example silane to obtain a silicon nanowire). These precursors are introduced into the reactor, most often mixed with other elements and in a gaseous form. They are then broken down under the effect of an energy source (heat, microwave, laser beam, etc.). The growth of nanowires is obtained when, rather than being deposited on a substrate, the chemical elements constituting the wires will preferentially dissolve in the metal particles, present on the substrate, which acts as a catalyst. The concentration of these elements in the nanoparticles increases to exceed the saturation threshold, which causes crystallization. This most often takes place on one face of the particle and, for semiconductor nanowires, generally on that in contact with the crystalline substrate. This contact surface constitutes the "growth germ", the size of which, defined by the diameter of the particle, regulates the diameter of the growing nanowire.

2.5 Electrospinning Method

Electrospinning [34] has been recognized by the scientific community as an operational technique allowing the spinning of nanofibers from solutions of different polymers. This technique is becoming very interesting, particularly in the field of textiles and innovative materials.

Using a conventional electrospinning process, a sol-gel or precursor solution is pumped through a thin nozzle with an internal diameter in the range of 400 to 700 micrometers (usually the needle is metallic). The nozzle acts as an electrode. A high electrical voltage, in the order of several thousand volts, is applied between this metal nozzle and a counter electrode (collector). The distance between the two electrodes is 10–40 cm in laboratory systems. Electrospinning allows having very long free-standing wires and their networks/clusters.

3 THEORETICAL INVESTIGATIONS OF NANOWIRES

With the large increase in computational power, important advances in electronic structure theory, and algorithmic development, theoretical studies became an essential tool to predict and understand experimental phenomena. In this optic, we will give some results obtained in some works that investigate the magnetic properties of nanowires by using Monte Carlo and DFT simulations.

To study the effect of some physical parameters (temperature, external magnetic field, crystal field, etc.) on the magnetic behavior of nanowires such as critical temperature, blocking temperature, and hysteresis behaviors, the authors use MFT, EFT, and the MCS combined to some physical models such as Ising and Heisenberg models. Most theoretical works are done with Monte Carlo simulation which has long been recognized as a powerful technique giving accurate results. The compensation temperature and the hysteresis loop present interesting magnetic properties that find their potential use in magnetic recording materials [19]. M. Boughrara et al. have studied the magnetic properties of cylindrical spin ½ Ising nanowire with core-shell structure [20], a mixed spin (½, 1) Ising nanowire with pure [21] and diluted surface [10]. The hysteresis behavior is examined and it is seen that the number of loops depends on the surface coupling and the interface exchange interaction values (Figure 6.1a). It is shown that, depending on the values of the exchange coupling in the core (R_I) and the shell (R_S) and the crystal field (D), these systems can exhibit compensation behavior (Figure 6.1b-c). Higher spin values of core-shell nanowire systems were also studied with Monte Carlo Simulation [22, 23]. For a spin 1 core-shell nanowire. It is observed that for the antiferromagnetic interface exchange interaction, the phase diagram exhibits the compensation temperature for a special range of surface exchange interactions. It is also found that the compensation and critical temperatures increase slightly when the absolute value of the interface exchange coupling increases. For a mixed spin (3/2, 1) nanowire with a core-shell structure, it is found that the compensation points can appear when the crystal field becomes larger than a critical value.

In another work, N. Zaim et al. [24] have investigated the blocking temperature of a nanowire which is composed of alternate layers (Figure 6.2a), R_A and R_{AB} are the exchange interactions in A layers and between A and B layers respectively, D_{xy} is the plane anisotropy. The authors studied the effect of the Hamiltonian parameters such as R_A, R_{AB}, and D_{xy} on the blocking temperature. By definition, the blocking temperature is a temperature that separates the ferromagnetic or the blocking phase and the superparamagnetic ones. It is shown that the blocking temperature increases with the increase of the value of R_B (Figure 6.2b) and R_{AB} (Figure 6.2c) and decreases with the increase of D_{xy} (Figure 6.2d).

Other work investigated the magnetic properties with Monte Carlo simulation of an ordered and disordered hexagonal array (with respect to the position of the nanowire in the array) composed of cobalt and

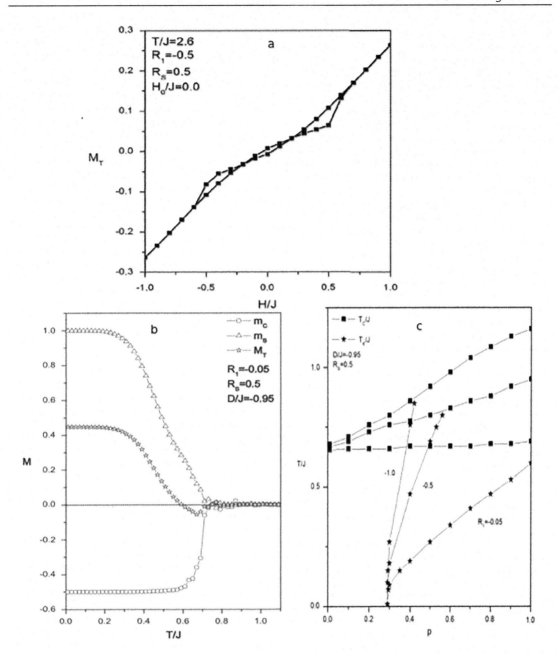

FIGURE 6.1 (a) Hysteresis loop for spin ½ nanowire [20], (b) compensation temperature for mixed spin ½ and 1 nanowire [21] and (c) compensation and critical temperature for mixed spin ½ and 1 nanowire with a diluted surface, p is the concentration of spin 1 at the shell [10]. Adapted with permission from References [20], [21], Source: Copyright (2014), and form Reference [10], Copyright (2015), Elsevier.

nickel nanowires (Ni_xCo_{1-x}) [25]. Firstly, the hysteresis behavior of single cobalt and nickel nanowires is studied. It is found that the coercive field of Co nanowire is: Hc = 415 Oe and for Ni nanowire, it is equal to 235 Oe. It is also shown that the saturation magnetization of Co nanowire (1400 emu/cm^3) is greater than that of the nickel one (490 emu/cm^3). For an array that contains a mixture of nickel and cobalt nanowires, even for disordered or ordered arrays, the saturation magnetization decreases as increasing the Ni

82 Applications of Low Dimensional Magnets

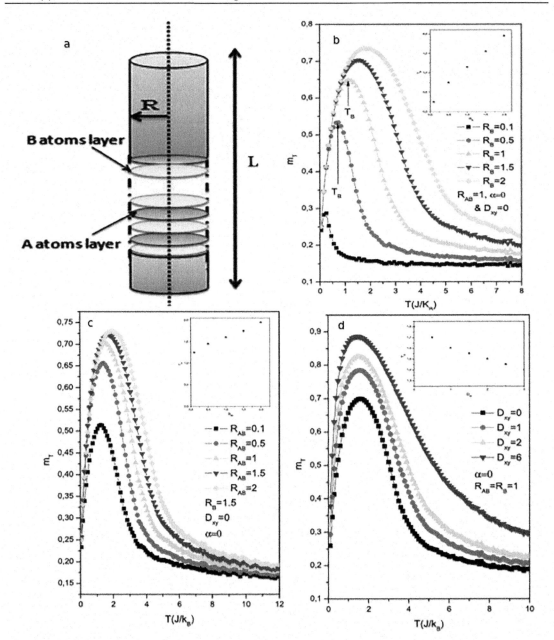

FIGURE 6.2 (a) Schematic representation of cylindrical nanowire. The effect of temperature on the Zero field cooled (ZFC) magnetization for different values of exchange interaction RB (b), for different values of RAB (c), and different values of the plane anisotropy Dxy (d) [24]. Adapted with permission from Reference [24].

Source: Copyright (2021), Springer Nature.

nanowires concentration (x) in the array. The coercivity and remanence decrease monotonically with the x concentration for the ordered array case. In contrast, for the disordered array, strange behavior is found for x<0.5, the coercivity and the remanence are approximately constant for this range of concentration.

The nanowires systems are also extensively studied by the Density functional theory [26–28]. Recently, Al. Rosa et al. [26] have investigated the Mn-doped ZnO nanowires by using the hybrid density functional theory. It is found that there is no effect of the site positions (Mn at the surface, at the subsurface, or inner sites) or the surface passivation on the total magnetic moments. The authors have also studied the Mn doping on the magnetic phases, and they have found even for two Mn atoms at the surface of one of them at the surface and the other at the subsurface, the two configurations favor an anti-ferromagnetic behavior and the magnetic moment is twice the single magnetic moment which is due to the negligible overlap between the two Mn states. They have also shown that the presence of surface dangling bonds stabilizes the anti-ferromagnetic phases in Mn-doped ZnO nanowires.

The magnetic properties of Mn-doped GaAs nanowires influenced by the surface dangling bonds (SDBs) are also studied in reference [27]. This work is also done with the DFT approach. The authors showed that the surface dangling bonds can induce surface magnetism and the Mn-doped GaAs with As-surface dangling bonds increase the critical temperature (Tc) above room temperature and can stabilize the ferromagnetic (FM) states. It is also found that due to the charge transfers between surface dangling bonds and Mn-3d orbitals, the magnetic moments present very abundant variation.

Jo et al. [28] have studied the effect of the Co concentration (x) on the magnetic moments of Fe_{1-x}-Co_x nanowires with a centered-staggered triangle (cst) and centered-staggered square (css) structures. They have obtained that the highest magnetic moment appears around x = 0.2 and decreases slowly with increasing Co concentration for the nanowire with CSS structure. It is also found that the magnetic moments of the studied system are 2.75 µB/atom for the range of x = 0.1–0.2 which is larger than 0.35µB/atom corresponding to the highest magnetic moment of Fe-Co bulk alloys. These results are comparable with the ones obtained by Barthélémy et al. [29]. The Slater-Pauling curve for the Fe-Co bulk alloys exhibits a peak around x = 0.35 and drops down quickly at x = 0.8 where the structure of the Fe-Co bulk alloys changes from bcc to fcc. The whole trend of the magnetic moments for the Fe-Co nanowires with CST and CSS structures is similar to that of Fe-Co bulk alloys. Regardless of the nanowire shape, the magnetic moments are mainly dependent on the Co concentration.

Based on the chain-of-spheres model with a symmetric fanning magnetization reversal mechanism, the coercivity of $Fe_{1-x}Co_x$ nanowires is also investigated. The obtained value of coercivity predicted by the symmetric fanning mechanism is in good agreement with the experimental results for the nanowire arrays embedded in porous templates. The free-standing Fe-Co nanowires have a large coercivity of about 2500 Oe which is in accordance with the value obtained of the nanowire arrays combined in porous templates.

Over the past years, the researchers have also focused on the magnetic behavior of magnetic multilayered structures for their novel multifunctional properties which may have great potential for future magneto-electronic device applications such as ultrahigh density memory devices and magnetic sensors [30, 31]. In this regard, a tremendous amount of multilayered magnetic nanowire studies have shown considerable promise. In fact, multilayered nanowires have been found to present a giant magnetoresistance effect at ambient temperature [32]. Wang et al. studied the magnetic properties of one-dimensional Fe/Cu multilayered nanowires with different widths of nonmagnetic Cu spacer. They suggested that the value of the interlayer exchange coupling leads to switch signs with the nonmagnetic Cu spacer thickness increases in the nanowire, and where the number of nonmagnetic Cu layers increases the magnitude of the interlayer exchange coupling value decreases significantly [13]. Furthermore, Pal et al. [14] have examined the role of nonmagnetic spacers on the magnetic properties of one-dimensional Ni/Cu and Ni/Al multilayered nanowires. They have reported that the change in the width of the nonmagnetic spacer leads to subtle changes in the magnetic properties of the nanowires. The results also indicate that Al spacer layers increases in Ni/Al nanowire lead to a nonmonotonic decrease in the magnetic moment per Ni atom (µav). It has been revealed that the dissimilar interfacial bonding gives rise to the difference in magnetic property between Ni/Cu and Ni/Al nanowire. Moreover, Panigrahi and coworkers [11] have controlled interlayer exchange coupling in one-dimensional Fe/Pt multilayered nanowire by using DFT. The authors have reported the magnetic moment of the interfacial Fe atom in the Fe/Pt multilayered nanowire is higher than that of the Fe atom away from the interface. The magnetic-moment enhancement at the interface is explained by the mechanism of multistep electron transfer between the layers and spin-flip within the layer. As a result, the nonmonotonous feature in interlayer exchange coupling is associated with the variation of the width of the nonmagnetic Pt spacer, the competition among short- and long-range direct exchanges, indirect Ruderman-Kittel-Kasuya-Yosida exchange, and superexchange.

Silicon nanowires are well known to possess a unique combination of their special characteristics such as one-dimensionality, high surface-to-volume ratio, tunable bandgap, and biocompatibility. Silicon nanowires are ideally suited for their widespread applications requiring such as solar cells, biological and chemical sensors, optoelectronics, and field-effect-transistors. To further expand and tailor their applications, the modification of the surface of silicon nanowires by chemical species is one of the feasible ways to attain this goal. The DFT approach is used to explore the magnetic properties of transition metal (TM = Sc, Ti, V, Cr, and Mn) atom inserted single and double one-dimensional styrene molecular wires confined on the hydrogen-terminated Si(100) surface (Si − [TM(styrene)]) [33]. The silicon-supported [TM (styrene)]$_\infty$ single and double molecular wires possess rich magnetic properties. Hence, due to the incomplete coordination, a significant increase is seen in the local magnetic moment of both Cr and Mn atoms in comparison with the gas phase TM–benzene wires. It is also shown that the [TM (styrene)]∞ single molecular wire with the Mn defect has uneven local magnetic moments on each Mn atom. The electronic structure and magnetic properties of silicon-supported molecular wires may be significantly affected by the TM defect.

4 EXPERIMENTAL INVESTIGATION OF NANOWIRES

To have a deeper insight into the structural and magnetic properties of the nanowires, the experimental investigations reveal very interestingly, especially the studies of the influence of the size of the nanowires (length and diameter) and the synthesis parameters such as the annealed temperature on their magnetic properties. To do that, the magnetic properties of some nanowires treated in some experimental studies will be presented. In this section, we highlight some particular results of cylindrical nanowires and core-shell ones. Based on the chronoamperometric technique of the electrochemical deposition. Maaz Khan et al. have successfully synthesized magnetic heterostructure nanowires with three different compounds (i.e., Au–NiO–Au, Ni–NiO–Ni, and Au–Ni–NiO) [15]. The magnetic hysteresis loops are measured for the different studied compounds (Figure 6.3). The results are obtained for three values of the temperature (2 K, 30 K, and 100 K). The Au/NiO/Au nanowire (Figure 6.3**a**) exhibits a linear magnetization at 30 K and 100 K which refers to antiferromagnetic/paramagnetic nature. For 2 K, a ferromagnetic behavior is observed. In these nanowires, NiO segment length is about 150 nm which indicates the frustrated spin at the surface is caused by a high surface to volume ratio. For 2 K, these frustrated spins are frozen. For the second composition (Ni/NiO/Ni), the ferromagnetic behavior is seen for 100 K and 30 K with a coercivity of 70 Oe and 78 Oe respectively (Figure 6.3**b**). For the third compound (Au/Ni/NiO), the ferromagnetic behavior is shown for the whole temperature range, which is attributed to the high Ni content in the compound in comparison to NiO (Figure 6.3**c**). The Zero field cooled (ZFC) and the field cooled (FC) curves are also measured. Typical ferromagnetic behavior is shown for Au/Ni/NiO nanowire.

Y.C. Zhang et al. studied Ni$_2$MnGa nanowires which are fabricated via the electrospinning method [34]. They have studied the effect of annealed temperature on critical behavior. Figure 6.4 shows the ZFC and FC curves. It can be seen that the annealed temperature has a great effect on the critical temperature; for 973 K annealed temperature (Figure 6.4a) the critical one is Tc = 235 K and for 1023 K (Figure 6.4b) one, Tc = 288 K. The difference in these results is due to the difference of the exchange interaction of the two samples which are annealed with two different temperatures. The obtained critical temperature T$_c$ of nanostructures is, in general, lower than that of their bulk, in this case, the obtained critical temperature (288 K) is lower than that given in [35] which is equal to 380 K. The low Tc in low-dimensional systems

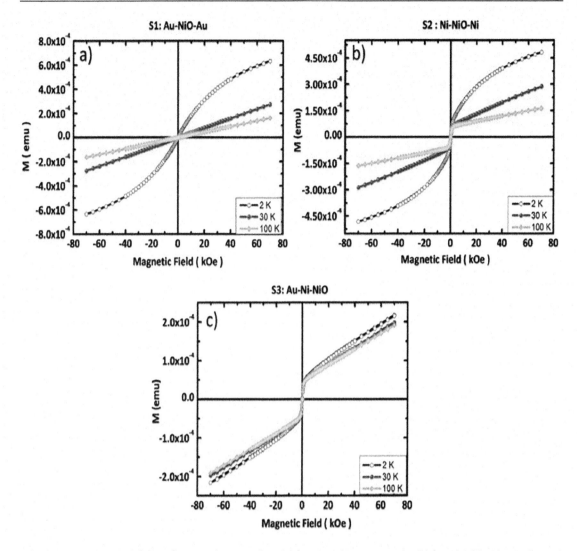

FIGURE 6.3 Hysteresis loop of magnetic nanowires: (a) for Au–NiO–Au sample, (b) for Ni–NiO–Ni sample, and for (c) Au–Ni–NiO sample recorded at 2 K, 30 K, and 100 K [15]. Adapted with permission from Reference [15].
Source: Copyright (2020), Elsevier.

has been attributed to finite size effects that are mainly due to the weak exchange interaction in grains and domains at the nanoscale. In this work, antiferromagnetic interactions of Mn-Mn can also lead to a decrease of Tc. Concerning the magnetization, it is also observed that for the 1023 K annealed temperature, the ZFC curves present a large maximum at higher temperature in this sample which is probably due to the occurrence of a non-equilibrium magnetic response, magnetic anisotropy, and a single domain structure or may be due to frustration and disorder.

Kirill V. Frolov et al. [36] have investigated a nanowire that is based on nickel and iron elements, they have used electrochemical deposition to synthesize $Ni_{1-x}Fe_x$. In this work, the magnetic properties of $Ni_{1-x}Fe_x$ were studied for different diameters (30 nm and 70 nm), for different concentrations (6.5% and 55% of Fe), and different orientations of an external magnetic field. It is proven that the saturation magnetization (Ms) of the samples with a high iron content (65%) is larger than the samples with a low ion iron (6.5%) and the remanent magnetization (Mr) and coercive field (Hc) values are highly dependent on the

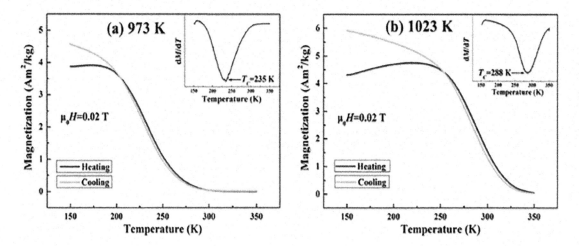

FIGURE 6.4 The temperature dependence of magnetization of the nanowire samples under different heat treatment conditions: (a) for 973 K and (b) for 1023 K annealed for 3 h. Insets show the dM/dT versus temperature curves that are used to calculate the critical temperature TC [34]. Adapted with permission from Reference [34].

Source: Copyright (2020), Elsevier.

diameter of samples. It is also shown that the orientation of the magnetic field, which is relative to the axis of the nanowire, affects the values of Ms, Mr, and Hc.

To investigate the effect of length on the magnetic behavior, A.H.A. Elmekawy et al. synthesized the iron hexagonal nanowire [37], the samples were obtained by templated electrodeposition into porous anodic aluminum oxide (AAO) and the hysteresis behaviors were studied for different directions of external magnetic field (perpendicular and parallel to the nanowire axis) and different lengths (7.4, 11.6 and 21.2 μm). It is shown that the length of nanowires plays a crucial role in determining magnetic behavior. It is found that the hysteresis behavior in presence of the magnetic field perpendicular to the axis wire is independent of the long wires; in contrast, where the magnetic field along the axis wire is applied, the hysteresis loop became rectangular under the length effect which is due to the shape anisotropy. For the applied field parallel to the axis wires, the coercivity H_c increases non-linearly with the nanowire length which is due to the competition between shape anisotropy and magnetostatic interactions. For perpendicular orientation the coercivity field increases with the length of wires, this increase even is four times smaller than that of parallel case.

In the same context, and to study the influence of the nanowire length on the magnetic behavior, N. Ghazkoob et al. have used AC pulse electrodepositing method to synthesize the Zinc ferrite nanowires ($ZnFe_2O_4$) for several lengths (1.2, 2.3, and 6.3 μm) [38]. In a part of this work, the authors studied the effect of nanowires length on the hysteresis behavior with a diameter of 38 nm and for a perpendicular external magnetic field to the nanowire axis. It is shown that the hysteresis loop appears only for the length of L = 6.3 μm. In the case of a parallel external magnetic field to the nanowire axis, it is found that the area of the hysteresis loop increases with the length of the wires. It is also found that the saturation magnetization Ms and the Hc increase with the increasing of the length, and this is due probably to the appearance of multi-domain instead of single-domain with the increasing of grain size and it is due also to the magnetostatic interaction between nanowires.

In this work, the authors also studied the effect of diameter on the magnetic characteristics, it is found that, while the field is parallel to the nanowire axis, the remanent and the saturation magnetizations and the magnetic coercivity (except for the diameter of 58 nm) decrease from 0.56 to 0.11 and from 902 to 197 Oe, respectively with increases in the diameter of the nanowire from 30 to 58 nm. It is also shown that with the increase of the diameter of the nanowires, the grain size increases also and there will be the possibility to convert the magnetic single domain (MSD) into a magnetic multi-domain (MMD) structure. For the energy required to rotate the magnetic moment, much greater is required than the energy of the magnetic domain movement. Therefore, zinc ferrite nanowires with small diameters present a greater magnetic coercivity, multi-domain structures become saturated with the magnetic domain movement, which requires less energy. Therefore, nanowires with large diameters have smaller magnetic coercivity. However, the transition from a MSD structure to a MMD structure has decreased the ratio Mr/Ms and increased the magnetostatic interaction. It is found that stronger magnetic anisotropy is related to a higher value of coercivity and loop squareness ratio.

Based on electrodeposition in the nanopores of alumina and polycarbonate membranes technique, the samples of cobalt nanowires were also prepared for different lengths [39]. It is shown that the length of the nanowire has a great effect on the magnetic parameters such as coercive field, the magnetization of saturation, and remanent magnetization. It is also found that the magnetic properties of the nanowire are strongly affected by the crystalline and shape anisotropies and the magnetostatic interaction between nanowires. In fact, the shape anisotropy affects the magnetic easy axis, for long nanowires (Length >> diameter), the easy axis is expected to be along the axis of the nanowire. In addition to that, it is proven that the length of the nanowires also affects the dipolar interaction and this induces a magnetic easy axis perpendicular to the nanowire axis [40]. The effect of dipolar interaction will be stronger than that of shape anisotropy and will lead to a decrease in coercivity and remanence for long nanowires.

The last part of this section will be focused on the magnetic behavior of nanowires with core-shell structures. Iron becomes a magical element in the nano field due to its use in wide fields of application such as medicine, environment, and technology which makes it to be a center of researchers' interest. However, one of the main problems researchers face when synthesizing iron-based nanostructures is that the samples burn when brought into contact with air, which is due to the high activity of iron. To prevent this situation, the solution is to cover the iron-based nanowire with an iron oxide layer which leads to a core-shell morphology [41–43]. This method has two advantages; it protects and stabilizes the Fe nanostructures.

To study the magnetic properties of such nanostructures with core-shell morphology. Xiaobing Cao et al. have synthesized Fe@α-Fe$_2$O$_3$ core-shell nanowires by using the chemical reaction of ferrous sulfate and sodium borohydride, as well as the post-annealing process in the air [41]. The magnetic properties are investigated and the hysteresis loop is measured for different synthesized parameters especially the time annealed temperature. In the same context, Fe-Fe$_3$O$_4$ core-shell nanowires were prepared by electrodeposition onto aluminum oxide templates (AOT) [42]. A group of Korean researchers has also synthesized and characterized a core-shell nanowire of Fe-FeO$_x$ [43]. In this last work, the nanowires were obtained by thermal conversion of the Fe nanowires by pulsed electrodeposition using an anodized aluminum oxide template.

The magnetic behaviors of Fe-Fe$_3$O$_4$ and Fe@α-Fe$_2$O$_3$ core-shell nanowires were investigated by studying the effect of annealed temperature-time on the hysteresis behavior. For the first nanowire (Fe@α-Fe$_2$O$_3$), it is shown that the saturation magnetization decreases and the coercive field increase with the increasing of the annealed temperature time. In fact, the increase of the magnetization saturation is due to the growth of antiferromagnetic (AFM) phase formed at the surface with annealing temperature time, this increase of the AFM phase at the surface induces more unidirectional anisotropy due to the exchange interaction between Fe core and α-Fe$_2$O$_3$ shell at the interface which certainly causes the increase of the coercive field. For Fe-Fe$_3$O$_4$ core-shell nanowire, it is found that the saturation magnetization decreases and the coercive field slightly decreases with the annealed temperature time. With the annealed temperature time, the Fe$_3$O$_4$ shell thicknesses increases to take maximum values (25 nm) after 72 hours, this result is obtained for the polycrystalline Fe core. Whereas for the single-crystal Fe core, the thickness of the shell does not exceed 12 nm. The increase of the shell thickness causes the reduction of the saturation magnetization which is due to the smaller magnetic moment of the Fe$_3$O$_4$ shell compared to the Fe core [44]. The reduction of coercivity is also due to the increase of the shell thickness that strongly depends on the shape anisotropy value of the core-shell nanowire, which is proportional to its magnetization [45]. In another work [46], the authors studied iron-iron oxide nanowires with a core-shell structure, and found that the saturation magnetization of the nanowires with a native oxide layer is higher than that of Fe for the oxidized sample (Figure 6.5).

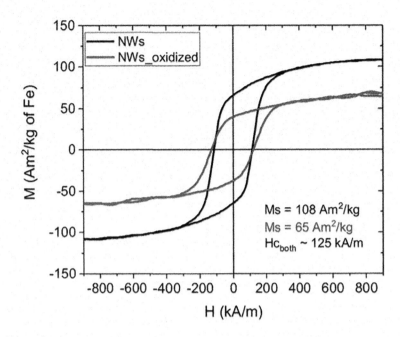

FIGURE 6.5 Magnetization curves as a function of the external magnetic field of nanowires with Fe core/shell with a native oxide shell (black curve) and nanowires with an oxidized shell (red curve) [46]. Figure adapted with permission from ref [46].

Source: Copyright the Authors (2000), some rights reserved; exclusive licensee [Springer Nature]. Distributed under a Creative Commons Attribution License 4.0 (CC BY) https://creativecommons.org/licenses/by/4.0/.

5 APPLICATIONS OF MAGNETIC NANOWIRES

5.1 Nanowires for Nonvolatile Memory Application

Non-volatile memories characterized by fast reads, write speeds and high storage density are the focus of much research into new materials in the nanostructured form [47]. The storage limit was firstly reported in 1997 (36 Gbit/in^2) and its main problem is the thermal stability of magnetically stored information.

Increasing the storage density of magnetic memories is possible with a decrease in the size of the bits storing information at the nanoscale. The use of non-continuous magnetic materials such as assemblies of nanoparticles or nanowires is one way to reduce the size of the bits. In fact, with recent technology, density storage has already reached densities one order of magnitude higher than the value shown above. Actually, some companies have achieved a density of 185 Tb/in^2 [48]. Some studies [49] suggest that the principal limitation is mainly determined by the maximum tolerable bit error rate and certain material parameters, including the saturation magnetization of the recording medium.

However, to enhance storage density, we need a fine isolated granular structure from the magnetic material, but the miniaturization is facing a big challenge which is the superparamagnetic limit. To resolve this challenge, the researchers proposed two ways, the first one consists of increasing the material effective anisotropy and the second one is to synthesize nanowires by increasing the thickness from nanodots to nanocylinders with constant surface area [50].

5.2 Nanowires for Biomedical Applications

Due to their sizes and functionalities, nanomaterials present great interest in the biomedical field. Their comparable size to those of cells, viruses, bacteria, etc., can allow them to serve many advantages in biomedical applications. To make these nanomaterials more suitable for use in the medical field, it is possible to set experimentally their shape and size.

In a previous work [51], the authors have shown that nanowires are more efficient than nanoparticles, especially in hyperthermia therapy. However, magnetic nanowires can be used in many applications such as drug delivery [52], hyperthermia therapy [53], magnetic resonance imaging (MRI) contrast agent [54], etc.

Concerning drug delivery, using magnetic nanowires and generally magnetic nanoparticles is a promising method for targeting drugs to a diseased area with a reduction of the negative effects caused by conventional drug delivery. In this case, the magnetic particle is coupled to a therapeutic agent that is injected into the blood vessels and is directed to the targeted location by an external magnetic field. In ref [55], the authors investigated an iron-based magnetic nanowire (FePd), which is synthesized and developed to have magnetic properties optimized for drug targeting. The nanowires have a high magnetic moment to reduce the field gradient required to capture them with a magnet. The choice of material and size of the nanowire and coating are such that they are dispersible in aqueous media, non-cytotoxic, easily phagocytosed, and non-complementary tactivators.

For hyperthermia therapy, the approach involved the injection of a fluid or gel containing magnetic nanowires distributed homogeneously in a tumor. By applying an oscillating magnetic field with a low frequency, the nanowires rotate around their center due to the torque generated by the shape anisotropy. The external rotation and friction of the fluids in the nanowire's boundary cause heating that can destruct the tumor.

Thanks to the magnetic resonance imaging method, the location, and migration of living cells in the human body can be tracked non-invasively in real-time over several days, this is of utmost importance for cancer treatment and medical treatment monitoring based on living cells, such as stem cell therapies. This technique uses magnetic nanowires as contrast agents. Nanowires with a Fe-core and a Fe-oxide shell are great materials that can be used as contrast agents [53]. In this work, the authors have shown that the iron nanowires with iron oxide shells present an excellent performance as a contrast agent. It was found that the performance of the nanowires as a contrast agent can be adapted by changing the thickness of the oxide shell and adding coating agents.

REFERENCES

[1] Nie, Z., Petukhova, A., Kumacheva, E., Properties and emerging applications of self-assembled structures made from inorganic nanoparticles. *Nature Nanotechnology*, 5 (2010) 15.

[2] Gleiter, H., Nanostructured materials: state of the art and perspectives. *Nanostructured Materials*, 6 (1995) 3.

[3] Song, L., Dou, K., Wang, R., Leng, P., Luo, L., Xi, Y., Kaun, Ch-Ch., Han, N., Wang, F., Chen, Y., Sr-doped cubic In2O3/rhombohedral In2O3 homojunction nanowires for highly sensitive and selective breath ethanol sensing: Experiment and DFT simulation studies. *ACS Applied Materials and Interfaces*, 12(1) (2019) 1270–1279.

[4] Qin, Y., Zhao, L., Jiang, Y., Modulation of Ag modification on NO2 adsorption and sensing response characteristics of Si nanowire: A DFT study. *Applied Surface Science*, 467 (2019) 37–44.

[5] Dikin, D. A., Chen, X., Ding, W., Wagner, G., Ruoff, R. S., Resonance vibration of amorphous SiO2 nanowires driven by mechanical or electrical field excitation. *Journal of Applied Physics*, 93(1) (2003) 226–230.

[6] Deka, S., Joy, P. A., Synthesis and magnetic properties of Mn doped ZnO nanowires. *Solid State Communications*, 142 (2007) 190.

[7] Jeon, B., Yoon, S., Yoo, B., Electrochemical synthesis of compositionally modulated FexPd1−x nanowires. *Electrochimica Acta*, 56 (2010) 401–405.

[8] Shamaila, S., Sharif, R., Riaz, S., Ma, M., Khaleeq-ur-Rahman, M., Han, X. F., Magnetic and magnetization properties of electrodeposited fcc CoPt nanowire arrays. *Journal of Magnetism and Magnetic Materials*, 320 (2008) 1803.

[9] Leite, V. S., Grandi, B. C. S., Figueiredo, W., Phase diagram of uniaxial antiferromagnetic small particles: Monte Carlo calculations. *Physical Review B*, 74 (2006) 094408.

[10] Boughrara, M., Kerouad, M., Zaim, A., Phase diagrams of ferrimagnetic mixed spin 1/2 and spin 1 Ising nanowire with diluted surface. *Physica A*, 433 (2015) 59.

[11] Panigrahia, P., Pati, R., Controlling interlayer exchange coupling in one-dimensional Fe/Pt multilayered nanowire. *Physical Review B* 79 (2009) 014411.

[12] Metropolis, N., Rosenbluth, A. W., Rosenbluth, M. N., Teller, A. H., Teller, E., Equation of state calculations by fast computing machines. *Journal of Chemical Physics*, 21 (1953) 1087.

[13] Wang, X., Zhi, M., Liang Cai, M., Ling, X., Influence of the thickness of nonmagnetic spacer on the magnetic properties of Fe/Cu multilayered nanowires. *Key Engineering Materials*. Trans Tech Publications Ltd, 787 (2018) 93–98.

[14] Pal, P. P., Pati, R., Magnetic properties of one-dimensional Ni/Cu and Ni/Al multilayered nanowires: Role of nonmagnetic spacers. *Physical Review B*, 77 (2008) 144430.

[15] Khan, M., Rafique, M. Y., Razaq, A., Mir, A., Karim, S., Anwar, S., Sultana, I., Fabrication of Au/Ni/NiO heterostructure nanowires by electrochemical deposition and their temperature dependent magnetic properties. *Journal of Solid State Chemistry*, 284 (2020) 121186.

[16] Balela, M. D. L., Yagi, S., Matsubara, E., Fabrication of cobalt nanowires by electroless deposition under external magnetic field. *Journal of the Electrochemical Society*, 158 (2011) 4.

[17] Ji, Guangbin, Tang, Shaolong, Xu, Baolong, Gu, Benxi, Du, Youwei, Synthesis of CoFe2O4 nanowire arrays by sol – gel template method. *Chemical Physics Letters*, 379 (2003).

[18] Liu, J. J., Yu, M. H., Zhou, W. L., Well-aligned Mn-doped ZnO nanowires synthesized by a chemical vapor deposition method. *Applied Physics Letters*, 87 (2005) 172505.

[19] Hatwar, T. K., Genova, D. J., Victora, F. L. H., Double compensation point media for direct overwrite. *Journal of Applied Physics*, 75 (1994) 6858.

[20] Boughrara, M., Kerouad, M., Zaim, A., Phase diagrams and magnetic properties of a cylindrical Ising nanowire: Monte Carlo and effective field treatments. *Journal of Magnetism and Magnetic Materials*, 368 (2014) 169.

[21] Boughrara, M., Kerouad, M., Zaim, A. J., The phase diagrams and the magnetic properties of a ferrimagnetic mixed spin 1/2 and spin 1 Ising nanowire. *Journal of Magnetism and Magnetic Materials*, 360 (2014) 222.

[22] Zaim, N., Zaim, A., Kerouad, M., Monte Carlo study of the random magnetic field effect on the phase diagrams of a spin-1 cylindrical nanowire. *Journal of Alloys and Compounds*, 663 (2016) 516.

[23] Feraoun, A., Zaim, A., Kerouad, M., Monte Carlo study of a mixed spin (1, 3/2) ferrimagnetic nanowire with core/shell morphology. *Physica B*, 445 (2014) 74.

[24] Zaim, N., Zaim, A., Kerouad, M., The blocking temperature of an amorphous alternate A and B layers cylindrical nanowire. *Journal of Superconductivity and Novel Magnetism*, 32 (2019) 3081.

[25] Castillo-Sepúlveda, S., Corona, R. M., Altbir, D., Escrig, J., Magnetic properties of mosaic nanocomposites composed of nickel and cobalt nanowires. *Journal of Magnetism and Magnetic Materials*, 416 (2016) 325.

[26] Rosa, A. L., Tacca, L. L., Frauenheim, Th., Lima, E. N., Electronic and optical properties of Mn impurities in ultra-thin ZnO nanowires: Insights from density-functional theory. *Physica E*, 109 (2019) 6.

[27] Zhang, Y., Chen, S. Z., Xie, Z. X., Yu, X., Deng, Y. X., Ning, F., Xu, L., Surface dangling bonds dependent magnetic properties in Mn-doped GaAs nanowires. *Physics Letters A*, 384 (2020) 126815.

[28] Jo, C., Lee, J. I., Jang, Y., Electronic and magnetic properties of ultrathin Fe–Co alloy nanowires. *Chemistry of Materials*, 17(10) (2005) 2667.

[29] Barthélémy, A., Fert, A., Petroff, F., Chapter 1 Giant magnetoresistance in magnetic multilayers. In *Handbook of Ferromagnetic Materials*. Amsterdam: North-Holland, 1999.

[30] Noh, S. J., Miyamoto, Y., Okuda, M., Hayashi, N., Kim, Y. K., Control of magnetic domains in Co/Pd multi-layered nanowires with perpendicular magnetic anisotropy. *Journal of Nanoscience and Nanotechnology*, 12 (2012) 428–432.

[31] Fedorov, F. S., Mönch, I., Mickel, C., Tschulik, K., Zhao, B., Uhlemann, M., Gebert, A., Eckert, J., Electrochemical deposition of Co(Cu)/Cu multilayered nanowires. *Journal of the Electrochemical Society*, 160 (2013) D13–D19.

[32] Tanase, M., Silevitch, D. M., Hultgren, A., Bauer, L. A., Searson, P. C., Meyer, G. J., Reich D. H., Magnetic trapping and self-assembly of multicomponent nanowires. *Journal of Applied Physics*, 91 (2002) 8549.

[33] Liu, X., Tan, Y., Li, X., Wu, X., Pei, Y., Electronic and magnetic properties of silicon supported organometallic molecular wires: a density functional theory (DFT) study. *Nanoscale*, 7(32) (2015) 13734–13746.

[34] Zhang, Y. C., Qin, F. X., Estevez, D., Franco, V., Peng, H. X., Structure, magnetic and magnetocaloric properties of Ni2MnGa Heusler alloy nanowires. *Journal of Magnetism and Magnetic Materials*, 513 (2020) 167100.
[35] Khovailo, V. V., Chernenko, V. A., Cherechukin, A. A., Takagi, T., Abe, T., An efficient control of Curie temperature TC in Ni – Mn – Ga alloys. *Journal of Magnetism and Magnetic Materials*, 272 (2004) 2067–2068.
[36] Frolov, K. V., Chuev, M. A., Lyubutin, I. S., Zagorskii, D. L., Bedin, S. A., Perunov, I. V., Lomov, A. A., Artemov, V. V., Khmelenin, D. N., Sulyanov, S. N., Doludenko, I. M., Structural and magnetic properties of Ni-Fe nanowires in the pores of polymer track membranes. *Journal of Magnetism and Magnetic Materials*, 489 (2019) 165415.
[37] Elmekawy, A. H. A., Iashina, E., Dubitskiy, I., Sotnichuk, S., Bozhev, I., Kozlov, D., Napolskii, K., Menzel, D., Mistonov, A., Magnetic properties of ordered arrays of iron nanowires: The impact of the length. *Journal of Magnetism and Magnetic Materials*, 532 (2021) 167951.
[38] Ghazkoob, N., ZargarShoushtari, M., Kazeminezhad, I., LariBaghal, S. M., Structural, magnetic and optical investigation of AC pulse electrodeposited zinc ferrite nanowires with different diameters and lengths. *Journal of Magnetism and Magnetic Materials*, 537 (2021) 168113.
[39] Lavin, R., Denardin, J. C., Cortes, A., Gomez, H., Cornejo, M., Gonzales, G., Magnetic properties of cobalt nanowire arrays. *Molecular Crystals and Liquid Crystals*, 521 (2010) 293–300.
[40] Han, G. C., Zong, B. Y., Luo, P., Wu, Y. H., Angular dependence of the coercivity and remanence of ferromagnetic nanowire arrays. *Journal of Applied Physics*, 93 (2003) 9202.
[41] Cao, X., Wang, W., Zhang, X., Li, L., Cheng, Y., Liu, H., Du, S., Zheng, R., Magnetic properties of fluffy Fe@α-Fe2O3 core-shell nanowires. *Nanoscale Research Letters*, 8 (2013) 423.
[42] Ivanov, Y. P., Alfadhel, A., Alnassar, M., Perez, J. E., Vazquez, M., Chuvilin, A., Kosel, J., Tunable magnetic nanowires for biomedical and harsh environment applications. *Scientific Reports*, 6 (2016) 24189.
[43] Lee, Ju Hun, Wu, Jun Hua, Lee, Ji Sung, Jeon, Ki Seok, Kim, Hae Ryong, Lee, Jong Heun, Suh, Yung Doug, Kim, Young Keun, Synthesis and characterization of Fe–FeOx core-shell nanowires. *IEEE Transactions on Magnetics*, 44 (2008) 11.
[44] Frison, R., Cernuto, G., Cervellino, A., Zaharko, O., Colonna, G. M., Guagliardi, A., Masciocchi, N., Magnetite – maghemite nanoparticles in the 5–15 nm range: Correlating the core – shell composition and the surface structure to the magnetic properties. *A Total Scattering Study, Chemistry of Materials*, 25 (2013) 4820.
[45] Del Bianco, L. A. J. M., Fiorani, D., Testa, A. M., Bonetti, E., Savini, L., Signoretti, S., Magnetothermal behavior of a nanoscale Fe/Fe oxide granular system. *Physical Review B*, 66 (2002) 174418.
[46] Martínez-Banderas, A. I., Aires, A., Plaza-García, S., Colás, L., Moreno, J. A., Ravasi, T., Merzaban, J. S., Ramos-Cabrer, P., Cortajarena, A. L., Kosel, J., Magnetic core – shell nanowires as MRI contrast agents for cell tracking, Journal of Nanobiotechnol, 18 (2020) 42.
[47] Prince, B. (Ed.), *Vertical 3D Memory Technologies*. John Wiley & Sons Ltd, Chichester, 2014.
[48] Sony creates an 185TB tape cartridge! *Storage Servers*, 01-May-2014.
[49] Evans, R. F. L., Chantrell, R. W., Nowak, U., Lyberatos, A., Richter, H.-J., Thermally induced error: Density limit for magnetic data storage. *Applied Physics Letters*, 100(10) 2012, 102402.
[50] Bochmann, S., Döhler, D., Trapp, B., Staňo, M., Fruchart, O., Bachmann, J., Preparation and physical properties of soft magnetic nickel-cobalt three-segmented nanowires. *Journal of Applied Physics*, 124 (2018) 163907.
[51] Lin, W. S., Lin, H. M., Chen, H. H., Hwu, Y. K., Chiou, Y., Shape effects of iron nanowires on hyperthermia treatment. *Journal of Nanomaterials*, 2013 (2013) 6.
[52] Liu, Y., Han, Q., Yang, W., Gan, X., Yang, Y., Xie, K., Xie, L., Deng, Y., Two-dimensional MXene/cobalt nanowire heterojunction for controlled drug delivery and chemo-photothermal therapy. *Materials Science & Engineering C*, 116 (2020) 111212.
[53] Lin, W. S., Lin, H. M., Chen, H. H., Hwu, Y. K., Chiou, Y. J., Shape effects of iron nanowires on hyperthermia treatment. *Journal of Nanomaterials*, 1 (2013) 1–6.
[54] Martínez-Banderas, A. I., Aires, A., Plaza-García, S., Colás, L., Moreno, J. A., Ravasi, T., Merzaban, J. S., Ramos-Cabrer, P., Cortajarena, A. L., Kosel, J., Magnetic core – shell nanowires as MRI contrast agents for cell tracking. *Journal of Nanomaterials*, 18 (2020) 42.
[55] Pondman, K. M., Bunt, N. D., Maijenburg, A. W., van Wezel, R. J., Kishore, U., Abelmann, L., ten Elshof, J., E., ten Haken, B., Magnetic drug delivery with FePd nanowires. *Journal of Magnetism and Magnetic Materials*, 380 (2015) 299.

Spin Torque Devices

M. Shakil, Halima Sadia, M. Isa Khan, M. Zafar

Contents

1 Introduction	93
2 History	94
3 Magnetoresistance Effect	95
4 Current Flow Perpendicular to Plane	96
5 Magnetic Tunnel Junctions	97
6 Spin-Transfer Torque-Magnetic Random Access Memory (STT-MRAM)	99
6.1 PMA Based Spin-Transfer Torque-Magnetic Random Access Memory	101
6.2 STT Based Memory Devices and Materials	102
7 Thermal Effects on MTJ	106
8 Advances and Prospects of STT-MRAM	107
9 STT-MRAM Comparison over Conventional Memory Devices	108
10 Current Challenges to STT-MRAM	109
10.1 Writing Improvements	109
10.2 Variation of MTJ Resistance	109
10.3 Thermal Stability	109
10.4 Demagnetizing Fields	110
11 Emerging Memory: Based on Spin Hall Effect	110
Conclusions	111
References	111

1 INTRODUCTION

Conventional electronic devices that carry charge/current also have spin but spin is randomly oriented. Hence these devices do not show spin behavior. In FM materials, electrons become spin-polarized (SP) and play a significant role over conventional electronic devices. In these devices, information is stored by switching magnetic states in a particular configuration. Switching occurs by applying a magnetic field that controls the material's coercive force. In 1996, it was found by Slonczewski [1] and Berger, that current flow through magnetic multi-layers could be more effective on the magnetic state. When current passes through FM material it becomes SP then it carries angular momentum. Thus current remains SP in neighboring non-magnetic (NM) layer, hence angular momentum possessed by this current makes interaction with magnetization in the subsequent magnetic layer. The SP current exerts STT on the magnetization in a subsequent device. If the current is large enough, it produces precession and reversal. Experimental demonstration of STT effect for magnetization [2,3] switching made it attractive for spintronic devices. In the existing MRAM implementation, the Oersted field is used for switching. These long-range fields mean that reduced fields are applied to neighboring bits. The possibility for Oersted field replacement by STT in MRAM has attracted significant attention in research.

Many device geometries have been observed with STT induced magnetizations effects such as lithographically defined point contact, mechanical point contact, manganite junctions, electro-chemically grown nanowires, tunnel junctions, semiconductor structures, and lithographically defined nano-pillars. These devices have two common particular characteristics, i) small cross-sectional area, and ii) magnetoresistive read out of magnetic state. These devices possess diverse computing functionality by increasing speed, density, non-volatility, and reliability. Giant magnetoresistance (GMR) was discovered and elaborated by Peter Grunberg and Albert Fert in 1988. This discovery drew much attention towards spin-based electronic devices. After 10 years of this discovery, the GMR effect was used in hard disk drives (HDD). There was a desperate increase in areal density owing to an increase of sensitivity of GMR read heads. Its considerable impact on technologies was recognized by Nobel Prize in 2007 [4]. The basic phenomenon behind the GMR effect is that; the resistance of the FM-spacer layer-FM system is dependent on the relative orientation of magnetization directions of FM material. This effect is assessed by magnetoresistance ratio (*MR*) which is a percentage of change between parallel (P) and anti-parallel (AP) orientation. When FM layers coupled anti-ferromagnetically (AFM) then sample possess relatively more resistance. By externally applied magnetic field, magnetization can be aligned P, and in such cases, the sample possesses less resistance. At intermediate angles resistance can be maintained at intermediate level. If metallic spacer is replaced by dielectric spacer in GMR, then amplified version of *MR* is called tunnel magnetoresistance (TMR). A better choice of dielectric spacer and FM electrodes can produce TMR over 600% at room temperature (RT) [5]. MTJs employing phenomena of TMR have been used in hard drives to empower higher sensitivity readback of data. GMR and TMR are exploited by the spin-based current.

STT phenomena can be used as a write mechanism in magnetic random access memories (MRAMs) based on MTJ. Initial execution of MRAM required the application of a conventional magnetic field to switch the bit-state. But it is proved inherently unscalable as increase of write field/current decreases the bit size [6]. STT is alternative approach as write current scales with the size of a cell. This combined effect of STT and MRAM makes STT-MRAM, the frontrunners of non-volatile data storage technology. STT-MRAM has been faced significant engineering challenges during its development. As total angular momentum transferred by a single electron is smaller. So, a large current density is required for a significant STT effect in traditional FM materials. MTJ requirement for STT-MRAMs is higher TMR ratio, low switching current, and higher thermal stability (Δ) $\left(\Delta = \dfrac{E}{k_B T}\right)$. One solution to minimize the current density for switching can be availed by using of materials with perpendicular magnetic anisotropy (PMA) while maintaining high data-retention capability. Usage of high PMA materials in MTJs maintaining high TMR is the current focus of research to unlock potential usage of STT based technologies. In 2001 two groups [7, 8] independently reported that Fe/MgO/Fe MTJ would have high TMR values. Later on, this prediction was verified on RT devices. TMR ratio of other MTJs have also been observed for instance CoFe|MgO|CoFe [9] and Co|SrTiO$_3$|Co [10]. Many other FM/oxide interfaces like Fe/Al$_2$O$_3$, Co/AlO$_x$, Fe/CrO$_2$, Fe/MgO, Co/MgO, and CoFeB/MgO have been studied to obtain higher TMR [11]. Among them, CoFeB has shown high SP with PMA. Nowadays MTJs built on sputtered amorphous CoFeB with MgO oxide layer have revealed a record TMR value of 604 % at RT [5]. Recently, spin-orbit torque i.e., exhibit spin Hall effect (SHE) MRAM (SOT-MRAM) has also been investigated which had manifested fast switching at RT [12]. But, the SHE MTJ-hTron bitcell scheme faces certain challenges like the large size and slow operation [13].

2 HISTORY

The first practical MRAM was introduced in 1947 with the Williams Kilburn tube, which used electrically charged spots present as a bit on the cathode ray tube [14]. That was the stimulating point for the beginning of future RAMs. This data storage technology did not last long because it was not reliable. Data storage based on magnetic core memories had been emerged in the 1950s and last for two decades. A semiconductor technology named dynamic random access memory (DRAM) was introduced in 1968. It was more compatible than previously integrated chips. Afterward, DRAM started used as standard memory technology. About a half-century later, magnetic memory was again considered critically as a versatile product for market demand. The initial step of MRAM technology was taken in the 1960s when it was recommended to put magnetoresistive elements in core memory instead of the toroid. Many issues had been resolved by MRAM such as non-volatility. Further research disclosed GMR phenomena in 1988 and then in the 1990s MTJ was introduced. After that, progress in research activities about MRAM declined again in 2000 when it was considered that there were scalability issues of magnetic field-based technology as it was not extendable to a smaller size which would result in higher density MRAM applications. But with the invention of the STT effect, it again aroused as a promising candidate in memory applications with greater switching ability, scalability, and reliability for the future horizon.

3 MAGNETORESISTANCE EFFECT

Memory devices based on the STT effect consist of spin valves as a building block. When an external magnetic field is not applied, magnetic layers' magnetization tends to align in a particular direction called easy axis which is collinear. Due to spacer layer types, spin valves are divided into two categories. Spin valves that exhibit insulator spacer layer demonstrate TMR as investigated by Julliere [15]. In 1975, Julliere demonstrated the TMR effect experimentally and obtained a 14% TMR value at 4K for Fe/Ge/Co tunnel junction [15]. On the other hand, the GMR effect is demonstrated when the spacer layer is NM metal. Switching based on spin transfer is firstly observed in GMR structures [8] and later in MTJs [16]. GMR effect was investigated by Fert and Grunberg independently and was awarded Noble Prize [17, 18]. The conductance of spin valves depends upon the relative alignment of magnetic layers. An important parameter to express spin valve is the *MR* ratio as given in eq. 1.

$$MR = \frac{G_P - G_{AP}}{G_{AP}} \times 100 \qquad (1)$$

Firstly, the measured *MR* ratio by Julliere and Fert was 14% and 80% in Fe/Ge-O/Co tunnel junctions and in Fe/Cr superlattices respectively [18] at temperature ≈ 4.2K. Afterward, several studies were carried out for an understanding of TMR and MTJ [19]. GMR can be understood from the concept that the scattering of electrons occurs when passed through the spin valve. When the configuration of the spin valve is P, electrons encounter smaller scattering. However large scattering occurs for AP configuration. TMR can be described by the spin filtering effect that originates owing to band structures of a component layer of the spin valve. For that particular reason, the probability of tunneling electrons across the insulator spacer layer (SL) relies on the layout of the spin valve based on TMR. Thus a higher *MR* ratio can be achieved by a TMR-based spin-valve as compared to GMR at RT. TMR-based spin-valve resistance is greater than GMR-based spin valve. Hence a magnetic configuration of the spin valve based on TMR can be electrically observed in CMOS circuits conveniently as compared to GMR based spin valve. TMR-based spin valves are more prominent for on-chip memory devices.

The earliest studies have also used aluminum oxide (AlO) as a tunnel barrier (TB). MTJ based on AlO developed in 1991 [20]. These tunnel barriers were very disordered and it was difficult to investigate them theoretically. Tunneling characteristics of Fe/MgO/Fe were calculated in 2001 by Butler et al., Mathon and Umerski [35, 36]. They investigated that electronic states and symmetry of the system make a large TMR possible. Emerging techniques have made both AlO and MgO barriers thin enough to meet the current densities requirement for magnetic switching with STT. Usage of the STT effect in tunnel junctions is more appropriate in many applications than metallic magnetic multilayers. Magnetic switching caused by STT effects can be more efficient as compared to magnetic switching caused by a current-induced magnetic field. Hence, magnetic memory devices can be produced which require a low current for switching and consequently have more energy efficiency and larger device density than devices based on field switching.

4 CURRENT FLOW PERPENDICULAR TO PLANE

The earliest work on GMR was performed for current flowing in a plane (CIP) of a multi-layered sample. In 1991 another geometry was found [21] in which the current flow perpendicular to plane (CPP) of sample and produce a large change in fractional resistance. CPP geometry has an unusual interest in present circumstances as STT effects are much more effective in CPP geometry as compared to CIP. In 1990, Parkin et al. discovered that oscillations of interlayer exchange coupling occur as a function of SL thickness [22]. Slonczewski [23] made calculations for interlayer exchange coupling when insulating TB used as an SL. At that time, there were no measurements for insulating TB. His work has two particular features to spin torque physics. Firstly, exchange coupling is calculated by observing spin current flowing across insulating TB. The spin current was also noticed even for zero applied bias across TB whenever electrodes magnetization is non-collinear and exchange coupling source is a transfer of angular momentum from SP current to each of the magnets. Secondly, additional coupling was also observed when the voltage applied across the TB. That was first observation of SST in multi-layered geometry for CPP. After that, there was little instant experimental investigation owing to rather primal technology for making MTJ at that time, and TBs provided at that time were too thick to allow for huge current densities required to STT powered magnetic dynamics.

Slonczewski [24] and Luc Berger [25] predicted that CPP in a multi-layer produces SST strong enough that it can reorient magnetization in one of the layers. The measurement of resistance changes in magnetic multilayered devices due to induced current firstly identified with spin-torque driven excitations by Tsoi et al., in 1998 for devices composed of mechanical point contact to metallic multilayer and then in manganite devices by Sun in 1999. After that, magnetization reversal by spin torque was observed in lithographically defined samples. At that time interest began to grow STT in metallic multilayers and raised interest in MTJ starting with the observation of TMR in 1995 at RT. One of the challenges for MTJ development is large critical current production required for STT switching in STT-MRAM. This critical current can be estimated by the relation given in eq. 2.

$$I_c = \frac{2e\pm}{\eta\hbar} M_s V \left(H_{k,\text{eff}} \right) \qquad (2)$$

α = Gilbert damping parameter
η = STT efficiency factor
M_s = saturation magnetization
V = magnetic volume

It is important to investigate materials that have lattice-matched BCC (001) with MgO and also possess lesser magnetization. Materials that have higher TMR also have more saturation magnetization such as Fe/MgO/Fe (001) with Ms ~ 1700 emu/cm³. Hence to reduce current densities (J_c as large as 107 A/cm²), materials with low saturation magnetization such as CoFeB, NiFe, and CoFe are being utilized. However, this low M_s free layer (FL) introduced the energy barrier decline (eq. 3).

$$E_B = M_s H_k V / 2 k_B T \qquad (3)$$

To tackle reduction in J_c, an alternative approach must be applied. On the other hand, I_c can also be minimized by establishing partial perpendicular anisotropy into FL for lowering the effective demagnetization field. When it goes to extreme value then the uniaxial anisotropy field perpendicular to the plane surpasses the total demagnetizing field and the FL easy axis becomes perpendicular. This perpendicular alignment (eq. 4) not only reduces J_c, but also enhances E_B.

$$I_c = \frac{2e\pm}{\eta\hbar} M_s V \left(H_{k\perp,\text{eff}}\right) \tag{4}$$

5 MAGNETIC TUNNEL JUNCTIONS

An MTJ consists of an ultra-thin insulating layer inserted between two FM layers. The tunneling resistance of MTJ relies upon the relative alignment of FM electrodes, as depicted in Figure 7.1 [26].

Tunneling resistance of P electrodes is represented by R_P (Figure 7.1(**a**)) which is lower as compared to the resistance produced from AP alignment (R_{AP}) of electrodes (Figure 7.1(**b**)). The corresponding effect is called TMR, which is of core significance in spintronic applications. The strength of the TMR effect is measured by the magnetoresistance (*MR*) ratio. The *MR* ratio can be indicated by expression (eq. 5)

$$MR = 2P_1P_2/(1-P_1P_2) \tag{5}$$

Where P_1 and P_2 represent SP of two FM electrodes at the Fermi level (E_F). Capping layer attached to one of the FM electrodes, if its nature is oxidizing. The density of states becomes SP at E_F hence,

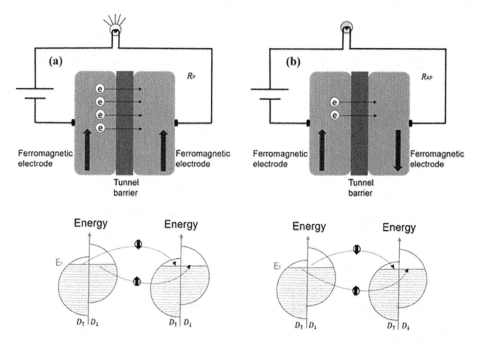

FIGURE 7.1 Schematic description of tunnel magnetoresistance effect in a magnetic tunnel junction. (a) The left side illustrates parallel magnetic state; (b) Right side illustrates anti-parallel magnetic state; E_F denotes Fermi level; R_P is the resistance of parallel magnetic state and R_{AP} is the resistance of anti-parallel magnetic state; majority spin and minority spin density shown by $D_\uparrow$ and $D_\downarrow$ respectively; black arrows denote magnetization direction; thin dotted black arrows denote tunneling transport of electrons.

Source: Adapted with permission from [26]. Copyright (2018) MRS Bulletin.

TMR also becomes changed owing to relative magnetic orientation. Practical MTJ structure is shown in Figure 7.2(a) [26]. The Upper FM layer is called FL while the lower FM layer is named as a reference layer (RL). The direction of magnetization of RL is pinned by applying external bias from the AFM layer (for example Ir-Mn (111)), as a part of SAF). SAF structure comprised of two AP-oriented FM layers and a SAF SL present between. When magnetization of FL is configured in such a way that it rotates continuously under the effect of the external applied magnetic field, then MTJ acts as a magnetic sensor device, for example, read-head of hard disk drive (HDD). If magnetization of FL is configured in such a way that it has two remnant states, then 1 bit of data is stored by MTJ in magnetic alignment form and thus works as a non-volatile MRAM as shown in Figure 7.2(b) [26]. Due to the TMR effect, resistance changes, and thus, the stored information is readable. For the writing function in MRAM, FL magnetization is switched by applying a magnetic field produced from electric current or by STT induction result from SP current. Today MgO is the most common TB [27]. However, other oxide materials like Al_2O_3 [28], ZnO [29], and $Mg_3B_2O_6$ [30] have also been used. Titanates [31] and ferrites [32] have also been used. Experimental findings on amorphous AlO_x showed that such materials exhibited a significant TMR ratio at RT [20], for Co/AlO_x/Co MTJ TMR value exceeded 15% at RT [33].

Later, extensive work was performed with AlO TB by using different electrodes with high SP. Miyazaki et al., [34] and Moodera et al., [33] made MTJ in 1995 by using amorphous AlO TB and two FM electrodes. They observed TMR at RT. *MR* ratio shown by these AlO based MTJs was up till 70–80% at RT which have commercialized the read heads for higher density HDDs and first-generation toggle MRAM. In 2004, using amorphous AlO_x TB and CoFeB electrodes, TMR ratio achieved up to 70% using a sputtering technique, and CoFeB composition was taken as $Co_{60}Fe_{20}B_{20}$. However, such *MR* ratio was not much higher for future devices. For instance, STT-MRAM needs those MTJs which have *MR* ratio >> 100% at RT.

MR ratio possessed by AlO based TB was less than 100% owing to its amorphous nature. No crystallographic symmetry exists in these TBs which causes the Δ_1 states (Bloch states) of DOS with different SP to tunnel through barrier and lead to tunneling SP. For transition metal FM it is below 0.5 at RT. Contrarily, TBs with crystalline symmetry such as MgO (001), anticipated tunneling through the barrier coherently by preserving the orbital symmetry. In 2001, MgO with a higher *MR* ratio was theoretically

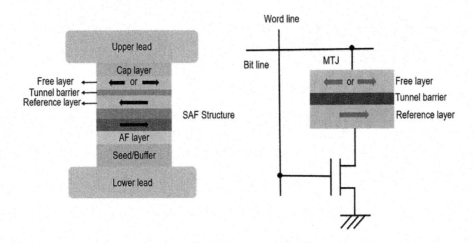

FIGURE 7.2 (a) Cross-sectional view of practical MTJ exhibiting in-plane magnetization (b) memory cell of MRAM. MOSFET denotes metal oxide semiconductor field-effect transistor; SAF denotes synthetic antiferromagnet; word line: a metal line that selects one column of transistors and makes it conductive; bit line: a metal line that provides voltage to one row of MTJ alignment. Adapted with permission from [26].

Source: Copyright (2018) MRS Bulletin.

predicted [7, 8]. However, an experimental prediction was made in 2004 [35]. Changing from AlO to MgO increased the read access time [36]. First-principles calculations have shown that only those Δ_1 states tunnel through MgO (001) barrier which has spherical symmetry [7]. Ishikawa et al., investigated properties of MTJ consisting of Co/Pt multilayer which act as recording layer where CoFeB layer inserted between MgO and recording layer to increase the TMR ratio. They indicated that placing the Ta layer between CoFeB and Co/Pt multilayer had improved the MTJ properties after annealing. After annealing at 300 °C, magnetic anisotropy energy per unit area obtained was over 1.2 mJ/m^2. Inserting CoFeB layer with 1.6 nm thickness, 91% TMR ratio and high thermal stability factor of 92 achieved in MTJ with diameter of 17 nm [37]. The Δ_1 states of body-centered cubic Fe (001) are 100% SP and a higher *MR* ratio (>>1000%) is anticipated theoretically [8]. Yuasa et al. assembled the fully epitaxial MTJs Fe (001)/MgO (001)/Fe (001) by employing molecular beam epitaxy and obtained a high *MR* ratio up to 180% at RT [35]. Parkin et al. fabricated MTJs with textured MgO (001) TB by employing sputter deposition and achieved a higher *MR* ratio up to 220% at RT [38].

MTJs based on MgO still have some problems for practical applications even though of their enormous *MR* ratios. Practical MTJ structure is shown in Figure 7.2(**a**) [26]. The problem here is that for the growth of MTJ film AFM/SAF multilayer is required, which is fcc (001) structure and possess threefold in-plane crystallographic symmetry while MgO (001) based MTJs possess fourfold in-plane symmetry. Hence owing to symmetry mismatch it cannot be grown on fcc (001). Djayaprawira et al. made the development of MTJs based on CoFeB/MgO/CoFeB to resolve crystal growth problem and an *MR* ratio greater than 200% was observed at RT [39]. The amorphous CoFeB layer was prepared when the concentration of B was kept higher than several atomic percent. Then, a highly textured layer of MgO (001) was grown on an amorphous CoFeB layer with the sputtering technique in the presence of an optimized environment. Once the annealing of MTJ reached above 250°C then CoFeB crystallized from MgO (001) interfaces because of the lattice matching between MgO (001) and body-centered cubic CoFeB (001) [40]. After annealing structure was textured CoFeB (001)/MgO (001)/CoFeB (001). Afterward utilizing this growth method, CoFeB/MgO/CoFeB MTJ can be fabricated having fourfold symmetry on practical bottom structures of any type of underlayer. Today CoFeB/MgO/CoFeB MTJ has gained significant attention towards practical applications, for instance, magnetic sensors and STT-MRAM.

6 SPIN-TRANSFER TORQUE-MAGNETIC RANDOM ACCESS MEMORY (STT-MRAM)

Silicon-based memory devices like DRAM, static random access memory (SRAM), and non-volatile flash store information in the form of electric charge on a memory cell that is electrically capacitive. The Digital state of the charged capacitor is "1" while for the uncharged capacitor is "0". MRAM is a crucial candidate for memory devices owing to its high speed, density, and non-volatility. On the other hand, STT-MRAM has attained remarkable attention from researchers due to its low write current and higher thermal stability. In non-volatile memory devices such as STT-MRAM, information is stored by the spin of an electron. A typical STT-MRAM consists of one transistor and one MTJ as shown in Figure 7.3(**a**)[26]. STT-MRAM technology was developed from theoretical investigations on SP current which had generated the torque and consequently alter the magnetic moment of a magnet by transfer of angular momentum. To perform read and write operations, the electrical current passed through the MTJ. First experimental verification regarding STT effect observed in nano-particles and spin valves. After that, the STT effect was observed in MTJs.

The first integrated STT-MRAM chip was made by Sony in 2005 using 4-kbit MTJ arrays. STT-MRAM consists of two FM electrodes usually CoFeB and one insulator layer present between two FM layers as shown in Figure 7.3(**b**). The Bottom FM layer has fixed magnetization named RL and the top FM layer consists of non-fixed magnetization that can be altered by employing SP current. When passed through FL, it

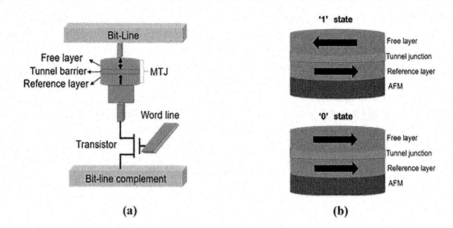

FIGURE 7.3 (a) Schematic description of bit cell of magnetic random access memory. (b) High and low resistance states of MTJ based on the relative orientation of magnetization. Adapted with permission from [26].
Source: Copyright (2018) MRS Bulletin.

absorbs the angular momentum of electrons, and consequently, magnetization flips. It is the reason that we call it STT. The stored information in MTJ can be sensed by the CMOS access transistor which forms the 1T-1MTJ structure. The magnetization direction of PMA STT-MRAM is perpendicular to the MTJ surface. MTJ shape for PMA STT-MRAM is circular while elliptical for in-plane STT-MRAM. When the direction of magnetization of both layers is similar, then MTJ carries low resistance. However, when the magnetization of both layers is AP then MTJ carries high resistance as shown in Figure 7.3(**b**). These two possibilities are represented by high (HI) and low (LO) states. These states maintain data encoding as

"0" state: $V_{BL,} = I_R * (R_L + R_T R)$ (6)
"1" state: $V_{BL,} = I_R * (R_H + R_T R)$ (7)

Where R_H and R_L represent high and low resistance of MTJ and I_R represent read current. Resistance of NMOS transistor represented by $R_T R$. Bit-line voltages at HI and LO states are represented by $V_{BL,}$ and V_{BL}. The voltage of MTJ is compared with a reference voltage to read STT-RAM. Which is defined as

$$Max\ (V_{BL,L}) < V_{ref} < Min\ (V_{BL,H})$$

When the magnetization direction of the FM layer in MTJ is switched from P to AP as indicated in Figure 7.4(**a**) [41] then the flow of electrons occurs from fixed layer towards FL. Further, the spinning of electrons takes place in a similar direction as the magnetization of the fixed layer prevails in the fixed layer and consequently, the SP current is achieved. Consequently, the SP current exerts STT effect on FL. When this SP current crossed the threshold value, the magnetization direction of the AP layer switches. The threshold current (J_{co}) of FM layer with given anisotropy defined by Slonczewski equation (eq. 8)

$$Jco \cong \frac{2\alpha e}{\eta \hbar}(M_s t)(2\pi M_{eff})$$ (8)

where M_{eff} is effective magnetization.

When switching occurs from P to AP layer as indicated in Figure 7.4(**b**) [41], then electrons move from FL to the fixed layer. Electrons whose spin is the same as fixed layer pass but become reflected at oxide layer boundary whose spin is opposite and injected into FL. These reflected electrons produce the STT effect in FL and switch the magnetization direction when the current exceeds the threshold value [41].

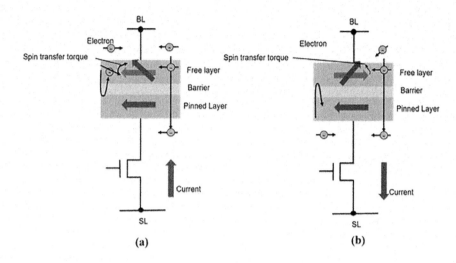

FIGURE 7.4 Memory cell schematic with MTJ. (a) switching from P to AP (b) switching from AP to P. Adapted with permission from [41].
Source: Copyright (2012) Elsevier.

6.1 PMA Based Spin-Transfer Torque-Magnetic Random Access Memory

PMA is significant in STT-MRAM as it requires a lower current for switching as compared to in-plane magnetic anisotropy for the same value of activation energy. Practical implementation of PMA materials took many years. The first result of integrated STT-MRAM based on PMA materials was published by Toshiba in collaboration with the National Institute of Advanced Industrial Science and Technology and Tohoku University Japan in 2008. They reported fast switching, low switching current, and good device data retention [42]. In 2010, researchers from IBM had reported a detailed evaluation of STT-MRAM based on PMA materials which were integrated into 4-kbit arrays. They illustrated low write error and narrow distribution of switching voltage, which satisfied the demand of 64-Mb chip for the first time [43]. Researchers from IBM and Tohoku University work independently and discovered that a good PMA can be obtained using a thin layer of CoFeB with TB MgO [44, 45]. This research established useful PMA materials required for MTJ, CoFeB/MgO. Here PMA arises from interfacial symmetry breaking called interfacial PMA. In CoFeB/MgO, PMA originates due to interfacial hybridization between $3d$–O $2p$ orbitals. In the earliest work, PMA had also been discovered using the Pt seed layer, but Pt presence prevented such layers to be useful for MTJ. Later on, FL based on CoFeB with double oxide interfaces was used to increase the PMA of FL which resulted in thermal stability as well. The oxide cap had also reduced the magnetic damping of FL by cutting off the spin pumping effect from heavy metal caps such as Pt, thus minimizing the switching current of a device [46].

PMA materials based on CoFeB are suitable for FL candidates for STT devices due to small Gilbert damping and greater tunneling SP. To drop-down stray dipole effects that act on FL, a SAF-free structure is mostly adopted. The FM materials of RL are normally PMA multilayers as well as Co|Ni, Co|Pd, and Co|Pt. These multilayers revealed a low SP and fcc structure that grew on (111) texture which makes it incompatible for the MgO barrier to obtain a greater TMR ratio. Higher TMR can be obtained by introducing a CoFeB layer at the interface and an SL of a low thickness (e.g., Ta) into RL based on multilayer. Here, SL acts as a transition layer which makes sure that the as-grown amorphous CoFeB layer has crystallized from the interface of MgO and becomes a bcc structure during the process of annealing to obtain high TMR. The main building blocks of STT-MRAM are MgO TB, iPMA FL based on CoFeB and SAF RL. Such a system of materials has also shown scaling potential lower than sub-20 nm [47, 48].

102 Applications of Low Dimensional Magnets

Fast switching in STT-MRAM is advantageous because it is directly proportional to switching current and inversely proportional to the product of resistance area (RA). To increase the speed of STT-MRAM, it is necessary to reduce RA which can be reduced by using thin MgO TB. It can also be obtained through high switching current I_c. But, it has also been reported that the desire for both high current and thin barriers may break the AlO TB and decrease the life time of MTJ [49]. The use of MgO is beneficial due to the large breakdown voltage and increased lifetime of MTJ. Moreover, MgO possesses a higher STT and TMR effect and induces good PMA. The perpendicular MTJs are significant for RT applications.

6.2 STT Based Memory Devices and Materials

Magnetization alignment of nano-magnet *m*, along its easy axis, is maintained by energy barrier "E_B" in the presence of thermal fluctuations as indicated in Figure 7.5(a) [50]. We suppose that to overcome E_B for switching from 0°–180° ($E_{B,1}$) is similar to switching from 180°–0°($E_{B,2}$). These are conditions when

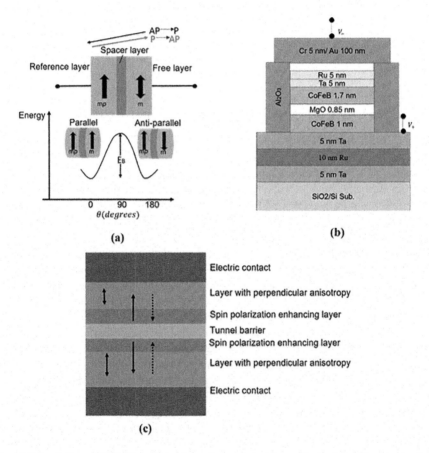

FIGURE 7.5 (a) Schematic vertical spin valve. Magnetoresistance type depends on the spacer layer. The free layer exhibits an anisotropy energy barrier and hence the spin valve is in either a parallel or anti-parallel arrangement. The top view represents the direction of the current flow for spin-valve switching. Adapted with permission from [50]. Copyright (2016) Proceedings of the IEEE. (b) Schematic structure of CoFeB/MgO p-MTJ annealed at 300°C. Adapted with permission from [44]. Copyright (2010) Nature materials. (c) Perpendicular MTJ cell. The thin FM layer possesses a high spin polarization factor; the thick FM layer possesses a higher perpendicular magnetocrystalline anisotropy. Tunnel barrier present between two FM layers. Adapted with permission from [41].

Source: Copyright (2012) Elsevier.

higher-order effects have been ignored. However, in the presence of such effects $E_{B,1} \neq E_{B,2}$. In that circumstance, E_B is smaller than $E_{B,1}$ and $E_{B,2}$. Production of E_B is realized by uniaxial anisotropy and its relation to volume V of nano-magnet is expressed by relation (eq. 9) [20].

$$E_B = K_{u,2}V = \mu_0 M_S V H_C/2 \tag{9}$$

Where $K_{u,2}$ is the uniaxial anisotropy constant of second order. M_S represents saturation magnetization and H_C is a magnetic anisotropy field. In the presence of thermal fluctuations, stability or lifetime of nano-magnet in relation to E_B is expressed by eq. 10 [51].

$$t_{LIFE} = t_0 \exp\left(\frac{E_B}{Tk_B}\right) \tag{10}$$

where E_B is the energy barrier, k_B is Boltzmann constant and t_0 is inverse attempt frequency that is ~ 1 ns. A nano-magnet $E_B \approx 40 k_B T, t_{LIFE}$ ~ 7.4 years.

As said above, MTJ consists of two FM layers (free and pinned) and an insulator layer sandwiched between them. Uniaxial anisotropy is established into FL to produce E_B that make sure of thermal stability. Switching of MTJ between the P and AP arrangement can be done by applying magnetic field which is also called field-induced magnetization (FIMS). So, its on-chip applications magnetic field (H) generated through current-carrying wires as (eq. 11)

$$H = 1/2\pi r \tag{11}$$

Distance present between nano-magnet and wire is denoted by r. Such a type of field is needed to overcome the critical field that is necessary to arrange the magnetic state of MTJ.

$$H > HC = \left(\frac{2E_B}{u_{0M_sV}}\right) \tag{12}$$

Critical current (I_c) required to produce a magnetic field is given in eq. 13.

$$I_c >= \left(\frac{4\pi E_B r}{u_{0M_sV}}\right) \tag{13}$$

As the size of MTJ is cut down then FL volume scales with r^2. Hence switching current increases. Additionally, the magnetic field is not well confined for switching of MTJ and it can affect current-carrying wires. Hence magnetic field switching is not scalable in MRAM. So, STT is an alternative approach.

Consider the flow of electrons from the pinned layer (PL) to FL. These magnetic layers SP the electrons. However, PL polarizes incoming electrons strongly in the direction of PL magnetization denoted by m_p. When these electrons tunnel through a barrier into FL, it applies torque and helps to orient magnetization m of FL with m_p. When electrons flow from FL to PL, FL tries to SP incoming electrons with m. However, electrons with the spin direction of m_p tunnel through easily across barriers while electrons with the spin opposite to m_p do not tunnel and remain in FL and apply torque on m. As a result, alignment of m occurs opposite to m_p. Hence switching direction is determined from the current flowing direction as shown in Figure 7.5 (b) [44]. Also, I_c is lowered to align parallel FL with PL, as spins injected into FL are mostly in the direction of m_p. When the alignment of FL is anti-parallelizing with PL then electrons coming in FL are not effectively SP. So, large I_c is the requirement to anti-parallelize FL with PL. FL magnetization dynamics in MTJ can be modeled by utilizing Landau–Lifshitz–Gilbert–Slonczewski (LLGS) equation (eq. 14).

$$\frac{\partial m}{\partial t} = -|\gamma| m \times H_{EFF} + \alpha\left(m \times \frac{\partial m}{\partial t}\right) + \text{STT} \tag{14}$$

Where the gyromagnetic ratio is denoted by γ, α is Gilbert's damping factor and H_{EFF} is the effective magnetic field. STT is generally given by eq. 15.

$$STT = \left| \frac{\gamma \hbar}{2\mu_0 e M_s V} \right| I \left(m \times \left[\eta (m_p \times m) - \eta' m_p \right] \right) \quad (15)$$

Here $\hbar$ denotes reduced Planck's constant, η and η' denotes the Slonczewski-like and field-like torques respectively and probably functions of m and m_p, e is electronic charge.

If FL possesses only uniaxial anisotropy then simplify the above equation such that η is constant and η' is equal to zero. η explains the degree of SP and depends upon the spin filtering effect. Substitute equation 15 into 14 to find out the condition for which STT switch MTJ configuration, which is as given (eq.16).

$$\left| \frac{\gamma \hbar}{2\mu_0 e M_s V} \right| I \geq \eta \alpha H_c \quad (16)$$

Using equations 12 and 16, critical current I_C for switching is obtained (eq. 17)

$$I_C = \frac{4 e \alpha E_B}{\eta \hbar} \quad (17)$$

At iso-E_B, I_C is constant in STT switching when MTJ scales down but rises for FMIS based MRAM. Hence STT-MRAM is more scalable.

Non-volatile memory devices require access time smaller than 10 ns. These devices have more endurance, scalable writing, and are well suited to CMOS technology. To meet these requirements MTJs with STT effect are being developed. MTJ cell structure as discussed above, consist of two FM electrodes. TMR ratio found by Julliere was 14% at temperature 4.2 K in Fe/Ge(-O)/Co. In 2001, Butler et al., [7] Mathon and Umerski [8] investigated that TMR can be obtained from 100 to 1000% on (001) oriented Fe/MgO/Fe MTJ based on first-principles calculations. Yuasa et al. proved it experimentally for (001) oriented epitaxial Fe/MgO/Fe MTJs made by molecular beam epitaxy [35, 52]. Parkin et al. also proved it for CoFe/MgO/CoFe MTJ made by a combination of molecular sputtering and ion beam [38]. The CoFe/MgO/CoFe/NiFe MTJs showed an MR ratio of about 62% at RT. Djayaprawira et al., [39] and Hayakawa et al., [53] showed a TMR ratio of more than 200% using in-plane MTJ for MgO barrier adopting the sputtering technique for $Co_{60}Fe_{20}B_{20}$ and $Co_{40}Fe_{40}B_{20}$ respectively. CoFeB/MgO MTJs are significant regarding mass production as they are deposited on thermally oxidized Si employing a conventional technique of sputtering and annealing to obtain higher TMR. Hayakawa et al., revealed a 260% TMR ratio for $Co_{40}Fe_{40}B_{20}$/MgO/$Co_{40}Fe_{40}B_{20}$ iMTJs [53]. Theoretical investigations have shown that TMR ratio up to 604% at RT observed for CeFeB/MgO MTJs [5, 54]. With the development of perpendicular MTJs, deposition of highly oriented bcc (001) CoFeB/MgO/CoFeB on previously used FM materials such as CoCrPt alloy or Co/Pt multilayer was not easy owing to different crystal structures of them such as fcc and hcp. TMR ratio obtained by Yoshikawa et al., [55] was higher than 100% for $L1_0$-ordered FePt/MgO/Fe/$L1_0$-ordered FePt. Ikeda et al. investigated that very thin film CoFeB with perpendicular anisotropy has a TMR ratio higher than 100% for CoFeB/MgO/CoFeB perpendicular MTJs. MTJs requirement for very large-scale integrated circuits (VLSI) are as follows:

- Very small feature size (F) that is in nm
- Small write current for STT (switching current I_C < FμA owing to transistor drivability)
- A higher TMR ratio is a requirement for quick sensing which is greater than 100%
- Recording layer needs higher thermal stability $\Delta = E_B/k_B T$ (> 60, here E_B is the energy barrier between P and AP arrangement, k_B represents Boltzmann constant and T is the absolute temperature.
- Ability to hold out against annealing at 350 to 410 °C without loss of higher TMR.

Expression for critical current (I_c) in in-plane MTJ and perpendicular-MTJ for coherent magnetization reversal is

$$\text{In-plane MTJ } I_c = J_c A = \alpha \frac{2e}{\hbar} \frac{M_s A t}{g(\theta)} \left(H_k + \frac{M_s}{\mu_0} \pm H_{ext} \right) \quad (18)$$

$$\text{Perpendicular-MTJ } I_c = J_c A = \alpha \frac{2e}{\hbar} \frac{M_s A t}{g(\theta)} \left(H_k - \frac{M_s}{\mu_0} \pm H_{ext} \right) = \alpha \frac{2e}{\hbar} \frac{M_s H_K^{eff}}{g(\theta)} V \quad (19)$$

Where J_c represents critical current density for switching, A represents area of junction, α represents magnetic damping constant, e denotes elementary charge, M_s designates saturation magnetization, t is the thickness of recording layer, reduced Planck's constant represented by $\hbar$, $g(\theta)$ is the relative angle of magnetization between reference and recording layer and also a function of spin polarization of tunneling current, H_K designates crystalline anisotropy, external field represented by H_{ext}, effective anisotropy field is represented by H_K^{eff}, the volume of the recording is denoted by V ($= At$) [8, 44, 56]. Perpendicular-MTJs obtain smaller I_c and large value of Δ as compared to in-plane MTJs, due to reduction in demagnetizing filed component and reduced area of junction ($A \propto kF^2$), however, Δ can be raised by utilizing higher magnetic anisotropy materials. It is the reason why STT-MRAM with perpendicular-MTJs is being developed. Many materials are explored to obtain perpendicular anisotropy as a recording layer, which includes transition metals/rare earth alloys [57, 58], and L1$_0$ – ordered (Co, Fe) – Pt and multilayers. Investigations also showed that 50 nm φ perpendicular-MTJs when Fe alloy doped by Pt and/or other elements followed up in a switching current of 10 μA and Δ of 32, while the TMR ratio demonstrated a smaller value of 23% [56].

It has been determined that traditional perpendicular materials do not meet up the requirement of VLSI applications. Afterward, it was thought to acquire a high TMR ratio a (100) CoFeB/MgO/CoFeB structure is required but adding that structure to electrode material which has perpendicular anisotropy and is contradictory with (001) structure causes destruction of (001) structure after annealing (thus lower TMR) or $M_s t$ value is much greater to observe switching. Ta/CoFeB/MgO with perpendicular magnetic easy axis stacked with an ultra-thin CoFeB layer was emerged in 2010, where interfacial anisotropy of CoFeB-MgO was accountable for the anisotropy. First-principles calculations investigated that interface magnetic anisotropy between Fe and O results in hybridization of Fe-3d and O-2p orbitals. Satisfactory characteristics of perpendicular-MTJ CoFeB/MgO were reported after initial investigations. So a higher TMR ratio can be obtained using a suitable material CoFeB/MgO in p-MTJ. A schematic description of perpendicular-MTJ CoFeB/MgO has shown in Figure 7.5(**b**). the junction diameter of CoFeB/MgO was a view observed by scanning electron microscope (SEM) and found 40 nm as shown in ref. [44]. The MTJ structures formed from the substrate side as Ta(5)/Ru(10)/Ta(5)/Co$_{20}$Fe$_{60}$B$_{20}$ (1.0)/MgO(0.85)/Co$_{20}$Fe$_{60}$B$_{20}$ (1.7)/Ta(5)/Ru(5) (numbers in parenthesis show nominal thicknesses in nm). TEM image of perpendicular-MTJ shown in ref. [5], when annealed at 300°C, which authenticated that the top and bottom layers of CoFeB and MgO have a continuous thickness in nm scale. Major and minor loops of out-of-plane R-H (resistance vs field) curves of perpendicular-MTJ CeFeB/MgO with 40 nm diameter when annealed at 300 °C are shown in ref. [44]. When a product of area and resistance (RA) is 18 Ωμm², the TMR ratio is 124%. Degradation in RA and TMR ratio was not noticed after annealing performed at 350 °C. A comparison of resistance vs current density (R-J) curves are demonstrated in ref. [44]. These are measured values at a current pulse duration of τ_P = 300 μs and in the absence of an externally applied magnetic field. Switching was noticed at a few current densities of mA/cm². For a single interface of CoFeB/MgO perpendicular-MTJ, Δ indicates constant value down to a critical junction size below which Δ tends to reduce. A constant value of Δ is owing to nucleation type reversal where a nucleation embryo having a size and length scale estimated by material parameters govern the magnetization reversal and set off highest achievable Δ of a stack. It was determined that Δ scales linearly by increasing thickness of recording layer [59]. But this happens only until size of junction is greater than critical size of nucleation embryo. Hence it is compulsory to increase the thickness of recording layer, increasing in-plane anisotropy while sustaining perpendicular easy axis. To acquire that, double interface of CoFeB/MgO structure assumed to rise interfacial anisotropy. It is found that double interfacial structure MgO/CoFeB/Ta/CoFeB/MgO in perpendicular-MTJs possessing synthetic FM reference layer, Δ increases by 60 at a dimension of 29 nm in diameter, while maintaining comparable intrinsic critical current density compared to the single interfacial recording structure CoFeB-MgO [37]. It is observed that the thickness of the CoFeB recording layer rises by reducing α. A fast speed of switching by STT, higher endurance, and low rate of write error was also observed in perpendicular-MTJ having a single interface CoFeB-MgO recording structure [60].

Theoretical investigations were made on new MTJ FeCo/$Mg_3B_2O_6$/FeCo using Kotoite ($Mg_3B_2O_6$) which may be a strong spin filtering candidate with decreased symmetry oxide. Usage of lower symmetry oxide region could lead to greater spin filtering and hopefully a more advanced device with higher TMR. Recently, high temperature annealing effect (350 °C–500 °C) on CoFeB/MgO/CoFeB MTJ investigated that was directly made on flexible polyimide substrate exhibiting good thermal tolerance. It was observed that with the increase of annealing temperature, TMR ratio raises up till ~200% at 450 °C annealing temperature. Such MTJ dependence on annealing temperature is alike to MTJ assembled on oxidized Silicon substrate. Scanning tunneling electron microscope image confirmed that CoFeB and MgO layers crystallization have improved which is considered a significant factor for increasing the TMR ratio [61]. Further investigation was made about the endurance of repeated stretching of flexible MTJ substrate. It was observed that the TMR ratio did not change during and after 1000 cycles of application of tensile strain greater than 1%. Hence flexible MTJ is a suitable candidate for future strain-sensing devices. This technology is significant for developing STT-MRAM. Recently a new spin-valve MTJ was fabricated and characterized utilizing *in situ* atomic layer deposition (ALD) technique of Al_2O_3 tunnel barrier with 0.55 nm thickness. Higher TMR values of ~77% and ~90% were achieved respectively at RT and at 100 K. These values are resembling to the best investigated values of MTJ using thermally oxidized AlO_x tunnel barrier.

STT-MRAM of 8 Mb, embedded in 28 nm logic platform having higher TMR ratio and retention time was reported as 10 years [62]. Chung et al., demonstrated 4 Gb STT-MRAM in 2016 having compact cell structure of 90 nm pitch by optimization of parasitic resistance. In this investigation, the improvement of write errors and read margin were also observed [63]. Recently, cryogenic properties of CoFeB/MgO perpendicular-MTJ were evaluated. Less error rate ($<10^{-4}$) was observed with switching temperature as low as 9 K and pulse duration 2 to 200 ns. Additionally, perpendicular-MTJs endurance was more than 10^{12} cycles with an amplitude of 0.85V and write pulses of 10 ns at 9 K. From 300 to 9 K, improvement in tolerance was observed by three orders of magnitude for the similar condition of stress. That work showed that CoFeB/MgO-based perpendicular-MTJ is practical for utilization in MRAM [64].

7 THERMAL EFFECTS ON MTJ

Critical current as shown in the above equation is the minimum current needed to switch FL. Moreover, switching was also observed for current lower than I_C. This random switching is caused by thermal effects which can be modeled as an effective magnetic field and are given as (eq. 20)

$$H_{Thermal} = \xi \sqrt{\frac{2\alpha k_B T}{|\gamma| M_s V \delta_t}} \tag{20}$$

Where ξ is a vector along with components and are independent Gaussian random variables, δ_t is the constant step of the time. There are two possible ways in which $H_{Thermal}$ can affect switching dynamics. Firstly, it has an effect because it is always presence. Secondly, it effects the initial relative angle present between m and m_p. The effect of each way is experienced differently. For precessional switching region (switching delay, $t_{SW} \leq 3$ ns), t_{SW} mostly depends on the initial angle between m and m_p, and θ (eq. 21)

$$t_{sw} \propto \frac{ln\left(\frac{\pi}{2\theta}\right)}{I-I_{C0}} \quad (21)$$

Where I_{C0} represents critical current switching delay which is usually of 1 ns. Thermal fluctuations in θ modeled as Boltzmann distribution eq. 22.

$$P(\theta) \propto \exp\left(\frac{E_B \cos^2\theta}{k_B T}\right) \quad (22)$$

In a precessional region of switching, the switching probability (P_{SW}) (eq. 23) of MTJ is estimated by using equations 21 and 22 as

$$P_{SW} \propto \exp\left(\frac{\Delta}{2}(1-\cos^2\phi)\right)(I-I_{C0})\sin^2\phi \quad (23)$$

Here $\phi = \left(\frac{\pi}{2}\right)\exp(-2\eta\mu_B/eM_sV)(I-I_{C0})t_{SW})$, μ_B is Bohr magneton

In the thermal activation region, $t_{SW} \geq 10$ ns, so P_{SW} is given as (eq. 24)

$$P_{SW} = 1 - \exp\left(-\frac{t}{\tau}\exp\left(-\Delta\left(1-\frac{I}{I_{C0}}\right)\right)\right) \quad (24)$$

Joule heating causes a decrease Δ when a current passes through MTJ. Thus, thermally assisted MRAM which utilizes the write current to heat the MTJ and operate STT-MRAM in the thermal activation region as suggested. However, such thermally assisted MRAM may not be considered suitable for memory applications that need write latency smaller than ~10 ns.

8 ADVANCES AND PROSPECTS OF STT-MRAM

Requirement of STT-MRAM to put back the higher density DRAM, the lateral size of MTJ bit must be smaller than ~ 15 nm. For sustaining the thermal stability of FL magnetization in these low-scale bits, the PMA energy of FL must be greater than ~ 2 MJ/m². A further requirement is of *MR* ratio which should be >> 300%. Moreover, the small Gilbert damping constant α ~ 0.001 and lower saturation magnetization is beneficial for the reduction of STT switching current density. For that purpose, CoFeB/MgO MTJ with in-plane magnetic anisotropy is not beneficial for ultrahigh-density STT-MRAM which possesses 10–15-nm MTJ diameter. Thus, it is necessary to develop perpendicular-MTJs with novel FM and TB materials. Further experimental attempts were made on FM materials called Heusler alloys (HAs) to enhance the *MR* ratio. Co_2YZ based alloys with $L2_1$ order and Mn-based alloys with $D0_{22}$ order which have shown 100% SP at the E_F. Another significant property of half-metallic (HM) HAs is that they exhibit a small Gilbert damping constant. Sakuraba et al. made attempt on Co_2MnSi based HAs and obtained an *MR* ratio of 570% at 2K temperature and TB used was amorphous AlO [65]. This giant *MR* ratio indicated that Co_2MnSi is HM even at low temperatures. However, the *MR* ratio was reduced to 67% at RT. After the discovery of MgO-based TB [35] HAs combined to MgO and *MR* ratio obtained was 2610% at 4.2 K and 429% at RT using off-stoichiometric $Co_2(Mn, Fe)Si$. HAs to exhibit a higher *MR* ratio at low temperatures as compared to CoFeB electrodes. However, the *MR* ratio at RT is still higher for the CoFeB electrode which is up to 600%. Suppression of thermal spin fluctuations at interface of electrode/tunnel barrier is a key to use HAs electrodes in practical applications. When the bit size of MTJ approaches to 10 nm, there is strong influence on magnetic and magneto-transport characteristics of individual MTJ bit due to the presence of grain and also the grain boundaries in polycrystalline MTJs which bring a substantial bit to bit variations in characteristics. This problem can be bypassed by using fully epitaxial MTJs using single-crystal barriers and by growing electrode layers on Si (001) substrates. Integration of such fully epitaxial MTJs is done on extensive scale integrated circuit chips by employing three-dimensional integration technology. FM materials possess a high damping factor when the thickness is reduced. However, a record low value of 0.0015 ± 0.0001 has been obtained in 53 nm thick Co_2FeAl with an optimal substrate temperature of 300 °C [66]. Such a low value inspired the researchers to use HAs in magnetic RAMs applications. Besides writing improvements, the use of HAs demonstrated reading enhancement as well owing to the greater magnetic moment, high SP, and high T_c [67]. Reported TMR ratio for $Co_2FeAl_{0.5}Si_{0.5}$ HA at RT was 386% [68]. Further, researchers predicted TMR even 1000% in the following 10 years [11].

9 STT-MRAM COMPARISON OVER CONVENTIONAL MEMORY DEVICES

Conventional memory devices use electron charges to store information. However, in magnetic RAMs information is stored in MTJ. SRAM has a high read and write speed but its volatile nature and large size make it expensive for embedded applications. On the other hand, DRAM requires constant refreshing and consumes more power. STT-MRAM is beneficial over conventional devices owing to their smaller cell size and low write current. STT-MRAM only conveys that current to a cell that is needed for writing. Writing current increases proportionally with cell size. Hence it is supposed that the STT-MRAM cell cloud should be lower than 20nm in the future. Cell size could be according to thermal stability requirements. In STT devices no current is required only the spin of an electron is changed. Therefore, these devices possess unlimited durability.

STT-MRAM cells go through write/erase cycles of more than 10^{15}. However, an extremum limit has not been found yet. The cell structure of STT-MRAM is alike to SRAM and DRAM but it occupies a smaller chip area. STT-MRAM could be set out as L2 cache substitution makes CPUs have greater memory and speed at no cost. Table 7.1 represents a comparison between STT-MRAM and other memory devices. Little voltage is required for read and write in STT-MRAM which reduces power consumption up to 75%. Usage of STT-MRAM in mobile phone technologies would reduce standby power and increase battery life. Recently, 8 Mb STT-MRAM was reported with a higher TMR ratio and embedded in a 28 nm

TABLE 7.1 Comparison of STT-MRAM with conventional memory technologies. Adapted from [6] copyright (2010) Proceedings of the IEEE.

	SRAM	DRAM	FLASH (NOR)	FLASH (NAND)	MRAM	STT-RAM
Non-volatile	No	No	Yes	Yes	Yes	Yes
Cell size ()	50–120	6–10	10	5	16–40	6–20
Read time (ns)	1–100	30	10	50	3–20	2–20
Write/erase time (ns)	1–100	50/50	1 μs/10 ms	1 ms/0.1 ms	3–20	2–20
Endurance	10^{16}	10^{16}	10^5	10^5	>10^5	>10^5
Write power	No	Low	Very high	Very high	High	Low
Other power consumption	Current leakage	Refresh current	None	None	None	None
High voltage required	No	2 V	6–8 V	16–20 V	3 V	< 1.5 V
	Existing products					Prototype

logic platform with a retention time of 10 years [62]. Another demonstration made by Chung et al., in 2016, a 4 GB STT-MRAM with compact cell structure with 90 nm pitch displayed by optimizing parasitic resistance. In such demonstrations, many factors were improved such as write rate and read margin. Owing to lower power consumption, non-volatility and more endurance of devices and easy fabrication process it has revolutionized the technology [6].

10 CURRENT CHALLENGES TO STT-MRAM

10.1 Writing Improvements

The critical current required for switching of FL magnetization is dependent on damping factor α. Thus, it is very crucial to search those materials which possess smaller damping factors to minimize the current needed for writing. Materials, for example, FePt or Co/Pd, were investigated for perpendicular-MTJ applications but they exhibited a higher damping constant which is not suitable. Conversely, interfacial perpendicular-MTJs based on CoFeB are important for STT-MRAM due to a low damping factor and lower writing current. Further, orthogonal spin transfer spin-valve devices have achieved a low writing error rate within a very limited pulse condition at 4 K [69]. Another way to increase the efficiency of STT is to find materials that have high SP. CoFeB is with reasonable SP used at the MTJ interface. HAs with much higher SP are also being searched for future candidates. HAs with high SP and high T_c make them remarkable for spintronic devices such as STT-MRAM and magnonic devices.

10.2 Variation of MTJ Resistance

Resistance of MTJ exponentially depends on the thickness of the insulator layer present between two FM layers. Investigations have found that a small difference in thickness of oxide layer between 14 Å to 14.1 Å increases 8% resistance. Moreover, magnetoresistance is dependent on the anisotropy of the cell. Hence small variations in geometry will affect the performance. This variation increases more as the oxide layer thins owing to the reduction of MTJ cells. It is the key parameter that limits the adaptation of STT-MRAM.

10.3 Thermal Stability

The logic value of STT-MRAM designates the direction of magnetization in FL. Sufficiently stable electronic spin can withstand thermal perturbation. Hence, low temperature is the requirement such that electronic spin in FL does not change. The electronic spin should be stable to bear thermal agitation. Thermally induced switching variation occurs randomly in MTJ and cannot be deterministically repeated. This major factor causes errors in STT-MRAM operation. The solution to this problem is to raise the working temperature of MTJ by introducing error elimination circuits to the design. Research is under focus to find alternatives that can optimize switching time and error rate in STT-MRAM. Moreover, errors occur due to the switching of magnetization direction in FL. Hence the error rate is reduced by flip optimization in the STT-MRAM device. Additionally, switching from "1" to "0" requires less switching current. The error rate is higher when switching is from "0" to "1" but a more reliable write rate for "1" to "0" switching [70]. Moreover, Sato et al. predicted that thermal stability rises in double interfacial structures by a factor of 1.9 from the highest value of perpendicular-MTJs with single CoFeB-MgO interfacial structure [71].

10.4 Demagnetizing Fields

FM materials film used in other contexts have magnetization in film planes to keep away surface magnetic poles. However, in MRAM and STT-MRAM magnetization direction perpendicular to the film plane is desirable due to decoupling of MTJ resistance from anisotropic restrictions. This perpendicular magnetization produces a demagnetizing field owing to surface magnetic poles. It is necessary to divert this demagnetizing field before direction switching. As a result, write current increases desirably. Current research investigated that demagnetizing fields have reduced due to perpendicular-MTJ design in STT-MRAM as in Figure 7.5(c) [41]. A perpendicular-MTJ is a bilayer coupled with a thin layer of maximum SP and a thick layer of large PMA. An effective magnetic field produced by these crystal structures that force magnetization along certain axes and cancel out a demagnetizing field. It has been found that the orientation of perpendicular domains in materials behaves independently of shape anisotropy, and it simplifies the process of manufacturing as it allows devices to fabricate in circular disks.

11 EMERGING MEMORY: BASED ON SPIN HALL EFFECT

Owing to some limitations of STT-MRAM, substitute memories based on the spin are also under investigation. An alternative approach is the MRAM based on the spin Hall effect (SHE) that is supposed to consume less power as compared to STT-MRAM. In the SHE phenomenon, spin-orbit coupling of electrons occurs and electrons with different spin deflect in separate paths, as a result, a refined spin current is produced. The first metallic material studied for SHE was Pt because it demonstrated the considerable value of such effect although at RT. The Pt possesses higher spin Hall conductivity at RT owing to higher spin-orbit coupling. In 2012, the SHE phenomenon for MRAM using β-Tantalum layer was investigated by Liu et al. [12]. An electric current passed through a thin layer of Ta that employs spin-torque switching owing to SHE phenomenon in adjoining FM layer CoFeB for in-plane and perpendicular magnetization at RT. Owing to the greater spin Hall angle of Ta, it was preferred over Pt. Higher ratio of spin current density to charge current density obtained which rises due to the SHE phenomenon. On the other hand, Ta does not produce damping in the FM layer, unlike Pt. The principle behind this is that in the Ta layer oscillating current produces spin current oscillations that apply spin-torque oscillations on magnetic moments of the FM layer. J. Kim et al. comparatively studied the performance of in-plane SHE MRAM and interface perpendicular STT-MRAM based cache memory utilizing identical material parameters and device dimensions with CoFeB as FL.

Simulations-based investigation showed that SHE MRAM exhibits better results such as write delay, threshold spin current, and failure rate (%) of retention regarding thermal stability. Y. Kim et al. suggested that SHE-based STT-bit cell that is beneficial for enhanced performance of on-chip memory, exhibiting 10 times less requirement of energy needed for writing and possess 1.6 times quick read time, as compared to the 1T1R (one transistor and one resistor) in-plane STT-MRAM. Apart from the lesser power utilization and enhanced switching speed, the usage of SOT-based three-terminal MTJ is also advantageous for separating the write and the read path. Reading reliability also improves by using SOT-MRAM. Moreover, to overcome the limitations of STT-MRAM, an extra terminal was introduced to SOT-MRAM that separates the read and write path these paths are at right-angle with respect to each. These terminals consist of a source line, bit line, a word line, and a write line. Access to the required cell during a read operation is obtained by using a word line. Current flow occurs between a source line and the write line during a write operation. FL magnetization affects the direction of the current and hence, the stored value in bit-cell. The MTJ resistance would be minimum for current flow from the source line towards the write line. For higher resistance of MTJ, the current is required to flow from the write line towards the source line [72]. Gambardella et al., and Miron et al., represented that current-induced magnetization occurs due to the Rashba effect [73] however, Liu et al. [34] accredited SHE for it. Nonetheless, disregarding the actual effect, the SOT is accountable for FL magnetization switching and consequently the name SOT-MRAM. Its advantages reflect that it could be a distinguished memory for future use. Though, still much exploration regarding SOT-MRAM is required to employ it practically.

CONCLUSIONS

In this chapter, the discussion was on materials exhibiting the STT effect and having applications in STT devices. From the above discussion, it is established that STT-based memory devices are beneficial over conventional memory devices. Current memory devices require better data retention, readability, writability, scaling, and easy fabrication. The magnetoresistance effect that enables reading of MRAM and writing methods such as writing based on magnetic field and writing based on STT effect also described. Different materials as building blocks of MTJs are elaborated along with their merits and demerits. Further, in-plane MTJs and perpendicular-MTJs are discussed. It was presented that CoFeB/MgO-based perpendicular-MTJs are significant for STT memory devices over in-plane MTJs as they require low write current, low damping constant and are thermally stable. From the literature review, it is investigated that CoFeB/MgO/CoFeB based MTJ is promising for STT-MRAM devices. However, HAs may also be used for future work due to high SP and high T_c in such FM materials. Additionally, SOT-based memory devices are also highlighted but many challenges are faced y them practically. Still, there is much to make investigations on them for future device

REFERENCES

1. P. Levin, *Periodic structure of spin-transfer current in ferromagnetic multilayers*. Physics Letters A, 2007. **360**(3): p. 467–471.
2. M. Tsoi, A. G. M. Jansen, J. Bass, W.-C. Chiang, M. Seck, V. Tsoi, and P. Wyder, *Excitation of a magnetic multilayer by an electric current*. Physical Review Letters, 1998. **80**(19): p. 4281.
3. E. B. Myersd, C. Ralphj, A. Katiner, N. Louieand, and R. A. Buhrman, *Current-induced switching of domains in magnetic multilayer devices*. Science, 1999. **285**(5429): p. 867–870.
4. S. M. Thompson, *The discovery, development and future of GMR: The Nobel Prize 2007*. Journal of Physics D: Applied Physics, 2008. **41**(9): p. 093001.
5. S. Ikeda, J. Hayakawa, Y. Ashizawa, Y. M. Lee, K. Miura, H. Hasegawa, M. Tsunoda, F. Matsukura, and H. Ohno, *Tunnel magnetoresistance of 604% at 300 K by suppression of Ta diffusion in Co Fe B/ Mg O/ Co Fe B pseudo-spin-valves annealed at high temperature*. Applied Physics Letters, 2008. **93**(8): p. 082508.
6. Stuart A. Wolf, Jiwei Lu, Mircea R. Stan, Eugene Chen, and Daryl M. Treger, *The promise of nanomagnetics and spintronics for future logic and universal memory*. Proceedings of the IEEE, 2010. **98**(12): p. 2155–2168.
7. W. H. Butler, X.-G. Zhang, T. C. Schulthess, and J. M. MacLaren, *Spin-dependent tunneling conductance of Fe| MgO| Fe sandwiches*. Physical Review B, 2001. **63**(5): p. 054416.
8. J. Mathon, and A. Umerski, *Theory of tunneling magnetoresistance of an epitaxial Fe/MgO/Fe (001) junction*. Physical Review B, 2001. **63**(22): p. 220403.
9. X.-G. Zhang, and W. Butler, *Large magnetoresistance in bcc Co/ Mg O/ Co and Fe Co/ Mg O/ Fe Co tunnel junctions*. Physical Review B, 2004. **70**(17): p. 172407.
10. J. P. Velev, K. D. Belashchenko, D. A. Stewart, M. van Schilfgaarde, S. S. Jaswal, and E. Y. Tsymbal, *Negative spin polarization and large tunneling magnetoresistance in epitaxial Co| SrTiO 3| Co magnetic tunnel junctions*. Physical Review Letters, 2005. **95**(21): p. 216601.
11. Atsufumi Hirohata, Hiroaki Sukegawa, Hideto Yanagihara, Igor Žutić, Takeshi Seki, Shigemi Mizukami, and Raja Swaminathan, *Roadmap for emerging materials for spintronic device applications*. IEEE Transactions on Magnetics, 2015. **51**(10): p. 1–11.

12. Luqiao Liu, Chi-Feng Pai, Y. Li, H. W. Tseng, D. C. Ralph, and R. A. Buhrman, *Spin-torque switching with the giant spin Hall effect of tantalum*. Science, 2012. **336**(6081): p. 555–558.
13. Minh-Hai Nguyen, Guilhem J. Ribeill, Martin V. Gustafsson, Shengjie Shi, Sriharsha V. Aradhya, Andrew P. Wagner, Leonardo M. Ranzani, Lijun Zhu, Reza Baghdadi, Brenden Butters, Emily Toomey, Marco Colangelo, Patrick A. Truitt, Amir Jafari-Salim, David McAllister, Daniel Yohannes, Sean R. Cheng, Rich Lazarus, Oleg Mukhanov, Karl K. Berggren, Robert A. Buhrman, Graham E. Rowlands, and Thomas A. Ohki, *Cryogenic memory architecture integrating spin hall effect based magnetic memory and superconductive cryotron devices*. Scientific Reports, 2020. **10**(1): p. 1–11.
14. F. C. Williams, and T. Kilburn, *Electronic digital computers*. Nature, 1948. **162**(4117): p. 487–487.
15. M. Julliere, *Tunneling between ferromagnetic films*. Physics Letters A, 1975. **54**(3): p. 225–226.
16. Yiming Huai, Frank Albert, Paul Nguyen, Mahendra Pakala, and Thierry Valet, *Observation of spin-transfer switching in deep submicron-sized and low-resistance magnetic tunnel junctions*. Applied Physics Letters, 2004. **84**(16): p. 3118–3120.
17. P. Nobel, *Nobel Prize in Physics-2005*, 2005.
18. M. N. Baibich, J. M. Broto, A. Fert, F. Nguyen Van Dau, F. Petroff, P. Etienne, G. Creuzet, A. Friederich, and J. Chazelas, *Giant magnetoresistance of (001) Fe/(001) Cr magnetic superlattices*. Physical Review Letters, 1988. **61**(21): p. 2472.
19. G. Autès, J. Mathon, and A. Umerski, *Strong enhancement of the tunneling magnetoresistance by electron filtering in an Fe/MgO/Fe/GaAs (001) junction*. Physical Review Letters, 2010. **104**(21): p. 217202.
20. T. Miyazaki, T. Yaoi, and S. Ishio, *Large magnetoresistance effect in 82Ni-Fe/Al-Al2O3/Co magnetic tunneling junction*. Journal of Magnetism and Magnetic Materials, 1991. **98**(1–2): p. L7–L9.
21. W. P. Pratt, Jr., S.-F. Lee, J. M. Slaughter, R. Loloee, P. A. Schroeder, and J. Bass, *Perpendicular giant magnetoresistances of Ag/Co multilayers*. Physical Review Letters, 1991. **66**(23): p. 3060.
22. S. Parkin, N. More, and K. Roche, *Oscillations in exchange coupling and magnetoresistance in metallic superlattice structures: Co/Ru, Co/Cr, and Fe/Cr*. Physical Review Letters, 1990. **64**(19): p. 2304.
23. J. C. Slonczewski, *Conductance and exchange coupling of two ferromagnets separated by a tunneling barrier*. Physical Review B, 1989. **39**(10): p. 6995.
24. J. Slonczewski, *Excitation of spin waves by an electric current*. Journal of Magnetism and Magnetic Materials, 1999. **195**(2): p. L261–L268.
25. L. Berger, *Effect of interfaces on Gilbert damping and ferromagnetic resonance linewidth in magnetic multilayers*. Journal of Applied Physics, 2001. **90**(9): p. 4632–4638.
26. Shinji Yuasa, Kazuhiro Hono, Guohan Hu, and Daniel C. Worledge, *Materials for spin-transfer-torque magnetoresistive random-access memory*. MRS Bulletin, 2018. **43**(5): p. 352–357.
27. X. Han, S. S. Ali, and S. Liang, *MgO (001) barrier based magnetic tunnel junctions and their device applications*. Science China Physics, Mechanics and Astronomy, 2013. **56**(1): p. 29–60.
28. J. D. R. Buchanan, T. P. A. Hase, and B. K. Tanner, *Determination of the thickness of Al2 O3 barriers in magnetic tunnel junctions*. Applied physics letters, 2002. **81**(4): p. 751–753.
29. Zhihuan Yang, Qingfeng Zhan, Xiaojian Zhu, Yiwei Liu, Huali Yang, Benlin Hu, Jie Shang, Liang Pan, Bin Chen, and Run-Wei Li, *Tunneling magnetoresistance induced by controllable formation of Co filaments in resistive switching Co/ZnO/Fe structures*. EPL (Europhysics Letters), 2014. **108**(5): p. 58004.
30. D. A. Stewart, *New type of magnetic tunnel junction based on spin filtering through a reduced symmetry oxide: FeCo| Mg3B2O6| FeCo*. Nano letters, 2010. **10**(1): p. 263–267.
31. Nuala M. Caffrey, Thomas Archer, Ivan Rungger, and Stefano Sanvito, *Prediction of large bias-dependent magnetoresistance in all-oxide magnetic tunnel junctions with a ferroelectric barrier*. Physical Review B, 2011. **83**(12): p. 125409.
32. A. V. Ramos, M.-J. Guittet, and J.-B. Moussy, *Room temperature spin filtering in epitaxial cobalt-ferrite tunnel barriers*. Applied Physics Letters, 2007. **91**(12): p. 122107.
33. J. S. Moodera, Lisa R. Kinder, Terrilyn M. Wong, and R. Meservey, *Large magnetoresistance at room temperature in ferromagnetic thin film tunnel junctions*. Physical Review Letters, 1995. **74**(16): p. 3273.
34. T. Miyazaki, and N. Tezuka, *Giant magnetic tunneling effect in Fe/Al2O3/Fe junction*. Journal of Magnetism and Magnetic Materials, 1995. **139**(3): p. L231–L234.
35. Shinji Yuasa, Taro Nagahama, Akio Fukushima, Yoshishige Suzuki, and Koji Ando, *Giant room-temperature magnetoresistance in single-crystal Fe/MgO/Fe magnetic tunnel junctions*. Nature Materials, 2004. **3**(12): p. 868–871.
36. Renu W. Dave, G. Steiner, J. M. Slaughter, J. J. Sun, B. Craigo, S. Pietambaram, K. Smith, G. Grynkewich, M. DeHerrera, J. Åkerman, and S. Tehrani, *MgO-based tunnel junction material for high-speed toggle magnetic random access memory*. IEEE Transactions on Magnetics, 2006. **42**(8): p. 1935–1939.

37. Hideo Sato, Michihiko Yamanouchi, Shoji Ikeda, Shunsuke Fukami, Fumihiro Matsukura, and Hideo Ohno, *MgO/CoFeB/Ta/CoFeB/MgO recording structure in magnetic tunnel junctions with perpendicular easy axis.* IEEE Transactions on Magnetics, 2013. **49**(7): p. 4437–4440.
38. S. S. Parkin, et al., *Giant tunnelling magnetoresistance at room temperature with MgO (100) tunnel barriers.* Nature Materials, 2004. **3**(12): p. 862–867.
39. David D. Djayaprawira, Koji Tsunekawa, Motonobu Nagai, Hiroki Maehara, Shinji Yamagata, and Naoki Watanabe, *230% room-temperature magnetoresistance in CoFeB/ MgO/ CoFeB magnetic tunnel junctions.* Applied Physics Letters, 2005. **86**(9): p. 092502.
40. S. Yuasa, and D. Djayaprawira, *Giant tunnel magnetoresistance in magnetic tunnel junctions with a crystalline MgO (0 0 1) barrier.* Journal of Physics D: Applied Physics, 2007. **40**(21): p. R337.
41. T. Kawahara, K. Ito, R. Takemura, and H. Ohno, *Spin-transfer torque RAM technology: Review and prospect.* Microelectronics Reliability, 2012. **52**(4): p. 613–627.
42. T. Kishi, H. Yoda, T. Kai, T. Nagase, E. Kitagawa, M. Yoshikawa, K. Nishiyama, T. Daibou, M. Nagamine, M. Amano, S. Takahashi, M. Nakayama, N. Shimomura, H. Aikawa, S. Ikegawa, S. Yuasa, K. Yakushiji, H. Kubota, A. Fukushima, M. Oogane, T. Miyazaki, and K. Ando, *Lower-current and fast switching of a perpendicular TMR for high speed and high density spin-transfer-torque MRAM.* In 2008 IEEE International Electron Devices Meeting, 2008. IEEE.
43. D. C. Worledge, G. Hu, P. L. Trouilloud, D. W. Abraham, S. Brown, M. C. Gaidis, J. Nowak, E. J. O'Sullivan, R. P. Robertazzi, J. Z. Sun, and W. J. Gallagher, *Switching distributions and write reliability of perpendicular spin torque MRAM.* In 2010 International Electron Devices Meeting. 2010. IEEE.
44. S. Ikeda, K. Miura, H. Yamamoto, K. Mizunuma, H. D. Gan, M. Endo, S. Kanai, J. Hayakawa, F. Matsukura, and H. Ohno, *A perpendicular-anisotropy CoFeB – MgO magnetic tunnel junction.* Nature materials, 2010. **9**(9): p. 721–724.
45. D. C. Worledge, G. Hu, David W. Abraham, J. Z. Sun, P. L. Trouilloud, J. Nowak, S. Brown, M. C. Gaidis, E. J. O'Sullivan, and R. P. Robertazzi, *Spin torque switching of perpendicular Ta| CoFeB| MgO-based magnetic tunnel junctions.* Applied Physics Letters, 2011. **98**(2): p. 022501.
46. Makoto Konoto, Hiroshi Imamura, Tomohiro Taniguchi, Kay Yakushiji, Hitoshi Kubota, Akio Fukushima, Koji Ando, and Shinji Yuasa, *Effect of MgO cap layer on Gilbert damping of FeB electrode layer in MgO-based magnetic tunnel junctions.* Applied Physics Express, 2013. **6**(7): p. 073002.
47. Woojin Kim, J. H. Jeong, Y. Kim, W. C. Lim, J. H. Kim, J. H. Park, H. J. Shin, Y. S. Park, K. S. Kim, S. H. Park, Y. J. Lee, K. W. Kim, H. J. Kwon, H. L. Park, H. S. Ahn, S. C. Oh, J. E. Lee, S. O. Park, S. Choi, H. K. Kang, and C. Chung, *Extended scalability of perpendicular STT-MRAM towards sub-20nm MTJ node.* In 2011 International Electron Devices Meeting. 2011. IEEE.
48. J. J. Nowak, R. P. Robertazzi, J. Z. Sun, G. Hu, J. H. Park, J. H. Lee, A. J Annunziata, G. P. Lauer, C. Kothandaraman, E. J. O'Sullivan, P. L. Trouilloud, Y. Kim, and D. C. Worledge, *Dependence of voltage and size on write error rates in spin-transfer torque magnetic random-access memory.* IEEE Magnetics Letters, 2016. **7**: p. 1–4.
49. R. Sbiaa, H. Meng, and S. Piramanayagam, *Materials with perpendicular magnetic anisotropy for magnetic random access memory.* Physica Status Solidi (RRL) – Rapid Research Letters, 2011. **5**(12): p. 413–419.
50. Xuanyao Fong, Yusung Kim, Rangharajan Venkatesan, Sri Harsha Choday, Anand Raghunathan, and Kaushik Roy, *Spin-transfer torque memories: Devices, circuits, and systems.* Proceedings of the IEEE, 2016. **104**(7): p. 1449–1488.
51. B. Behin-Aein, S. Salahuddin, and S. Datta, *Switching energy of ferromagnetic logic bits.* IEEE Transactions on Nanotechnology, 2009. **8**(4): p. 505–514.
52. Shinji Yuasa, Akio Fukushima, Taro Nagahama, Koji Ando, and Yoshishige Suzuki, *High tunnel magnetoresistance at room temperature in fully epitaxial Fe/MgO/Fe tunnel junctions due to coherent spin-polarized tunneling.* Japanese Journal of Applied Physics, 2004. **43**(4B): p. L588.
53. Jun Hayakawa, Shoji Ikeda, Fumihiro Matsukura, Hiromasa Takahashi, and Hideo Ohno, *Dependence of giant tunnel magnetoresistance of sputtered CoFeB/MgO/CoFeB magnetic tunnel junctions on MgO barrier thickness and annealing temperature.* Japanese Journal of Applied Physics, 2005. **44**(4L): p. L587.
54. J. Hayakawa, S. Ikeda, Y. M. Lee, F. Matsukura, and H. Ohno, *Effect of high annealing temperature on giant tunnel magnetoresistance ratio of CoFeB/ MgO/ CoFeB magnetic tunnel junctions.* Applied Physics Letters, 2006. **89**(23): p. 232510.
55. Masatoshi Yoshikawa, Eiji Kitagawa, Toshihiko Nagase, Tadaomi Daibou, Makoto Nagamine, Katsuya Nishiyama, Tatsuya Kishi, and Hiroaki Yod, *Tunnel magnetoresistance over 100% in MgO-based magnetic tunnel junction films with perpendicular magnetic L1 $ _ {0} $-FePt electrodes.* IEEE Transactions on Magnetics, 2008. **44**(11): p. 2573–2576.

56. Hiroaki Yoda, Tatsuya Kishi, Toshihiko Nagase, Masatoshi Yoshikawa, Katsuya Nishiyama, Eiji Kitagawa, Tadaomi Daibou, Minoru Amano, Naoharu Shimomura, Shigeki Takahashi, Tadashi Kai, Masahiko Nakayama, Hisanori Aikawa, Sumio Ikegawa, Makoto Nagamine, Junichi Ozeki, Shigemi Mizukami, Mikihiko Oogane, Yasuo Ando, Shinji Yuasa, Kei Yakushiji, Hitoshi Kubota, Yoshishige Suzuki, Yoshinobu Nakatani, Terunobu Miyazaki, and Koji Ando, *High efficient spin transfer torque writing on perpendicular magnetic tunnel junctions for high density MRAMs*. Current Applied Physics, 2010. **10**(1): p. e87–e89.

57. A. Canizo Cabrera, Che-Hao Chang, Chih-Cheng Hsu, Ming-Chi Weng, C. C Chen, C. T. Chao, J. C. Wu, Yang-Hua Chang, and Te-Ho Wu, *Perpendicular magnetic tunneling junction with double barrier layers for MRAM application*. IEEE Transactions on Magnetics, 2007. **43**(2): p. 914–916.

58. H. Ohmori, T. Hatori, and S. Nakagawa, *Perpendicular magnetic tunnel junction with tunneling magnetoresistance ratio of 64% using MgO (100) barrier layer prepared at room temperature*. Journal of Applied Physics, 2008. **103**(7): p. 07A911.

59. H. Sato, M. Yamanouchi, K. Miura, S. Ikeda, R. Koizumi, F. Matsukura, and H. Ohno, *CoFeB thickness dependence of thermal stability factor in CoFeB/MgO perpendicular magnetic tunnel junctions*. IEEE Magnetics Letters, 2012. **3**: p. 3000204.

60. C. Yoshida, and T. Sugii, *Reliability study of magnetic tunnel junction with naturally oxidized MgO barrier*. In 2012 IEEE International Reliability Physics Symposium (IRPS). 2012. IEEE.

61. S. Ota, A. Ando, T. Sekitani, T. Koyama, and D. Chiba, *Flexible CoFeB/MgO-based magnetic tunnel junctions annealed at high temperature (≥ 350° C)*. Applied Physics Letters, 2019. **115**(20): p. 202401.

62. Y. J. Song, J. H. Lee, H. C. Shin, K. H. Lee, K. Suh, J. R. Kang, S. S. Pyo, H. T. Jung, S. H. Hwang, G. H. Koh, S. C. Oh, S. O. Park, J. K. Kim, J. C. Park, J. Kim, K. H. Hwang, G. T. Jeong, K. P. Lee, and E. S. Jung, *Highly functional and reliable 8Mb STT-MRAM embedded in 28nm logic*. In 2016 IEEE International Electron Devices Meeting (IEDM). 2016. IEEE.

63. S.-W. Chung, T. Kishi, J. W. Park, M. Yoshikawa, K. S. Park, T. Nagase, K. Sunouchi, H. Kanaya, G. C. Kim, K. Noma, M. S. Lee, A. Yamamoto, K. M. Rho, K. Tsuchida, S. J. Chung, J. Y. Yi, H. S. Kim, Y. S. Chun, H. Oyamatsu, and S. J. Hong, *4Gbit density STT-MRAM using perpendicular MTJ realized with compact cell structure*. In 2016 IEEE International Electron Devices Meeting (IEDM). 2016. IEEE.

64. Lili Lang, Yujie Jiang, Fei Lu, Cailu Wang, Yizhang Chen, Andrew D. Kent, and Li Ye, *A low temperature functioning CoFeB/MgO-based perpendicular magnetic tunnel junction for cryogenic nonvolatile random access memory*. Applied Physics Letters, 2020. **116**(2): p. 022409.

65. Y. Sakuraba, M. Hattori, M. Oogane, Y. Ando, H. Kato, A. Sakuma, and T. Miyazaki, *Giant tunneling magnetoresistance in Co2Mn Si/ Al – O/ Co2Mn Si magnetic tunnel junctions*. Applied Physics Letters, 2006. **88**(19): p. 192508.

66. Sajid Husain, Serkan Akansel, Ankit Kumar, Peter Svedlindh, and Sujeet Chaudhary, *Growth of Co2 FeAl Heusler alloy thin films on Si (100) having very small Gilbert damping by Ion beam sputtering*. Scientific reports, 2016. **6**(1): p. 1–11.

67. I. Galanakis, and P. H. Dederichs, *Half-metallicity and Slater-Pauling behavior in the ferromagnetic Heusler alloys*. In Half-metallic Alloys. 2005, Springer: p. 1–39.

68. N. Tezuka, N. Ikeda, F. Mitsuhashi, and S. Sugimoto, *Improved tunnel magnetoresistance of magnetic tunnel junctions with Heusler Co2FeAl 0.5 Si 0.5 electrodes fabricated by molecular beam epitaxy*. Applied Physics Letters, 2009. **94**(16): p. 162504.

69. G. E. Rowlands, C. A. Ryan, L. Ye, L. Rehm, D. Pinna, A. D. Kent, and T. A. Ohki, *A cryogenic spin-torque memory element with precessional magnetization dynamics*. Scientific Reports, 2019. **9**(1): p. 1–7.

70. R. Maddah, S. M. Seyedzadeh, and R. Melhem. *CAFO: Cost aware flip optimization for asymmetric memories*. In 2015 IEEE 21st International Symposium on High Performance Computer Architecture (HPCA). 2015. IEEE.

71. H. Sato, M. Yamanouchi, S. Ikeda, S. Fukami, F. Matsukura, and H. Ohno, *Perpendicular-anisotropy CoFeB-MgO magnetic tunnel junctions with a MgO/CoFeB/Ta/CoFeB/MgO recording structure*. Applied Physics Letters, 2012. **101**(2): p. 022414.

72. Rajendra Bishnoi, Mojtaba Ebrahimi, Fabian Oboril, and Mehdi B. Tahoori, *Architectural aspects in design and analysis of SOT-based memories*. In 2014 19th Asia and South Pacific Design Automation Conference (ASP-DAC). 2014. IEEE.

73. Ioan Mihai Miron, Kevin Garello, Gilles Gaudin, Pierre-Jean Zermatten, Marius V. Costache, Stéphane Auffret, Sébastien Bandiera, Bernard Rodmacq, Alain Schuhl, and Pietro Gambardella, *Perpendicular switching of a single ferromagnetic layer induced by in-plane current injection*. Nature, 2011. **476**(7359): p. 189–193.

Nanosensors Based on Magnetic Materials

8

Kumar Navin, Rajnish Kurchania

Contents

1 Introduction 115
2 Theoretical Background and Physics 118
   2.1 Hall Effect 118
   2.2 Magnetoresistance 118
       2.2.1 Anisotropic Magnetoresistance 118
       2.2.2 Giant Magnetoresistance 119
       2.2.3 Tunneling Magnetoresistance 119
   2.3 Nuclear Magnetic Resonance (NMR) 120
   2.4 Magneto-Optic Kerr Effect (MOKE) 120
3 Magnetic Nanosensors 120
   3.1 Gas Sensors 120
       3.1.1 Gas Sensor Based on Magnetic Properties of Materials 122
       3.1.2 Hall Effect Gas Sensors 122
       3.1.3 MOKE Based Gas Sensors 124
       3.1.4 Ferromagnetic Resonance-Based Gas Sensors 125
       3.1.5 Magnetostatic Spin Wave-Based Gas Sensor 125
   3.2 Biosensors 127
       3.2.1 Magnetic Nanoparticles-Based Sensors 129
       3.2.2 Magnetoresistance Based Sensors 131
       3.2.3 NMR Based Sensors 133
4 Conclusions 134
References 134

1 INTRODUCTION

Information is available in different forms that need to be usefully detected, translated, and processed are the primary focus of emerging science and technology. This information has a wide range of applications in different aspects of human life, requiring a very sensitive, robust, and miniature system. The most commonly used terms for the detection and conversion/translation of signals from one form to another are sensors, transducers, and actuators. A transducer is broadly defined as a device that converts energy from one form to another form. Actuators are a type of transducer that converts energy into motion. Sensors are also a type of transducers that receives a signal and converts it into a readable format.

A sensor consists of a sensitive medium or layer that responds to the external stimuli followed by a transducer that translates and processes the signal into electrical form. Figure 8.1 shows the schematic representation of the operation of the sensors and their different variations used for detection applications. Sensors are broadly classified into two categories based on the requirement of energy source for their operation as active and passive sensors. The active sensors required an external energy source (thermistors), whereas no external energy source is required for passive sensors (such as thermocouples. piezoelectric sensors). In addition, sensors are also categorized based on their applications such as mechanical, optical, magnetic, thermal, chemical, biological, acoustic, etc. which are summarized in Table 8.1 with different measurement applications [1]. Different static and dynamic parameters are used to characterize the effectiveness of the sensors, such as accuracy, precision, error, resolution, sensitivity, selectivity, reproducibility, stability, and noise. A brief definition of these parameters is summarized in Table 8.2.

FIGURE.8.1 Schematic representation of the working of a sensor.

TABLE 8.1 Different types of sensors and their measurement parameters [2].

SENSOR TYPES	MEASUREMENT PARAMETERS
Mechanical sensors	Position, acceleration, moment, torque, stress, strain, force, pressure, etc.
Optical sensors	Absorption, reflection, scattering, luminescence, fluorescence, refractive index, etc.
Magnetic sensors	Magnetic field, susceptibility, flux, permeability, etc.
Thermal sensors	Temperature, specific heat, thermal conductivity, flux, etc.
Electrical sensors	Charge, current, potential, conductivity, dielectric properties, etc.
Chemical sensors	Chemical identity (liquid and gaseous phase), concentration, pH, states, chemical interaction, etc.
Biological sensors	Biological identity, concentration, interactions, growth, and behavior, etc.
Acoustic sensors	Wave amplitude, velocity, polarization, etc.

TABLE 8.2 Different Parameters to Represent Sensor Characteristics [3]

PARAMETERS	DEFINITION AND REPRESENTATION
Accuracy (%)	How correct is the sensor output representing the actual value?
Error (%)	Difference between the actual value of the quantity being measured and the actual value obtained from the sensor.
Precision	The number of decimal places to which a measurand can be reliably measured.
Resolution	The smallest incremental change in the measurand will result in a detectable increase in the output signal.
Sensitivity	It is the ratio between the incremental change in the sensor output and its incremental change in the measurand at the input.
Selectivity	A sensor's ability to measure a single component in the presence of others is known as its selectivity.
Noise	Random fluctuation in the output signal without any change in the measurand.
Repeatability	Repeatability is the ability of the sensor to produce the same response for successive measurements of the same input under similar conditions.
Reproducibility	The ability of sensors to reproduce responses after a measurement condition has been changed.
Detection limit	The smallest magnitude of the measurand that can be measured with a sensor.
Response time	Time taken by a sensor to arrive at a stable value.

The term "nanosensors" is broadly defined as the sensor that utilizes the properties of nanomaterials to enhance signal detection. Nanosensors is a multidisciplinary and broad area of research, still in the developing stage with a wide range of applications. Nanomaterials are generally defined as materials with at least one dimension is confined in the range of 1 to 100 nm range. As the size of the materials is reduced to the nanoscale, the surface-to-volume ratio increases drastically as compared to the bulk materials. The very high surface-to-volume ratio of the nanomaterials leads to several unique properties which enhance its performance in sensors. Table 8.3 shows the different types of nanostructures defined based on their confinement direction.

The recent advance in the synthesis and manufacture of nanomaterials allows the construction of miniaturized devices, such as laboratory-on-chip (LOC) based devices. LOC devices provide a new possibility for cost-effective, easy, and fast sensor devices, especially for biological and chemical detection. Magnetic nanomaterials show their possible application for the design of miniaturized nanosensor devices due to their smaller size and excellent magnetic properties. The magnetic properties of nanomaterials are significantly different from their bulk counterparts due to finite-size effects. For example, ferromagnetic/ferrimagnetic nanoparticles show "superparamagnetic" behavior with negligible coercivity. These are suitable for the possible applications in magnetic resonance imaging and magnetic fluid hyperthermia. A brief overview of the different mechanisms used for the design of the nanosensors based on the magnetic nanomaterials is explained below.

TABLE 8.3 Different Types of Nanomaterials and their Properties.

NANOMATERIALS	CONFINEMENT DIRECTION	STRUCTURE
0D nanostructure	x, y, z direction	Nanoparticles, Nanoclusters
1D nanostructure	y, z direction	Nanorods, Nanowires, Nanotubes
2D nanostructure	z direction	Nanosheets, Thin films, Nanomesh
3D nanostructure		Nanopillars array, 3D architectures, Dendritic structures

2 THEORETICAL BACKGROUND AND PHYSICS

2.1 Hall Effect

In the Hall effect, a voltage difference is produced in a current-carrying conductor in the transverse direction when the magnetic field is applied to the direction perpendicular to it. The Hall voltage (V_H) and Hall coefficient (R_H) is calculated as

$$V_H = \frac{B.I}{\rho.w} \text{ and } R_H = \frac{V_H.w}{B.I} \qquad (1)$$

where, B, I, ρ, and w are the magnetic field, current, charge density, and width of the specimen, respectively [Figure 8.2(**a**)].

Hall effect has a wide range of applications for sensing applications because the Hall voltage is directly proportional to the magnetic field. Hall effect has a wide range of applications in switches, magnetic microscopy, automotive, and biomedical applications. The sensor response is calculated by measuring the change in Hall voltage due to interaction with sensing identity. For example, for gas sensing applications the sensor response is calculated as [4]

$$S_O = (V_g - V_a)/V_a \text{ and } S_R = (V_g - V_a)/V_a \qquad (2)$$

where, S_O, S_R, V_g, and V_a are sensor responses for oxidizing, reducing, Hall voltage in a gaseous environment, and Hall voltage in air, respectively. The main advantages of the Hall effect sensors are

(i) Accuracy, reliability, and reproducibility of the sensors.
(ii) Sensitive to the static magnetic field.
(iii) These are solid-state devices with no moving parts, so these are immune to environmental conditions such as humidity, dust, and vibrations as compared to the optical and mechanical sensors.
(iv) These sensors can work over a wide range of frequencies.

The most commonly used materials used for the fabrication of Hall effect sensors are Gallium arsenide (GaAs), Indium phosphide (InP), Indium arsenide (InAs), and Graphene.

2.2 Magnetoresistance

Magnetoresistance is defined as the change in resistivity of the materials in the presence of the external magnetic field. There are different types of magnetoresistive effects such as anisotropic magnetoresistance (AMR), giant magnetoresistance (GMR), and tunneling magnetoresistance (TMR) which have been used for the sensor technology

2.2.1 Anisotropic Magnetoresistance

Anisotropic magnetoresistance is a phenomenon in which the electrical resistivity of the material depends on the relative angle between the direction of current and the direction of magnetization [Figure 8.2(**b**)] [5]. It originates due to the spin-orbit interaction of the material with a magnetic field. The AMR effect is positive when the current direction and magnetic field direction are parallel with each other due to the enhanced probability of the s-d scattering of electrons. Mathematically, the AMR effect is represented as

$$AMR(\%) = \frac{\rho_{//} - \rho_\perp}{\rho_\perp} \tag{3}$$

where, $\rho_{//}$ and $\rho_\perp$ represents resistivity of the material measured when current and a magnetic field is parallel and perpendicular to each other.

2.2.2 Giant Magnetoresistance

The giant magnetoresistance effect is observed in the multilayer ferromagnetic (FM)/non-magnetic (NM) multilayer system [Figure 8.2(c)]. GMR originates due to the spin-dependent scattering of the conduction electrons near the interface due to the different magnetic orientations of the different layers. The adjacent magnetic layer can undergo a ferromagnetic or antiferromagnetic interaction depending on the thickness of the non-magnetic spacer layer. Therefore, the resistance is greater due to the pronounced spin-dependent scattering when its orientation is antiparallel. Thus, resistance can be switched by applying an external magnetic field, which changes the magnetic configuration of the multilayer system. GMR effect is used to design spin valve structure which has a wide range of commercial applications in GMR-based sensors, data storage, and MEMS devices [Figure 8.2(e)]. The total resistance of the GMR system is expressed as [6]

$$R = R_o + \Delta R_{GMR} Sin^2(\theta/2) \tag{4}$$

where, R_0, ΔGMR, and θ represent ferromagnetic resistance, enhancement factor due to GMR, and relative angle of orientation of magnetization in adjacent layers. GMR effect strongly depends on the device geometry of the system which can be either current in plane (CIP) or current perpendicular to plane configuration (CPP).

2.2.3 Tunneling Magnetoresistance

The tunneling magnetoresistance effect is based on the spin-polarized tunneling of the conduction electrons in a magnetic tunnel junction based on ferromagnetic metal/insulator/ferromagnetic metal junction based multilayer system [Figure 8.2(d)]. The device geometry of the TMR-based device is similar to the GMR effect except for the intermediate layer between two ferromagnetic layers. The relative change in resistance due to the TMR effect is given as [6]

$$TMR(\%) = \frac{2P_1 P_2}{1 - P_1 P_2} \tag{5}$$

where P_1 and P_2 are the spin polarization of the two ferromagnetic layers defined in terms of the spin-dependent density of states (D) of the materials as-

$$P = \frac{D_\uparrow(E_F) - D_\downarrow(E_F)}{D_\uparrow(E_F) + D_\downarrow(E_F)} \tag{6}$$

AMR-based materials are most commonly used for the design of memory and angle sensors, but the value of AMR as compared to GMR and TMR effects limits its commercial applications [7]. The ferromagnetic layer of these structures is composed of ferromagnetic metals and alloy-based films (Fe, Co, Ni, Fe Co, and FeNi), while the non-magnetic layers are made from Cu, Cr, Ag, Ru, and Au [5]. In TMR structure, a few nanometers thick Al_2O_3 and MgO are used as a tunneling barrier. For GMR and TMR precise control thickness and interface of the multilayer are crucial factors for its efficient operation. Several thin film deposition techniques have been developed for the preparation of multilayer structure including electron beam evaporation, sputtering, pulsed layer deposition, molecular beam epitaxy.

2.3 Nuclear Magnetic Resonance (NMR)

NMR is a technique based on the analysis of magnetic properties of the nuclei of the atom to exploit the chemical properties of the materials. In NMR, certain nuclei absorb some electromagnetic radiation and reemit when placed in a strong magnetic field [8]. It is a very fast and non-destructive technique suitable for biomedical samples because it does not require any specific sample preparation. In general, the working of NMR involves three steps: alignment of nuclear spins by applying a constant magnetic field (B), perturbation of this alignment by using an alternating magnetic field (RF pulse), and detection of the NMR signals. In conventional NMR devices, a strong magnet is required which limits its portable and on-site applications. The use of magnetic nanomaterials makes the NMR system portable, cheaper, flexible, and suitable for on-site applications such as NMR-based sensing devices. These magnetic nanoparticles work as proximity sensors to amplify the molecular response.

2.4 Magneto-Optic Kerr Effect (MOKE)

The Magneto-optic Kerr effect is based on the observation of the change in polarization and intensity of the reflected light from a magnetized surface [9]. It is a surface-sensitive technique and the signal is independent of the contribution of the substrate. Figure 8.2(**f**) shows the schematic of the experimental setup used for the MOKE measurement. It is used for the sensing application by measuring the change in magnetic properties (saturation, coercivity, and squareness of the hysteresis loop) of the magnetic film due to interaction with the sensing materials (such as gaseous molecule for gas sensor application).

3 MAGNETIC NANOSENSORS

The structure, properties, and applications of nanosensors based on magnetic nanomaterials for gas sensor, biosensors, and sensors for environmental monitoring have been discussed in the following sections.

3.1 Gas Sensors

A gas sensor is a device that generates the signals in terms of the detectable change in electrical, optical, magnetic, thermal, and electrochemical properties when gas molecules are adsorbed on the active surface material. These gas sensors have a wide range of applications to detect various harmful gases (such as SO_2, NO_2, NH_3, CO_2, CH_4, etc.) in various industrial, domestic, and environmental monitoring applications [10]. The key parameters used to evaluate the performance and effectiveness of a gas sensor are low cost, good selectivity, reproducibility, quick response time, less sensitivity to moisture and temperature fluctuations, and room temperature operation. The integration of exceptional physical, chemical, optical, electrical, and magnetic properties of nanomaterials with sensors increases its sensing performance in terms of size, stability, sensitivity, accuracy, precision, and resolution.

Magnetic gas sensor technology is based on the detection of changes in magnetic properties of the materials as a result of the interaction of the gaseous atoms. The magnetic properties of the materials such as the Hall effect, change in magnetization, magnetoresistance, magneto-optical Kerr effect, and

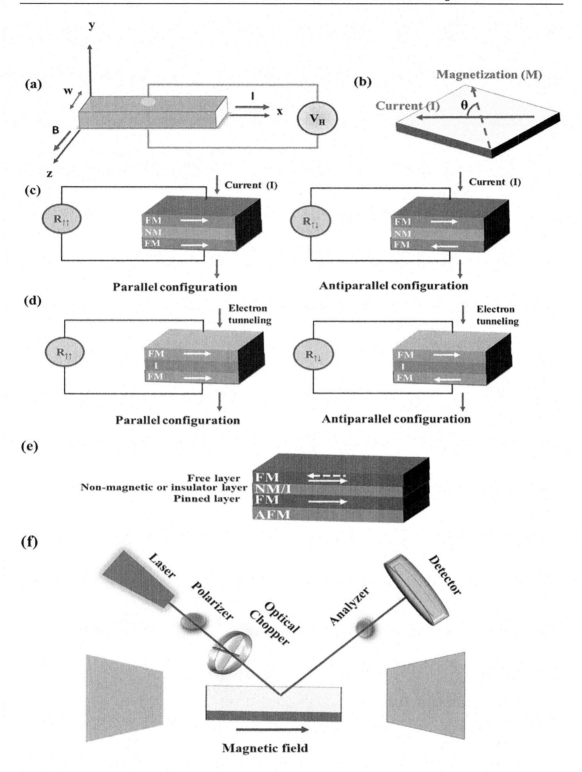

FIGURE 8.2 Schematic representation of the (a) Hall effect, (b) Anisotropic magnetoresistance (AMR), (c) Giant magnetoresistance (GMR), (d) Tunneling magnetoresistance (TMR), (e) Spin valve structure., and (f) Magneto-optical Kerr effect (MOKE).

ferromagnetic resonance are the properties that can be used as a probe for sensing application. These sensors have several advantages over other sensors such as chemical or electrochemical-based gas sensors due to their selectivity and stability to reproduce signals in adverse operating environmental conditions such as temperature, pressure, moisture, etc. Some advantages of the magnetic gas sensors are [11]:

(a) Magnetic gas sensors can work without any electrical contact with the active materials. It increases the performance of the sensors by increasing the signal-to-noise ratio and risk factor of explosion or degradation of contacts in highly reactive gaseous environments.
(b) The operating range of the sensors can be easily tuned from low to high-temperature regions by selecting the appropriate Curie temperature of the magnetic materials.
(c) Faster response of the variation in magnetic parameters as compared to electrical or chemiresistive sensors.

Different types of magnetic materials such as ferromagnetic and paramagnetic materials, dilute magnetic semiconductors, ferrites, transition metal-based alloys, and composites have been used in the form of thin films and nanoparticles for the fabrication of gas sensors [12–16].

3.1.1 Gas Sensor Based on Magnetic Properties of Materials

The most commonly used magnetic gas sensors are based on the monitoring of the change in magnetic parameters of the materials [Saturation magnetization (M_s), Remanent magnetization (M_r), and Coercivity (H_c)] as a result of interaction with gaseous atoms. Figure 8.3(**a**) shows the schematic design of the gas sensor based on the magnetic properties of the materials. It consists of a magnetic measurement system such as a vibrational sample magnetometer (VSM) that produces a variable magnetic field around the sample, Gas flow setup, and heating arrangements. When a magnetic sample interacts with the gaseous molecules, it changes the surface chemistry of the sample by oxidation or reduction of the sample. It results in the change in overall magnetic proprieties (H_c, M_r, and M_s) of the material which can be measured.

Punnose et al. have reported a novel gas sensing method based on the measurement of the change in magnetic properties of the nanomaterials [17]. They have studied the H_2 gas sensing properties of the $Sn_{0.95}Fe_{0.05}O_2$ dilute magnetic semiconductor (DMS) nanoparticles (20–70 nm) above 475 K. A large change in magnetic properties (H_c, M_r, and M_s) of the nanoparticles has been observed in the presence of H_2 gas which changes the oxygen stoichiometry in the sample. They have also explored the H_2 gas sensing properties of the hematite (Fe_2O_3) nanoparticles [18]. They have experimentally observed excellent sensing properties of the hematite nanoparticles with high sensitivity (better than the explosive limit). The impurity phase formation due to gas-sample interaction results in the change in magnetic properties of the antiferromagnetic Fe_2O_3 nanoparticles. Glover et al. have studied the gas sensing properties of the $CoFe_2O_4$ nanoparticles (size = 10nm, surface area = 150 m^2/g) for the SO_2 gas sensing [19]. A large decrease in remnant magnetization (23%), saturation magnetization (20%), and coercivity (9%) were observed due to the chemisorption of the gaseous atoms on the sample surface. Gas sensors based on magnetic properties have several advantages in terms of sensitivity and response but the cost of the experimental setup limits its applications.

3.1.2 Hall Effect Gas Sensors

Hall effect-based gas sensors are another important category of magnetic gas sensors which is based on the measurement of the change in Hall voltage of the materials in response to the interaction with gaseous molecules. Figure 8.3(**b**) shows the basic design of Hall effect-based gas sensors proposed by Hammond and his coworkers [20]. It consists of a tin oxide thin film (0.5 mm × 2.0 mm × 1100 Å) deposited on Si substrate with an insulator SiO_2 thin film (1μm). A platinum electrode (3000Å) was deposited on the film surface for the measurement of conductivity, Hall voltage, and temperature of the film surface. An electromagnet is used to produce a magnetic field perpendicular to the metal oxide thin films. Hall voltage,

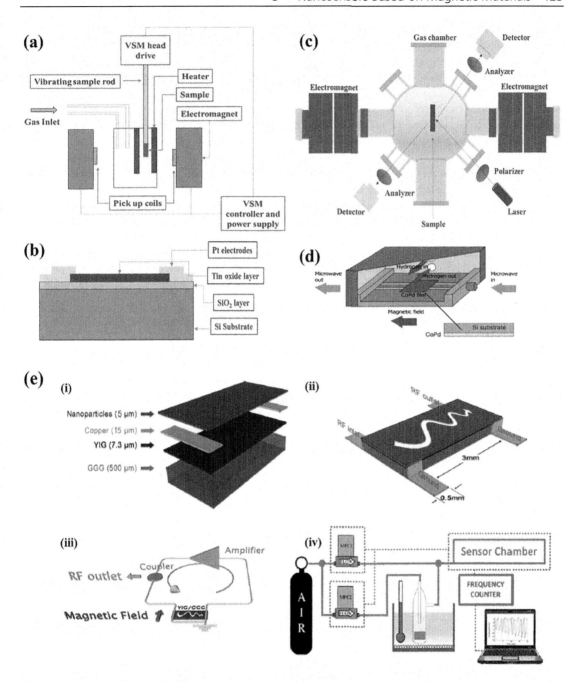

FIGURE 8.3 Schematic representation of the magnetic gas sensors based on (a) change in magnetic properties of a material, (b) Hall effect, (c) magneto-optical Kerr effect (MOKE), (d) ferromagnetic resonance (FMR) based [Adapted with permission from Ref. [13]. Copyright (2019) Hydrogen Energy Publications LLC. Elsevier Ltd.], and (e) Magnetostatic spin-wave (MSSW-DL) (i) composition of different layers, (ii) geometry of the device, (iii) oscillator controlled by MSSW-DL, (iv) set-up for gas sensing measurement. Adapted with permission from Ref. [15].

Source: Copyright (2015) The Royal Society of Chemistry.

conductivity, electron density, Hall mobility was measured in air and the presence of H$_2$ gas. A small change in Hall mobility from 0.9 to 1.83 cm^{2}V^{-1}s^{-1} was observed when exposed to the H$_2$ atoms.

Lin et al. have reported the NO$_2$ gas sensing properties of the WO$_3$ based on the Hall effect mechanism [4]. A WO$_3$ thin film was deposited on the Al$_2$O$_3$ substrate by using the screen-printing process. Au electrode was deposited on the substrate to measure the Hall coefficient in the presence of a magnetic field. It shows the possibility of WO$_3$ prepared by screen printing for a low-cost NO$_2$ sensor with good reproducibility and room temperature operation. The low sensitivity of the sensor for low concentration of the NO$_2$ concentration is the main drawback of such type of sensor. Shaeifi et al. have also studied the Galinstan-based (Ga$_2$O$_3$) Hall effect sensor for the detection of NO$_2$, NH$_3$, and CH$_4$ gases [21]. They have prepared Galinstan thin film on SiO$_2$/Si substrate and deposited gold electrode for the measurement. The sensor shows its highest response with a lower detection limit as 1ppm and 20 ppm for the NO$_2$ and NH$_3$ gases respectively with stable operation at 100 °C. Das et al. have studied the H$_2$ gas sensing properties of the CoPd thin film based on the extraordinary Hall effect (EHE) [22]. Co$_x$Pd$_{(100-x)}$ thin film was deposited on GaAs substrate by rf sputtering method for different concentrations of Co. Palladium is used for the alloy due to the high dissolution of H$_2$ atoms by making palladium hydride. Hydrogen sensitivity was observed for the Co concentration in the range of 20%<x<50% with the strongest response near x = 38%. The high sensitivity of H$_2$ sensing was observed for the CoPd alloy by tuning Co concentration and magnetic field.

3.1.3 MOKE Based Gas Sensors

Magneto-optical Kerr effect (MOKE) based magnetic sensors are another important category of magnetic field-based gas sensors. High surface sensitivity, room temperature measurement with the low magnetic field, and independence from the substrate contribution are the major advantages of the MOKE-based gas sensors. Ciprian and his coworkers have studied the gas sensing properties of the Co/ZnO nanorod-based hybrid nanostructure by using MOKE magnetometry [12]. A Co thin film (100nm) was deposited on an Al$_2$O$_3$ substrate by RF magnetron sputtering. ZnO nanorods are grown on Co film by using RF sputtering with a few angstroms thick catalyst (Sn) layer [Figure 8.4]. A ZnO nanorod of different sizes S1 (length = 100–200 nm) and S2 (length = 1 mm, diameter = 20–30 nm) was prepared by deposition for different time

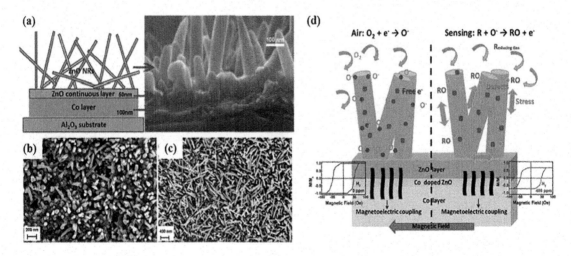

FIGURE 8.4 (a) Schematic representation of ZnO nanorods/Co hybrid composite structure with cross-sectional SEM, (b-c) Morphology of S1 and S2, (d) Magnetic gas sensing process. Adapted with permission from Ref. [12].

Source: Copyright (2016) The Royal Society of Chemistry.

duration. The ZnO/Co hybrid structure has been tested for the sensing of H_2, CO, and C_3H_6O gases. It has been observed that the gaseous molecules interact with ZnO nanorods resulting in an increase in internal stress by creating oxygen vacancies. The stress developed in the nanorods is transduced to the Co film by a decrease in magnetization.

A $[Pd/Fe]_2$ and $[Co_{40}Pd_{60}/Cu]_{10}/Fe$ multilayer films have also been explored for the H_2 gas sensing on the MOKE-based detection mechanism [14, 23, 24]. $[Co_{40}Pd_{60}(2nm)/Cu(2nm)]_{10}/Fe(7 nm)$ multilayer GMR-like structure has been deposited on Si(100) substrate by magnetron structure with Cu as a spacer and Fe as a buffer layer [23]. The experimental setup used for gas sensing was demonstrated in Figure 8.3(c). Magnetic and transport properties measurement shows that the structural defects and charge transfer due to H_2 absorption affects the spin-dependent transport. It results in the fluctuation in the magnetic properties of the materials which can be detected by MOKE signals. $[Pd/Fe]_2$ multilayer film was deposited on MgO (001) substrate by electron beam evaporation [24]. As the spacer layer of the Pd is sensitive to the H_2 due to the formation of Pd-hydride, the interlayer coupling is also becoming sensitive to the H_2 concentration. This multilayer system works like a GMR- type sensor and changes in magnetic properties can be detected by MOKE magnetometer.

3.1.4 Ferromagnetic Resonance-Based Gas Sensors

Ferromagnetic resonance (FMR) is another important mechanism used for the design of magnetic gas sensors. It is based on the FMR measurement for the detection of small changes in magnetic anisotropy especially perpendicular magnetic anisotropy (PMA) in ferromagnetic/non-magnetic (heavy metal such as Pd, or Pt) as a result of the exposure to the gaseous environment. FMR depends on the static magnetic field and the strength of PMA of the materials that occur at microwave frequencies. Several reports are available on the exploration of FMR in a non-magnetic metal/ferromagnetic metal (Pd/Co) multilayer film for gas sensing applications. Figure 8.3(d) shows the basic setup used for the FMR-based gas sensor. A multilayer thin film sample is placed on a stripline through which a radio frequency signal is transmitted (0.5–100 GHz). The FMR absorption is observed by measuring the transmitted power at a given frequency and magnetic field.

Lueng et al. have studied the FMR properties of the Co_xPd_{1-x} (x = 0.65, 0.39, 0.24, and 0.14) for hydrogen gas sensing. For x = 0.39, and x = 0.24, shows the promising applications for hydrogen gas sensing in a broad concentration range from 0.05% to 100% [25]. A large shift in FMR peak (1.7 times) has been observed with an H_2 concentration change from 10% to 67% for x = 0.24. Causer et al. have studied the H_2 gas sensing mechanism of Co/Pd film by using FMR response in combination with neutron reflectivity [26]. They observed that the Pd layer expands due to the formation of hydride when exposed to the H_2 gas. The plane magnetic moment of the Co/Pd film increases and interfacial PMA is reduced which affects the FMR frequency.

3.1.5 Magnetostatic Spin Wave-Based Gas Sensor

The variation in magnetic properties of the magnetic nanoparticles due to interaction with gaseous atoms has been explored for the design of high sensitivity, rapid response, with good reproducibility. The change in magnetic properties of the nanoparticles is explored by using a magnetostatic spin wave (MSW) oscillator. Figure 8.3(e) shows the design of MSW based gas sensor containing a thin film of magnetic nanoparticles coated on YIG films. Matagui et al. have explored the gas sensing properties of $CuFe_2O_4$ nanoparticles for the sensing of volatile organic compounds (Benzene, Isopropanol, Toluene, Ethanol, Xylene, Dimethylformamide) by using an MSW oscillator with high sensitivity and good reproducibility [15]. They have also studied the role of $CuFe_2O_4$, $MnFe_2O_4$, $ZnFe_2O_4$, and $CoFe_2O_4$ nanoparticles for the sensing of volatile organic compounds [27]. The high sensitivity of the magnetic static wave with a small change in magnetic properties of the nanoparticles due to gaseous interactions allows the detection of low concentrations of gaseous atoms.

Thus, different mechanisms, designs, and working of magnetic gas sensors have been discussed in this section. Table 8.4 summarizes different types of magnetic gas sensors and their properties reported by different research groups. The advantages and disadvantages of different gas sensing mechanisms have been summarized in Table 8.5. Thus, the magnetic gas sensor with nanomaterials shows the possibility

TABLE 8.4 Different Types of Magnetic Gas Sensors and Their Properties

S. NO.	WORKING PRINCIPAL	SENSING MATERIALS	SENSING GASES	PROPERTIES	REF.
1.	Magnetic properties of the DMS nanoparticles (M_s, M_r, and H_c)	$Sn_{0.95}Fe_{0.05}O_2$ nanoparticles	H_2 gas sensing	High sensitivity and measurement without electrical contacts	[17]
2.	Magnetic properties of the manganite nanoparticles (M_s, M_r, and H_c)	Fe_2O_3 nanoparticles	H_2 gas sensing	High sensitivity	[18]
3.	Magnetic properties of the ferrite nanoparticles (M_s, M_r, and H_c)	$CoFe_2O_4$ nanoparticles (10 nm)	SO_2 gas sensing	Large change in magnetic properties (23% in M_r, 20% in M_s, 9% in H_c) due to chemisorption	[19]
4.	Hall effect sensor	WO_3 thin film	NO_2 gas sensing	Low cost, good reproducibility, room temperature operation	[28]
5.	Hall effect sensor	Ga_2O_3 thin film	NO_2, NH_3, and CH_4 gas sensing	Good sensitivity for low concentration of gases	[21]
6.	Hall effect sensor	$Co_xPd_{(100-x)}$ polycrystalline thin film with different Co concentrations on GaAs substrate	H_2 gas sensing	H_2 sensitivity was observed for the 20%<x<50% with the strongest response near x = 38%. Sensitivity depends on Concentration and magnetic field.	[22]
7.	Magneto-optical Kerr effect	ZnO/Co hybrid nanostructure	H_2, CO, and C_3H_6O gas sensing	Fast, sensitive, scalable, low cost, portable, and room temperature operation	[12]
8.	Magneto-optical Kerr effect	$[Pd/Fe]_2$ multilayer thin film on MgO substrate by electron beam evaporation	H_2 gas sensing	GMR type sensors for H_2 sensing	[24]
9.	Magneto-optical Kerr effect	$Co_{40}Pd_{60}/Cu$ deposited on Si substrate	H_2 gas sensing	Low-pressure H_2 detection sharp and reproducible	[29]
10.	Ferromagnetic Magnetic Resonance	Co_xPd_{1-x} (x = 0.65, 0.39, 0.24, and 0.14)	H_2 gas sensing	Hydrogen gas sensing in a broad concentration range from 0.05% to 100% for x = 0.24 and 0.39.	[25]
11.	Ferromagnetic Magnetic Resonance	CoPd film	H_2 gas sensing	FMR combined with neutron reflectivity to explore the H_2 gas sensing mechanism	[26]
12.	Magnetostatic surface spin wave oscillator	$CuFe_2O_4$ nanoparticles	Volatile organic compounds	high sensitivity, rapid response, with good reproducibility	[15]
13.	Magnetostatic surface spin wave oscillator	$CuFe_2O_4$, $MnFe_2O_4$, $ZnFe_2O_4$ and $CoFe_2O_4$ nanoparticles	Volatile organic compounds	high sensitivity, rapid response, with good reproducibility	[27]

TABLE 8.5 Advantages and Disadvantages of the Different Magnetic Properties Used for Gas Sensing.

S. NO.	SENSOR PROPERTIES	ADVANTAGES	DISADVANTAGES
1.	Magnetic properties	• Highly sensitive to reducing or oxidizing gases	• High cost • Complex setup design
2.	Hall effect sensor	• Less sensitive to the gas concentration • High surface area nanostructured favors the enhanced sensing performance	• Sensitivity depends on the operating temperature • Sensing response depends on the type of material (n or p-type) • Performance is highly dependent on the film composition, thickness, and field
3.	Magneto-optical Kerr effect	• High surface sensitive properties • MOKE signal is independent of substrate contributions	• Tricky analyzer angle adjustment
4.	Ferromagnetic resonance effect	• High S/N ratio	• Proper adjustment of the magnetic field is needed
5.	Magnetostatic surface spin wave oscillator	• Detect low concentration of gases • Accurately measures weak variations • Good reproducibility	• High cost • Multistep complicated setup

of a gas sensing mechanism with high sensitivity, quick response, and reproducibility. The design of flexible, wearable, and self-powered magnetic gas sensors is the requirement of the future development of magnetic gas sensor technology.

3.2 Biosensors

A biosensor or biological sensor is an analytical device used for the detection of a biological molecule. It contains a biological receptor as a sensing element for the detection of specific biomolecules and a transducer and signal processing system. The application of nanomaterials improves the stability, cost, sensitivity, and selectivity of the sensors. A biological element interacts with the analyte and generates a biological signal which is further converted into an electrical signal by a transducer for display. Different types of biosensors have been developed based on their applications such as glucose sensors, immune sensors, biocomputers, biochips, etc. American biochemist L. L. Clarke has discovered (1962) the first biosensor to detect oxygen levels in the blood [30]. After that different generations of biosensors have been developed in terms of the attachment of the bioreceptors, recognition of the biomolecules, transducer, and electronics to improve the sensitivity limit of biosensors. In addition, recent progress is focused on the integration of nanomaterials with biosensor technology to improve its performance [31].

Figure 8.5 shows the block diagram of the biosensors including receptors or sensors, transducers, and electronic or signal conditioning circuits. Sensors interact with biological elements and the signal generated in response is further amplified and processed for display. A biosensor is generally designed to detect the concentration and nature of the analyte which includes nucleic acid, specific proteins, viruses, bacteria, etc. The bioreceptors (antibodies, nucleic acids, etc.) communicate through the analyte to generate signals which are further sensed by a transducer and processed. The primary requirement of the biosensors is that the biological molecules should be properly immobilized with a transducer so that they can be utilized a number of times over a long period. Different types of biosensors have been developed such as electrochemical biosensors, physical biosensors, piezoelectric biosensors, thermometric biosensors,

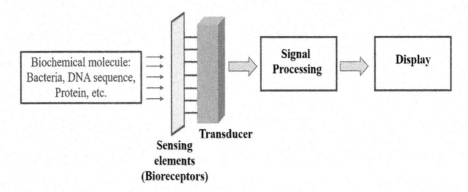

FIGURE 8.5 Block diagram of the working of biosensors.

optical biosensors, and magnetic biosensors. These biosensors have a wide range of applications in diagnosis, drug delivery, health monitoring, biomedicine, food safety and processing, environmental monitoring, defense, and security. But most of the commercial applications of biosensors are based on diagnostic applications, especially glucose sensing for diabetic patients. In this section, a design, working principle, and applications of magnetic nanobiosensors have been discussed.

Recently developed biosensor instrumentation is divided into two broad categories [32]:

(1) Rapid measurement of biological interaction accurately by using sophisticated and expensive high-quality machines.
(2) A less expensive, portable device suitable for mass production requiring an easy and less specialized operation.

The need for cost-effective, less invasive, mass-produced, wearable biosensors is highly desirable to balance increased healthcare spending and consumer demands. There are two broad categories of biomolecules transduction that can be label-based or label-free detection mechanisms. Label-based detection methods are based on the immunoassay, which is an antigen-antibody reaction. In label-based techniques, enzyme-linked immunosorbent assay (ELISA) or Radioimmunoassay (RIA) is used to detect the presence of antigen or antibody in the sample. It is a very sensitive biochemical technique that used a fluorescent marker to detect the binding of the antibody with the fluorophore-tagged linker molecules. It is a time-consuming process that required skilled supervision for the accuracy of the analysis. The label-free sensing techniques don't require tagging a molecule to detect it. It combines advanced fabrication techniques with nanomaterials to provide easy, rapid, and point of care diagnostics with high accuracy.

There are a variety of nanomaterials used for biosensing applications this includes nanoparticles, nanowires, nanotubes, nanomembranes, etc. For example, magnetic nanoparticles have been explored for diagnostic and therapeutic applications [33, 34]. Haghinaz et al. have explored the possible applications of $La_{0.7}Sr_{0.3}MnO_3$ nanoparticles for magnetic resonance imaging (MRI) and magnetic fluid hyperthermia applications [34]. Among different biosensors, magnetic biosensors are particularly important due to several advantages such as [35]

(a) Long life: Magnetic probes are more stable than other sensing probes (such as fluorescent tags).
(b) Noise: Magnetic biosensors have fewer background noise effects.
(c) Sensitivity: Magnetic assay shows higher sensitivity which allows detection at lower sample concentration.
(d) The biological environment can be easily regulated by controlling the external magnetic field.

The working principle of most of the magnetic biosensors is based on the magnetic nanoparticles, Magnetoresistance (AMR, GMR, TMR), and nuclear magnetic resonance (NMR) [36]. The advantages and disadvantages of different techniques used for the magnetic Nanosensors are summarized in Table 8.6.

3.2.1 Magnetic Nanoparticles-Based Sensors

Magnetic nanoparticles have a wide range of applications in biosensors. Generally, metal nanoparticles (Fe, Co, and Ni) and magnetic nanoparticles based on metal oxides with various sizes and morphology have been synthesized for biosensor applications. These magnetic nanoparticles have a very large surface area and can be easily functionalized through antibodies and aptamers (single-stranded DNA or RNA). These surface-functionalized magnetic nanoparticles have potential applications in immunosensors. The size distribution, morphology, and surface functionalization contribute to the reproducibility of the results [37].

Magnetic particle imaging (MPI) is an emerging tool of imaging based on the detection of iron oxide nanoparticles traces in a living organism with high contrast and sensitivity. It allows getting information from the cells, blood, and is used for the targeted drug delivery application [38]. Magnetic particle spectroscopy (MPS) is a similar technique as MPI in which topographic images can be reconstructed by analyzing a magnetic response of the nanoparticles [39]. In MPS, the magnetic nanoparticles are periodically magnetized by applying an external sinusoidal magnetic field, and response is monitored by pickup coils. The mono and dual-frequency modes are used for the excitation of magnetic nanoparticles. The magnetic moment of the nanoparticles is aligned with the magnetic field by Neel or Brownian relaxation process, which is opposed by thermal energy. The relaxation process is also depending on the physical condition of the magnetic nanoparticles such as temperature, viscosity, and biochemical interactions. In general, iron oxide nanoparticles (magnetite/maghemite) coated with a capping layer and functionalized with specific biomolecules (proteins, antibodies, etc.) are used for the MPS technique. MPS has a wide range of applications in cell analysis, biological assay with functionalized nanoparticles, magnetic imaging, and hyperthermia [40, 41]. The high sensitivity of the detection is achieved by MPS due to negligible background noise generated in biological samples.

Magnetic resonance imaging (MRI) is a powerful diagnostic tool for biomedical applications. Magnetic nanoparticles play an important role as a contrast agent in MRI because of their magnetic properties, biocompatibility, and flexibility in modification of the surface properties by functionalization to control kinetics and functionalities. Iron oxide and gadolinium nanoparticles are the most commonly used contrast agents in MRI [42]. Figure 8.6(a) shows the design and working of a magnetic resonance

TABLE 8.6 Advantages and Disadvantages of Different Techniques Used for the Design of Magnetic Nanosensors [36]

S. NO.	METHOD	ADVANTAGES	DISADVANTAGES
1.	Giant Magnetoresistance (GMR)	• High sensitivity • Portable device • Mass production capability	• Well trained technicians are needed • Time consuming • High fabrication cost of the biosensors
2.	Magnetic Tunnel Junction (MTJ)	• High sensitivity • Mass production capability	• Well trained technicians are needed • High noise • Hard to acquire a linear response • The complicated and costly fabrication process • Time consuming
3.	Magnetic particle spectroscopy (MPS)	• High sensitivity • Low cost per test • Portable device available	• Well trained technicians are needed • Time consuming • less sensitive
4.	Nuclear Magnetic Resonance (NMR)	• Portable device available	• Well trained technicians are needed • Time consuming • Medium sensitivity

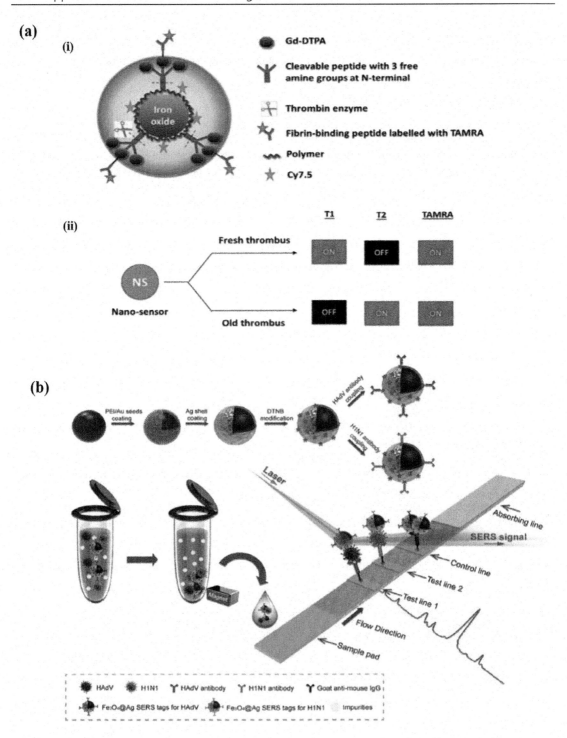

FIGURE 8.6 (a) Design and working of the functionalized iron oxide nanoparticles for the MRI T1/T2 switchable nanosensors. Adapted with permission from Ref. [43]. Copyright (2018) The Royal Society of Chemistry, (b) Schematic representation of the (i) Synthesis of antibody modified magnetic nanoparticles (Fe3O4@Ag), (ii) Magnetic SERS Strip for Detecting Two Respiratory Viruses. Adapted with permission from Ref. [44].

Source: Copyright (2019) The Royal Society of Chemistry.

nanosensor based on iron oxide nanoparticles for detecting and discriminating cardiovascular disease (thrombosis) [43]. The sensor can switch between T1 and T2 signals depending on the age of the thrombus and the presence or absence of thrombin at the thrombus site.

In the current scenario of pandemic (COVID-19), the early detection of respiratory viruses is the key factor for the prevention in spreading and initiation of the treatment. Magnetic nanoparticles are playing an important role in the early detection of such viruses in a short duration with high sensitivity. Wang et al. have reported a surface-enhanced Raman scattering (SERS) based lateral flow immunoassay strip for the detection of influenza A H1N1 virus and human adenovirus simultaneously [44]. A schematic description of the working of the device is represented in Figure 8.6(**b**). The Ag coated Fe_3O_4 nanoparticles functionalized with target antibodies and SERS dye are used as a magnetic tag for the specific recognition and SERS detection of the viruses. It can replace the colloidal gold nanoparticles-based technique as a powerful tool for virus detection due to its very high sensitivity and high throughput.

3.2.2 Magnetoresistance Based Sensors

Magnetoresistive sensors are an interesting area of research for biomedical and life sciences applications due to their high sensitivity with low magnetic fields [36]. Figure 8.7 shows the principal and working of the highly sensitive giant magnetoresistance (GMR) based biosensor for the detection of influenza A

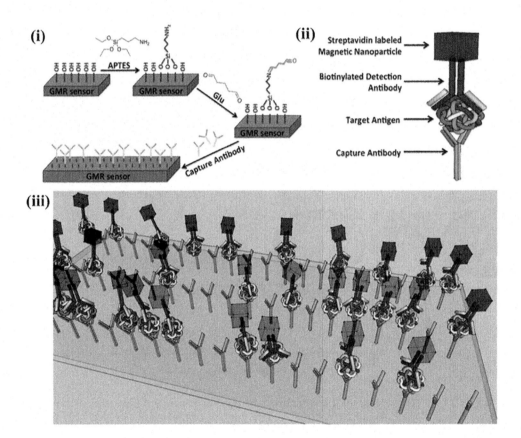

FIGURE 8.7 Schematic representation of (i) GMR biosensor functionalization, (ii) Sandwich structure, and (iii) Influenza A virus detection. Adapted with permission from Ref. [45].

Source: Copyright (2016) Krishna, Wu, Perez, and Wang. Distributed under a Creative Commons Attribution License (CC BY).

virus [45]. Superparamagnetic nanoparticles or beads functionalized with target antibodies are used as a marker for the detection. When magnetic nanoparticles are attached to the sensor surface its stray field changes the resistive of the sensor which can be detected by the GMR sensor. GMR spin-valve multilayer structure [Ta(50Å)/NiFe(20Å)/CoFe (10Å)/Cu(33Å)/CoFe (25Å)/IrMn (80Å)] was deposited by magnetron sputtering on Si/SiO$_2$ surface. A capture antibody was attached to the GMR surface. It can detect viral concentration in the range of 1.5×10^2 to 1.0×10^5 TCID$_{50}$/ml which is better than ELISA-based techniques.

Wu et al. have designed a portable, sensitive, and point of care device for the detection of influenza A virus, known as Z-lab [46]. It is a GMR-based sensor device that provides accurate and affordable detection of the virus in remote locations without the requirement of any laboratory skills. Figure 8.8 shows the design and working of the Z-lab device. It consists of a magnetic sandwich assay which is bound to the GMR sensor surface if the virus is present in the sample. For the food safety applications, Wu et al. have reported TMR-based biosensors for the detection of Escherichia coli O157:H7 bacteria within a 5h time duration with a concentration range of 10^5-10^6 CFU/ml [47]. High sensitivity and rapid detection make it suitable for food safety applications. Mak et al. have reported the application of a spin valve sensor (8 × 8 array) for the simultaneous detection of multiple mycotoxins with high sensitivity [48].

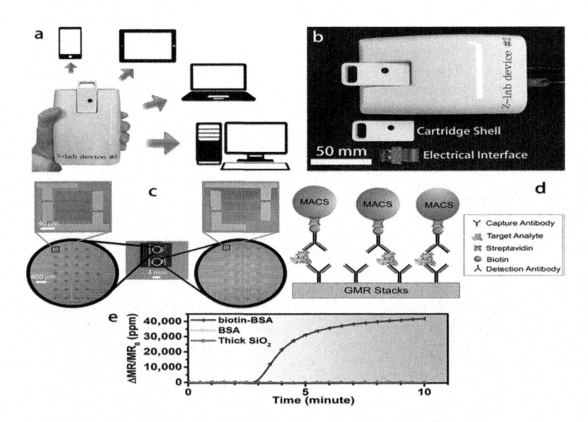

FIGURE 8.8 (a) Z-lab platform for magnetic sandwich assay (i) portable device by data can be collected by the wireless network through electronic gadgets, (ii) Components of the device, (iii) GMR chip, (iv) Magnetic sandwich assay, and (v) real-time data. Adapted with permission from Ref. [46].

Source: Copyright (2017) American Chemical Society.

3.2.3 NMR Based Sensors

Nuclear magnetic resonance is a powerful analytical tool for the detection of bacteria in a biological sample. A portable, rapid, and accurate measurement is highly desired for molecular analysis of the sample. Lee et al. have reported a portable, rapid, multichannel, and miniaturized diagnostic magnetic resonance (DMR) based biosensor system for the point of care analysis of the biological sample as shown in Figure 8.9 [49]. It uses magnetic nanoparticles as a proximity sensor for the amplification of molecular interactions. There are four main components of the DMR system: a micro-NMR chip with microcoils, small permanent magnets, NMR electronics, and a microfluidic network for sample handling. The micro-coil is prepared by electroplating with a photolithography technique used for the RF signal generation and detection. It is assembled on a printed circuit board with other electronic components and generates a pulse sequence to measure longitudinal (T_1) and transverse (T_2) relaxation times. The cross-linked iron oxide magnetic nanoparticles are used for the NMR signal measurement. It is a portable, low-cost, and high throughput device which provides a point of care detection of the biological samples with high sensitivity and selectivity.

Magnetic biosensor based on the NMR provides a solution to address the food safety concern for public health and safety. Luo et al. have also demonstrated a portable NMR-based biosensor device to detect the presence of food-borne bacteria (Escherichia coli O157:H7) [50]. Figure 8.10(i) shows the schematic of the working of NMR based pathogen detection system. Iron oxide nanoparticles synthesized by the sol-gel method are conjugated with antibodies for pathogen detection. The magnetic nanoparticles-pathogen conjugates are washed and filtered to reduce the interference from unbound magnetic nanoparticles. After filtration, the number of magnetic nanoparticles is proportional to the pathogen concentration. The presence of magnetic nanoparticles in the sample produces disturbance in the local magnetic field [Figure 8.10(ii)]. The concentration of the pathogens in the sample is detected from the NMR signal by measuring the water proton's spin-spin relaxation time.

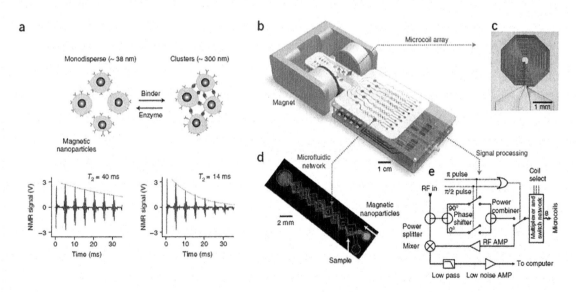

FIGURE 8.9 Structure and working of the diagnostic magnetic resonance (DMR) (a) Principal of proximity assay using magnetic nanoparticles, (b) device design of DMR system, (c) Micro RF coil, (d) microfluidic network, (e) NMR electronic circuits. Adapted with permission from Ref. [49].

Source: Copyright (2008) Springer Nature.

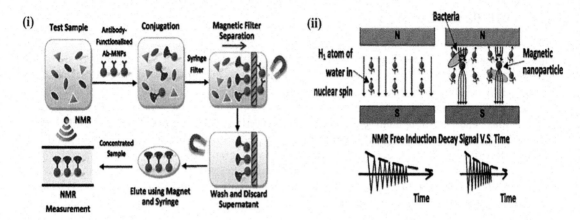

FIGURE 8.10 (i) Schematic representation of the NMR-based biosensing mechanism for pathogen detection, (ii) Magnetic nanoparticles as a proximity biomarker for NMR detection. Adapted with permission from Ref. [50].

Source: Copyright (2016) Yilun Luo and Evangelyn C. Alocilja. Distributed under a Creative Commons Attribution License (CC BY 4.0).

4 CONCLUSIONS

In this chapter, we have discussed the principle, design, and operation of nanosensors based on magnetic nanomaterials for applications in gas sensors, biosensors, and environmental safety applications. Nanosensors based on magnetic materials show several advantages compared to other sensors in terms of sensitivity, miniaturization, reproducibility, cost-effectiveness, fast response with portable design. Further development of flexible, portable, self-powered sensors is highly desirable to provide a larger-scale point-of-care solution. It will provide a new perspective for the advancement of existing work for future development.

REFERENCES

[1] A. Munawar, Y. Ong, R. Schirhagl, M.A. Tahir, W.S. Khan, S.Z. Bajwa, Nanosensors for diagnosis with optical, electric and mechanical transducers, *RSC Adv.* 9(12) (2019) 6793–6803.

[2] R. Abdel-Karim, Y. Reda, A. Abdel-Fattah, Nanostructured materials-based nanosensors, *J. Electrochem. Soc.* 167(3) (2020) 037554.

[3] J. Fraden, *Handbook of Modern Sensors: Physics, Designs, and Applications*, Springer (2004).

[4] J. Lin, W. Xie, X. He, H. Wang, A novel NO_2 gas sensor based on Hall effect operating at room temperature, *Appl. Phys. A* 122(9) (2016) 1–6.

[5] G. Florio, *Applications of Magnetic Materials*, Elsevier (2021).

[6] X. Zhang, W. Butler, *Theory of Giant Magnetoresistance and Tunneling Magnetoresistance, Handbook of Spintronics*, Springer (2014).

[7] S. Yang, J. Zhang, Current progress of magnetoresistance sensors, *Chemosensors* 9(8) (2021) 211.

[8] H. Günther, *NMR Spectroscopy: Basic Principles, Concepts and Applications in Chemistry*, John Wiley & Sons (2013).

[9] O. Borovkova, H. Hashim, M. Kozhaev, S. Dagesyan, A. Chakravarty, M. Levy, V. Belotelov, TMOKE as efficient tool for the magneto-optic analysis of ultra-thin magnetic films, *Appl. Phys. Lett.* 112(6) (2018) 063101.

[10] K. Xu, C. Fu, Z. Gao, F. Wei, Y. Ying, C. Xu, G. Fu, Nanomaterial-based gas sensors: A review, *Instrum. Sci. Technol.* 46(2) (2018) 115–145.
[11] P.V. Shinde, C.S. Rout, Magnetic gas sensing: Working principles and recent developments, *Nanoscale Adv.* 3(6) (2021) 1551–1568.
[12] R. Ciprian, P. Torelli, A. Giglia, B. Gobaut, B. Ressel, G. Vinai, M. Stupar, A. Caretta, G. De Ninno, T. Pincelli, New strategy for magnetic gas sensing, *RSC Adv.* 6(86) (2016) 83399–83405.
[13] C. Lueng, P. Lupo, T. Schefer, P. Metaxas, A. Adeyeye, M. Kostylev, Sensitivity of ferromagnetic resonance in PdCo alloyed films to hydrogen gas, *Int. J. Hydrog. Energy* 44(14) (2019) 7715–7724.
[14] J.-Y. Liang, Y.-J. Chou, C.-W. Yin, W.-C. Lin, H.-J. Lin, P.-W. Chen, Y.-C. Tseng, Realization of an H2/CO dual-gas sensor using CoPd magnetic structures, *Appl. Phys. Lett.* 113(18) (2018) 182401.
[15] D. Matatagui, O. Kolokoltsev, N. Qureshi, E. Mejía-Uriarte, J. Saniger, A magnonic gas sensor based on magnetic nanoparticles, *Nanoscale* 7(21) (2015) 9607–9613.
[16] J. Lin, Z. Chen, X. He, W. Xie, Detection of H2S at room temperature using ZnO sensors based on Hall effect, *Int. J. Electrochem. Sci.* 12 (2017) 6465–6476.
[17] A. Punnoose, K. Reddy, J. Hays, A. Thurber, M.H. Engelhard, Magnetic gas sensing using a dilute magnetic semiconductor, *Appl. Phys. Lett.* 89(11) (2006) 112509.
[18] A. Punnoose, K. Reddy, A. Thurber, J. Hays, M.H. Engelhard, Novel magnetic hydrogen sensing: A case study using antiferromagnetic haematite nanoparticles, *Nanotechnology* 18(16) (2007) 165502.
[19] T.G. Glover, D. Sabo, L.A. Vaughan, J.A. Rossin, Z.J. Zhang, Adsorption of sulfur dioxide by CoFe2O4 spinel ferrite nanoparticles and corresponding changes in magnetism, *Langmuir* 28(13) (2012) 5695–5702.
[20] Z. Torabi, A. Mohammadi Nafchi, The effects of SiO2 nanoparticles on mechanical and physicochemical properties of potato starch films, *J. Chem. Health Risks* 3(1) (2013).
[21] M. Shafiei, F. Hoshyargar, N. Motta, A.P. O'Mullane, Utilizing p-type native oxide on liquid metal microdroplets for low temperature gas sensing, *Mater. Des.* 122 (2017) 288–295.
[22] S. Das, G. Kopnov, A. Gerber, Detection of hydrogen by the extraordinary Hall effect in CoPd alloys, *J. Appl. Phys.* 124(10) (2018) 104502.
[23] J.-Y. Liang, Y.-C. Pai, T.-N. Lam, W.-C. Lin, T.-S. Chan, C.-H. Lai, Y.-C. Tseng, Using magnetic structure of Co$_{40}$Pd$_{60}$/Cu for the sensing of hydrogen, *Appl. Phys. Lett.* 111(2) (2017) 023503.
[24] C.-C. Hsu, P.-C. Chang, Y.-H. Chen, C.-M. Liu, C.-T. Wu, H.-W. Yen, W.-C. Lin, Reversible 90-degree rotation of Fe magnetic moment using hydrogen, *Sci. Rep.* 8(1) (2018) 1–12.
[25] C. Lueng, P.J. Metaxas, M. Sushruth, M. Kostylev, Adjustable sensitivity for hydrogen gas sensing using perpendicular-to-plane ferromagnetic resonance in Pd/Co Bi-layer films, *Int. J. Hydrog. Energy* 42(5) (2017) 3407–3414.
[26] G.L. Causer, M. Kostylev, D.L. Cortie, C. Lueng, S.J. Callori, X.L. Wang, F. Klose, In operando study of the hydrogen-induced switching of magnetic anisotropy at the Co/Pd interface for magnetic hydrogen gas sensing, *ACS Appl. Mater. Interfaces* 11(38) (2019) 35420–35428.
[27] D. Matatagui, O. Kolokoltsev, N. Qureshi, E. Mejía-Uriarte, C. Ordoñez-Romero, A. Vázquez-Olmos, J. Saniger, Magnonic sensor array based on magnetic nanoparticles to detect, discriminate and classify toxic gases, *Sens. Actuators B Chem.* 240 (2017) 497–502.
[28] J.Y. Lin, Y.A. Zhang, L.J. Wang, T.L. Guo, WO3-based sensor based on Hall Effect for NO2 detection: Designed and investigation, *Adv. Mater. Res., Trans. Tech. Publ.* (2011) 1042–1046.
[29] S. Zhang, N. Pan, Supercapacitors performance evaluation, *Adv. Energy Mater.* 5(6) (2015) 1401401.
[30] S. Vigneshvar, C. Sudhakumari, B. Senthilkumaran, H. Prakash, Recent advances in biosensor technology for potential applications – an overview, *Front. Bioeng. Biotechnol.* 4 (2016) 11.
[31] M. Holzinger, A. Le Goff, S. Cosnier, Nanomaterials for biosensing applications: A review, *Front. Chem.* 2 (2014) 63.
[32] A.P. Turner, Biosensors: Sense and sensibility, *Chem. Soc. Rev.* 42(8) (2013) 3184–3196.
[33] N.D. Thorat, O. Lemine, R.A. Bohara, K. Omri, L. El Mir, S.A. Tofail, Superparamagnetic iron oxide nanocargoes for combined cancer thermotherapy and MRI applications, *Phys. Chem. Chem. Phys.* 18(31) (2016) 21331–21339.
[34] R. Haghniaz, K.R. Bhayani, R.D. Umrani, K.M. Paknikar, Dextran stabilized lanthanum strontium manganese oxide nanoparticles for magnetic resonance imaging, *RSC Adv.* 3(40) (2013) 18489–18497.
[35] V. Nabaei, R. Chandrawati, H. Heidari, Magnetic biosensors: Modelling and simulation, *Biosens. Bioelectron.* 103 (2018) 69–86.
[36] K. Wu, R. Saha, D. Su, V.D. Krishna, J. Liu, M.C.-J. Cheeran, J.-P. Wang, Magnetic-nanosensor-based virus and pathogen detection strategies before and during covid-19, *ACS Appl. Nano Mater.* 3(10) (2020) 9560–9580.
[37] Y. Zhang, J. Xu, Q. Li, D. Cao, S. Li, The effect of the particle size and magnetic moment of the Fe3O4 superparamagnetic beads on the sensitivity of biodetection, *AIP Adv.* 9(1) (2019) 015215.

[38] X. Zhu, J. Li, P. Peng, N. Hosseini Nassab, B.R. Smith, Quantitative drug release monitoring in tumors of living subjects by magnetic particle imaging nanocomposite, *Nano Lett.* 19(10) (2019) 6725–6733.
[39] K. Wu, D. Su, R. Saha, J. Liu, V.K. Chugh, J.-P. Wang, Magnetic particle spectroscopy: A short review of applications using magnetic nanoparticles, *ACS Appl. Nano Mater.* 3(6) (2020) 4972–4989.
[40] K. Wu, J. Liu, R. Saha, D. Su, V.D. Krishna, M.C.-J. Cheeran, J.-P. Wang, Magnetic particle spectroscopy for detection of influenza A virus subtype H1N1, *ACS Appl. Mater. Interfaces* 12(12) (2020) 13686–13697.
[41] A.V. Orlov, J.A. Khodakova, M.P. Nikitin, A.O. Shepelyakovskaya, F.A. Brovko, A.G. Laman, E.V. Grishin, P.I. Nikitin, Magnetic immunoassay for detection of staphylococcal toxins in complex media, *Anal. Chem.* 85(2) (2013) 1154–1163.
[42] A. Avasthi, C. Caro, E. Pozo-Torres, M.P. Leal, M.L.J.S.-M.N.F.E. García-Martín, B. Applications, Magnetic nanoparticles as MRI contrast agents, *Top. Curr. Chem.* (2020) 49–91.
[43] H.T. Ta, N. Arndt, Y. Wu, H.J. Lim, S. Landeen, R. Zhang, D. Kamato, P.J. Little, A.K. Whittaker, Z.P. Xu, Activatable magnetic resonance nanosensor as a potential imaging agent for detecting and discriminating thrombosis, *Nanoscale* 10(31) (2018) 15103–15115.
[44] C. Wang, C. Wang, X. Wang, K. Wang, Y. Zhu, Z. Rong, W. Wang, R. Xiao, S. Wang, Magnetic SERS strip for sensitive and simultaneous detection of respiratory viruses, *ACS Appl. Mater. Interfaces* 11(21) (2019) 19495–19505.
[45] V.D. Krishna, K. Wu, A.M. Perez, J.-P. Wang, Giant magnetoresistance-based biosensor for detection of influenza A virus, *Front. Microbiol.* 7 (2016) 400.
[46] K. Wu, T. Klein, V.D. Krishna, D. Su, A.M. Perez, J.-P. Wang, Portable GMR handheld platform for the detection of influenza A virus, *ACS Sens.* 2(11) (2017) 1594–1601.
[47] Y. Wu, Y. Liu, Q. Zhan, J.P. Liu, R.-W. Li, Rapid detection of Escherichia coli O157: H7 using tunneling magnetoresistance biosensor, *AIP Adv.* 7(5) (2017) 056658.
[48] A.C. Mak, S.J. Osterfeld, H. Yu, S.X. Wang, R.W. Davis, O.A. Jejelowo, N. Pourmand, Sensitive giant magnetoresistive-based immunoassay for multiplex mycotoxin detection, *Biosens. Bioelectron.* 25(7) (2010) 1635–1639.
[49] H. Lee, E. Sun, D. Ham, R. Weissleder, Chip – NMR biosensor for detection and molecular analysis of cells, *Nat. Med.* 14(8) (2008) 869–874.
[50] Y. Luo, E.C. Alocilja, Portable nuclear magnetic resonance biosensor and assay for a highly sensitive and rapid detection of foodborne bacteria in complex matrices, *J. Biol. Eng.* 11(1) (2017) 1–8.

Role of Magnetic Nanomaterials in Biomedicine

9

Bhagavathula S. Diwakar, D. Chandra Sekhar,
Venu Reddy, P. Bhavani, Ramam Koduri, S. Srinivasarao

Contents

1	Introduction	137
2	Magnetic Nanoparticles for Biomedical Applications	138
	2.1 Magnetic Bioseparation	138
	2.1.1 Examples of Magnetic Bioseparation	139
	2.2 Magnetic Hyperthermia	140
	2.3 Targeted Drug Delivery	141
	2.4 Magnetofection Agents	143
	2.5 Magnetic Resonance Imaging (MRI)	143
	2.5.1 T1-Contrast Agents for MRI	144
	2.5.2 T2/T2*- Agents for MRI	144
References		144

1 INTRODUCTION

Nanotechnology is a science branch that studies nanometer-sized objects and their applications. It has amazing applications in various aspects of materials, devices, and systems. Currently, nanomaterials-based device production technology is one of the largest progressed technologies in scientific research and applications. At present, nanoparticles are also being used for commercial applications unlike in the past where they have been limited studies only to their physical and chemical properties. Out of various types of nanomaterials, magnetic nanomaterials have many different applications in numerous aspects. Magnetic nanoparticles (MNPs) include Fe, Co, Ni, Mn, Gd, and their alloys, oxide compounds, cation complexes which are magnetic elements and also polymers, etc., that show ferromagnetism, paramagnetism, or even superparamagnetism. The characteristics of nanoparticles largely depend on their chemical composition, crystal structure, size, and shape, and also raw materials for production. Along with the fundamental effects like the quantum size influence, the nanoparticles also have special magnetic properties like superparamagnetism, high coercivity, low curie temperature, and high magnetic susceptibility [1–4].

DOI: 10.1201/9781003196952-9

Nanoparticles are also found applications in catalysis, mineralogy, data storage, environmental science in the concentration of pollutants, and also in biomedicine due to their brilliant properties in chemistry and physics [5,6].

Nanoparticles provide numerous opportunities in biomedicine and its technology. They are (1) Nanoparticles have a size difference that is smaller than or similar to the size of cells (10–100 μm), viruses (20–450 nm), proteins (5–50 nm), and DNA double helix (2 nm width) i.e., they can "get close" to a biological entity of concern. The nanometer size prefers the interaction among nanoparticles and bioentities to improve the ability of biomolecules to coat on the surface of nanoparticles hence providing a controllable "tagging". (2) The magnetic nanoparticles are easily regulated by an external magnetic field. It is used to improve transport and immobilization applications of biological objects labeled with magnetic nanoparticles in human cells. Hence, they can help to deliver an object like an anticancer drug, to a targeted area of the body, like a tumor. (3) Due to the difference in the external magnetic field, the magnetic nanoparticles receive energy from the excitation field and transfer heat to the target area. Therefore, magnetic nanoparticles are useful for chemotherapy and radiation therapy because heated tissue can effectively damage tumor cells.

In addition, various functional molecules or units such as enzymes, antibodies, cells, DNA, and RNA can be easily associated with nanoparticles due to their magnetic properties, biocompatibility, biodegradability, and functional groups. Anyway, the surface of magnetic nanoparticles can be changed as per the requirements. Changing the surface of magnetic nanoparticles comprise nonpolymer, immobilization of an organic and inorganic molecule, and targeted ligand modification [7]. Commonly used modifiers include lipids, polyethylene glycol (PEG), gluconic acid, polyvinylpyrrolidone (PVP), aliphatic acid, polyvinyl alcohol (PVA), peptides, gelatin, chitosan, methylsilane, and liposomes [8]. Such agents increase the biocompatibility of magnetic nanoparticles, inhibit aggregation, inhibit protein uptake, increase the duration in circulation, reduce toxic nature, and improve targeting. Due to this, the MNPs are being largely helpful in biomedicine. The magnetic nanoparticles also have other uses like probing and separation of cells. The main disadvantage of the functionalized magnetic nanoparticles is their spherical shape and this is overcome by making cylindrical shapes through the electrodeposition technique [9]. The surface functionalization of nanomaterials helps to bind various ligands to various regions of the nanoparticle surface. This property is used to improve magnetic nanoparticles that can respond to weaker magnetic fields which are used in nanoparticle self-assembling. It allows restricting cell assembly in different forms. In this chapter, we focused on the applications of MNPs in biomedicine, including (i) the magnetic separation methods, (ii) hyperthermia treatment, (iii) targeted drug delivery, (iv) magnetofection agents, and (v) magnetic resonance imaging (MRI).

2 MAGNETIC NANOPARTICLES FOR BIOMEDICAL APPLICATIONS

2.1 Magnetic Bioseparation

In the magnetic bioseparation technique, employing an applied magnetic field, the bioentities are isolated from the bio-environment. Currently, magnetic nanomaterials with superparamagnetic nature are helpful for bioseparation as they can be effortlessly magnetized with an applied magnetic field. When this field is withdrawn, captured bio-entities are dispersed instantly into the solution. The bioseparation technique consists of two steps: (1) Conjugation and labeling of bioentities with magnetic nanomaterials; (2) Separation of labeled bioentities from the solution with a magnetic separator. This process is usually applied in isolation and purification of various kinds of bioentities, for instance, cells, bacteria, proteins, and nucleic acids, etc. Magnetic separation is similarly used in addition to optical probing to accomplish enzymatic immunoassays with magnetic enzymes. As a result, this magnetic separation method is soon being investigated for a variety of separation applications.

   a. **Separation of Proteins:** Proteins are made up of 22 different amino acids that control hydrophilicity, hydrophobicity, polarity, and non-polarity. Magnetic nanoparticles (MNPs) have been associated with various protein copolymers using a variety of methods like ligand binding, hydrophobic and electrostatic interactions [10]. The electrostatic properties of the amino acids are important for biological separation because they are present as zwitterions consisting of equal surface charges (+ve & −ve) at their isoelectric point (pI), and the net surface charge can vary with the pH of the environment. As a result, the charged protein attaches to the reversely charged MNP [11].
   b. **DNA Separation:** DNA-based MNPs isolation depends on the same type of interaction with proteins that, the charge distribution on the surface of a DNA molecule regulates its attachment to the corresponding MNP [12]. In a physiological environment, DNA molecules are −ve charged due to their phosphate backbone. Even though the environment is acidic, the proton is positively charged because it binds to the phosphate group [13]. One study used these electrostatic interactions to isolate salmon sperm DNA using mesoporous silica magnetite nanocomposites fabricated using the matrix method [14]. At pH ~ 7.4 and high salt concentrations, the nanocomposite has a positive charge, which ultimately promotes electrostatic interactions with the negatively charged phosphate backbone of DNA, ensuring successful separation. Nearly 100% of the DNA was obtained from the nanocomposite surface compared to less than 10% from the magnetite core.
   c. **Separation of Other Biomolecules:** In addition to proteins and DNA, MNPs were used in isolating numerous biomolecules like cells, bacteria, genes, and viruses because of the adaptability of functional groups which helped to change their surfaces [15]. Anti-CD3 monoclonal antibody bio-conjugated to core/shell (Fe_3O_4/Au) MNPs were efficient to isolate T-cells successfully up to 98.4% [16].

2.1.1 Examples of Magnetic Bioseparation

In biomedicine, it is important to identify, purify, and evaluate biomolecules or other applications that often require isolation from the environment. Among the available methods, magnetic separation is convenient [17]. Today, analysis of MNPs shows many effective methods for modulating and investigating the interactions between synthesized MNPs and biological molecules [18]. In general, two methods are frequently used to isolate proteins. (i) Conjugation of MNPs to specific ligand molecules and (ii) conjugation of MNPs to antibodies. Initial protein isolation was performed using Ni-nitrilotriacetic acid (Ni-NTA) MNPs to isolate His-tag proteins from cell lysates.

Sun et al. demonstrated decorated FePt MNPs with Ni-NTA complexes using mercapto alkanoic acid [19]. The resulting MNP isolates His-tag proteins of *E. coli* lysates with high capacity and selectivity. Zheng et al. prepared magnetic microspheres of phenylboronic acid Fe_3O_4 @ polydopamine (Fe_3O_4 @ PDAPBA) to evaluate the selectivity and binding capacity of magnetic microspheres Fe_3O_4 @ PDAPBA using standard glycoproteins and non-glycoproteins. The results show that the adsorption capacity of 160 mg/g and 140 mg/g for the standard glycoproteins obtained from the bovine liver, namely ovalbumin and catalase, which was about 38 times higher than for the non-glycoprotein lysozyme, myoglobin, and Ribonuclease A and bovine hemoglobin were 25, 57, 47, and 20 mg/g respectively [20]. Bucak et al. improved phospholipid-coated MNPs for protein extraction from protein mixtures. These proteins have high adsorption capacities up to 1200 mg protein/ml of adsorbent [21].

Another isolation method involves antibody-linked MNPs, which are mainly used in immunoassays. Matsunaga et al. discovered the @MNP antibody for a fully automated immunoassay to assay human insulin [22]. They developed a guaranteed automated sandwich immunoassay to recommend antibody-protein sterile magnetic nanoparticles for the accurate identification of human insulin. Isolation of DNA also facilitates a similar process for the isolation of proteins. The attachment of DNA to MNPs depends on the DNA molecules and the surface properties of the MNPs. Biao et al. improved silica-coated nanoparticles for the isolation of bacterial plasmid DNA from bacteria cultures. These particles have a +ve surface charge at neutral pH, increasing the potential for DNA to bind to MNPs. MNPs respond strongly and systematically to magnetic fields, providing high isolation yields and purity of plasmid DNA. Zhao et al prepared water-dispersed magnetic nanoparticles covered by salicylic acid. These MNPs were originally isolated from solution and attached to mammalian cells for attachment to genomic DNA by extraction and isolation method. Yongjun et al. found a fast identification process of *Pseudomonas aeruginosa*, based on magnetic separation. They used polymerase chain reaction to efficiently amplify the biotin-labeled UTP DNA fragment of the gyrB gene and identified *P. aeruginosa* with a detection limit of 7.5 fM gyrB fragment [23]. Xi et al. reported that *hepatitis B* from pure protein (detection limit: 0.1 ng/ml) using a chemiluminescent aptasensor constructed according to a magnetic separation system and immunoassay [24].

2.2 Magnetic Hyperthermia

The demand for nanomedicine has skyrocketed in recent years. Although side effects of anticancer drugs are known, there is a great demand for higher specificity and efficacy. The paradox of high drug concentration at the target site and in the least-damaged normal tissue can be effectively solved with the help of nanoparticles. Another point is the "intelligence" of nanoparticles, which can release drugs according to external signals, temperature rise, or physiological conditions of cancer cells. Although nanoliposome formulations have already entered the market, they do not yet have previously described properties. However, this is the right path because side effects are limited in clinical practice, and there is ultimately a serious economic debate about the use of nanotechnology in the development of anticancer drugs.

Nanocarrier drug targeting systems are especially paving the way for traditional anticancer drugs that are on the verge of losing patent protection. Connecting to nanoparticles and discovering new specificities and new properties leads to entirely new products and new patent portfolios.

Nanotechnology has the potential to revolutionize tumor detection and treatment. It is well known that early tumor diagnosis is necessary even before anatomical abnormalities are discovered. Non-invasive detection of tumors at an early stage to determine the precise relationship between cancer biomarkers and clinical pathology and to maximize therapeutic efficacy has been a major challenge in cancer detection. For breast cancer, for example, a molecular imaging target should be able to accurately identify tumors with a diameter of less than 0.3 mm (about 1000 cells) for accurate diagnosis.

Targeted and localized delivery is a major challenge in cancer treatment. Effective treatment of cancer requires the ability to selectively attack tumor cells while protecting normal tissues from the increased burden of drug toxicity. However, since many anticancer drugs are simply designed to kill cancer cells and are often less specific, distributing them to normal, healthy organs or tissues is undesirable due to their potential side effects.

As a result, systemic use of these drugs usually leads to serious side effects in other tissues such as myelosuppression, cardiomyopathy, neurotoxicity, etc., basically limiting the maximum dose of the drug. Together with this, rapid clearance and widespread distribution to non-target organs and tissues require large doses of the drug, which may be uneconomical and also difficult due to non-specific toxicity. Risky cycles of high doses along with toxicity are the major drawbacks of current cancer treatments. In most cases, drug toxic side effects in patients appear long before tumors appear.

Magnetic hyperthermia is a cancer treatment process in which magnetic nanoparticles generate heat by the action of interchanging magnetic fields of appropriate amplitude and frequency, and MNPs fixed inside or near the tumor raise the tumor temperature. It kills tumor cells by necrosis when the temperature exceeds 45% and increases the effectiveness of chemotherapy when the temperature reaches near 42%. The concept of magnetothermal therapy using small particles was projected by Gilchrist et al., in 1957, but was mainly associated with inaccurate temperature measurements and poor parameters of AC magnetic fields in animal models [25]. Later, Jordan and colleagues established that iron oxide nanoparticles exhibited significantly higher specific absorption rates (SAR [W/g]) in clinically acceptable H°f combinations compared to multidomain particles [26].

It was made a big step in cancer treatment using hyperthermia techniques since then. The size and the composition of nanoparticles are important for hyperthermia techniques. Magnetic nanoparticles widely used in hyperthermia were mainly composed of Fe-O nanoparticles, but nanoparticles of other metals, such as cobalt, iron, or FeCo, increase magnetization and maximum SAR values compared to iron oxide. These metal nanoparticles have been rarely used because of their toxicity. Of the many research groups around the globe testing and developing the technology, Germany is the only country, testing autothermal therapy at the clinic of the Charite Medical University, a radiation therapy clinic in Berlin [27].

The following are a few recent and important experiments. Le Renard and colleagues examined the use of implant formation *in situ* including superparamagnetic iron oxide nanoparticles (SPIONs) as a less invasive treatment of tumors using local magnetic field-induced hyperthermia. They are developing injectable gels that trap magnetic particles in cancer cells. Furthermore, studies in mice show that thermoreversible hydrogels are inadequate and that alginate hydrogels induce robust implants localized at the tumor periphery, but organogel compositions show different microstructures. It is of clinical importance to produce co-solvent formulations containing up to 20 wt %. For magnetic microparticles with minimal toxicity and better tumor implantation [28]. Autothermal therapy can kill tumor cells alone by leaving the surrounding healthy tissue. If heat release occurs in the theoretical model assuming non-interacting nanoparticles, the basic idea of "no interaction" is violated, and consequently the interaction with biological systems and resulting heat release cannot be predicted. Experimental data from Dennis and colleagues show that the collective behavior of the kilohertz mode of magnetite nanoparticles interaction was achieved by spacing and anisotropy results in significant thermogenesis which leads to complete regression of mammary tumors in mice. Takada et al. discovered a chemotherapeutic immunotherapy (CTI) strategy to better treat melanoma by binding N-propionylcysteaminylphenol (NP-CAP) to the superficial layer of magnetite nanoparticles (NP-CAP/M). NP-CAP/M significantly inhibited tumor growth rechallenge and increased the lifespan of mice by using AMF alone instead of control magnetite or using an alternating magnetic field (AMF) without AMF [29]. Lately, Kim and his colleagues from Argonne, USA, described cell interactions with magnetic microdiscs defined as lithography with a ground state of spin vortices. Their study also exhibited that spin vortex-mediated stimulation caused 2 effects: disruption of cell membrane integrity and programmed cell death [30]. For example, the application of tens of hertz alone in 10 minutes in a lower frequency field was sufficient to cause equivalent in vitro damage to cells at 90° owing to the high temperature generated by magnetic vortex microdisk. This indicates a positive effect of tumor treatment based on nanoscale-thick magnetic vortex microdisks.

Micron-sized needles comprising magnetic nanoparticles were investigated as the best system to transfer heat to tumor material for hyperthermia treatment. The limitation of Nee'l relaxation for MNPs was described by Hergt et al. for complete models with hysteresis [31]. Chen et al. were synthesized nano magnetite and monitored for heating performance under existing conditions using blood sugar (dextrose), electrolyte (commercial saline), protein (bovine serum albumin, BSA), and viscosity (glycerol). The SAR values were not affected by protein and blood glucose, but the SAR values have fallen significantly in the electrolyte atmosphere due to Na+ precipitation. For viscous media by glycerol, the result was a decrease in SAR value by increasing glycerol concentration. This process is a fast and inexpensive method to evaluate magnetic nanoparticles to generate analytical information about the effectiveness of physiological interventions, which can significantly reduce cost and time when testing *in vitro* or animals [32] Thiopronin-stabilized nanoparticles were able to kill bacteria at from 6.25 to 50 mg/ml concentrations. This application to kill bacteria by magnetic hyperthermia will find many uses in the future to treat a wide range of infectious diseases.

2.3 Targeted Drug Delivery

The use of MNPs as carriers in drug delivery was invented in the late 1970s by Widder, Senja, and his colleagues [33]. The principle behind the idea is that targeted chemotherapeutic or therapeutic nucleic acid molecules are bound or encapsulated in the core of a magnetic particle. With an additional layer of polymer or metallic coating, this magnetic particle core enhances its functionality. After functionalization, the conjugate of magnetic particles and the therapeutic agent is administered orally or intravenously into the bloodstream. With the support of an external magnetic field, the composition of magnetic particles and the therapeutic agent is directed to the targeted site and then the drug is distributed [34].

Since this idea was expressed by Senya et al., there are many reports in the field of self-targeted drug delivery. Sun et al. utilized bacterial magnetosomes (BMs) as self-targeting drug carriers and observed antitumor properties of doxorubicin (DOX)-loaded BMs (DBMs) in EMT6 & HL60 cell lines. This study also demonstrated the healing potential of DBM in targeted therapy for liver cancer. Polyak et al explored whether cell transport strategies exist. It has been proposed that self-targeting the endothelial cells to the steel surface of intra-arterial stents is enhanced by the mechanism (i) generates cells that are sensitive to magnetic fields and (ii) magnetic moments within the MNP and field gradients around the steel stent induced by the strong magnetic field thus targeting, MNP-loaded cells on the stent wire.

In vitro studies have shown MNP-mounted bovine aortic endothelial cells (BAECs) can be targeted magnetically to stent steel wires. According to *in vivo* study, BAECs transduced with adenovirus expressing MN-loaded luciferase (Luc) targeted a stent placed in the carotid artery of mice subjected to an external magnetic field and expressed Luc than nonmagnetic controls. This was much higher. Cao and colleagues purified silicate-coated iron-carbon composite particles for use as drug carriers in targeted therapy. The 200–300 nm composite particles were efficiently obtained using a high-energy planetary ball mill and a method of reducing hydrogen. Composites have the benefits of activated carbon and magnetic Fe, so the drug adsorption and desorption cappotential are excellent, and the magnetic goal is strong. In *in vivo* experiments, the adsorbed composite particle (TcO4) Tc99 showed significant biodistribution in the left side of the liver of the pig under the influence of an external magnetic field. After the doxorubicin-adsorbed complex particles were injected into the artery, the quantity of doxorubicin in the liver tissue was 23.8 times more in the target region than in the non-target region. They also hypothesized that composite particles could penetrate through the wall around the stroma of tissues & liver cells under the control of an external magnetic field in the target region. This means that Si-coated Fe-C composite particles can be widely applied to types of tumors as effective drug carriers.

Benerjee and Chen have created a novel multifunctional magnetic nanocarrier to simultaneously perform cancer treatment and probe, programmed to respond to environmental stimuli (e.g., pH values), doxorubicin (DOX), and adipic acid created by combining dehydrazide implanted gumarabic MNPs (ADHGAMNP) modified through hydrolytically degradable pH-sensitive hydrazone bonds. A subsequent nanocarrier DOXADHGAMNP with an average diameter (~)13.8 nm could deliver about 6.52 mg of bound DOX. It exhibited pH-induced DOX release in an acidic atmosphere (pH 5.0), but was moderately stable at pH (pH 7.4). The sensitivity of this novel nanocarrier to pH was also confirmed by studying the zeta potential and optical density of plasmons at various pH values.

This nanocarrier was versatile and was used to develop a drug delivery vehicle that combines target with sensory perception and therapy [35]. However, another magnetic nanocarrier for targeted drug delivery was recently discovered by Banerjee et al. In their study, they replaced magnetic nanoparticles with gum arabic and inserted 2-hydroxypropylcyclodextrin (HCD) into them employing hexamethylene diisocyanate (HMDI) as a linker. As a result of characterizing the properties such as size, hydrodynamic diameter, and incorporation of HCD, the resulting magnetic nanocarrier displayed excellent loading capacity for an anticancer drug, trans-retinoic acid (ATRA). The enhancement in biocompatibility and specific targeting and complexation with hydrophobic drugs have made nanosystems a stimulating opportunity to target hydrophobic drugs [36].

Targeted delivery of drugs to the brain, that is, the use of magnetic nanocarriers through the vascular endothelium at the dense junction of the blood-brain barrier in brain tumor patients, is of great biomedical importance. Focus on the interaction between freely circulating magnetic nanoparticles and the blood-brain barrier, including Stepp et al. To overcome this problem, they discovered an in vivo model that quantitatively measures the fluctuations in the velocity and volume of cerebrovascular blood flow when exposed to nanoparticles that circulate under self-control.

Data from these trials suggest that the gradient magnetic field drug targeting method is suitable for treating cancer in regions located deep in the body (head, neck, chest, arms, legs) with a space of 8.7 cm.

The results showed that technology using a gradient magnetic field and magnetite nanoparticles can be used as a practical tool for the subcutaneous administration of anticancer drugs in the human body.

2.4 Magnetofection Agents

Magnetofection is a method of transfection with magnetic nanoparticles using a magnetic field that presents in nucleic acids to various target cells. Their concept is to use nucleic acids to bind magnetic nanoparticles coated with lipids or surfactants to form composites, which are then enriched & injected into various cells with an additional magnetic field, like a neodymium magnetic device (Nd–Fe–B).

Under the impact of a magnetic field, a vector with nucleic acid enriches substantially rapidly and is introduced into the cell as a gene vector at a very high dose. After 10 years of improvement, magnetofection methods have become widespread using all classes of nucleic acids. Non-viral transfection systems convert other viruses (adenoviruses and retroviruses) into other cell types as primary cells, suspension cells, and adherent cells [37,38]. Researchers reported enhanced gene delivery by using superparamagnetic nanoparticles (SPIONs) covered with polyethyleneimine (PEI) in a constant, pulsed magnetic field. They investigated the effect of the magnetic field to significantly increase the transfection efficiency by more than 40 times compared to cells that were not opened to the magnetic field.

In vivo gene transfer in synovial cells in both static and pulsed magnetic fields after intra-articular addition of SPIONS associated with a plasmid containing a reporter gene encoding a fluorescent protein. and resulting gene expression was studied. From these results, intra-articular addition of functionalized SPIONS caused mild to moderate synovitis, and the evidence for gene expression was inconclusive [39]. Lee et al. completed a non-viral delivery method based on PEI coated on the surface of bacterial magnetic nanoparticles (BMP). The ability to transfect the BMPsPEI/DNA complex was greater than that of the PEI/DNA complex because it protected the BMPsPEIDNA nano complex from the action of DNase. However, the BMPsPEI complex showed minimal toxicity to cells in vitro compared to PEI [40].

Jahnke et al. In a phase I study [41], found a treatment plan following immune system activation by using magnetofection to deliver the 3 feline cytokine genes to cancer cells. Mikhailyk et al provided protocol regarding the design and conduct of nucleic acid transfer experiments in vitro adhesion and suspension cell cultures using magneofection techniques.

2.5 Magnetic Resonance Imaging (MRI)

MRI displays images of the interior of a subject in a non-invasive method based on the properties of proton nuclei in water and lipids. MRI is broadly used for visualizing physiological changes in living materials of the body. Compared to computed tomography (CT), MRI has the following advantages: (i) uses non-ionizing radio frequency (RF) signals for imaging and is the best match for uncalcified body tissue (ii) detect different tissue features (iii) generate cross-sectional images from all images except for inclined planes using various scan parameters (iv) for further detection of tumors (v) ideal for multiple studies in a row in a short time (vi) T1-weighted MRI images, T2- and T2-weighted*, providing several contrast mechanisms. T1 relaxation is the time constant for a nuclear spin to return to equilibrium.

When the proton-nucleus changes from a higher-energy state to a lower-energy state, it is associated with a loss of energy by the adjacent nuclei. Relaxation T1 is considered as a longitudinal yield of the pure magnetic field in the basic state of the maximum length with the main magnetic field. T1 is typically about 1 second for tissue. Relaxation T2 is the decay time constant of a signal. The T2 relaxation happens during the rotation of the proton nucleus in higher or lower energy states, where energy is exchanged, but no energy is transferred to the surrounding lattice. Resulted magnetic moments interact with each other to reduce magnetization or decay after the core releases its extra energy. T2 is generally no more than 100 ms for normal tissue. T2 * is the moment when the lateral magnetization drops to 37% of the initial value. Magnetic field and occurs in all magnets. T2 * is determined by the uneven B0 and loss of transverse magnetization at a rate more than T2.

2.5.1 T1-Contrast Agents for MRI

The gadolinium-based MRI contrast agent, prepared from relatively simple chemical composition, is capable of providing sufficient contrast enhancement for a variety of applications with various sizes and chemical properties [42]. Gd components based on different-sized contrast agents exhibit different functions. For example (1) polyamidoamine (PAMAM)-based small-size dendrimer core-based contrast agent with a size of less than 3 nm easily penetrates the vascular endothelium, resulting in rapid perfusion; (2) about 3–6 nm wavelength contrast agents, rapidly expelled from the body through the kidneys, and has a positive effect on the improvement of the contrast agent function in the kidneys (3) 7–12 nm contrast agent is used as an intravenous contrast agent because contrast agent remains in the bloodstream (4) ~12–15 nm contrast agents, readily recognized by the reticuloendothelial system (RES) than fecal matter. (5) ~15–20 nm contrast agents work differently in the body.

However, various contrast agents according to the Gd component have many functions as follows (i) hydrophilic contrast agents are utilized for lymphatic imaging (2) Targeting antibodies, receptors, DNA, functional peptides can act as tumor-specific agents with diagnostic and therapeutic functions for gadolinium components. Gd-based formulations have a variety of uses as contrast agents for MRI; (3) Certain organs such as liver, spleen, lungs, kidneys, brain, and lymph node system. Earlier results targeting and surveillance node imaging have determined that Gd-based contrast agents can be further used in preclinical studies or clinical practice. A further example of a gadolinium T1 contrast agent is solid Gd_2O_3 nanoparticles. These luminescent hybrid Gd_2O_3 nanoparticles coated with a polydioxanone shell layer contain carboxylated PEG and organic fluorophores covalently bonded for immobilization on the surface of inorganic nanoparticles. These studies suggest that these particles cause improvements in MRI-positive contrast agents compared to commonly used positive contrast agents such as GdDOTA in MRI clinical practice.

2.5.2 T2/T2*- Agents for MRI

The T2/T2* agent was first improved by Ohgushi in 1978. The T2/T2* ultra-small superparamagnetic iron oxide nanoparticles (USPION) for contrast agents were introduced primarily by Weisleder and colleagues in the 1990s [43]. USPION generally induces potent transverse and longitudinal effects *in vivo* when related to Gd-based contrast agents. As n example, USPION formulation (NC100150) at 7 μmol iron/kg induced a signal drop nearly equivalent to that of a standard bolus injection of 0.2 mmol/kg gado-diamide to track cerebral blood volume (CBV) and cerebral blood measurements flow (CBF) [44]. So far, USPION has been widely used by targeted or non-targeted methods in different biological or clinical applications, such as small target molecules, targeted receptor, and magnetically labeled cells [45], atherosclerotic plaque, cellular inflammation, tissue inflammation, reticuloendothelial system (RES) including liver, spleen, islets, lungs and lymph nodes; perfusion imaging of brain, myocardium, cartilage, and kidney; MR angiography; and imaging of tumor vessels. Currently published literature, the number of T2 contrast agent nanoparticles for biomedical imaging in animal models is very large, for example, the Dai group reports FeCo core/single graphite shell nanoparticles and the Chun group reports spinel ferrite nanoparticles other than USPIO. Early *in vivo* results demonstrated long-term positive contrast enhancement in vascular MRI in a rabbit model. These nanoparticles with an inner shell serve as agents for complex diagnostic and therapeutic applications.

REFERENCES

1. J. Frenkel and J. Dorfman, "Spontaneous and induced magnetisation in ferromagnetic bodies", *Nature* 126, 1930, 274–275.
2. X. Battle and A. Labarta, "Finite-size effects in fine particles: Magnetic and transport properties", *J. Phys. D: Appl. Phys.* 35, 2002, R15–R42.

3. M. Chen, J. P. Liu and S. J. Sun, "One-step synthesis of FePt nanoparticles with tunable size", *J. Am. Chem. Soc.* 126, 27, 2004, 8394.
4. M. Vazquez, C. Luna, M. P. Morales, R. Sanz, C. J. Serna and C. Mijangos, "Magnetic nanoparticles: Synthesis, ordering and properties", *Phys. B: Condens. Matter* 354, 2004, 71–79.
5. M. Takafuji, S. Ide, H. Ihara and Z. H. Xu, "Preparation of poly(1-vinylimidazole)-grafted magnetic nanoparticles and their application for removal of metal ions", *Chem. Mater.* 16, 10, 2004, 1977.
6. J. Jeong, T. H. Ha and B. H. Chung, "Enhanced reusability of hexa-arginine-tagged esterase immobilized on gold-coated magnetic nanoparticles", *Anal. Chim. Acta* 569, 2006, 203.
7. M. Arruebo, M. Valladares and A. Gonzalez-Fernandez, "Antibody-conjugated nanoparticles for biomedical applications", *J. Nanomater.* 2009, Article ID 439389.
8. A. K. Gupta and M. Gupta, "Synthesis and surface engineering of iron oxide nanoparticles for biomedical applications", *Biomaterials* 26, 2005, 3995.
9. P. C. Lin, P. H. Chou, S. H. Chen, H. K. Liao, K. Y. Wang, Y. J. Chen and C. C. Lin, "Ethylene glycol-protected magnetic nanoparticles for a multiplexed immunoassay in human plasma", *Small* 2, 4, 2006, 485.
10. H. Churchill, H. Teng and R. M. Hazen, "Correlation of pH dependent surface interaction forces to amino acid adsorption: Implications for the origin of life", *Am. Min.* 89, 2004, 1048.
11. J. Lambert, "Adsorption and polymerization of amino acids on mineral surfaces: A review", *Orig. Life Evol. Biosph.* 38, 2008, 211–242.
12. J. Woda, B. Schneider, K. Patel, K. Mistry and H. M. Berman, "An analysis of the relationship between hydration and protein-DNA interactions", *Biophys. J.* 75, 1998, 2170–2177.
13. M. Pavlin and V. B. Bregar, "Stability of nanoparticle suspensions in different biologically relevant media", *Dig. J. Nanomater. Biostructures.* 8, 2012, 1389.
14. K. A. Melzak, C. S. Sherwood, R. F. B. Turner and C. A. Haynes, "Driving forces for DNA adsorption to silica in perchlorate solutions", *J. Colloid Interface Sci.* 181, 1996, 635–644.
15. C. M. Earhart, C. E. Hughes and R. S. Gaster, "Isolation and mutational analysis of circulating tumor cells from lung cancer patients with magnetic sifters and biochips", *Lab Chip*, 14, 2014, 78–88.
16. Y.-R. Cui, C. Hong, Y.-L. Zhou, Y. Li, X.-M. Gao and X.-X. Zhang, "Synthesis of orientedly bioconjugated core/shell Fe 3O4@Au magnetic nanoparticles for cell separation", *Talanta*, 85, 2011, 1246–1252.
17. L. Lu, X. Wang, C. Xiong and L. Yao, "Recent advances in biological detection with magnetic nanoparticles as a useful tool", *Sci. China Chem.* 58, 2015, 793.
18. H. Gu, K. Xu, C. Xu and B. Xu, "Biofunctional magnetic nanoparticles for protein separation and pathogen detection", *Chem. Commun.* 9, 2006, 941.
19. S. Sun, C. Murray, D. Weller, L. Folks and A. Moser, "Monodisperse FePt nanoparticles and ferromagnetic FePt nanocrystal superlattices", *Science* 287, 2000, 1989.
20. J. Zheng, Z. Lin, L. Zhang and H. Yang, "Polydopamine-mediated immobilization of phenylboronic acid on magnetic microspheres for selective enrichment of glycoproteins and glycopeptides", *Sci. China Chem.* 58, 2015, 1056.
21. S. Bucak, D. A. Jones, P. E. Laibinis and T. A. Hatton, "Protein separations using colloidal magnetic nanoparticles", *Biotechnol. Prog.* 19, 2003, 477.
22. T. Tanaka and T. Matsunaga, "Fully automated chemiluminescence immunoassay of insulin using antibody–protein A–bacterial magnetic particle complexes", *Anal. Chem.* 72, 2000, 3518.
23. Y. Tang, Z. Li, N. He, L. Zhang, C. Ma, X. Li, C. Li, Z. Wang, Y. Deng and L. He, "Nonporous silica nanoparticles for nanomedicine application", *J. Biomed. Nanotechnol.* 9, 2013, 312.
24. Z. Xi, R. Huang, Z. Li, N. He, T. Wang, E. Su and Y. Deng, "Selection of HBsAg-specific DNA aptamers based on carboxylated magnetic nanoparticles and their application in the rapid and simple detection of hepatitis B virus infection", *ACS Appl. Mater. Interfaces* 7, 2015, 11215.
25. A. Jordan, R. Scholz, P. Wust, H. Fahling and R. Felix, "Magnetic fluid hyperthermia (MFH): Cancer treatment with AC magnetic field induced excitation of biocompatible superparamagnetic nanoparticles", *J. Magn. Magn. Mater.* 201, 1999, 413.
26. A. Jordan, P. Wust, H. Fahling, W. John, A. Hinz and R. Felix, "Inductive heating of ferrimagnetic particles and magnetic fluids: physical evaluation of their potential for hyperthermia", *Int. J. Hyperther.* 9, 1993, 51.
27. A. Jordan, R. Scholz, K. Maier-Hauff, M. Johannsen, P. Wust, J. Nadobny, H. Schirra, H. Schmidt, S. Deger, S. Loening, W. Lanksch and R. Felix, "Presentation of a new magnetic field therapy system for the treatment of human solid tumors with magnetic fluid hyperthermia", *J. Magn. Magn. Mater.* 225, 2001, 118.
28. P. E. Le Renard, O. Jordan, A. Faes, A. Petri-Fink, H. Hofmann, D. Rufenacht, F. Bosman, F. Buchegger and E. Doelker, "The in vivo performance of magnetic particle-loaded injectable, in situ gelling, carriers for the delivery of local hyperthermia", *Biomaterials* 31, 2010, 691.

29. T. Takada, T. Yamashita, M. Sato, A. Sato, I. Ono, Y. Tamura, N. Sato, A. Miyamoto, A. Ito, H. Honda, K. Wakamatsu, S. Ito and K. Jimbow, "Growth inhibition of re-challenge B16 melanoma transplant by conjugates of melanogenesis substrate and magnetite nanoparticles as the basis for developing melanoma-targeted chemo-thermo-immunotherapy", *J. Biomed. Biotechnol.* 2009, Article ID 457936.
30. D. H. Kim, E. A. Rozhkova, I. V. Ulasov, S. D. Bader, T. Rajh, M. S. Lesniak and V. Novosad, "Biofunctionalized magnetic-vortex microdiscs for targeted cancer-cell destruction", *Nat. Mater.* 9, 2010, 165.
31. R. Hergt, S. Dutz and M. Zeisberger, "Validity limits of the Néel relaxation model of magnetic nanoparticles for hyperthermia", *Nanotechnology* 21, 2010, 015706.
32. S. Chen, C. L. Chiang and S. Hsieh, "Simulating physiological conditions to evaluate nanoparticles for magnetic fluid hyperthermia (MFH) therapy applications", *J. Magn. Magn. Mater.* 322, 2010, 247.
33. A. Senyei, K. Widder and G. Czerlinski, "Magnetic guidance of drug-carrying microspheres", *J. Appl. Phys.* 49, 1978, 3578.
34. S. C. McBain, H. H. P. Yiu and J. Dobson, "Magnetic nanoparticles for gene and drug delivery", *Int. J. Nanomed.* 3, 2008, 169.
35. S. S. Banerjee and D. H. Chen, "Multifunctional pH-sensitive magnetic nanoparticles for simultaneous imaging, sensing and targeted intracellular anticancer drug delivery", *Nanotechnology* 19, 2008, 505104.
36. S. S. Banerjee and D. H. Chen, "Grafting of 2-hydroxypropyl-β-cyclodextrin on gum arabic-modified iron oxide nanoparticles as a magnetic carrier for targeted delivery of hydrophobic anticancer drug", *Int. J. Appl. Ceram. Tec.* 7, 2010, 111.
37. F. Scherer, M. Anton, U. Schillinger, J. Henkel, C. Bergemann, A. Kruger, B. Gansbacher and C. Plank, "Magnetofection: enhancing and targeting gene delivery by magnetic force in vitro and in vivo", *Gene Ther.* 9, 2002, 102.
38. C. Plank, M. Anton, C. Rudolph, J. Rosenecker and F. Krötz, "Enhancing and targeting nucleic acid delivery by magnetic force", *Expert Opinion on Biological Therapy* 3, 2003, 745.
39. L. Q. Galuppo, S. W. Kamau, B. Steitz, P. O. Hassa, M. Hilbe, L. Vaughan, S. Koch, A. Fink-Petri, M. Hofman, H. Hofman, M. O. Hottiger and B. von Rechenberg, "Gene expression in synovial membrane cells after intraarticular delivery of plasmid-linked superparamagnetic iron oxide particles – A preliminary study in sheep", *J. Nanosci. Nanotechnol.* 6, 2006, 2841.
40. X. Li, B. Wang, H. Jin, W. Jiang, J. Tian, F. Guan and Y. Li, "Bacterial magnetic particles (BMPs)-PEI as a novel and efficient non-viral gene delivery system", *J. Gene Med.* 9, 2007, 679.
41. A. Jahnke, J. Hirschberger, C. Fischer, T. Brill, R. Kostlin, C. Plank, H. Kuchenhoff, S. Krieger, K. Kamenica and U. Schillinger, "Intra-tumoral gene delivery of feIL-2, feIFN-γ and feGM-CSF using magnetofection as a neoadjuvant treatment option for feline fibrosarcomas: A Phase-I Study", *J. Vet. Med. A* 54, 2007, 599.
42. H. Kobayashi and M. W. Brechbiel, "Nano-sized MRI contrast agents with dendrimer cores", *Adv. Drug Deliver. Rev.* 57, 2005, 2271.
43. R. Weissleder, G. Elizondo, J. Wittenberg, C. A. Rabito, H. H. Bengele and L. Josephson, "Ultra small super paramagnetic iron oxide: Characterization of a new class of contrast agents for MR imaging", *Radiology* 175, 1990, 489.
44. E. X. Wu, H. Tang and J. H. Jensen, "Applications of ultrasmall superparamagnetic iron oxide contrast agents in the MR study of animal models", *NMR Biomed.* 17, 2004, 478.
45. M. Lewin, N. Carlesso, C. H. Tung, X. W. Tang, D. Cory, D. T. Scadden and R. Weissleder, "Tat peptide-derivatized magnetic nanoparticles allow in vivo tracking and recovery of progenitor cells", *Nat. Biotehcnol.* 18, 2000, 410.

Recent Advances in Carbon-Based Nanomaterials for Spintronics

10

Trupti K. Gajaria, Narayan N. Som, Shweta D. Dabhi

Contents

10.1	Introduction	147
10.2	Experimental and Theoretical Tools for Spintronics Investigation	149
10.3	The Effects of Dimensional Confinement on Spintronic Properties	149
	10.3.1 Graphene-Based 2D Materials	149
	10.3.2 Carbon-Based One-Dimensional Materials	152
	10.3.3 Carbon-Based Quantum Dots	152
10.4	Effect of External Parameters Strain/Pressure on Spintronic Properties	153
10.5	Effect of Defects and Doping	155
10.6	Carbon-Based Devices and Applications	157
10.7	Conclusive Remarks and Future Scope	157
References		158

10.1 INTRODUCTION

The 21st century has witnessed the revolutionary advancement in cutting-edge technologies and the relevant materials utilized for transferring technology to real-life application/s. Ever since the beginning of the digital era, the storage, transfer, and processing of data have become the center of attention, and the development of nano-materials with robust properties suitable for accomplishing data-oriented tasks has become a top priority for the tech-tycoons. In this context, the famous Moore's law that predicted the number of transistors on a chip will almost double every two years [1] can hold true only if the size of the components gets reduced significantly. This implicates more components on a single chip meaning super-fast functioning and so will be the overall performance of the device. It can be presumed that there should be a limit for getting more and more transistors embedded on a single chip, as the smallest possible part would be an electron which is unphysical. However, considering the advancement in design and development, the group led by professor Ali Javey successfully created a MoS_2 based transistor with a 1 nm gate made-up of CNT [2], which indicates a possibility of constructing ultra-small devices on a single chip in near future. Although the development and scale-up are still under process, reminiscent of the famous Feynman quote: *"There're still plenty of rooms at the bottom!"* The major issue with miniaturization and compacting is that under a high density of transistors, the components will be so close to each other that the electrons instead of remaining in the gate will directly tunnel through one to another components which needs to be avoided to have an off state [3]. Such rigorous issues serve as directives for further investigations thereby reducing the intricacy of the phenomenon together with shedding light on the root cause of the peculiar dynamics within the material.

As the carrier transport and dynamics are solely governed by the electrons, it is necessary to first understand the fundamental properties and the factors controlling the same. The electron being the indivisible unit of matter possesses three properties, mass, charge, and spin. The mass of an electron is almost 1,836 times smaller than that of a nucleon, and its arrangement in form of orbital configuration along with specific symmetrized crystal lattice makes it crucial for controlling carrier dynamics within the material. As the motion of electronic charge governs the transport within the electronic devices, the controlled motion of spin together with charge is the foundation of the spintronics [4]. This implies that along with the charge, one exploits electron spin also like a degree of freedom and, therefore, the current generated under such circumstances is generally known as "spin-polarized current" or "spin current".

The era of spintronics started back in the 1960s when the research group headed by Dr. Leo Asaki at IBM who later shared the 1973 Nobel Prize for his discovery of the electron tunneling effect [5]. Following the fundamental research in this area and the development of quantum structures like superlattices inspired researchers to study magnetic multi-layered materials that finally led to the discovery of giant-magnet-resistance (GMR) in 1988 [6,7]. GMR being the foundation of the spintronics was first observed when alternate thin layers of Fe (ferromagnetic) and Cr (non-magnetic) were stacked together and subjected to an external magnetic field [8]. Such configuration showed critical modulation in the electrical resistance of the system when influenced by external magnetic fields. Further, it showed that the coupling between the layers in the presence of a magnetic field plays a key role in aligning the magnetic moments in parallel (anti-parallel) configuration to cause enormous reduction (enhancement) in the electrical resistance of the system [9]. The major applications of GMR include reading the data from hard disks, magnetic sensors, and magnetoresistive random-access memory (RAM) [10–15]. Followed by such groundbreaking discoveries, many other phenomena like anisotropic magnetoresistance (AMR), tunneling magnetoresistance (TMR), colossal magnetoresistance (CMR), spin injection, spin-transfer torque, etc., were reported.

Most importantly, the era of spintronics received an enormous boost when such phenomena were observed in non-magnetic semiconducting materials like graphene [16,17]. Carbon is the most abundant material on earth with its presence ranging from the earth's crust to all living beings to the atmosphere indicating its utilization in various energy applications. Narrowing down to the solid carbon materials, there are two stable allotropes of carbon, diamond, and graphite. Out of which the diamond is considered as the hardest crystal ever found naturally on earth. Graphite, on the other hand, received more attention after the successful exfoliation in the form of a single layer in 2004 [18]. Other nano allotropes of carbon like Buckyball/Fullerene and CNTs also received overwhelming recognition due to their dramatic carrier dynamics and exotic properties like targeted drug-delivery, quantum size effect dependent fluorescence, ultra-high thermal conductivity, supra-hard structure, ultrafast carrier dynamics, and many other properties important for the latest fields like spintronics. It is interesting to note that the researchers were able to induce magnetism and superconductivity within such materials despite lacking the d and/or f orbital electrons.

The objective of the present chapter is to weave the development and contributions of carbon-based nanomaterials in the domain of spintronics and to shed light on the advancements in the field. The proceeding section briefly describes the experimental and theoretical techniques utilized for investigating such materials. Followed by this, we individually review the contributions from all three categories of the carbon nanomaterials to the field of spintronics and the summary has been quoted in the conclusion section.

10.2 EXPERIMENTAL AND THEORETICAL TOOLS FOR SPINTRONICS INVESTIGATION

The investigation of spintronic properties requires a keen understanding of spin dynamics and the coupled role-play between electronic charge and spin. The spin of the electron was first measured by Stern and Gerlach in 1921 [19]. Following this, numerous eminent researchers contributed to the field. Finally, the application of the quantum mechanical treatment with the inclusion of relativistic effects became one of the foundations for unraveling the microscopic dynamics within complex materials. The very first spin effect field transistor constructed using metal-oxide-semiconductor technology was proposed by Datta and Das in 1989 [20]. Prior to investigating the initial step requires preparing a magnetic material. Besides metallic magnetic and non-magnetic coupled configurations, the present era is dominated by carbon-based semiconductors due to the advancement in structural and electronic manipulative techniques like introducing defects, doping, and configurational modifications. The performance of a spintronic device can be estimated through its efficiency of generating spin current [21]. Besides, the spin-injection expressing the capacity for transferring the spin from magnetic to the non-magnetic regime is a crucial factor affecting the performance of the device. The carbon-based system being organic possess low spin-orbit coupling (SOC) and weak hyperfine interaction are one of the factors affecting spin relaxation [22]. As observed the anisotropy caused by SOC results in the existence of magnetism even under monolayer configuration though with a lower magnitude of Curie temperature as compared to its bulk phase [23]. Experimental realization of anomalous Hall Effect (AHE), spin-Hall effect (SHE), magneto-optical Kerr effect (MOKE), spin-transfer torque (STT), etc., helps to quantify the candidature of material for a spintronic application.

Theoretical treatments like density functional theory (DFT) [24] on the other end are widely used for predicting almost all ground and excited-state properties of the materials ranging from metals to semiconductors to insulators and semi-metals. As the name suggests, the DFT solely depends on the electron density and as the investigation for spintronic materials requires to take into account the spin angular momentum, the so-called spin-polarized DFT calculations the energy functional depends on both minority as well as majority spin densities, and thus accounts for spin-dependent properties. The sum of both gives the charge density while the difference of the same should yield the spin density. A theoretical framework like linear response [25] and non-equilibrium Green's function [26] in conjunction with the DFT gives reliable insight into the spin dynamics within the materials.

10.3 THE EFFECTS OF DIMENSIONAL CONFINEMENT ON SPINTRONIC PROPERTIES

0D, 1D, or 2D nanomaterials can be constructed by restricting the motion of a charge carrier in either three/two or one orthogonal directions of the material. The 0D usually covers materials like quantum dots, Buckyball, and nanoparticles with extremely low diameters (< 100 nm), whereas the 1D nanomaterials have different geometries like nanowires (NWs), nanotubes (NTs), nanoribbons (NRs), etc.; whereas, the 2D materials may possess either of the mono/multilayer/heterostructured configurations. The quantum size effects get more pronounced when the size of the material approaches near the de-Broglie wavelength of the electron.

10.3.1 Graphene-Based 2D Materials

The discovery of Graphene leads to the origin of two-dimensional materials and it gains a lot of attention due to its tunable electronic properties besides their density of states acquiring constant energy variation due to the confinement effect [27]. Graphene and its 2D allotropes such as graphene, graphyne, graphitic carbon nitride, and others have been widely employed in spintronics, because of their spin transport properties and efficient spin with alteration in their properties [28–30]. Especially, Graphene has attracted researchers and scientists for spintronic devices as it is first used as a medium to transport spin information over a very long distance at room temperature due to its strong suppression of backscattering, long spin relation time, and high diffusion length [31]. This uniqueness of it is due to weak SOC and high electron mobility which leads to longer spin relation and spin diffusion length respectively along with the gate-tunable carrier concentration and absence of hyperfine interaction. One of the difficulties is that graphene cannot generate spin current and so it is not suited for direct application in spintronics devices alone. However, adding magnetism to graphene and controlling its properties through the proximity effect offers it many unique features such as induced SOC and exchange interaction making it appropriate for spintronic devices [32]. In this section, we are going to discuss detailed studies about the graphene spintronic fields including spin injection, proximity effect, twistronics, enhancing the SOC effects by forming heterostructure and tuning spin relation time [33,34].

Most of the studies focus on the enhancement of SOC present within graphene as its hindrance to the control and tunability of the spintronic device. David et. al, induced the SOC through the virtual band to band tunneling by forming the bilayer heterostructure of graphene and transition metal dichalcogenide (TMDC) [33]. In this approach, they have derived the relation between the intrinsic property of considered TMDC and induced SOC in graphene using twistronic motivating from previous studies of density functional theory and tried to fill the gap between the experimental and theoretical results that is the mismatch between the layers of the heterostructure. They have successfully derived the Hamiltonian for the Dirac point of graphene under the influence of the top layer of TMDC (MoS_2). In this approach, the tunneling is considered from the orbital to the energy band with the perception of the Dirac point lining with the bands of TMDC while considering TMDC valance and conduction contribution of near chalcogenides layers for tunneling. They have shown that with the fixed tunneling matrix satisfied with satisfied quasimomentum conservation similar to bilayer graphene and Figure 10.1(**a**) shows the twisted graphene over TMDC monolayer and (**b**) Backfolded TMDC BZ vectors satisfying the quasimomentum conservation for the rotated Dirac point of graphene. They have validated their approach by calculating valley Zeeman and the Rashba type SOC shown to be well-matched with the previous report. Their studies confirm that induced SOC depends on the twist angle and induced SOC even disappears with twisting due to the presence of Dirac point within the TMDC bandgap. The maximum induces valley Zeeman was observed at a favorable position of Dirac Point that is close to the conduction band of TDMC. Recently, graphene has been shown to generate spin current while stacking with metallic $2H-TaS_2$ [34] which induces Rashba SOC in graphene by forming heterostructure. Cho et al. confirming their experimental finding with the tight-binding model and DFT calculation, that is, the spin-splitting of 5-layer graphene-TaS_2 Dirac – likes states with a spin gap of about 70 meV. The formation of spin-helical 2D Dirac-like fermions was observed by them which resides at the interface of $5LG/2H-TaS_2$ heterostructure and their work motivates the use of nonmagnetic van der Waals materials for the generation and manipulation of spin current of graphene. At the same time, another group Hoque and co-workers [35], studied the proximity-induced spin-galvanic effect in graphene with semimetal $MoTe_2$ van der Waals heterostructure with seat room temperature, a phenomenon govern by spin-charge conversion. In this work, the experimental spin-galvanic signals are explained using theoretical calculations based on spin-orbit induced spin-splitting in the graphene bands of the heterostructure. They observed with the implication of the gate electric field alters the spin properties that is spin-orbit interaction (SOI) and Rashba effect of the graphene-$MoTe_2$ heterostructure, therefore they applied electrified on heterostructure using density functional theory to explore the cause. On applying different strength of electric field on hetero-structure, they found that splitting of the low energy bands in the Gr-$MoTe_2$ interface was tuned and splitting of Rasbha SOI is large in case of the negative field as compared to positive filed. Spin galvanized effect in proximities graphene is due to inverse Rashba-Edelstein effect in the context of proximity induced SOI from $MoTe_2$. In a nut-shell, Graphene develops a strong SOI and a spin texture with spin-split conduction and valence bands as a consequence of the proximity effect, making it excellent for tuning with a gate voltage. Indeed, the heterostructure of graphene with TMDC induced SOC, spin current in the graphene monolayer, besides the spin lifetime anisotropy of

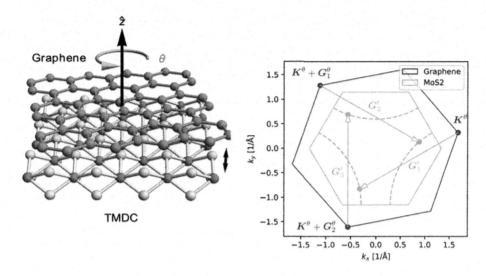

FIGURE 10.1 (Left) 3D view of graphene on top of monolayer TMDC. θ represents the twist angle between graphene and the TMDC layer, and d⊥ is the perpendicular distance between graphene and the upper (closest) chalcogen layer of the TMDC. (Right) Shows the backfolded TMDC BZ vectors satisfying the quasi-momentum conservation for the rotated Dirac point of graphene K^θ. The dashed lines indicate the full paths of the back-folded vectors in the range of twist angles θ ∈ [0,π/3]. Moreover, $G^\theta_{1,2}$ are rotated reciprocal lattice vectors of graphene, while $G'_{1,2,3}$ are reciprocal lattice vectors of the TMDC. As an example, here we have shown in orange the BZ of MoS_2 (with lattice constant aT = 3.15 Å). Adapted with permission from reference [33].

Source: Copyright (2019) American Physical Society.

graphene also shown to be prolonged with the proximity effect using the first principle-based model and it can be used as experimental evidence for observing the strong spin-valley locking for both the graphene-TMDC and TMDC heterostructures [33,35,36].

The realistic quantum spin dynamic modeling is found to be applicable for spin relaxation mechanism dominating by SOC effect together with anisotropy. They observed the anisotropy of TMDC was weakly impacted due to the electron-electron or phonon-electron coupling. Not only has the formation of heterostructure made the graphene properties suitable for applicable spintronic but the heterojunction of graphene with Ferromagnetic metals (FM) is one of the ways to realize efficient spin injection into graphene mainly depend upon the interfacial hybridization and magnetic exchange interaction [37]. Liu et, al., underline the fact to realize a better graphene-based transistor, the existence of a dead magnetic layer at the interface of FM/graphene may be hindrance for applicability. Therefore, they come up with the unique a specially designed FM1/FM2/SC structure (FM1 = 30 MLs Ni, FM2 = 1 ML Fe, and SC = graphene) to simulates the realistic FM/graphene interface of the proposed graphene-based transistor using Density functional theory and X-ray magnetic circular dichroism (XMCD). The Fe/graphene interface have been found to stabilize with the inclusion of Ni layer atop Fe film and obtained magnetic moment of Fe ML on graphene order of 1.23 μB/atom and validated the experimental results with density functional theory. This works motivates to interface engineering of heterojunction Transition metal ML over graphene for spintronic technology. One of the major drawbacks of graphene is that it cannot generate spin current alone, however recent study shows that bi-layer graphene by the proximity of an interlayer antiferromagnet (CrSBr), there is a large, induced exchange interaction is observed which give rise to spin-charge coupling in it [32]. Ghiasi et al. [32] observed the spin polarization of conductivity (up to 14%) and a spin dependent Seebeck effect in the magnetic graphene. This offers ultrathin magnetic memory and sensory devices based on magnetic graphene.

One of the recent articles by Choudhuri et al. [7] suggests carbon nitride as a promising material for spintronics applications. However, this system is non-magnetic and experimentally able to induce magnetism into it with doping of transition metal. The other allotropes of graphene such as Twin T-graphene, T-graphene (TG), PHE-graphene, THD-graphene, R-graphyne (RGY), and a-graphyne have been predicted theoretically. Recently, Bhattacharya et, al., [38] showed Twin T-graphene with nitrogen doping exhibits bipolar magnetic semiconductors nature and it is a potential candidate for spintronics devices.

10.3.2 Carbon-Based One-Dimensional Materials

Quantum materials, the nanomaterials showing remarkable quantum effects under confined dimensions are of crucial importance for nanodevices utilized in cutting-edge technologies. Especially, when any two of the three dimensions are confined for electronic motion, the unique van Hove singularities are observed, which are usually responsible for the exotic properties of one-dimensional (1D) nanomaterials. 1D nanomaterials can be synthesized using various methods like for CNTs, chemical vapor deposition, arc discharge, etc. have received significant acceptance, whereas the defect-free crystalline NWs growth can be achieved via DC arc discharge [39], molecular beam epitaxy [40], etc. Further, for multi-walled CNTs, catalytic chemical vapor deposition (CCVD) is largely utilized [41]. CNTs can be viewed as the rolled-over sheets of graphene that show diverse properties depending on the rolling direction. CNTs with different chirality can be made either single-walled or multi-walled [42,43]. There exist many ways of controlling the charge and spin transport through the CNTs; for example, rotation, subjecting to a magnetic field, edge-passivation, chemical functionalization, creation of defects/doping, and many more [44,45]. Besides this, the twist-dependent modification in the properties of the material has become the locus of the device design and development [46].

In this regard, one of the reports [47] suggests the chemical functionalization of CNT results in giant spin signals showing feasible spin transport guided via MWCNTs. The article [47] sheds light on how the weak SOC and absence of hyperfine interactions within the functionalized MWCNTs aid in transporting spin currents across millimeter distances, which can be considered as a milestone of the spintronics journey as the prior findings could transport the spin currents across micro-meter distances only. Another 1D form of graphene that can be made via limiting one of the lateral dimensions of graphene sheet to have a thin strip-like structure is referred as nano-ribbon. Following the numerous applications of graphene nanosheets, the graphene nanoribbon (GNR) fetched the attention of the device fabrication industries through its unique and exotic carrier dynamics and geometric configuration. Depending on the edge termination of graphene, the GNR can be classified either as *zigzag* or *armchair* [46]. Furthermore, doping, defect, and edge functionalization in carbon-based NRs have been observed to show remarkable spintronic properties. It is shown that switching between the ferromagnetic (FM) and anti-ferromagnetic (AFM) states can be achieved using modulating the edge termination [48], and terminating both edges at phosphorous atoms, the NR mimics the behavior of z-GNR. Furthermore, passivating the NRs with hydrogen resembles the phosphorene-like properties and provides an opportunity to electrostatically modify the electronic properties of the system [48].

10.3.3 Carbon-Based Quantum Dots

Carbon-based nanostructures namely graphene quantum dot (GQD), carbon nanoparticles, Fullerene, etc. with the tiny size of a few nanometers possess extraordinary properties due to the quantum confinement effect in all three dimensions and are also known as 0D materials. QDs generally possess many intriguing properties as compared to their bulk counterparts. Their availability in different shapes like spherical, triangle, square, hexagonal, etc., and sizes with tunable semiconducting nature make them desirable for diverse applications. The possibility to make carbon-based quantum dots to be magnetic in nature has grabbed the attention of the scientific and technological communities. As mentioned earlier, the carbon does not contain d- or f- electrons so the resultant magnetic moment would not be persistent. It was observed that the z-graphene flakes terminated with hydrogen atoms possess spin-ordered electronic states giving upswing to magnetic quantum phenomena. Though C60 or Pd do not show any ferromagnetic behavior in normal conditions, a system made up of Pd/C60 bilayer exhibits ferromagnetic characteristics with high Curie temperature (>500 K) [49]. Ferromagnetic behavior has been observed when irradiated by light on C60 due to oxidation [50]. Room temperature vertical spin valves constructed using bathocuproine has been studied by Gobbi et al. [51]. In this device, transport takes place through the organic layer; even at room temperature, the electron spin coherence is preserved up to 60 nm distance.

Intrinsic magnetism of graphene quantum dots is experimentally studied by Y. Sun et al. [16] Due to edge passivation, Curie like paramagnetism has been observed with local moment 1.2 µB at 2 K. Triangular graphene flakes with zigzag edges have been studied by first-principles calculation [52] with a major focus on geometry and size-dependent transport properties of graphene flakes that can be useful for spintronic device applications. By varying the bias voltage, the transition from FM to AFM state can also be achieved. Controlling the quantum state of an electron spin at room temperature has been reported by B. Náfrádi et al. [53] for conducting carbon nanospheres. Semiconducting z-graphene nanoflakes (with tunable energy band gap) show strong edge magnetism due to the localized edge states [54]. These graphene nanoflakes with zigzag edges show FM, AFM, or NM behavior depending on the size of the nanoflakes. Moreover, the bandgap can also be controlled with the system size and hence make them strong candidates for future spintronic devices.

10.4 EFFECT OF EXTERNAL PARAMETERS STRAIN/ PRESSURE ON SPINTRONIC PROPERTIES

The strain plays a vital role in altering the fundamental properties of materials, it is considered an alternative approach to tailoring the band gap of graphene and its allotrope for Spintronic applications. Strain is an avoidable parameter that arises due to the mismatch between the hetero-structure and substrates. Moreover, single or multiple layers can certainly adapt to the geometry of the patterned substrate underneath them, resulting in layer deformation that is subjected to a strain field [27]. The stranstronic leads to the emergence of a strong pseudogauge field which in turn manipulates the band nature and even affects the Dirac electron of graphene [55]. Banerjee et. al., carried out the periodic modulated strain up to 10% on graphene to study its effect on Dirac electrons. Pseudogauge potential is produced by this lattice strain, which may be seen as rippling graphene at the high strain in Figure 10.2.

In Figure 10.2, strain deforms the hexagonal structure of the Brillouin zone, and this strain gradient also gives rise to pseudomagnetic field which is perpendicular to the graphene plane. The superlattice region of short and long C-C bond length is viewed as locally dense and rare regions respectively, this altering zones of oppositely pedagogue fields energies at the interface between them (Figure 10.2(**d**)). These fields are spatial derivatives of potentials that increase with the strain gradient. They observed that the pseudomagnetic fields also form the oppositely propagating valley-Hall edge state in rippled graphene and higher density of states at positive biases on flat terrace usually observed graphene grown on metallic substrate Cu. Moreover, the realization of magnetic graphene due to the creation of a pseudomagnetic field is subjected to high strain without the distortion in the geometry is a difficult task except for the rippled graphene. Thereafter, Alimohammadian et. al. applied temperature and magnetic fields simultaneously on graphene and observed weak ferromagnetic behavior in graphene flakes experimentally by vibrating sample magnetometer (VSM) which is an important aspect for spintronic [56].

Moreover, Graphene exhibits spin-polarized edge states in form of z-GNR. One of the edges of GNR is spin-polarized with FM states while the other side is AFM and can be half-metallic with a proper implication of a strong electric field. However, strong electric fields limit this process, thereafter Zhang et al. [57], realized the half-metallicity of GNR through in-plane bending leading to inhomogeneous

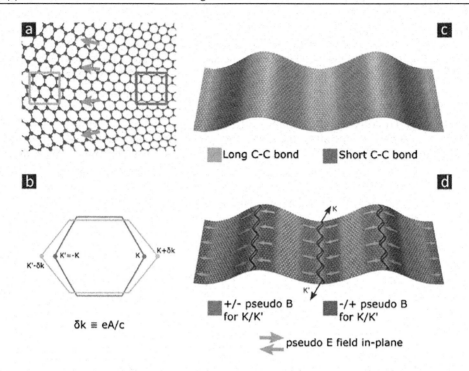

FIGURE 10.2 (a) High (low) density of carbon atoms and, hence, electrons are created in regions marked by light-blue (yellow) regions due to a strain gradient. This inhomogeneous charge distribution results in an electric field (green arrows). (b) Stretching of bonds cause the Dirac cones at K and K' points to shift symmetrically (yellow) from their original unstrained positions (light-blue) in the reciprocal space. As a momentum shift δk can be interpreted as generating a pseudovector potential term eA/c18 (where e is the electronic charge and c is the velocity of light), this creates pseudomagnetic fields with opposite signs at the two valleys. (c) The strain associated with rippling creates rare (yellow) and dense (turquoise) regions in the graphene, effectively acting as two different materials in a superlattice. (d) Pseudofields form near the interfaces of these "materials", both electric (green arrows) and magnetic (red/blue regions indicating the ±ẑ field direction for pseudospin up electrons respectively; pseudospin down are flipped). The up and down magnetic fields are separated by only a few nanometers, on the same order as the magnetic length, making the individual Landau levels to interact. LDOS peaks are maximized at the ripple crests and troughs, where valley polarized snake states (violet curved lines) are also expected to form due to the reversal of the pseudospin dependent pseudomagnetic fields across these lines. Adapted with permission from Reference [55].

Source: Copyright (2020) American Chemical Society.

strain that is compressive and tensile repose of NR. This strain-induced half metallicity in the z-GNR has been studied with the use of the generalized Bloch theorem coupled with self-consistent charge density-functional tight-binding (SCC-DFTB). The considered maximum strain in the in-plane bending deformation is up to 5%. If the strain is applied individually that is tensile and compressive then band gaps in typical covalent semiconductors vary linearly and sensitively with strain, with the energies of the band edge states moving in opposing directions. However, with their method, using in-plane bending, a relatively low-level strain is required to achieve half-metallic states. This method is similar to the in-plane bending of graphene over a flexible substrate.

At the same time Ni et al., [58] investigated spin calorintronic properties of the nanobubble at the edge of z-GNR under strain influence using first-principle calculation combined with the non-equilibrium Green's function approach. The geometry strain highly influences spin polarization, magnetoresistance, and Seebeck coefficients of nanobubble (NB)-GNR. They observed NB-NGR can generate a perfect bidirectional spin-polarized current under the temperature gradient. This straintronic extended to other graphene allotropes one of such is γ-graphyne nanoribbon (γ-GYNR), recently Li et al. investigated spin-dependent properties of (γ-GYNR) between two gold electrodes using DFT [59]. They have been studied in three distinctive ways: the molecular junction without strain and scattering region is flat (M1); the second molecular junction appears to be curved in the x-axis with a U-curved structure (M2) and the third molecular junction holds an S-curved structure and has twice the scattering region as compared to M1 with the strain effect it can be bent into the opposite direction of +x and −x-axis. M1 is observed to be semiconductor in nature, the more transmission peaks near Femi level are observed in M2 under stain effect while in M3 spin-splitting phenomenon under strain effect. M3 junction exhibit magnetic due to asymmetric S-shape structure and change in electric dipole under strain make it more suitable for device application. The structural distortion-induced strain in graphene has been investigated experimentally by Hsu et. al., they have strain-induced giant pseudo-magnetic fields and global valley polarization by direct STM/STS studies. They demonstrate that nanoscale strain engineering can provide a controlled method for changing topological states in monolayer graphene and can induce giant pseudo-magnetic fields (up to 800 T) with desired spatial distributions by global inversion symmetry breaking due to graphene over nanostructure [60].

10.5 EFFECT OF DEFECTS AND DOPING

For spintronic device application, inducing of magnetic moment is a crucial and important factor, and this alteration in magnetic properties of materials such as graphene which diamagnetic nature can lead to meet the demand of magnetic storage with help of two-dimensional thin magnets [28]. There are many several studies on magnetic graphene due to the functionalization graphene with hydrogen atoms called as graphone, heavy adatoms, vacancy defects, and molecular doping [61–63]. The vacancy defect is shown to be effective for having ferromagnetic carbon systems such as Graphite, single-walled nanotube, carbon nanofoam, and fullerenes [64]. It is noteworthy that using DFT, it is shown that in graphite and diamond the magnetization depends on the spatial distribution and increasing the vacancies in these materials leads to a decrease in the magnetization. The induced magnetization is observed together in the vacancy with the presence of a nitrogen atom near the vacancy position of graphite [61]. The increase in vacancies leads to the formation of dangling electrons but all are occupied in strengthening the sp^2 (C-C) bonds in graphite resulting in no electron available for the localized spin-polarized level.

However, nitrogen doping inspired much research to explore the magnetism in graphene, graphdiyne, and zig-zag GNR [65,66]. However, graphene is derived from graphite, shows vacancies induced magnetic moment, and is confirmed with magnetotransport measurements and spin-polarized density-functional theory calculations by Chen et. al. [67], they have calculated charge distribution for the spin-up and spin-down states for graphene with two A sublattice vacancies and one B sublattice vacancy using DFT, as shown in Figure 10.3. The vacancy in graphene leads to Jhon-Teller distorted triangle was observed with magnetic moment 1.5 μ_B and the localized Vσ state is split into highly localized Vσ1, Vσ2, and Vσ3 states (Figure 10.3) due to this distortion with single carbon vacancy. The increment in the magnetic moment was observed and it was about 1.7 μ_B with two A sublattice vacancies which arises due to the strong interaction between the local moments through the wave functions of Vσ states whereas the lower value in the case of a single vacancy was due to AFM coupling between the localized and the itinerant band spins and confirms with experimental results.

Non-carbon atom doping of the graphene lattice has been found as a promising method for imprinting magnetic ordering into graphene, which is desirable for spintronic, optoelectronic, and magnetooptical applications [65]. Błonski et al. observed the magnetic induction in graphene with nitrogen doping which is concentration-dependent, below 5% of nitrogen doping, graphene sustains its non-magnetic

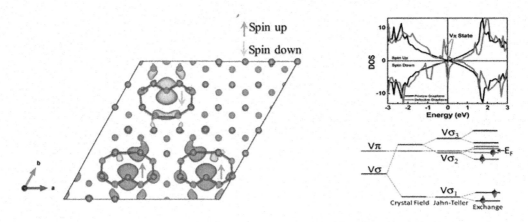

FIGURE 10.3 The charge distribution for the spin-up and spin-down states for graphene with two A sublattice vacancies and one B sublattice vacancy (left side) and the density of states of pristine and vacancy introduced graphene with a schematic diagram of band splitting shown below. The figures are reproduced from reference [67].

Source: Copyright (2014) Royal Society of Chemistry. The article was printed under a CC-BY license.

behavior while for 5.1% of doping, it acquires ferromagnetic behavior at 65K. The number of induced paramagnetic centers increased as the concentration of nitrogen was increased and it became firmly embedded in the crystal lattice of graphene, eventually forming magnetically active motifs with conduction electrons providing interaction pathways between them and establishing long-range magnetic ordering when the temperature was decreased. These results are also validated using DFT calculations. Inspired by the nitrogen doping introduced magnetism in the NM materials, Zhang et al. investigated magnetic characteristics of graphdiyne powder by magnetic moment measurement with temperature changing from 2 K to 300 K. and they observed graphdiyne exhibit large saturation moment Ms = 0.51 emu/g at 2K due to the sp-hybridization caused by the formation of acetylenic bonds resulting in a locally increased magnetic moment and pyridine nitrogen is more beneficial to enhance magnetism of graphdiyne [66].

Nitrogen doping plays different roles in graphene, graphdiyne, and even z-GNR which are antiferromagnetically coupled to opposite sides of each other, therefore one N-doping energetically prefers to be localized at the ribbon edge, whereas, two N-doping each on the opposite side leads to nonmagnetic behavior. As Nitrogen doping induces electron carriers, whereas Boron induces holes carriers; recently B-doped armchair GNR (7AGNR), which has a width of seven C atoms deposited on an Au(111) substrate, to overcome the drawback of the previous study that is the valence and conduction bands of the B-doped 7AGNR are revealed to contain a contribution from the dopant states. They have investigated the spin-polarized electron transport property of Fe-codoped 7AGNR junction. They observed the spin polarization obtained for the codoped junction and only Fe- functionalized is 0.96 and 0.7 respectively and act as a spin filter for the codoped system [68]. Similar, to the observation with B-doped graphene, and N-doped graphene, both dopants are also studied for graphitic heptazine carbon nitride (gh-C_3N_4), where it is shown that low concentration leads to the loss of the half-metallicity nature as concentration plays an important role in inducing magnetism and half-metallicity. They have also investigated the strain effect for B-doped gh-C_3N_4 and observed the loss of half-metallicity at 5% and 1.5% of compressive and tensile strains, respectively. In the case of graphyne nanoribbon, medium-high concentration Cobalt doping leads to FM nature which can act as a perfect spin-filter [7,69]. We observed that the non-carbon doping and vacancy indeed include magnetism in non-magnetic carbon materials and this modulation in properties makes it suitable for spintronic devices.

10.6 CARBON-BASED DEVICES AND APPLICATIONS

Due to high carrier mobility and low SOC, graphene is an exceptional material for lateral spin transport. The spin degree freedom of electrons has been utilized in spintronics for data storage and logic gate devices. Carbon-based materials have potential applications for transistors, capacitors, logic gates, memory devices, and spin valves. The characteristic parameters related to electronic transport can be affected by magnetization which leads to anomalous behavior in spin transport. Utilizing the spin component in the electronic devices; the future generation nanodevices are made to reduce the power consumption as well as to increase the memory and processing capabilities. A smart insight of an electron spin state is via a quantum bit (qubit) as they can be switched between the spin-up and spin-down quantum states. In this section, electronic devices made up of carbon-based materials viz. semiconductor QD, NT/NW, graphene allotropes have been discussed.

To obtain the magnetic edges, Y. Sun proposed GQDs passivated by hydroxyl group with zigzag edges [16] which can be useful for spintronics applications. Two weakly coupled QDs are constructed in series to form a double QD spin-valve such that individual split gates are magnetized in parallel or antiparallel direction. Spin valves fabricated using CNTs show very sharp resistance switching. They have also reported magnetization reversals using the AMR experiments [70]. CNTs functionalized with DNA can be useful as a source and detector of spin-polarized charge carriers due to the helical structure of DNA [71]. Further, spin polarization increases with the length of the CNT. Transistors scaled up to 5-nm Gate length show excellent performance reaching to subthreshold swing value of 70 mV/dec under 0.1 V bias [72]. The scalability perspective of CNT-based field-effect transistors in the limit of sub-5 nm Gate length has been studied using density functional theory calculations [73].

Bullard et al. have proposed spintronic devices made up of topological frustration. Triangular zigzag graphene nanoflakes (or triangulene) like structures possessing free radicals have a net spin and potential applications in logic devices and optoelectronics [74]. Further, scaling of the spin can be obtained with the size of the triangular graphene nanoflakes as the number of free radicals also changes with the number of benzene rings. Thus, desired properties can be achieved by tuning the geometrical properties. Besides this, Zhou et al. [75] have prepared graphene-passivated cobalt electrodes which can be used for vertical spin valves. It also reports the fabrication of a fullerene C60 based organic spin valve. There are many opportunities utilizing carbon-based materials for spintronics device fabrications.

10.7 CONCLUSIVE REMARKS AND FUTURE SCOPE

In summary, we revisited the outstanding research findings of inducing magnetism in non-magnetic semiconductors with the main focus on carbon-based nanomaterials. We witnessed that a lot of efforts have been made to introduce magnetism in graphene together with enhancement in SOC through unique approaches such as spin-injection, proximity effect, twistronics, tuning spin relaxation time, etc. These approaches are indeed inducing strong SOC effects under certain conditions and are suitable for spintronic devices. The study of twisting the angle between the graphene and TMDC showed that the SOC effects can be significantly affected and even can vanish in graphene as a function of the twist angle. Li et. al. also showed the dependence twisting angle and observed the maximum enhancement of the spin-splitting of graphene with 20° angle [36]. The rotation also changed SOC type from Zeeman type to Rashba type and spin-splitting is found to be sensitive towards the gate induced potential. Both studies show that controlling of twisting of hetero-structure might give more rooms to explore the materials for the betterment of the device performance.

The heterostructure with the semi-metals and metallic TMDC motivates the use of nonmagnetic van der Waals materials for the generation and manipulation of spin current of graphene, which can be extended further under the influence of non-carbon dopants or strain. Apart from the Graphene its other 2D allotropes, are shown to have potentials for spintronic devices, but their realization for the device is yet not fully advanced. However, graphene confined zig-zag nanoribbon and graphene quantum dot (GQD) with zigzag edges passivated by hydroxyl groups [16] have possibly substantial spin-polarized edge states. Generally, most of the GQD and CQD are non-magnetics, therefore their applicability to the spintronic device is low. The induction of magnetism into a non-magnetic material can be achieved via functionalization/doping or applying strain/pressure on the material. The most inspiring approach for the synthesized GQD (with the average diameter of ca. 2.04 nm) was shown to exhibit the purely Curie-like paramagnetism with the local moment of $1.2\,\mu_B$ at 2 K by Sun et. al. They observed that passivating the zigzag edge of GQD by hydroxyl group can be a promising method to get the edged magnetic GQD and can be an alternative candidate for the spintronic device.

Among the many other dopants, N and B have been widely used for induction of magnetism in carbon nanomaterials, it is interesting to note that Graphene and graphitic heptazine carbon nitride (gh-C3N4) exhibit non-magnetic nature at low concentration whereas at higher concentrations, they exhibit magnetic behavior. Even codoping with transition metal dopants may lead to interesting results as observed in case of 7AGNR. Indeed, the functionalization, straining and doping induced the magnetism in non-magnetic materials have open-up the door for utilization of these materials for spintronic applications. The studies show that induction of magnetism depends on the concentration and lots of factors are needed to underline while fabricating devices, strain is unavoidable factor which also affects the properties of device. This chapter will be helpful to experimentalist as well as theoretical researchers for the aspect of magnetic properties and device fabrication.

REFERENCES

[1] G. W. Jeon, K. W. Lee, and C. E. Lee, *Layer-Selective Half-Metallicity in Bilayer Graphene Nanoribbons*, Sci. Rep. **5**, 9825 (2015).

[2] M. Perucchini, E. G. Marin, D. Marian, G. Iannaccone, and G. Fiori, *Physical Insights into the Operation of a 1-Nm Gate Length Transistor Based on MoS2 with Metallic Carbon Nanotube Gate*, Appl. Phys. Lett. **113**, 183507 (2018).

[3] R. Hanson, L. P. Kouwenhoven, J. R. Petta, S. Tarucha, and L. M. K. Vandersypen, *Spins in Few-Electron Quantum Dots*, Rev. Mod. Phys. **79**, 1217 (2007).

[4] I. Žutić, J. Fabian, and S. Das Sarma, *Spintronics: Fundamentals and Applications*, Rev. Mod. Phys. **76**, 323 (2004).

[5] D. N. Langenberg, *The 1973 Nobel Prize for Physics*, Science. **182**, 701 (1973).

[6] G. Binasch, P. Grünberg, F. Saurenbach, and W. Zinn, *Enhanced Magnetoresistance in Layered Magnetic Structures with Antiferromagnetic Interlayer Exchange*, Phys. Rev. B **39**, 4828 (1989).

[7] I. Choudhuri, P. Bhauriyal, and B. Pathak, *Recent Advances in Graphene-like 2D Materials for Spintronics Applications*, Chem. Mater. **31**, 8260 (2019).

[8] E. Dagotto, *Brief Introduction to Giant Magnetoresistance (GMR)* (Springer, Berlin, Heidelberg, 2003), pp. 395–405.

[9] S. S. P. Parkin and D. Mauri, *Spin Engineering: Direct Determination of the Ruderman-Kittel-Kasuya-Yosida Far-Field Range Function in Ruthenium*, Phys. Rev. B **44**, 7131 (1991).

[10] J. M. Daughton, *GMR Applications*, J. Magn. Magn. Mater. **192**, 334 (1999).

[11] C. Reig, M.-D. Cubells-Beltrán, and D. Ramírez Muñoz, *Magnetic Field Sensors Based on Giant Magnetoresistance (GMR) Technology: Applications in Electrical Current Sensing*, Sensors **9**, 7919 (2009).

[12] A. Kurnicki, *Needle Type GMR Sensor in Biomedical Applications*, Przegląd Elektrotechniczny **89**(3b), 297 (2013).

[13] E. T. Enikov, G. Edes, J. Skoch, and R. Anton, *Application of GMR Sensors to Liquid Flow Sensing*, J. Microelectromechanical Syst. **24**, 914 (2015).

[14] V. D. Krishna, K. Wu, A. M. Perez, and J.-P. Wang, *Giant Magnetoresistance-Based Biosensor for Detection of Influenza a Virus*, Front. Microbiol. **7**, 1 (2016).
[15] I. Bakonyi and L. Péter, *Electrodeposited Multilayer Films with Giant Magnetoresistance (GMR): Progress and Problems*, Prog. Mater. Sci. **55**, 107 (2010).
[16] Y. Sun, Y. Zheng, H. Pan, J. Chen, W. Zhang, L. Fu, K. Zhang, N. Tang, and Y. Du, *Magnetism of Graphene Quantum Dots*, NPJ Quantum Mater. **2**, 5 (2017).
[17] M. Pizzochero and E. Kaxiras, *Imprinting Tunable π-Magnetism in Graphene Nanoribbons via Edge Extensions*, J. Phys. Chem. Lett. **12**, 1214 (2021).
[18] K. S. Novoselov, A. K. Geim, S. V. Morozov, D. Jiang, Y. Zhang, S. V. Dubonos, I. V. Grigorieva, and A. A. Firsov, *Electric Field Effect in Atomically Thin Carbon Films*, Science. **306**, 666 (2004).
[19] W. Stern, and O. Gerlach, *Experimental Determination of the Magnetic Moment of Silver Atom*, Zeit. Für Phys. **8**, 110 (1921).
[20] S. Datta and B. Das, *Electronic Analog of the Electro-optic Modulator*, Appl. Phys. Lett. **56**, 665 (1990).
[21] A. Hirohata, K. Yamada, Y. Nakatani, I.-L. Prejbeanu, B. Diény, P. Pirro, and B. Hillebrands, *Review on Spintronics: Principles and Device Applications*, J. Magn. Magn. Mater. **509**, 166711 (2020).
[22] S. Sanvito, *Molecular Spintronics*, Chem. Soc. Rev. **40**, 3336 (2011).
[23] S. P. Gubin, Y. A. Koksharov, G. B. Khomutov, and G. Y. Yurkov, *Magnetic Nanoparticles: Preparation, Structure and Properties*, Russ. Chem. Rev. **74**, 489 (2005).
[24] W. Kohn, A. D. Becke, and R. G. Parr, *Density Functional Theory of Electronic Structure*, J. Phys. Chem. **100**, 12974 (1996).
[25] D. Ködderitzsch, K. Chadova, and H. Ebert, *Linear Response Kubo-Bastin Formalism with Application to the Anomalous and Spin Hall Effects: A First-Principles Approach*, Phys. Rev. B – Condens. Matter Mater. Phys. **92**, 184415 (2015).
[26] J. Taylor, H. Guo, and J. Wang, *Ab Initio Modeling of Quantum Transport Properties of Molecular Electronic Devices*, Phys. Rev. B – Condens. Matter Mater. Phys. **63**, 245407 (2001).
[27] E. Blundo, E. Cappelluti, M. Felici, G. Pettinari, and A. Polimeni, *Strain-Tuning of the Electronic, Optical, and Vibrational Properties of Two-Dimensional Crystals*, Appl. Phys. Rev. **8**, 021318 (2021).
[28] W. Han, R. K. Kawakami, M. Gmitra, and J. Fabian, *Graphene Spintronics*, Nat. Nanotechnol. **9**, 794 (2014).
[29] M. R. Rezapour, C. W. Myung, J. Yun, A. Ghassami, N. Li, S. U. Yu, A. Hajibabaei, Y. Park, and K. S. Kim, *Graphene and Graphene Analogs toward Optical, Electronic, Spintronic, Green-Chemical, Energy-Material, Sensing, and Medical Applications*, ACS Appl. Mater. Interfaces **9**, 24393 (2017).
[30] E. C. Ahn, *2D Materials for Spintronic Devices*, NPJ 2D Mater. Appl. **4**, 17 (2020).
[31] N. Tombros, C. Jozsa, M. Popinciuc, H. T. Jonkman, and B. J. Van Wees, *Electronic Spin Transport and Spin Precession in Single Graphene Layers at Room Temperature*, Nature **448**, 571 (2007).
[32] T. S. Ghiasi, A. A. Kaverzin, A. H. Dismukes, D. K. de Wal, X. Roy, and B. J. van Wees, *Electrical and Thermal Generation of Spin Currents by Magnetic Bilayer Graphene*, Nat. Nanotechnol. **16**, 788 (2021).
[33] A. David, P. Rakyta, A. Kormányos, and G. Burkard, *Induced Spin-Orbit Coupling in Twisted Graphene-Transition Metal Dichalcogenide Heterobilayers: Twistronics Meets Spintronics*, Phys. Rev. B **100**, 85412 (2019).
[34] L. Li, J. Zhang, G. Myeong, W. Shin, H. Lim, B. Kim, S. Kim, T. Jin, S. Cavill, B. S. Kim, C. Kim, J. Lischner, A. Ferreira, and S. Cho, *Gate-Tunable Reversible Rashba-Edelstein Effect in a Few-Layer Graphene/2H-TaS2 Heterostructure at Room Temperature*, ACS Nano **14**, 5251 (2020).
[35] A. M. Hoque, D. Khokhriakov, K. Zollner, B. Zhao, B. Karpiak, J. Fabian, and S. P. Dash, *All-Electrical Creation and Control of Spin-Galvanic Signal in Graphene and Molybdenum Ditelluride Heterostructures at Room Temperature*, Commun. Phys. **4**, 124 (2021).
[36] Y. Li and M. Koshino, *Twist-Angle Dependence of the Proximity Spin-Orbit Coupling in Graphene on Transition-Metal Dichalcogenides*, Phys. Rev. B **99**, 075438 (2019).
[37] W. Q. Liu, W. Y. Wang, J. J. Wang, F. Q. Wang, C. Lu, F. Jin, A. Zhang, Q. M. Zhang, G. Van Der Laan, Y. B. Xu, Q. X. Li, and R. Zhang, *Atomic-Scale Interfacial Magnetism in Fe/Graphene Heterojunction*, Sci. Rep. **5**, 11911 (2015).
[38] D. Bhattacharya and D. Jana, *Twin T-Graphene: A New Semiconducting 2D Carbon Allotrope*, Phys. Chem. Chem. Phys. **22**, 10286 (2020).
[39] A. Pak, A. Ivashutenko, A. Zakharova, and Y. Vassilyeva, *Cubic SiC Nanowire Synthesis by DC Arc Discharge Under Ambient Air Conditions*, Surf. Coatings Technol. **387**, 125554 (2020).
[40] Y. Wu, B. Liu, Z. Li, T. Tao, Z. Xie, K. Wang, X. Xiu, D. Chen, H. Lu, R. Zhang, and Y. Zheng, *Synthesis and Properties of InGaN/GaN Multiple Quantum Well Nanowires on Si (111) by Molecular Beam Epitaxy*, Phys. Status Solidi **217**, 1900729 (2020).
[41] S. P. Patole, P. S. Alegaonkar, H.-C. Lee, and J.-B. Yoo, *Optimization of Water Assisted Chemical Vapor Deposition Parameters for Super Growth of Carbon Nanotubes*, Carbon N. Y. **46**, 1987 (2008).

[42] E. T. Thostenson, Z. Ren, and T.-W. Chou, *Advances in the Science and Technology of Carbon Nanotubes and Their Composites: A Review*, Compos. Sci. Technol. **61**, 1899 (2001).
[43] P. Nikolaev, M. J. Bronikowski, R. K. Bradley, F. Rohmund, D. T. Colbert, K. Smith, and R. E. Smalley, *Gas-Phase Catalytic Growth of Single-Walled Carbon Nanotubes from Carbon Monoxide*, Chem. Phys. Lett. **313**, 91 (1999).
[44] M. M. Cunha, J. R. F. Lima, F. Moraes, S. Fumeron, and B. Berche, *Spin Current Generation and Control in Carbon Nanotubes by Combining Rotation and Magnetic Field*, J. Phys. Condens. Matter **32**, 185301 (2020).
[45] C. Li, E. T. Thostenson, and T.-W. Chou, *Sensors and Actuators Based on Carbon Nanotubes and Their Composites: A Review*, Compos. Sci. Technol. **68**, 1227 (2008).
[46] S. R. Das and S. Dutta, *Edge State Induced Spintronic Properties of Graphene Nanoribbons: A Theoretical Perspective*, in *Carbon Nanomaterial Electronics: Devices and Applications* (Springer, Singapore, 2021), pp. 165–198.
[47] R. Bonnet, P. Martin, S. Suffit, P. Lafarge, A. Lherbier, J.-C. Charlier, M. L. Della Rocca, and C. Barraud, *Giant Spin Signals in Chemically Functionalized Multiwall Carbon Nanotubes*, Sci. Adv. **6**, 1 (2020).
[48] L. Cao, Y. S. Ang, Q. Wu, and L. K. Ang, *Electronic Properties and Spintronic Applications of Carbon Phosphide Nanoribbons*, Phys. Rev. B **101**, 035422 (2020).
[49] S. Ghosh, S. Tongay, A. F. Hebard, H. Sahin, and F. M. Peeters, *Ferromagnetism in Stacked Bilayers of Pd/C60*, J. Magn. Magn. Mater. **349**, 128 (2014).
[50] Y. Murakami and H. Suematsu, *Magnetism of C60 Induced by Photo-Assisted Oxidation*, Pure Appl. Chem. **68**, 1463 (1996).
[51] X. Sun, M. Gobbi, A. Bedoya-Pinto, O. Txoperena, F. Golmar, R. Llopis, A. Chuvilin, F. Casanova, and L. E. Hueso, *Room-Temperature Air-Stable Spin Transport in Bathocuproine-Based Spin Valves*, Nat. Commun. **4**, 2794 (2013).
[52] H. Şahin, R. T. Senger, and S. Ciraci, *Spintronic Properties of Zigzag-Edged Triangular Graphene Flakes*, J. Appl. Phys. **108**, 074301 (2010).
[53] B. Náfrádi, M. Choucair, K.-P. Dinse, and L. Forró, *Room Temperature Manipulation of Long Lifetime Spins in Metallic-like Carbon Nanospheres*, Nat. Commun. **7**, 12232 (2016).
[54] W. Hu, Y. Huang, X. Qin, L. Lin, E. Kan, X. Li, C. Yang, and J. Yang, *Room-Temperature Magnetism and Tunable Energy Gaps in Edge-Passivated Zigzag Graphene Quantum Dots*, NPJ 2D Mater. Appl. **3**, 17 (2019).
[55] R. Banerjee, V. H. Nguyen, T. Granzier-Nakajima, L. Pabbi, A. Lherbier, A. R. Binion, J. C. Charlier, M. Terrones, and E. W. Hudson, *Strain Modulated Superlattices in Graphene*, Nano Lett. **20**, 5, 3113 (2020).
[56] M. Alimohammadian and B. Sohrabi, *Observation of Magnetic Domains in Graphene Magnetized by Controlling Temperature, Strain and Magnetic Field*, Sci. Rep. **10**, 21325 (2020).
[57] D. B. Zhang and S. H. Wei, *Inhomogeneous Strain-Induced Half-Metallicity in Bent Zigzag Graphene Nanoribbons*, NPJ Comput. Mater. **3**, 32 (2017).
[58] Y. Ni, G. Deng, J. Li, H. Hua, and N. Liu, *The Strain-Tuned Spin Seebeck Effect, Spin Polarization, and Giant Magnetoresistance of a Graphene Nanobubble in Zigzag Graphene Nanoribbons*, ACS Omega **6**, 15308 (2021).
[59] Y. Li, X. Li, S. Zhang, L. Cao, F. Ouyang, and M. Long, *Strain Investigation on Spin-Dependent Transport Properties of γ-Graphyne Nanoribbon Between Gold Electrodes*, Nanoscale Res. Lett. **16**, 5 (2021).
[60] C. C. Hsu, M. L. Teague, J. Q. Wang, and N. C. Yeh, *Nanoscale Strain Engineering of Giant Pseudo-Magnetic Fields, Valley Polarization, and Topological Channels in Graphene*, Sci. Adv. **6**, 19, 1–8 (2020).
[61] Y. Zhang, S. Talapatra, S. Kar, R. Vajtai, S. K. Nayak, and P. M. Ajayan, *First-Principles Study of Defect-Induced Magnetism in Carbon*, Phys. Rev. Lett. **99**, 107201 (2007).
[62] V. Shukla, *Observation of Critical Magnetic Behavior in 2D Carbon Based Composites*, Nanoscale Adv. **2**, 962 (2020).
[63] S. C. Ray, N. Soin, T. Makgato, C. H. Chuang, W. F. Pong, S. S. Roy, S. K. Ghosh, A. M. Strydom, and J. A. McLaughlin, *Graphene Supported Graphene/Graphane Bilayer Nanostructure Material for Spintronics*, Sci. Rep. **4**, 3862 (2015).
[64] P. Chen and G. Zhang, *Carbon-Based Spintronics*, Sci. China Physics, Mech. Astron. **56**, 207 (2013).
[65] P. Błoński, J. Tuček, Z. Sofer, V. Mazánek, M. Petr, M. Pumera, M. Otyepka, and R. Zbořil, *Doping with Graphitic Nitrogen Triggers Ferromagnetism in Graphene*, J. Am. Chem. Soc. **139**, 3171 (2017).
[66] M. Zhang, X. Wang, H. Sun, N. Wang, Q. Lv, W. Cui, Y. Long, and C. Huang, *Enhanced Paramagnetism of Mesoscopic Graphdiyne by Doping with Nitrogen*, Sci. Rep. **7**, 11535 (2017).
[67] J. J. Chen, H. C. Wu, D. P. Yu, and Z. M. Liao, *Magnetic Moments in Graphene with Vacancies*, Nanoscale **6**, 8814 (2014).
[68] S. Tsukamoto, V. Caciuc, N. Atodiresei, and S. Blügel, *Spin-Polarized Electron Transmission Through B-Doped Graphene Nanoribbons with Fe Functionalization: A First-Principles Study*, New J. Phys. **22**, 063022 (2020).

[69] J. Pan, S. Du, Y. Zhang, L. Pan, Y. Zhang, H. J. Gao, and S. T. Pantelides, *Ferromagnetism and Perfect Spin Filtering in Transition-Metal-Doped Graphyne Nanoribbons*, Phys. Rev. B–Condens. Matter Mater. Phys. **92**, 205429 (2015).

[70] H. Aurich, A. Baumgartner, F. Freitag, A. Eichler, J. Trbovic, and C. Schönenberger, *Permalloy-Based Carbon Nanotube Spin-Valve*, Appl. Phys. Lett. **97**, 153116 (2010).

[71] M. W. Rahman, K. M. Alam, and S. Pramanik, *Long Carbon Nanotubes Functionalized with DNA and Implications for Spintronics*, ACS Omega **3**, 17108 (2018).

[72] C. Qiu, Z. Zhang, M. Xiao, Y. Yang, D. Zhong, and L.-M. Peng, *Scaling Carbon Nanotube Complementary Transistors to 5-Nm Gate Lengths*, Science. **355**, 6322, 271 (2017).

[73] L. Xu, J. Yang, C. Qiu, S. Liu, W. Zhou, Q. Li, B. Shi, J. Ma, C. Yang, J. Lu, and Z. Zhang, *Can Carbon Nanotube Transistors Be Scaled Down to the Sub-5 Nm Gate Length?*, ACS Appl. Mater. Interfaces **13**, 31957 (2021).

[74] Z. Bullard, E. C. Girão, J. R. Owens, W. A. Shelton, and V. Meunier, *Improved All-Carbon Spintronic Device Design*, Sci. Rep. **5**, 7634 (2015).

[75] G. Zhou, G. Tang, T. Li, G. Pan, Z. Deng, and F. Zhang, *Graphene-Passivated Cobalt as a Spin-Polarized Electrode: Growth and Application to Organic Spintronics*, J. Phys. D. Appl. Phys. **50**, 095001 (2017).

Rare Earth Manganites and Related Multiferroicity

11

Suresh Chandra Baral[†], P. Maneesha[†], Ananya T. J[††], Srishti Sen[††], Sagnika Sen[††], Somaditya Sen[*], E. G. Rini[*]

[†]Equal first authorship, [††]Equal contribution, [*]Corresponding authors

Contents

1	Introduction	164
2	Synthesis of Multiferroic $RMnO_3$	165
	2.1 Polycrystalline Ceramics	165
	2.2 Single Crystal	167
	2.3 Thin Films	167
3	Physical Properties of Manganites	168
	3.1 Crystallographic Structure	168
	3.1.1 X-Ray and Neutron Diffraction Studies	168
	3.1.2 Importance of Structural Distortion in Multiferroicity	170
	A-Site Ionic Size	171
4	Multiferroicity in $RMnO_3$	171
	4.1 Multiferroicity in Orthorhombic $RMnO_3$	171
	4.2 Multiferroicity in Hexagonal $RMnO_3$	172
5	Multiferroicity of Doped $RMnO_3$	174
	5.1 Divalent Elements Doped $RMnO_3$ ($RA'MnO_3$, A' = Ca, Ba, Sr)	175
	5.2 Rare Earth Elements Doped $RMnO_3$ ($RR'MnO_3$, R' = Rare Earth)	175
6	Applications of Rare-Earth Manganites	176
	6.1 Applications in Electronic Devices	176
	6.2 Environmental and Medical Applications	177
	6.3 Magneto-Caloric Refrigerator	177
	6.4 Solid Oxide Fuel Cells (SOFC)	177
7	Conclusions and Outlook	177
8	Acknowledgments	178
	References	178

DOI: 10.1201/9781003196952-11

1 INTRODUCTION

Perovskite manganite materials are oxides of manganese with the general formula $R_{1-x}A_xMnO_3$ where A is mostly rare earth or alkali/alkaline earth metal and R is a trivalent rare-earth element. These are extensively studied due to their fascinating properties such as colossal magnetoresistance (*CMR*), magnetocaloric and multiferroic effects. These are used in potential applications, e.g., spintronics, Ferro-electromagnets, magnetic refrigeration, magnetic storage media, magnetic sensors, etc. Multiferroic properties are reported in several manganites, which deal with exciting physics of co-existence of two or more of the primary ferroic properties, e.g., ferromagnetic, ferroelectric, ferroelastic, ferrotoroidic coexist in the same crystallographic phase. In multiferroic materials, a switchable electric polarization (P) results in (anti)ferroelectricity, and the magnetization (M) results in (anti)ferromagnetism or strain (ϵ) leads to Ferro elasticity with decreasing temperature.

Localized spins or magnetic moments are present in a material and are responsible for the type of magnetism. The d or f orbitals of transition-metals at B-sites or rare-earth ions at A-sites which are partially filled with localized electrons exhibit magnetic moments. Magnetic order can be induced with the exchange interaction of these localized moments. On the other hand, a ferroelectric material needs an empty d orbital for the spatial-inversion symmetry breaking via cation off-centering for inducing a spontaneous polarization in ferroelectric materials like *BaTiO₃*. In *RMnO₃*, because of the hybridization of transition metal *Mn-3d* and the *O-2p* states, manganites don't meet the condition of "d^0-ness". Hence, these rare multiferroic materials face a contradiction of the microscopic origin of magnetism and ferroelectricity. Hence, an alternative mechanism for ferroelectricity for single-phase multiferroic *RMnO₃* must exist which is different from traditional ferroelectric materials. In addition, they have no single electron pair (s^2), which may cause inversion symmetry to be lost through mixing with an excited (s^1) (p^1) state. Therefore, *RMnO₃* with a noncentrosymmetric and polar structure may have different physical mechanisms. The reason for this multiferroicity in these materials may be caused by the hexagonal structure ($P6_3/m$) having unusual five- and seven-fold coordination polyhedral about the *Mn* and *R* ions.

Depending on the microscopic origin mechanisms of ferroelectricity, the multiferroic materials can be classified into two types:

Type 1: The magnetism and the ferroelectricity emerge independently apart from a weak coupling.
Type 2: Ferroelectric state will induce (due to electronic or ionic displacements) as the magnetic order breaks the inversion symmetry. The ferroelectric polarization is a by-product of another driving order parameter, e.g., spin order.

Type 1 multiferroics exhibit ferroelectricity usually at a temperature higher than the room temperature and a low-temperature magnetic order. Type 2 exhibits the coexistence of the magnetic and ferroelectric orders at the same temperature.

Hexagonal *RMnO₃* is a good example of Type 1 multiferroics. On the other hand, in Type 2 multiferroics, spatial inversion symmetry breaking induces simultaneous ferroelectricity in magnetically ordered materials. Examples of Type 2 include transverse spin spirals (realized, e.g., in *TbMnO₃*) and orthorhombic *RMnO₃* (R = Ho to Lu) with E-type magnetic structures. Geometric constraints and competing magnetic interaction result in strong magnetic frustration and that leads to the above-mentioned inversion symmetry breaking magnetic orders. Hence there is a competition for the ground state due to the close energy of different magnetic orders. This is the reason why small perturbations like physical pressure, doping, external magnetic or electric fields, can affect the type 2 multiferroics and show extreme sensitivity to these changes. This makes them attractive materials for applications in magnetoelectric sensors or a new type of memory element.

RMnO₃ are unique strongly electron correlated systems. In these materials, physical properties such as magnetism, conductivity, and lattice distortion, are entangled with each other. In *RMnO₃* materials,

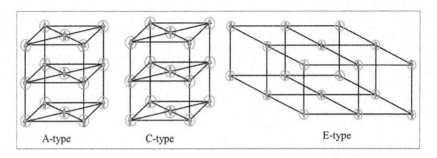

FIGURE 11.1 AFM ordering: A-type, C-type, and E-type.

the exciting multiferroicity and the magnetoresistance are due to the coupling of spin, charge, and orbital motion. These materials have rich phase diagrams and also exhibit interesting physical phenomena due to the mutual interaction of Mn and rare-earth ions.

The simple $RMnO_3$ structure can exist in either hexagonal or orthorhombic phases at ambient pressure. For larger R ions (*La, Pr, Nd, Sm, Eu, Gd, Tb,* and *Dy*), $RMnO_3$ prefers the orthorhombic structure with space group *Pbnm*, while with small R ions (e.g., *Ho, Er, Yb*) a hexagonal structure with space group $P6_3cm$ is preferred. Multiferroic property is shown by $RMnO_3$ for smaller rare earth ionic radii R = *Gd, Tb, Dy, Ho, Er, Tm, Yb* and for doped $RMnO_3$ with formula $R_{1-x}A_xMnO_3$ with A = *Ca, Sr, Mg, Pb*, etc. The Mn^{3+} moments will form an incommensurate sinusoidal ordering below a certain temperature T_N for the smaller rare-earth ion $RMnO_3$. The spins are aligned in these $RMnO_3$ perovskites according to some definite patterns or orders. Based on the radius of R^{3+}, the spin configuration represents either an A-type AFM ordering i.e., along (001), or an E-type AFM ordering [which is a mixture of both A-type (001) and C- type (110)] at a lower temperature. Hence, changes in the ionic radii generate modifications in the magnetic properties of these materials.

2 SYNTHESIS OF MULTIFERROIC $RMNO_3$

The literature is observed to detail the synthesis of most $RMnO_3$ materials with emphasis on a few. The synthesis of these materials is diverse in nature and requires a detailed discussion. Amongst the various materials, *$LaMnO_3$, $PrMnO_3$, $YbMnO_3$, $EuMnO_3$, $GdMnO_3$, $TbMnO_3$, $DyMnO_3$, $HoMnO_3$, $PmMnO_3$, $LuMnO_3$, $ErMnO_3$,* and *$NdMnO_3$* are the prime manganite materials. The synthesis of these materials can be categorized according to the form of the end product in terms of polycrystalline ceramic materials, single crystals, and thin films. Different types of precursors ranging from oxides, nitrates, carbonates, citrates, oxalates have been used as precursors using different synthesis routes like solid-state sintering, sol-gel route, hydrothermal synthesis, etc. The following paragraphs detail a few examples.

2.1 Polycrystalline Ceramics

$LaMnO_3$: $LaMnO_3$ is an important starting material in the field of perovskite $RMnO_3$. Multiferroicity is observed only in substituted $LaMnO_3$. As in this chapter we are dealing with multiferroic perovskites we will concentrate on the doped $LaMnO_3$.

$La_{1-x}Sr_xMnO_3$: Dabrowski et al. synthesized polycrystalline $La_{1-x}Sr_xMnO_3$ (x = 0.2–1.0) using the solid-state reaction method with La_2O_3, $SrCO_3$ and MnO_2 as precursors [1].

La$_{1-x}$Ca$_x$MnO$_3$: He et al. prepared *La$_{1-x}$Ca$_x$MnO$_3$* ($0 \leq x \leq 1$), by the solid-state reaction method with precursors as *La$_2$O$_3$*, *CaCO$_3$* and *MnO$_2$* [2].

YbMnO$_3$: Huang et al. synthesized high-quality *YbMnO$_3$*, via a wet-chemical route using *Yb$_2$O$_3$*, *MnCO$_3$*, and *[CH$_2$N(CH$_2$CO$_2$H)$_2$]$_2$*. Single-phase powder of the polymorph was synthesized through high-pressure (HP) annealing, calcination in air at 800 °C, and then further fired at 1200 °C. Then they prepared a powder of hexagonal h-*YbMnO$_3$* packed into a gold capsule and treated in a cubic anvil-type HP equipment (at Tokyo Tech) at 5 GPa and 1100 °C for 30 min [3].

EuMnO$_3$: Das et al. synthesized *EuMnO$_3$* nanoparticles using a modified hydrothermal method with precursors as *Eu (NO$_3$)$_3$•5H$_2$O, Mn (NO$_3$)$_3$•H$_2$O* and citric acid, to procure single-phase crystalline orthorhombic crystals, calcined at 750 °C [4].

Eu$_{1-x}$Na$_x$MnO$_3$: Sol-gel prepared *Na*-doped *EuMnO$_3$* (x = 0.0, 0.1, 0.2) compounds were reported by Nandy et al. using *Eu$_2$O$_3$* and *Mg (CH$_3$COO)$_2$•4H$_2$O* consequently sintered at 500 °C, 800 °C and 1000 °C [5].

GdMnO$_3$: Pal et al. reported the solid-state synthesis of polycrystalline *GdMnO$_3$* using *Gd$_2$O$_3$* and *Mn$_2$O$_3$* calcined at 1000 °C and further sintered at 1350 °C [6].

Gd$_{1-x}$Eu$_x$MnO$_3$: Single phase *Eu* doped *GdMnO$_3$* (x = 0.2 and 0.8) were produced by Ibrahim et al. through the solid-state reaction route, using precursors *Gd$_2$O$_3$, Mn$_2$O$_3$*, and *Eu$_2$O$_3$* which were calcined at 600 °C for 10 h followed by heating at different temperatures between 1000–1350 °C [7].

TbMnO$_3$: Cui et al. prepared ceramic samples of *TbMnO$_3$* via a solid-state reaction using *Tb$_2$O$_3$*, calcined at 1250 °C [8].

DyMnO$_3$: Polycrystalline *DyMnO$_3$* was prepared by Chen et al. via the citrate-gel process, by dissolving *Dy$_2$O$_3$* and *MnO* in aqueous solutions of citric and nitric acids, calcined at 600 °C in air to remove organic residues, and then further calcined 950 °C [9].

Dy$_{1-x}$K$_x$MnO$_3$: Yadagiri et al. synthesized bulk compounds of *K*-doped *DyMnO$_3$* (x = 0.1, 0.2 & 0.3) through the solid-state technique using *Dy$_2$O$_3$, K$_2$CO$_3$* and *MnO$_2$*, calcined at 1000 °C and further heated at 1300 °C [10].

HoMnO$_3$: Polycrystalline hexagonal *HoMnO$_3$* were prepared by Wu et al. via the solid-state technique using oxide precursors [11]

Ho$_{1-x}$Dy$_x$MnO$_3$: Magesh et al. synthesized polycrystalline *Dy* doped *HoMnO$_3$* (x = 0.1, 0.2 & 0.3) via a solid-state route using *Dy$_2$O$_3$, Mn$_2$(CO$_3$)$_3$*, calcined at 1350 °C [12].

ErMnO$_3$: *ErMnO$_3$* was synthesized by Massa et al. by the sol-gel citric acid route using *Er$_2$O$_3$* and *MnCO$_3$* as citrate precursors. The solution was dried at 120 °C, decomposed at 600 °C followed by treatment at 700 °C [13].

Er$_{1-x}$Ca$_x$MnO$_3$: Carron et al. obtained polycrystalline *La$_{1-x}$Ca$_x$MnO$_3$* (x = 0–1.0) doped *ErMnO$_3$* samples by sol-gel synthesis using citrate precursors and were treated at 1100 °C for 12 h [14].

TmMnO$_3$: Bulk *TmMnO$_3$* samples were prepared by Araujo et al. via a solid-state reaction using *Tb$_2$O$_3$* and *MnCO$_3$* as precursors. The powders were calcined in a conventional oven at 1200 °C for which a hexagonal phase was obtained. The phase could be later converted into orthorhombic through HP/HT treatment [15].

PrMnO$_3$: Similar to *LaMnO$_3$*, the pure *PrMnO$_3$* shows multiferroicity only after partial replacement of *Pr* by other elements.

Pr$_{1-x}$Sr$_x$MnO$_3$: Saw et al. prepared polycrystalline *Pr$_{1-x}$Sr$_x$MnO$_3$* ($0.2 \leq x \leq 0.40$) using the nitrate route with *Pr$_6$O$_{11}$, SrCO$_3$*, and *MnO$_2$* as precursors, and calcined several times between 800 °C and 1200 °C [16].

Pr$_{1-x}$Ca$_x$MnO$_3$: Polycrystalline samples of *Pr$_{0.6}$Ca$_{0.4}$MnO$_3$* were synthesized by Yang et al. via the solid-state reaction method which is sintered at 1450 °C [17]

Pr$_{1-x}$Ba$_x$MnO$_3$: Panwar et al. prepared polycrystalline samples of *Pr$_{1-x}$Ba$_x$MnO$_3$* (x = 0.33–0.80) using the conventional solid-state reaction route, and calcined between 900 °C and 1100 °C followed by sintering at 1260 °C [18].

Nd$_{0.5}$Sr$_{0.5}$MnO$_3$: Nanocrystalline samples of *Nd$_{0.5}$Sr$_{0.5}$MnO$_3$*, of average particle size 30 and 55 nm were prepared by the sol-gel technique, annealed at temperatures 700 °C and 800 °C [19].

2.2 Single Crystal

The continuous, homogeneous, and highly organized structure of single crystals make it one of the most significant classes of materials, having unique characteristics. Conventionally three techniques of single crystal growth have been used. The most prevalent process is growth from the melt, which relies on the solidification and crystallization of a melted substance. In the melt growth process, the material is melted prior to growth, which allows ease in achieving large single crystals in a shorter interval in comparison to other growth techniques. The Zone-movement method, Verneuil method, and the Floating zone method are the most common melt growth techniques. Similarly, instead of the material being melted in the first place, a suitable solvent is used to dissolve the material to be crystallized. The third technique is vapor-phase growth, which is more often used to fabricate thin single-crystal films on substrates instead of bulk single crystals. Single crystals in the vapor phase can be synthesized using a sublimation process, a gas phase reaction, and a transport reaction. Solid-state conversion of polycrystalline materials to single crystals has gained much attention in the scientific community. In the following paragraph, some examples are mentioned to demonstrate the versatility of single-crystal synthesis of different $RMnO_3$ materials.

$EuMnO_3$: Cu-doped $EuMnO_3$ was synthesized by Yang et al. after homogenized grinding of Eu_2O_3, CuO, and $MnCO_3$ for about 30 minutes followed by 6 h of calculations at 800 °C and 900 °C and further heat treatments at 1150 °C [20].

$TbMnO_3$: Single crystals of $TbMnO_3$ of dimensions $2 \times 2 \times 1$ mm³ were grown at the University of Oxford by Forrest et al. using a flux growth method [12].

$DyMnO_3$: Milov et al. synthesized $DyMnO_3$ orthorhombic perovskite single crystals by zone melting with optical heating in air [13].

$HoMnO_3$: Lee et al. synthesized rod-like $HoMnO_3$ single crystals of dimensions up to 2×2×7mm³, utilizing the conventional Bi_2O_3 flux method [21].

$ErMnO_3$: $ErMnO_3$ single crystals were grown by Vermette et al. through a high-temperature solution growth method using $PbF_2/PbO/B_2O_3$ flux, further annealed for 24 h at 1120 °C in oxygen atmosphere [22].

$YbMnO_3$: The orthorhombic phase of $YbMnO_3$ was acquired by Duttine et al. after heating the hexagonal phase of polycrystalline $YbMnO_3$ (heated up to 950 °C and then further heated at 1100 °C for 48h in air, in air, for 12h) under high pressure (5 GPa), which was obtained by solid-state reaction [23].

$La_{0.7}Sr_{0.3}MnO_3$: Martin et al. prepared $La_{0.7}Sr_{0.3}MnO_3$ single crystals of size 3–4 mm in diameter and 8 cm in length, using the floating zone (FZ) method [20].

$Pr_{1-x}Sr_xMnO_3$: Single crystals of $Pr_{1-x}Sr_xMnO_3$ (x = 0.22, 0.24, 0.26) were grown by a floating zone technique, using irradiative heating (Markovich et al.) [17].

$Nd_{0.5}Sr_{0.5}MnO_3$: Venkatesh et al. produced single crystals of $Nd_{0.5}Sr_{0.5}MnO_3$ using an infrared image furnace by the floating zone technique [24].

2.3 Thin Films

$GdMnO_3$: Strained $GdMnO_3$ thin films of various thicknesses ranging from 10 to 110 nm were grown by Li et al. on $SrTiO_3$ substrates, displaying nano-scale twin-like domains. Substrate temperature of 800 °C and an oxygen partial pressure of 10 Pa were used for depositions [25].

$TbMnO_3$: Lee et al. used a pulsed laser deposition method to grow 50 nm thick $TbMnO_3$ thin films. The $TbMnO_3$ target was prepared from the conventional solid-state reaction using Tb_2O_3 and Mn_2O_3 powders. Its base vacuum pressure was maintained at around 10^{-6} Torr [26].

$DyMnO_3$: Bulk $DyMnO_3$ ceramic was synthesized sintered at a temperature of 1400 °C for 12 h and further calcined in the temperature range of 1000–1200 °C, through a solid-state reaction process, which was used as a target in pulsed laser deposition for the film fabrication. The DMO thin film was deposited on LAO substrates by the PLD technique [27].

HoMnO₃: Shimamoto et al. produced *HoMnO₃* films of thickness ranging from 20–400 nm, fabricated by pulsed laser deposition. Pulsed beams were focused onto a *HoMnO₃* target for which the temperature was maintained at 780 °C [28].

ErMnO₃: Jang et al. prepared *ErMnO₃* films using a pulsed laser deposition method using *Pt/Al₂O₃* substrates. The 20-nm-thick *Pt/Al₂O₃* substrates were fabricated using a dc-magnetron sputtering technique, at an *Ar* pressure of 5 mTorr and a deposition temperature of 200 °C [29].

LaSrMnO₃: Malik et al. had grown a $La_{0.6}Sr_{0.3}MnO_3$ thin film with a thickness of 100 nm on *LaAlO₃* substrate by pulsed laser deposition [30]. Another method was used by Meda et al. to prepare $La_{0.6}Sr_{0.3}MnO_3$ films of thickness 300 nm, where deposition took place via the liquid delivery- metal-organic chemical vapor deposition method [31].

La$_{1-x}$Ca$_x$MnO₃: Films of $La_{0.72}Ca_{0.28}MnO_3$ of thickness 100 to 900 Å were prepared by *Liang* et al. on *LaAlO₃* single-crystal substrates through the process of sputtering, with pressure maintained at the order of 1 x 10⁵ Pa [32].

(PrSr)MnO₃: Liu et al. prepared epitaxial thin films of *(PrSr)MnO₃* (x = 0.3), of thickness 500–600 Å, which were developed on NGO substrate using an oxygen plasma-assisted home-made MBE system, equipped with the RHEED monitoring system. Substrate temperatures were maintained at 650–800 °C [33].

(PrCa)MnO₃: Epitaxial *(PrCa)MnO₃* (x = 0.3) thin films were developed on *SrTiO₃* substrates by Moon et al. through the pulsed laser deposition using eclipse methods. The films of 100-nm thickness were produced at 1023 K under a 150 mTorr oxygen environment and then cooled to room temperature in a 300 Torr oxygen atmosphere [34].

NdSrMnO₃: Chiu et al. prepared epitaxial $Nd_{0.5}Sr_{0.5}MnO_3$ thin films of a total thickness of 16.5 nm, which were deposited on STO substrates by pulsed laser deposition. The substrate temperature was maintained at 750 °C and the oxygen background pressure was kept at 160 mTorr [35].

3 PHYSICAL PROPERTIES OF MANGANITES

3.1 Crystallographic Structure

The ideal perovskite (AMX_3) belongs to the cubic structure with a *Pm-3m* space group. The A-cation is surrounded by twelve X anions. The *M*-cation is octahedrally coordinated by six *X* ions. The *X* anions are coordinated by two *M*-cations and four *A*-cations. In this section, x-ray diffraction (XRD) and neutron diffraction studies will be covered.

3.1.1 X-Ray and Neutron Diffraction Studies

The XRD studies can be performed using several types of diffractometers ranging from desktop versions to a proper laboratory XRD system in the lab. In general, the common x-ray source is a *Cu-K$_α$* source with wavelength 1.54 Å but there are exceptions where other sources like *Ag* and *Mo* are used with much lower wavelengths. High-resolution XRD can also be performed at a synchrotron beamline where a different technique is used to choose the right wavelength for a specific measurement. However, certain structural aspects, like magnetic structure and O-Wykoff positions cannot be determined by x-ray sources even at the beamline. To extract this information, one needs to perform neutron diffraction at specific labs. In the following section, an attempt has been made to look into the various studies performed on the *RMnO₃* compounds to emphasize the importance of the materials.

Neutrons have a magnetic moment and it gets scattered by both the nucleus and the resultant magnetic moment of electrons. To study the ordering of the spins whether is of Ferro-, Antiferro- and ferrimagnetic nature, neutron diffraction studies are the precise technique. Hence, neutron diffraction helps in determining the magnetic structure and dynamical properties of many important materials

XRD results for different $RMnO_3$: XRD studies of multiferroic manganites like $EuMnO_3$, $GdMnO_3$, $TbMnO_3$, $HoMnO_3$, $DyMnO_3$ and $YMnO_3$ are being discussed as examples in this section.

$EuMnO_3$: A PANalytical X'PERT PRO instrument was used to collect XRD of $EuMnO_3$, equipped with Cu-K_α radiation (λ = 1.5406 Å) for 2θ, 10^0–80^0. The XRD pattern matches nicely with the orthorhombic $EuMnO_3$ with $Pbnm$ symmetry (JCPDS le no. 261126).

Mota et al. performed high-pressure XRD data on the Extreme Conditions Beamline (ECB) at the European Synchrotron Radiation Facility (ESRF) on the ID27 high-pressure with a $Pbnm$ symmetry [4]. Their findings were that at high pressure too no structural phase transitions were observed [36].

$Eu_{1-x}Na_xMnO_3$: Powder X-ray diffraction (XRD) patterns of $Eu_{1-x}Na_xMnO_3$ (x = 0.0, 0.1 and 0.2) were collected from a Bruker D8 advance diffractometer in the 2θ range 20°–80°. A distorted orthorhombic $Pbnm$ is observed for all x [5].

$GdMnO_3$: The X-ray powder diffraction (XRD) pattern of $GdMnO_3$ was collected from a PANalytical x-ray diffractometer. The XRD pattern of the pellets sintered at 1350°C for 24 h in air, in a reduced oxygen atmosphere (95% N_2 and 5% H_2) for 15 min, 30 min, 45 min, and 1 h reveals a pure orthorhombic phase with $Pbnm$ symmetry [6].

$TbMnO_3$: Bridges et al. observed an orthorhombic single phase of $TbMnO_3$ by X-ray diffractometer. Aliouane et al. performed a neutron diffraction experiment using the flat cone neutron diffractometer E2 at the Hahn-Meitner Institute's BENSC facility and carried out polarized neutron diffraction as a function of temperature and magnetic field. These measurements aided information regarding the magnetic structure and contributing magnetic species i.e., Tb or Mn. $Pbnm$ orthorhombic phase is observed from NPD data. Reflections belonging to A- and G- modes were observed by cooling below T_N (~41 K) and the F- and C-modes reflections below T_S (28 K). A cycloidal spin order for Mn was observed. The final transition happens at T_N ~7 K where the spin order becomes independent at Tb [37].

$DyMnO_3$: Mota et al. observed orthorhombic $Pnma$ space group from XRD data [36]. Neutron powder diffraction (NPD) experiments on both $DyMnO_3$ (annealed at 1200 °C in an oxygen atmosphere for 4 weeks) and $DyMnO_3$ (annealed at 1200 °C in oxygen atmosphere) were performed at the high-intensity neutron diffractometer WOMBAT [139] at the OPAL reactor at ANSTO using a wavelength of 2.41 Å from a Ge (113) monochromator. NPD patterns for both samples within the 2θ region (16–40°) reveal orthorhombic $Pbnm$ phase and paramagnetic phase at 300 K. Paul et al. observed the Mn sinusoidal contribution at 18 K; the Mn and induced Dy spin-spiral phase at ≈ 12 K; and the contributions from the commensurate Dy, incommensurate Mn, and partially incommensurate Dy phase at ≈ 1.3 K. Similar results were observed for both the samples.

$ErMnO_3$: High-pressure XRD experiments were conducted at the 4W2 beamline of the Beijing Synchrotron Radiation Facility (BSRF) by angle-dispersive measurements with a wavelength of 0.6199 Å. The x-ray beam was focused in the horizontal and vertical direction to a 26 x 8 μm^2 (FWHM) spot using Kirkpatrick-Baez mirrors. XRD pattern at ambient pressure reveals a hexagonal $ErMnO_3$ and the space group is $P6_3cm$. Two new diffraction peaks appear at 20.2 GPa indicating a new phase. These two peaks can be attributed to the (111) and (022/211) reflections of the orthorhombic structure revealing that $ErMnO_3$ undergoes a hexagonal to orthorhombic phase transition at 20.2 GPa. The phase transition takes place gradually with increasing pressure and is completed at ~57 GPa. However, the phase transition is kinetically hindered above 39 GPa [38].

$HoMnO_3$: The synchrotron-based high-pressure XRD measurements up to ~25GPa carried out at the beamline X17C at the National Synchrotron Light Source (NSLS), in Brookhaven National Laboratory (BNL) revealing an orthorhombic structure $Pbnm$ phase for the entire pressure range [11]. A focused monochromatic X-ray beam of λ = 0.4066 Å was used as a source. The data were collected with a Rayonix 165 charge-coupled device (CCD) detector.

$LaSrMnO_3$: The XRD data of $La_{1-x}Sr_xMnO_3$ was recorded using Philips X'Pert MRD Diffractometer by Meda et al. The XRD experiment was operated at 40 kV and 45 mA using Cu Kα monochromatic radiation. XRD pattern reveals orthorhombic $Pbnm$ phase for x [31].

LaCaMnO₃: The room temperature X-ray diffraction data of $La_{1-x}Ca_xMnO_{3+\delta}$ ($0 \leq x \leq 1$) reported by He et al. reveals that all samples are single-phase orthorhombic structures with the *Pnma* space group [2].

PrSrMnO₃: Ajay et al. reported X-ray diffraction (XRD) patterns of $Pr_{1-x}Sr_xMnO_3$ ($0.20 \leq x \leq 0.40$). For $x = 0.20$, 0.33 and 0.40, orthorhombic symmetry with the *Pnma* space group was observed. For $x = 0.25$, a rhombohedral symmetry with the $R\bar{3}c$ space group was observed [16].

3.1.2 Importance of Structural Distortion in Multiferroicity

Tilting and distortion of the BO_6 octahedra and cationic displacements can lead to the formation of distortion of an ideal perovskite structure. Electronic instabilities such as Jahn-Teller distortion leads to such octahedral distortion and cationic displacements, whereas the smaller size of *A*-site cation yields octahedral tilting. Such deformations often lead to the separation of the positive and negative charge centers in the crystalline structure. This leads to the generation of ferroelectricity of other ferroic properties that make these materials multiferroic. Hence, to understand the mechanism of multiferroicity the distortions in the lattice should be understood in detail.

Tolerance factor: Distortions in the perovskites and hence the structural symmetry of the crystals can be estimated in terms of tolerance factor (t), defined by, $t = (R_A+R_O)/\sqrt{2}(R_B+R_O)$ where R_A, R_B, and R_O are the ionic radii of *A*, *B* and *O* ions. The structure becomes cubic for $t = 1$, rhombohedral for $t < 1$, and orthorhombic for $t \sim 0.8$. Deviation of t from 1 results in lowering the symmetry by tilting and rotation of the octahedra. The *A-site* cation tries to fill the Cubo-octahedral interstices of the oxygen sublattice created by the *B* site octahedra. The rotation or tilting of the octahedra is to facilitates the volume reduction of the Cubo-octahedral. The minimum energy configuration dictates the in-phase or anti-phase tilting.

Glazer notation describes the octahedral tilting distortions in perovskites. The tilting of one octahedron causes the neighboring octahedra to be tilted in perovskites. For small angles of tilt, the component tilts can be taken to be about the [100], [010], [001] pseudo cubic axes. The magnitudes of tilt are represented symbolically by a set of three letters which refer to the axes in the order [100], [010], [001] pseudo cubic axes. The unequal tilt is denoted by *abc*. Equality of tilts is represented by repeating the letter, e.g., *abb* means equal tilts about [010] and [001] with a different tilt about [100]. Glazer used superscripts +, – or 0 to represent the same tilt, opposite tilt, or no tilt of successive octahedra along an axis.

The figure shows Glazer representation of multiferroic *TbMnO₃* using CIF (COD 1525688 ID). The structure was observed along the equivalent pseudo cubic axis for orthorhombic *Pbnm* i.e., [110] and [1$\bar{1}$0] axes. Antiphase tilting of successive octahedra was observed down these axes and an in-phase tilting was observed along the [001] axis. Hence, the Glazer representation is $(a^-a^-c^+)$ for the *Pbnm* phase of *TbMnO₃*.

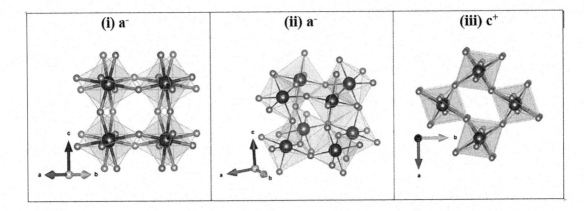

FIGURE 11.2 The Glazer representation ($a^-a^-c^+$) for the *Pbnm* phase of *TbMnO₃*.

A-Site Ionic Size

In $GdFeO_3$ a type of distortion is observed in which the corner-sharing octahedra tilts around the *b*-axes in alternating directions, and rotates around the *c*-axis in alternating directions. In orthorhombic manganites, such a distortion increases with reducing the ionic radius of the *R* ion. Hence, the ionic radius can dictate the magnitude of the *Mn-O-Mn* superexchange angle and hence the magnetism. The departure of the crystal structure from an ideal cubic perovskite leads to differences in the *R-O* and *Mn-O* bond lengths which is manifested only by lattice modifications in terms of cooperative rotation and tilting of the MnO_6 octahedra. For smaller *R* ionic radius (*Gd–Yb*) orthorhombic $RMnO_3$, a significant $GdFeO_3$-like distortion and staggered orbital ordering take place, resulting in a ferromagnetically coupled nearest-neighbor and antiferromagnetically coupled next-nearest-neighbor Mn^{3+}. This results in the frustration of Mn^{3+} spins depending on the *R* ions. As a result, the magnetic structure turns into a noncollinear spiral spin order for *R = Gd – Dy* and E-type antiferromagnetic order for *R = Ho – Yb*. The octahedral rotation increases with decreasing ionic radius of *R*. The lattice is further distorted by the Jahn-Teller JT distortion of the MnO_6 octahedra. The JT distortion increases with decreasing radius of *R* (where *R = La-Dy*). However, in hexagonal $RMnO_3$ the structural distortions result in asymmetric coordination of oxygen atoms around the rare earth ion.

4 MULTIFERROICITY IN RMNO$_3$

In the $RMnO_3$ with large ionic radii elements, i.e., *R = La* to *Dy*, the *Pbnm* orthorhombic phase is the stable phase. As the *R* ionic radii decrease for *R = Ho to Lu* the hexagonal phase appears. *Tb* and *Dy* manganites can be epitaxially stabilized in the hexagonal phase. The metastable orthorhombic phase of smaller rare earth ionic radii manganites can also be synthesized with high-pressure synthesis or other unconventional methods. The multiferroicity can be induced in $RMnO_3$ with larger *R* ions (*La, Pr, Nd, Sm*) by doping of the *R*-site with elements such as alkaline earth elements or other *R*-ions, whereas, in smaller *R* ions (*Eu, Ho, Er, Tb, Dy, etc.*) the multiferroicity arises inherently due to its hexagonal structure. Compared to the hexagonal phases the commensurate spiral and *E-AF* structures of orthorhombic $RMnO_3$ exhibit a weak magnetoelectric coupling. Hence, hexagonal manganites are also attracting great research interest as a promising platform for multiferroic properties. An extra degree of freedom for tuning the magnetoelectric coupling can be achieved by novel vortex-like ferroelectric domain walls which are locked to the antiphase structural domain walls. Because of tilting and Jahn-Teller distortions of the MnO_6 octahedra, the *Mn* 3d orbitals exhibit a staggered type ordering. Modifications of the *Mn* electronic structure are expected as the strain-induced additional MnO_6 distortions will affect the orbital ordering.

The multiferroic properties of $RMnO_3$ can be understood from measurements such as dc or ac magnetization, heat capacity, dielectric properties, electrical conductivity, etc. by detecting the thermodynamic nature of the phase transitions.

4.1 Multiferroicity in Orthorhombic RMnO$_3$

As discussed above under special conditions or by larger size ion-doping only a metastable orthorhombic phase of smaller rare earth ionic radii manganites can be attained. An inherent weak multiferroicity due to inherent structural properties is observed for smaller *R (Eu, Ho, Er, Tb, Dy, etc.)*. Table 11.1 represents lattice parameters and magnetic ordering temperatures of various orthorhombic $RMnO_3$. The following examples are listed for a better understanding:

$TbMnO_3$: The orthorhombic (*Pbnm*) $TbMnO_3$ has been a key material in studies of spin-driven ferroelectrics. Due to the antisymmetric Dzyaloshinskii–Moriya interaction, the ferroelectric polarization

TABLE 11.1 Lattice Parameters and Magnetic Ordering Temperatures of Orthorhombic $RMnO_3$

$RMNO_3$	A (Å)	B (Å)	C (Å)	$T_N(K)$	REFERENCES
$TbMnO_3$	5.293	5.838	7.403	41	[37]
$HoMnO_3$	5.248	5.839	7.356	42	[39]
$TmMnO_3$	5.231	5.814	7.323	44	[15]
$LuMnO_3$	5.198	5.788	7.297	40	[40]

P is induced along the c axis by the inversion symmetry breaking due to a cycloidal spiral spin order. The cycloidal spiral spin order in $TbMnO_3$ arises as a result of the competition between ferromagnetic (FM) in-plane nearest neighbor interaction (J_1) and antiferromagnetic in-plane next-nearest neighbor (J_2) interactions of Mn moments. However, this competition which can result in a spin-ordered state in this material is very low and can be modified by replacing Tb with other rare-earth elements. By substituting smaller ions such as Ho and Y (that is, applying chemical pressure), the Mn–O–Mn bond angles can be decreased leading to an 'E-type' antiferromagnetic (E-AFM) structure. Additionally, applying physical pressure leads to structural modifications such as shortening of interatomic distance which can be another tool to tune the magnetic competition, spin state, and resulting spin-driven ferroelectricity in this material.

Orthorhombic $HoMnO_3$: The multiferroic properties of $HoMnO_3$ arise in both orthorhombic structure as well as in hexagonal structure. The o-$HoMnO_3$ is a distorted perovskite material having the space group $Pbnm$. The multiferroic behavior of orthorhombic $HoMnO_3$ is a type-II multiferroic in nature, where the ferroelectricity in these materials is induced by the Mn antiferromagnetic order, and not by that of the R ions. There are two anomalies that are observed at $T_N = 42\ K$ and at $T_{Ho} = 15\ K$ in this material. Mn-Mn and Ho-Mn exchange striction mechanisms coexist in this system. These mechanisms can lead to both ferroelectric and antiferroelectric orders along an axis (Mn-Mn) and the c axis (Ho-Mn). The Ho-Mn interaction produces the c-axis polarization, while the Mn-Mn exchange striction induces the a-axis antiferroelectric order.

Orthorhombic $TmMnO_3$: In bulk orthorhombic-$TmMnO_3$, antiferromagnetism sets in at about $42K$ and evolves into an E-type AFM order at $T_N \approx 32\ K$. In o-$TmMnO_3$ the ferroelectric polarization arises from collinear Mn^{3+} magnetic order. The large rotation of the oxygen octahedra around the Mn^{3+} ions is expected to result in appreciable antiferromagnetic superexchange interactions along the a-axis through pairs of oxygen anions that compete with the ferromagnetic interactions in the ac plane.

Orthorhombic $LuMnO3$: Orthorhombic $LuMnO_3$ with perovskite-type structure can be synthesized by high-pressure methods. The crystal structure of this metastable polymorph of $LuMnO_3$ shows a strong Jahn–Teller distortion of the MnO_6-octahedra resulting in orbital ordering. The Mn-atoms order antiferromagnetically in an E-type structure below the Neel temperature $T_N \sim 40\ K$.

4.2 Multiferroicity in Hexagonal $RMnO_3$

The multiferroicity in hexagonal structure arises due to the strong coupling of the magnetic order to the lattice or phonons that was experimentally verified by Raman experiments which showed an enhancement of the phonon frequencies of two modes that modulate the Mn-Mn interaction, below the Neel temperature. All hexagonal manganites show antiferromagnetic (AFM) order of the Mn^{3+} spins below their respective Neel temperatures, T_N. Table 11.1 shows the lattice constants and the Neel temperatures of nine hexagonal $RMnO_3$ (R = Dy to Yb). The AFM transitions happen below 100K.

As ferroelectric and magnetic orders coexist below the Neel temperature, there is a need for a detailed analysis of how the two order parameters interact with each other and how different physical properties

TABLE 11.2 Lattice Parameters and Magnetic Ordering Temperatures of Hexagonal $RMnO_3$

RMNO₃	A (Å)	C (Å)	T_N(K)	REFERENCES
HoMnO₃	6.142	11.42	76	[41]
LuMnO₃	6.046	11.41	90	[42]
ErMnO₃	6.112	11.40	79	[41]
TmMnO₃	6.092	11.37	84	[41]
YbMnO₃	6.062	11.36	87	[41]
DyMnO₃	6.182	11.45	57	[43]

might be affected by their coupling. The AFM order parameters and *c*-axis ferroelectric polarization cannot be linearly coupled due to symmetry. There may be substantial anomalies of dielectric quantities at the magnetic phase transitions due to higher-order couplings mediated through the strong spin-lattice interaction.

The detailed mechanism and origin of this multiferroicity in different rare-earth manganites will be discussed below.

Hexagonal HoMnO₃: From the XPS study, it was observed that displacements of ion pairs happen at the three magnetic phase transitions of *HoMnO₃*. The most pronounced anomalies can be observed in the dielectric properties data as the temperature decreased, three distinct and sharp anomalies could be observed at T_N = 76 K, T_{SR} = 32.8 K, and T_{Ho} = 5.4K (where T_{SR} is the spin reorientation temperature and T_{Ho} is the ferrimagnetic transition temperature of *Ho*). In hexagonal manganites, the kink of $\varepsilon(T)$ at T_N is a characteristic sign of frustrated magnetic order. In contrast, the sharp peak at T_{SR} was only seen in *HoMnO₃* and is an unusual occurrence where Mn^{3+} spins rotate 90°, causing the transition. An explanation for the Mn^{3+} spin rotation may come from the increased magnetic fluctuations of the Ho^{3+} subsystem upon decreasing temperature. The sudden change of the elastic moduli at T_N and T_{SR} are indications of the importance of the spin-lattice coupling in hexagonal manganites, particularly in *HoMnO₃*.

ErMnO₃: *ErMnO₃* is another compound belonging to the hexagonal rare-earth manganite family. The ferroelectric transition temperature of *ErMnO₃* is around 588K. There are two transitions related to magnetic structures that are observed in these materials from neutron studies: T_{Er} ~10K and T_N ~80K. The AFM magnetic order of *ErMnO₃* in zero magnetic field sets in below T_N with the magnetic space group $P6_3cm$, which is similar to other $RMnO_3$ rare-earth manganites.

Hexagonal TmMnO₃: The magnetic order of spins sets in at T_N = 84K and the symmetry was determined as $P6_3cm$. *TmMnO₃* has a magnetically induced electric polarization that is substantially higher than in any other heavy rare-earth manganites with commensurate magnetic order. anomalies in the temperature dependence of the lattice constants were observed at the magnetic phase transitions that are evidence for strong coupling effects between the chemical and magnetic lattices. Also, according to electric hysteresis loop and dielectric data, hexagonal *TmMnO₃* ceramic exhibits ferroelectricity at room temperature and shows ferroelectric-paraelectric transition at around 621K.

YbMnO₃: The bulk magnetic and dielectric properties of *YbMnO₃* reveal the onset of magnetic order at T_N = 90K. Also, a second phase transition was found below 5 K and at magnetic fields above 30 kOe a metamagnetic transition was detected in field-dependent magnetization $P6_3cm$ magnetic symmetry below T_N is favored based on SHG optical measurements. The low-temperature transition involves Yb^{3+} moments. The *f*-moments of the *Yb* on 4*b* sites are systematically polarized according to the $P6_3cm$ antiferromagnetic structure. The phase transition at T_{Yb} involves the order of the Yb^{3+} moments on 2*a* sites antiparallel to the 4*b* moments and a reorientation of the Mn^{3+} spins resulting in the ferrimagnetic order of the *f*-moments according to the $P6_3cm$ magnetic symmetry.

DyMnO₃: *DyMnO₃* has a metastable hexagonal structure. *DyMnO₃* is ferroelectric in its hexagonal form, like all other hexagonal manganites. With the largest ion, Dy^{3+} has the most expanded structure with lattice constant a and c greater than all other hexagonal $RMnO_3$. Magnetic exchange couplings among the different ions are expected to be reduced, and $DyMnO_3$ (T_N = 57K) has the lowest Neel temperature among the $RMnO_3$ compounds. Noteworthy features include a sharp peak in the ac magnetic susceptibility at T_{Dy} = 6K and a remanent magnetization below this temperature, suggesting a ferrimagnetic state at low T. External magnetic fields along the c-axis enhance the low-temperature ferrimagnetic phase. Also, T_N = 68 K and T_{Dy} = 8K were obtained from magnetic and heat capacity data. The magnetic symmetry in the temperature range between T_N and T_{Dy} was determined as $P6_3cm$, with the Dy^{3+} moments all parallel to the c-axis.

There is an interesting difference between the symmetry of $DyMnO_3$ and those of $ErMnO_3$, $TmMnO_3$, and $YbMnO_3$, but it coincides with the symmetry of $HoMnO_3$ below the spin reorientation temperature, T_{SR}. The coupling between $3d$ spins and $4f$ moment is of the biquadratic form. This biquadratic interaction can trigger the simultaneous order of the Mn^{3+} spin and Dy^{3+} moment systems with different magnetic symmetries making the material multiferroic.

5 MULTIFERROICITY OF DOPED $RMNO_3$

The ordering phenomena including spin-, charge-, and orbital ordering, all coupled to lattice properties make the doped rare-earth manganites special compounds with multiple functionalities. These properties lead to a double exchange interaction and formation of the Jahn-Teller polaron, colossal magnetoresistance near the Curie temperature, dense granular magnetoresistance, and optically-induced magnetic phase transitions. The most interesting property exhibited by these doped rare-earth manganites is the multiferroicity which comes from the variable valence states of Mn due to multivalent doping at A site and B site of the $RMnO_3$ perovskite structure. In the larger rare earth orthorhombic manganites which are not inherently multiferroic, one can induce multiferroicity by doping other elements in the crystal structure. Also, in smaller rare earth ions which are having inherent multiferroicity, the ferroic properties can be enhanced with doping. The rare earth in the A site can be modified ($RA'MnO_3$) with divalent elements A' = Ba, Sr, Mg, etc., or monovalent elements A' = Na, K or rare earth elements ($RR'MnO_3$ R' = rare earth) of comparable radii. As compared to the parent $RMnO_3$ materials the doped $R_{1-x}A_xMnO_3$ shows a wide range of electrical and magnetic properties due to the coupling interaction of Mn ions (Mn^{3+} and Mn^{4+}) through a double exchange mechanism. The magneto transport properties are also due to Jahn-Teller (JT) distortion and grain boundary (GB) effects. In the structural analysis section, it was mentioned how doping of cations having different sizes at the A site induces multiferroicity due to JT distortion of MnO_6 octahedron. The coupling of structural, electronic, and magnetic properties provides these compounds with high flexibility for attaining specific custom-designed functionalities. Disorder generation at the R and TM sites, by systematic control of particle size, hydrostatic pressure, magnetic and electric field, etc., can strongly influence the charge, orbital and spin ordering [44].

Divalent elements doped rare-earth manganites are one of the interesting areas of research to explore the multiferroicity exhibited by these materials. The morphologies, chemical compositions, magnetic phase structures, and grain size distributions greatly influence the physical and chemical properties of rare-earth manganite nanoparticles. The thermal history during the synthesized procedure also affects the properties exhibited by these materials. However, the alkali metal-doped $RMnO_3$ does not show multiferroicity but it exhibits excellent properties in the field of the magnetocaloric effect. Although A site, B site, and A and B site modifications of $RMnO_3$ are possible which can exhibit multiple functionalities, here in this chapter based on rare-earth manganites multiferroic compounds we are focusing only on the A site modification of $RMnO_3$ which exhibit multiferroicity.

5.1 Divalent Elements Doped RMnO₃ (RA'MnO₃, A' = Ca, Ba, Sr)

The competing magnetic orders, metal-insulator transition, and colossal magnetoresistance (CMR) behavior are the main attractions of the doped manganites with compositional formula, $R_{1-x}A_xMnO_3$, where R is a trivalent rare-earth cation and A is a divalent alkaline earth element. The valence of the Mn ion can be tuned between 3+ and 4+ by the replacement of trivalent rare-earth by divalent alkaline earth. These mixed-valence Mn^{3+}/Mn^{4+} ions are the main reason for the magneto transport properties of manganites. The formation of small lattice polarons in the paramagnetic state can induce large resistance. Below a particular temperature, T_{CO}, the Mn^{3+} and Mn^{4+} ions order in a regular pattern; the phenomenon is called charge ordering, e.g. *(La, Ca) MnO₃*. These charge-orders can be site-centered or bond-centered. The site-centered charge centering can be attributed to spherical charge distribution around the nucleus while the bond-centered charge ordering can be due to electron density concentrated between two nuclei. Different spatial regions contain different electronic orders making these materials electronically inhomogeneous. When the A-site cation in the doped *RMnO₃* is large or has a large e_g bandwidth, electron hopping from Mn^{3+} to Mn^{4+} takes place through the double exchange mechanism. The correlation of magnetization and resistivity or magneto resistivity is through this double exchange interaction. Ba-doped *LaMnO₃* *(La₁₋ₓBaₓMnO₃)* exhibits ferromagnetism above room temperature and a significant magnetoresistance value. In the *Ca*-doped *PrMnO₃* *(Pr₁₋ₓCaₓMnO₃ x = 0.3, 0.4)* dielectric constant anomalies can be seen around the charge-ordering or the antiferromagnetic transition temperatures. In these materials, magnetic fields impose a significant effect on dielectric properties which reveals magneto-electric coupling. *Nd₀.₆Sr₀.₄MnO₃* exhibits room temperature semiconducting properties and by applying a low magnetic field a transition from semiconductor behavior to Ohmic resistivity is observed. For semiconductor phases, the static resistance of *Nd₀.₆Sr₀.₄MnO₃* drastically increases (1000 times) which results in negative magnetoresistance (MR) [45].

5.2 Rare Earth Elements Doped RMnO₃ (RR'MnO₃, R' = Rare Earth)

Doping rare earth elements in the *A*-site of *RMnO₃* leads to interesting properties which are absent in undoped *RMnO₃* samples. When a rare earth ion is doped in *RMnO₃* due to the difference in the ionic radii, bond lengths and bond angles change which modifies exchange interactions. Due to these changes electrical and magnetic properties of these doped samples are modified.

For Gadolinium doped *TbMnO₃* *(Tb₁₋ₓGdₓMnO₃)*, a region with a stable spiral spin ordering was observed in the ab plane with polarization along the a-axis. This type of ordering has been previously seen only in *GdMnO₃* with large applied magnetic fields. *TbMnO₃* is ferroelectric and has incommensurate antiferromagnetic (AFM) ordering whereas *GdMnO₃* is paraelectric and has canted AFM (weakly ferromagnetic) states. For the mixed system, *Tb₁₋ₓGdₓMnO₃* anticorrelation between the (weak) ferromagnetism and the ferroelectricity was observed near the phase boundaries.

A rich phase diagram is observed with varying compositions for the series *Eu₁₋ₓYₓMnO₃*. A weak ferromagnetic and weak ferroelectric phase was observed for x = 0.2 *(Eu₀.₈Y₀.₂MnO₃)*. This was not observed for *EuMnO₃*. Ho, doped *DyMnO₃*, *(Dy₁₋ₓHoₓMnO₃, x = 0, 0.1)* exhibits a ferroelectric transition around 18 K. An additional transition due to polarization flop occurs at 12 K. The magnetoelectric coupling strength of 10% in *Dy₁₋ₓHoₓMnO₃* is two orders stronger in magnitude as compared to the hexagonal manganites. This can be reduced by the suppression of the spiral magnetic ordering. In addition, the reduction in the AFM ordering and ferroelectric ordering temperatures can also be possible by the modification of the spiral ordering [46]. Although *SmMnO₃* is not a ferroelectric compound, Y-doped *SmMnO₃* *(Sm₁₋ₓYₓO₃)* exhibits an electric polarization that induces ferroelectricity in addition to the exciting magnetism and hence is a multiferroic. A similar magnetoelectric coupling is also observed for *Gd₁₋ₓYₓMnO₃* (x = 0.1–0.4).

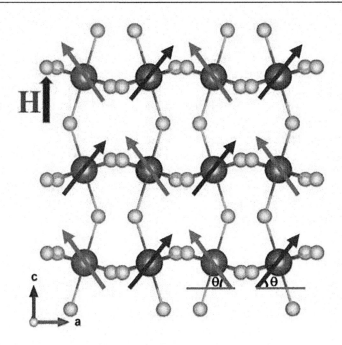

FIGURE 11.3 Canted Antiferromagnetic spin structure.

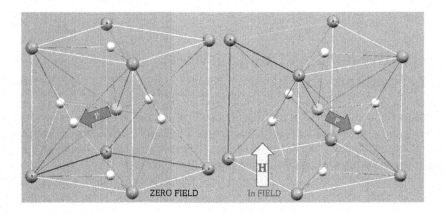

FIGURE 11.4 Polarization flop as a result of magnetic field (H).

6 APPLICATIONS OF RARE-EARTH MANGANITES

6.1 Applications in Electronic Devices

Application of the magnetic field in $RMnO_3$ results in a huge change in resistivity that invites great excitement in the field of multiferroics and the fabrication of novel electronic devices. The colossal magnetoresistance (CMR) effect could be the key to the next generation of magnetic memory devices, magnetic-field sensors, or transistors. Making use of the CMR effect in a film or using a spin valve structure, rare earth manganites are used as Magnetic field sensors. It is used in electric field effect devices using a ferroelectric gate. Another application is bolometric uncooled infrared (IR) sensors using the metal-insulator transition at the Curie temperature. It finds application in High-Temperature Superconductor (HTS)–CMR devices like flux-focused magnetic transducers and spin-polarized quasi-particle injection devices.

6.2 Environmental and Medical Applications

Epitaxial $Pr_{1-x}Ca_xMnO_3$ perovskite thin films reveal electro-catalytic water splitting which leads to oxygen evolution. Photocatalytic behavior is observed in $Nd_{0.6}Sr_{0.4}MnO_3$ and varies with the annealing temperature. Efficient solar-driven thermochemical CO_2 dissociation can be done with the redox properties generated by Mn^{4+}/Mn^{3+} pairs due to the doping of divalent elements in $RMnO_3$. These can also be used as a redox material in solar thermochemical fuel production through H_2O/CO_2 splitting. Unique redox properties are observed in substituted $LaMnO_3$. In general, due to their magnetic properties, $RMnO_3$ has received massive interest in the field of hyperthermia. The transition temperature (T_c) can be tuned in half-metallic ferromagnetic $La_{1-x}Sr_xMnO_3$ (LSMO). The overheating risk during hyperthermia treatment can be avoided due to the tunability of T_C of LSMO materials [47]

6.3 Magneto-Caloric Refrigerator

Magnetic refrigeration has the potential to reduce energy use by 30%, popularly known as the magnetocaloric effect (MCE). It finds application in heating or cooling technology as a potent alternative to conventional vapor compression. $RMnO_3$ exhibits MCE (entropy changes with magnetic fields). These can be used as an alternative for refrigerant material in magnetic refrigeration. Doped manganites like $La_{1-x}M_xMnO_3$, with $M = Na, K$, and Ag, exhibit MCE. For these materials change in entropy is maximum for different ranges of magnetic fields at different temperatures.

6.4 Solid Oxide Fuel Cells (SOFC)

$RMnO_3$ is an excellent example of electrode material in SOFCs, especially in Sr- doped $LaMnO_3$. The chemical stability and the thermal expansion coefficient are suitable as cathode material for high temperature (~1000 °C) SOFCs. Other manganites like $PrMnO_3$ and $NdMnO_3$ also exhibit electronic transport properties of SOFC that are used in low-temperature regimes usually less than 900 °C. Substitution of strontium in $NdMnO_3$ and $PrMnO_3$ enhances the properties of promising cathode material for SOFCs. Modification of the structure of the SOFC material may introduce distribution of strain energy over a larger thickness which can modify or accommodate the strain at the interface and overcome the lattice mismatch. This in turn can modify the performance of the SOFCs.

7 CONCLUSIONS AND OUTLOOK

$RMnO_3$ compounds are one of the most interesting regimes in the field of multiferroic materials. We have included most of the exciting details in the field of rare earth-manganites like structure, synthesis, multiferroicity, and various applications. The different origins of multiferroicity in orthorhombic and hexagonal $RMnO_3$ and the modification of the structure for the enhancement of the multiferroicity are undergoing intensive research. The application of doped and undoped $RMnO_3$ in the next generation of nanoelectronics needs significant effort and progress in these research areas. Most of the physical properties of these materials originate from the interactions between spin, charge, orbital, and lattice degrees of freedom. Still, the full understanding of these interaction mechanisms, the complexity of structure, and its allied properties related to the structure are not detailed in the literature. We believe that this chapter will give insight into the field of rare-earth manganites and their structure correlated properties and multiferroicity. The various applications of these unique materials discussed above are just a few examples of the wide application domain of the rare earth manganites.

8 ACKNOWLEDGMENTS

The first authors *SCB* acknowledges the Department of Science and Technology (DST, Govt. of India) for Inspire fellowship (*IF190617*) and *MP* acknowledges Government of India for Prime Minister Fellowship (PMRF – 2101307). The corresponding authors *SS* acknowledges the Department of Science and Technology (DST, Govt. of India) for research grant under the AMT project (*DST/TDT/AMT/2017/200*), and *EGR* acknowledges the Department of Science and Technology (DST, Govt. of India) for financial support under the Women Scientist Scheme-A (*SR/WOS-A/PM-99/2016 (G)*).

REFERENCES

[1] B. Dabrowskia, S. Kolesnika, A. Baszczukb, O. Chmaissema, T. Maxwella, and J. Mais, "Structural, transport, and magnetic properties of RMnO3 perovskites (R = La, Pr, Nd, Sm, 153Eu, Dy)," *Journal of Solid State Chemistry*, vol. 178, no. 3, pp. 629–637, 2005.

[2] Qinglin He, Xiantu Zhang, Haoshan Hao, and Xing Hua, "High-temperature electronic transport properties of La1-xCaxMnO3+δ ($0.0 \leq x \leq 1.0$)," *Physica B: Condensed Matter*, vol. 403, no. 17, pp. 2867–2871, Aug. 2008.

[3] Y. H. Huang, H. Fjellvag, M. Karppinen, B. C. Hauback, H. Yamauchi, and J. B. Goodenough, "Crystal and magnetic structure of the orthorhombic perovskite YbMnO3," *Chemistry of Materials*, vol. 18, no. 8, pp. 2130–2134, Apr. 2006.

[4] Raja Dasab and Pankaj Poddar, "Observation of exchange bias below incommensurate antiferromagnetic (ICAFM) to canted A-type antiferromagnetic (cAAFM) transition in nanocrystalline orthorhombic EuMnO 3," *RSC Advances*, vol. 4, no. 21, pp. 10614–10618, Feb. 2014.

[5] Anshuman Nandy, Tanusree Kar, Satya Ranjan Bhattacharyya, Dipankar Das, and Swapan Kumar Pradhan, "Alteration of magnetic behavior and microstructural distortion of EuMnO3 by partial substitution of Eu with monovalent Na," *Journal of Alloys and Compounds*, vol. 715, pp. 214–223, 2017.

[6] A. Pal and P. Murugavel, "Impact of cationic vacancies on the physical characteristics of multiferroic GdMnO3," *Journal of Applied Physics*, vol. 123, no. 23, p. 234102, Jun. 2018.

[7] Jamal-Eldin F. M. Ibrahim, Ayhan Mergen, E. İlhan Sahin, and Haythem Basheer, "The effect of europium doping on the structural and magnetic properties of GdMnO3 multiferroic ceramics," *Acerp.ir*, vol. 3, no. 4, pp. 1–5, 2017.

[8] Yimin Cui, Rongming Wang, and Jianqiang Qian, "Effect of cation addition on dielectric properties of TbMnO3," *Physica B: Condensed Matter*, vol. 392, no. 1–2, pp. 147–150, Apr. 2007.

[9] J. M. Chen, J. M. Lee, T. L. Chou, S. A. Chen, S. W. Huang, and H. T. Jeng, "Pressure-dependent electronic structures in multiferroic DyMnO3: A combined lifetime-broadening-suppressed x-ray absorption spectroscopy and ab initio electronic structure study," *Journal of Chemical Physics*, vol. 133, no. 15, p. 154510, Oct. 2010.

[10] K. Yadagiri and R. Nithya, "Structural and micro-Raman studies of DyMnO3 with potassium substitution at the Dy site," *RSC Advances*, vol. 6, no. 98, pp. 95417–95424, 2016.

[11] T. Wu, H. Chen, P. Gao, T. Yu, and Chen Z Liu, "Pressure dependent structural changes and predicted electrical polarization in perovskite RMnO3," *Journal of Physics Condensed Matter Physics*, vol. 28, no. 5, p. 056005, Jan. 2016.

[12] J. Magesh, P. Murugavel, R. V. K. Mangalam, K. Singh, Ch. Simon, and W. Prellier, "Strong enhancement of magnetoelectric coupling in Dy 3+ doped HoMnO 3," *Applied Physics Letters*, vol. 101, no. 2, p. 022902, Jul. 2012.

[13] N. E. Massa, del Campo Leire, Holldack Karsten, Ta Phuoc Vinh, Echegut Patrick, and Kayser Paula, "Far- and mid-infrared emission and reflectivity of orthorhombic and cubic ErMn O3: Polarons and bipolarons," *Physical Review B*, vol. 98, no. 18, p. 184302, Nov. 2018.

[14] L. Martin Carron, A. de Andres, M. J. Martínez Lope, M. T. Casais, and J. A. Alonso, "Raman phonons as a probe of disorder, fluctuations, and local structure in doped and undoped orthorhombic and rhombohedral manganites," *Physical Review B – Condensed Matter and Materials Physics*, vol. 66, no. 17, pp. 1–8, 2002.

[15] B. S. Araujo, A. M. Arevalo Lopez, C. C. Santos, J. P. Attfield, C. W. A. Paschoal, and A. P. Ayala, "Spin-phonon coupling in the incommensurate magnetic ordered phase of orthorhombic TmMnO3," *Journal of Physics and Chemistry of Solids*, vol. 154, p. 110044, Jul. 2021.

[16] Ajay Kumar Saw, Ganesha Channagoudra, Shivakumar Hunagund, Ravi L. Hadimani, and Vijaylakshmi Dayal, "Study of transport, magnetic and magnetocaloric properties in Sr2+ substituted praseodymium manganite," *Materials Research Express*, vol. 7, no. 1, pp. 1–14, 2019.

[17] Yang Huali, Liu Yiwei, Zhang Jiandi, Zhang Xiangqun, Cheng Zhaohua, and Xie Yali, "Anisotropic field-induced melting of orbital ordered structure in Pr0.6Ca0.4MnO3," *Physical Review B – Condensed Matter and Materials Physics*, vol. 91, no. 17, p. 174405, May 2015.

[18] Neeraj Panwar, S. K. Agarwal, G. L. Bhalla, D. Kaur, and D. K. Pandya, "Structural, electrical and magnetic properties of Pr1-xBa xMnO3 (x = 0.33–0.80)," *International Journal of Modern Physics B*, vol. 21, no. 15, pp. 2647–2656, Jun. 2007.

[19] Anis Biswasa and I. Das, "Magnetic and transport properties of nanocrystalline Nd0.5 Sr0.5 MnO3," *Journal of Applied Physics*, vol. 102, no. 6, p. 064303, 2007.

[20] A. M. Yang, Y. H. Sheng, M. A. Farid, H. Zhang, X. H. Lin, G. B. Li, L. J. Lui, and J. H. Lin, "Copper doped EuMnO3: Synthesis, structure and magnetic properties," *RSC Advances*, vol. 16, no. 17, pp. 13928–13933, 2016.

[21] N. Lee, Y. J. Choi, M. Ramazanoglu, W. Ratcliff, V. Kiryukhin, and S. W. Cheong, "Mechanism of exchange striction of ferroelectricity in multiferroic orthorhombic HoMnO3 single crystals," *Physical Review B – Condensed Matter and Materials Physics*, vol. 84, no. 2, p. 020101, Jul. 2011.

[22] J. Vermette, S. Jandl, and M. M. Gospodinov, "Raman study of spin-phonon coupling in ErMnO3," *Journal of Physics Condensed Matter*, vol. 20, no. 42, p. 425219, Oct. 2008.

[23] Duttine Mathieu, Wattiaux Alain, Balima Felix, Decorse Claudia, Moutaabbid Hicham, and D. H. Ryan, "Modulated magnetic structure in 57Fe doped orthorhombic YbMnO3: A Mössbauer study," *AIP Advances*, vol. 9, no. 3, p. 035008, Mar. 2019.

[24] R. Venkatesh, "Study of the magnetic behavior of single-crystalline Nd 0.5Sr0.5MnO3," *Journal of Applied Physics*, vol. 99, no. 8, p. 08Q311, 2006.

[25] Li Xiang, Lu Chengliang, Dai Jiyan, Dong Shuai, Chen Yan, and Hu Ni, "Novel multiferroicity in GdMnO3 thin films with self-assembled nano-twinned domains," *Scientific Reports*, vol. 4, p. 7019, Nov. 2014.

[26] Jung Hyuk Lee, Daesu Lee, Tae Won Noh, Pattukkannu Murugavel, Jae Wook Kim, and Kee Hoon Kim, "Formation of hexagonal phase of TbMnO3 thin film and its multiferroic properties," *Cambridge.org*, vol. 22, no. 8, pp. 2156–2162, Aug. 2007.

[27] Weitian Wang, "Epitaxial growth and in-plane dielectric properties of DyMnO3 by pulsed-laser deposition," *International Journal of Modern Physics B*, vol. 33, no. 28, p. 1950333, Nov. 2019.

[28] K. Shimamoto, Y. W. Windsor, Y. Hu, M. Ramakrishnan, A. Alberca, and E. M. Bothschafter, "Multiferroic properties of uniaxially compressed orthorhombic HoMnO3 thin films," *Applied Physics Letters*, vol. 108, no. 11, p. 112904, Mar. 2016.

[29] Seung Yup Jang, Daesu Lee, Jung Hyuk Lee, Pattukkannu Murugavel, and Jin Seok Chung, "Ferroelectric properties of multiferroic hexagonal ErMnO3 thin films," *Journal of the Korean Physical Society*, vol. 55, no. 2 PART 1, pp. 841–845, 2009.

[30] Malik Iftikhar Ahmed, Huang Houbing, Wang Yu, Wang Xueyun, Xiao Cui, and Sun Yuanwei, "Inhomogeneous-strain-induced magnetic vortex cluster in one-dimensional manganite wire," *Science Bulletin*, vol. 65, no. 3, pp. 201–207, Feb. 2020.

[31] Lamartine Meda, Klaus H. Dahmen, Saleh Hayek, and Hamid Garmestani, "X-ray diffraction residual stress calculation on textured La2/3Sr1/3MnO3 thin film," *Elsevier*, vol. 263, no. 1–4, pp. 185–191, Mar. 2004.

[32] Yuan Chang Liang, "Correlation between lattice modulation and physical properties of La0.72Ca0.28MnO3 films grown on LaAlO3 substrates," *Journal of Crystal Growth*, vol. 303, no. 2, pp. 638–644, May 2007.

[33] H. Wang and H. M. H.-J. T. H. T. Y. Guojun Liu, "Growth of PrSrMnO3-like thin films on NGO (1 1 0) substrates by plasma assisted MBE," *Journal of Crystal Growth*, vol. 227–228, pp. 960–965, Jul. 2001.

[34] Fujimoto Masayuki, Koyama Hiroshi, Nishi Yuji, Suzuki Toshimasa, Kobayashi Shinji, and Tamai Yukio, "Crystallographic domain structure of an epitaxial (Pr0.7Ca 0.3) MnO3 thin film grown on a SrTiO3 single crystal substrate," *Journal of the American Ceramic Society*, vol. 90, no. 7, pp. 2205–2209, Jul. 2007.

[35] I. Ting Chiu, Alexander M. Kane, Rajesh V. Chopdekar, Peifen Lyu, Apurva Mehta, and Chris M. Rouleau, "Phase transitions and magnetic domain coexistence in Nd0.5Sr0.5MnO3 thin films," *Journal of Magnetism and Magnetic Materials*, vol. 498, p. 166116, Mar. 2020.

[36] D. A. Mota, A. Almeida, V. H. Rodrigues, M. M. R. Costa, P. Tavares, and P. Bouvier, "Dynamic and structural properties of orthorhombic rare-earth manganites under high pressure," *Physical Review B – Condensed Matter and Materials Physics*, vol. 90, no. 5, p. 054104, Aug. 2014.

[37] N. Aliouane, O. Prokhnenko, R. Feyerherm, M. Mostovoy, J. Strempfer, and K. Habicht, "Magnetic order and ferroelectricity in RMnO3 multiferroic manganites: Coupling between R- and Mn-spins," *Journal of Physics Condensed Matter*, vol. 20, no. 43, p. 434215, Oct. 2008.

[38] Lin Chuanlong, Liu Jing, Li Xiaodong, Li Yanchun, Chu Shenqi, and Xiong Lun, "Phase transformation in hexagonal ErMnO3 under high pressure," *Journal of Applied Physics*, vol. 112, no. 11, p. 113512, Dec. 2012.

[39] S. M. Feng, Y. S. Chai, J. L. Zhu, N. Manivannan, and Y. S. Oh, "Determination of the intrinsic ferroelectric polarization in orthorhombic HoMnO3," *New Journal of Physics*, vol. 12, p. 092902, Jul. 2010.

[40] H. Okamoto, N. Imamur, B. C. Hauback, M. Karppinenb, H. Yamauchib, and H. Fjellvaga, "Neutron powder diffraction study of crystal and magnetic structures of orthorhombic LuMnO3," *Solid State Communications*, vol. 146, no. 3–4, pp. 152–156, Apr. 2008.

[41] B. Lorenz, "Hexagonal manganites—(RMnO3): Class (I) multiferroics with strong coupling of magnetism and ferroelectricity," *ISRN Condensed Matter Physics*, vol. 2013, pp. 43, 2013.

[42] D. G. Tomuta, S. Ramakrishnan, G. J. Nieuwenhuys, and J. A. Mydosh, "The magnetic susceptibility, specific heat and dielectric constant of hexagonal YMnO3, LuMnO3 and ScMnO3," *Journal of Physics: Condensed Matter*, vol. 13, no. 20, p. 4543, May 2001.

[43] Naoki Kamegashira, Hirohisa Satoh, and Satoshi Ashizuka, "Synthesis and crystal structure of hexagonal DyMnO3," *Materials Science Forum*, vol. 449, pp. 1045–1048, 2004.

[44] Vijay B. Shenoy and C. N. R Rao, "Electronic phase separation and other novel phenomena and properties exhibited by mixed-valent rare-earth manganites and related materials," *Philosophical Transactions of the Royal Society A: Mathematical, Physical and Engineering Sciences*, vol. 366, no. 1862, pp. 63–82, Jan. 2008.

[45] I. A. Abdel-Latif, Adel A. Ismail, Houcine Bouzid, and A. Al-Hajry, "Synthesis of novel perovskite crystal structure phase of strontium doped rare earth manganites using sol gel method," *Journal of Magnetism and Magnetic Materials*, vol. 393, pp. 233–238, Nov. 2015.

[46] P. M. J. Magesh, R. V. K. Mangalam, K. Singh, Ch. Simon, and W. Prellier, "Ferroelectric ordering and magnetoelectric effect of pristine and Ho-doped orthorhombic DyMnO3 by dielectric studies," *Journal of Applied Physics*, vol. 118, no. 7, p. 074102, Aug. 2015.

[47] Weiren Xia, Zhipeng Pei, Kai Leng, and Xinhua Zhu, "Research progress in rare earth-doped perovskite manganite oxide nanostructures," *Nanoscale Research Letters*, vol. 15, no. 1, pp. 1–55, Jan. 2020.

Magnetic Nanofillers-PVDF Nanocomposite Laminated Structures for Broad-Band Electromagnetic Shielding Applications

12

Soumyaditya Sutradhar

Contents

1	Introduction	182
2	Classification of Magnetic Nanofillers	183
	2.1 Ferromagnetic Materials	183
	2.2 Ferrimagnetic Materials	183
	2.3 Hexaferrite Materials	184
3	Importance of Magnetic Nanofillers in EMI Shielding	184
4	Advantages of PVDF Matrix	185
	4.1 Chemical Stability	185
	4.2 Piezoelectric Response	186
	4.3 Tunable Flexibility and Dimension	186
5	Method of Magnetic Nanofillers-PVDF Nanocomposite Preparation	187
6	Important Features of Magnetic Nanofillers-PVDF Nanocomposite Systems	188
	6.1 Structural Property	188
	6.2 Microstructural Property	188
	6.3 Thermal Property	189
	6.4 Chemical Property	190
	6.5 Magnetic Property	190
	6.6 Dielectric Property	191
	6.7 EMI Shielding Effectiveness Property	191
	6.7.1 Shielding Effectiveness Due to Reflection	192
	6.7.2 Shielding Effectiveness Due to Absorption	192

DOI: 10.1201/9781003196952-12

	6.7.3 Shielding Effectiveness Due to Multiple Reflections	192
	6.7.4 Total Shielding Effectiveness	193
7	EMI Shielding Applications of Magnetic Nanofillers-PVDF Nanocomposite Materials	193
	7.1 EM Pollution Reducer	193
	7.2 Coating Material for Electrical Cable	194
	7.3 Stealth Technology	194
8	Conclusion	195
References		195

1 INTRODUCTION

Nowadays, technology has advanced at its best and it reaches almost at its peak. The advancement of technology has improved the quality of human life and it brings up the ultimate coziness of livelihood to us by reducing the burdens from human shoulder at their workplace and home [1–5]. Not only has it reduced the workload from us, but it also shortens the gap between people all around the world by providing a very fast mode of communication using cutting-edge communication devices. People have already embossed their footsteps over the surface of the moon, and they have sent their representatives/devices to mars and near the sun. All these have been done successfully by us only due to the advancement of technology. The advancement of technology is not only helping to improve the quality of human life but at the same time, it also extends the long existence of human life by felicitating the technology-based advanced medical facilities to us [6–8]. All these show that the advancement of technology is a precious gift of science to us. In our childhood age, we have read an article called "Technology for Mankind" by Jacob Bronowski where the author has extended his views on the advantages of modern technology for mankind [9].

Nowadays, the improvement of technology, as well as its usefulness in almost every part of our life, shows how relevant the article was. However, we should be concerned about the other side of technological advancement. We shouldn't forget that technological advancement can also be a curse to mankind as it can bring up some of the very serious challenges to us if it is not used appropriately or prudently. It causes a great impact/threat on the environment and society. In recent times, the whole world is suffering from three major issues due to this high-end success of technological advancement and these have made us very much concerned about the environment and society all around us. These burning issues are water pollution, air pollution, and global warming. The social activists are regularly bringing these burning issues to our notice and knowledge and the scientists/technologists are devoting their knowledge, wisdom, and skill to cutting down these problems by finding suitable solutions. Fortunately, people are now very much concerned about these issues, but I think we are not acquainted with how many of us are aware of the fourth one, which has been evolved and extended only due to the extensive use of very modern and sophisticated electromagnetic devices/gadgets, an essential gift of science and technology to mankind. This fourth one, which can affect our environment, society, and even each of us almost equally and seriously, like the previously mentioned issues, is called electromagnetic (EM) pollution.

The essential and high demand for high-frequency electromagnetic devices in our everyday life has manifested this EM pollution. As the beneficiary of these EM devices, the human race, animals, and plants are also facing some serious problems due to the excessive exposure of this EM pollution, as it causes eye problems, it also affects different parts of the brain, creating many problems to pregnancy, etc. It also degrades the nutrient quality of food products. The uncontrolled EM radiation mostly in the microwave region of frequency (0.3 to 300 GHz) can also affect plants by producing abnormal chromosomal activity, DNA destruction, a decrease of growth, malfunctioning of seed germination, and many others. Radiation can also affect the immune system of living objects [9–15]. Now the time has come when we must take this problem very seriously and we need to combat this electromagnetic pollution. Also, the superior shielding effectiveness can reduce the Radar Cross-Section (RCS) when the material is applied over the surface of the flying object to make them fifth generation stealth fighter aircraft. Herein, through this article I have tried to find out the solution of the problem related to the electromagnetic/microwave pollution and stealth technology by the fabrication of magnetic nanofillers-PVDF nanocomposite material with superior electromagnetic interference (EMI) shielding effectiveness (SE) in the radio frequency (RF)/GHz frequency range and the related information of these state-of-art nanocomposite materials to my beloved readers.

2 CLASSIFICATION OF MAGNETIC NANOFILLERS

Now, to fabricate the state-of-art magnetic nanofillers-PVDF nanocomposite material the selection of magnetic material plays the most crucial role. So, a brief knowledge of various magnetic nanomaterials is necessary in this regard. These magnetic nanofillers with a significantly high magnetic moment are useful for the fabrication of resultant nanocomposite structures of high shielding effectiveness. The magnetic nanofillers with sufficient magnetic moment and different chemical compositions can be classified into different categories, which are mentioned below.

2.1 Ferromagnetic Materials

Ferromagnetic materials consist of permanent atomic magnetic moments and they have a very high tendency to align themselves in a particular direction even when the external magnetic field is not present. They possess spontaneous magnetization and a very high value of internal molecular magnetic field due to quantum mechanical exchange forces. Examples of ferromagnetic materials include cobalt, iron, nickel, gadolinium, dysprosium, permalloy, magnetite, etc. Some of the key advantages of ferromagnetic materials such as high permeability, high resistance, low hysteresis loss, low coercivity, high temperature, and chemical stability make them important in the field of science and technology. Properties of ferromagnetic materials are very interesting for different research and technological applications. Some of the important properties of ferromagnetic materials are mentioned below.

(i) The ferromagnetic substances are strongly attracted by the magnetic field.
(ii) These substances show permanent magnetism even in the absence of a magnetic field.
(iii) The ferromagnetic substance changes to paramagnetic when it is heated at a certain high temperature more specifically above Curie temperature.

These ferromagnetic materials find their applications in transformers, electromagnets, magnetic tape recording, hard drives, generators, telephones, loudspeakers, electric motors, hard disks, magnetic storage devices, etc.

2.2 Ferrimagnetic Materials

A ferrimagnetic material or ferrite (example: magnetite, Fe_3O_4) is a low-cost, corrosion-resistant, oxide ceramic material that consists large fraction of iron (III) oxide (Fe_2O_3) as the main component material. Additional metallic elements, such as strontium, barium, manganese, nickel, cobalt, copper, and zinc are also found in the composition of ferrites (examples: $NiFe_2O_4$, $CoFe_2O_4$, $Ni_{0.5}Zn_{0.5}Fe_2O_4$, $Ni_{0.4}Zn_{0.4}Cu_{0.2}Fe_2O_4$, $Mn_{0.4}Zn_{0.4}Cu_{0.2}Fe_2O_4$, etc.). The structure of ferrite is spinel in nature (AB_2O_4), and it consists of A-site (tetrahedral site) and B-site (octahedral site) respectively. Modulation of the magnetic property of ferrite or ferrimagnetic materials can be done suitable by the percentage variation of iron and other metallic elements at A-site (tetrahedral site) and B-site (octahedral site) respectively. These ferrite materials are ferrimagnetic in nature i.e., these materials can be magnetized or attracted to a magnet and the magnetic property of these materials can be modulated based on the cation distribution over A-site (tetrahedral site) and B-site when all other physical parameters are kept constant. Most of the ferrites are insulating in nature and this particular property of ferrite makes them most useful in different applications such as magnetic cores for transformers to suppress eddy currents. Other well-known applications of ferrite materials include magnetic recording tapes, radar-absorbing materials in stealth aircraft, microphones and loudspeakers, small motors for cordless appliances, and automobile applications.

2.3 Hexaferrite Materials

In the family of ferrite or ferrimagnetic materials, hexagonal ferrites or hexaferrites are the most important magnetic materials whose importance in research and commercial field is growing exponentially day by day. Some common examples of hexaferrites are:

(i) M-type ferrites, such as $BaFe_{12}O_{19}$ (BaM or barium ferrite), $SrFe_{12}O_{19}$ (SrM or strontium ferrite), and cobalt–titanium substituted M ferrite, Sr- or $BaFe_{12-2x}Co_xTi_xO_{19}$ (CoTiM).
(ii) U-type hexaferrites ($Ba_4Me_2Fe_{36}O_{60}$), such as $Ba_4Co_2Fe_{36}O_{60}$, or Co_2U.
(iii) W-type hexaferrites ($BaMe_2Fe_{16}O_{27}$), such as $BaCo_2Fe_{16}O_{27}$, or Co_2W
(iv) X-type hexaferrites ($Ba_2Me_2Fe_{28}O_{46}$), such as $Ba_2Co_2Fe_{28}O_{46}$, or Co_2X.
(v) Y-type hexaferrites ($Ba_2Me_2Fe_{12}O_{22}$), such as $Ba_2Co_2Fe_{12}O_{22}$, or Co_2Y.
(vi) Z-type hexaferrites ($Ba_3Me_2Fe_{24}O_{41}$), such as $Ba_3Co_2Fe_{24}O_{41}$, or Co_2Z.

The high magnetic permeability and moderate permittivity of these hexaferrite materials make them a good choice for permanent magnets, high-density magnetic recording and data storage materials and they are also highly suitable for microwave devices like resonance isolators, filters, circulators, phase shifters, etc. In addition to all these applications, the hexaferrite materials have excellent chemical stability, mechanical hardness, and low eddy current loss at high frequencies. Some of the hexaferrite materials (especially the X-type and U-type hexaferrites) are very good electromagnetic wave absorbers and are useful for EMC, RAM, and stealth technology.

3 IMPORTANCE OF MAGNETIC NANOFILLERS IN EMI SHIELDING

Magnetic nanoparticles are the most important candidates in high-frequency electromagnetic device applications. The high magnetic permeability, moderate permittivity, and tunable loss factors of most of the ferrite and hexaferrite nanoparticles in their ferri-/ferro-magnetic state make these magnetic nanoparticles the most significant material in the high-frequency region. In this direction, the most important applications include a transformer core, a memory device, sensor, and electromagnetic interference (EMI) shielding effectiveness in the broad-band microwave region. Ferrite and hexaferrite nanomaterials consist of large magnetic moments, i.e., these materials contain a significant number of magnetic dipoles. Also, the polarization effects at their lattice sites are responsible for the generation of electric dipolar polarization inside their lattice structure. Now, the interactions between the magnetic field vector of the incident EM wave with magnetic dipoles and the electric field vector of the incident EM wave with electric dipoles present in most of the magnetic nanoparticles (ferrite and hexaferrite nanoparticles) generate the EMI shielding effectiveness due to absorption. The flipping of both types of dipoles (magnetic and electric) at microwave frequency region under the influence of magnetic field vector and electric field vector of incident electromagnetic radiation generate heat energy. Therefore, the interactions between EM radiation with magnetic dipoles and electric dipoles and thereby the resonance oscillation of these dipoles are responsible for this energy conversion from microwave energy to heat energy. This heat energy will be absorbed by the materials themself or this energy will be radiated off by the materials to their surroundings leading to the reduction or attenuation of the energy of incident electromagnetic radiation which is commonly known as EMI shielding due to absorption.

4 ADVANTAGES OF PVDF MATRIX

Poly(vinylidene fluoride) (PVDF) is a thermoplastic fluoropolymer that can be prepared by the polymerization of vinylidene difluoride. Also, PVDF is a non-reactive, semi-crystalline, highly flexible, lightweight, semi-transparent polymer material with large pyroelectric and piezoelectric behavior. These properties of PVDF make it the most interesting one in the field of polymer research among all the available polymers [16–18]. Also, PVDF is a very well-known polymer material for its various polymorphism and five different crystalline phases such as α-, β-, γ-, δ- and ε-phases are available [19–21]. Among these five different crystalline phases, α-, β- and γ-phase polymorphs are the most common types available therein. Here α-phase is the non-polar phase, β- and γ-phases are the polar phases of PVDF, and the β-phase is the most interesting one for different technological applications as it constitutes the effective piezoelectric, pyroelectric, and ferroelectric responses in PVDF [22]. The most common way to obtain the polar β-phase of PVDF from its non-polar α-phase is the mechanical stretching method of PVDF [23]. But this method is not suitable for the preparation of laminated nanocomposite materials because the stretching process restricts the loading of a high amount of nanofillers and/or leads to uncontrolled reconfigurations and the agglomeration of the nanofillers [24]. Although the development of PVDF reinforced laminated nanocomposites have been considered as the subject of intensive research by various researchers but the architectural difficulties related to the formation of the desired phase of PVDF make the research work very challenging to the researchers. In this direction, improved β-phase crystallization of nanofillers loaded PVDF nanocomposites can be obtained by a simple and cost-effective solution casting method [25]. In this method, the production of nanofillers-PVDF nanocomposite materials with nearly uniform distribution of the nanofillers in the matrix of PVDF can be obtained and it also helps to develop/modulate many important and interesting properties. The addition of nanofillers in the matrix of PVDF can enhance its performance or it can provide new responses, by capitalizing the size, shape, and properties of the nanofillers [26]. Also, the processability and mechanical property of PVDF is an advantage compared to bare ferrites nanoparticles, which makes the PVDF-based nanocomposite systems self-standing and flexible. Moreover, a sufficiently high magnetic permeability can be achieved within the PVDF nanocomposite materials by the incorporation of the magnetic nanoparticles, and finally, the magnetic, dielectric, and microwave absorption effects can be observed in such laminated PVDF nanocomposite materials for different technological applications [27].

4.1 Chemical Stability

Poly(vinylidene fluoride) (PVDF) is a chemically inert material at room temperature and it shows chemical inertness to a wider range of chemicals that includes minerals and organic acids, aliphatic and aromatic hydrocarbons, alcohols, oxidants, and halogenated solvents. At a comparatively higher temperature, this PVDF can dissolve in various organic solvents such as amines and esters. The chemical inertness of PVDF at room temperature makes it the most suitable candidate for the design and fabrication of corrosion-resistive coating material over different health care equipment and high-frequency electromagnetic devices to be performed the role of an EMI shielding jacket.

4.2 Piezoelectric Response

Poly(vinylidene fluoride) (PVDF) is a very important material from the piezoelectric point of view. The poor electrical conductivity and high dielectric constant make this PVDF the most suitable piezoelectric material for different applications such as sensors, nanogenerators, etc. The presence of electroactive β-phase in PVDF can be found with its all-trans (TTTT) conformation which makes the β-phase of PVDF the most suitable one to convert internal elastic energy to dielectric energy when the external force is applied. PVDF has the piezoelectric anisotropy, due to which different piezoelectric characteristics can be found in each direction.

4.3 Tunable Flexibility and Dimension

PVDF bead is a hard material, so the modulation of flexibility and dimension of them is quite difficult. But the flexibility and dimension of PVDF or PVDF-based nanocomposite materials can be modulated suitably by one of the easiest synthesis processes called the solution casting method. Now, various factors are there to customize the flexibility and dimension of PVDF or PVDF-based nanocomposite materials in the solution casting method. These factors include PVDF dissolution temperature, amount of solvent, shape, and size of the mold, soaking temperature, etc. All these factors will help immensely to develop the PVDF or PVDF nanocomposite materials with a high degree of flexibility, large surface area, and controllable thickness so that the resultant nanocomposite can be used in different applications. The given Figure 12.1 shows the self-standing behavior and flexibility of some representative magnetic nanofillers-PVDF nanocomposite materials.

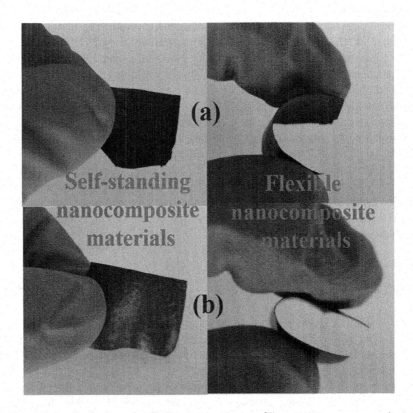

FIGURE 12.1 Image of self-standing and flexible magnetic nanofillers-PVDF nanocomposite materials.

5 METHOD OF MAGNETIC NANOFILLERS-PVDF NANOCOMPOSITE PREPARATION

Magnetic nanoparticles (ferrites and hexaferrites nanoparticles) with a significant amount of magnetic moment are very important for the usefulness of the resultant nanocomposite materials in shielding effectiveness study. For the desired composition of ferrites (examples: $NiFe_2O_4$, $CoFe_2O_4$, $Ni_{0.5}Zn_{0.5}Fe_2O_4$, $Ni_{0.4}Zn_{0.4}Cu_{0.2}Fe_2O_4$, $Mn_{0.4}Zn_{0.4}Cu_{0.2}Fe_2O_4$, etc.) and hexaferrites (examples: Co_2U, Co_2W, Co_2X, Co_2Y, Co_2Z, etc.) the stoichiometric amount of all the precursor materials will be taken in a beaker and a complete homogeneous solution would be prepared by ethanol. One mole citric acid will be dissolved in 60 ml ethanol and this homogeneous solution will be added to the above salt solution under vigorous stirring conditions. Thereafter this prepared solution will be sonicated for two hours at about 60 °C. After the sonication, the solution will be vigorously stirred at a regular interval to get a heavily dense form of the solution. The solution will be placed into the oven at 80 °C and the prepared solution will be dried slowly for over two days to get the dried form of the solution. Finally, the as-prepared powder sample will be sintered at various temperatures (should be greater than the phase forming temperature of the respective magnetic nanoparticles) for 6 hours to obtain the magnetic nanoparticles with a desired magnetic moment. After the formation of magnetic nanomaterials, the magnetic nanofillers-PVDF laminated nanocomposite structures can be prepared by a very simple solution casting method. In this process, solid PVDF pallets will be dissolved in DMF with magnetic stirring and the temperature of the stirrer needs to be set at 70 °C. The continuation of this process for a certain interval of time helps to dissolve PVDF completely into the DMF solution. The complete dissolution of solid PVDF pallet into DMF helps us to get the thick and transparent gel. After the formation of PVDF gel, a certain weight percentage of magnetic nanofillers of the high magnetic moment will be added to the solution of PVDF and the whole solution will be taken for ultra-sonication. The ultra-sonication needs to continue for two hours to obtain the homogeneous mixing of the magnetic nanofillers in the solution of PVDF. Finally, magnetic nanofillers embedded PVDF materials will be obtained by casting the whole mixture in a well cleaned and moisture-free glass plate and evaporating the solvent at 110 °C [28–30]. The detailed preparation of magnetic nanofillers embedded PVDF nanocomposite materials has been displayed in Figure 12.2.

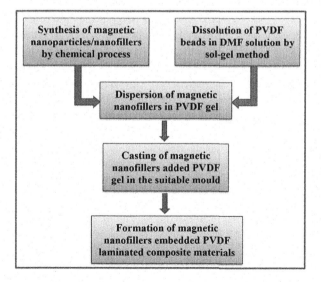

FIGURE 12.2 Flowchart of the synthesis of magnetic nanoparticles/nanofillers embedded PVDF laminated nanocomposite materials.

6 IMPORTANT FEATURES OF MAGNETIC NANOFILLERS-PVDF NANOCOMPOSITE SYSTEMS

As per the EMI shielding applications are concerned, these magnetic nanofillers embedded PVDF nanocomposite materials are the most important candidates due to their pertinency in both efficiency and applicability. The tailor-made applications of these nanocomposite structures with high EMI shielding effectiveness have attracted the interest of both research and industrial sectors. Many articles so far have been published on the EMI shielding effectiveness of these magnetic nanoparticles embedded PVDF nanocomposite structures in the broadband GHz frequency region [28–30]. Also, these nanocomposite materials can be used more conveniently over the surface of different geometrical structures as a coating/paint material in comparison to the well-known EMI shielding materials having powder structures [31]. So, the large EMI shielding capacity in the broadband GHz frequency region and the applicability of these lightweights, flexible, and laminated nanocomposite structures as a coating/paint material make them most appreciable candidates in different EMI shielding applications that include, EM pollution reduction, defense, telecommunications, medical equipment and many more [29]. Modulation of structural, microstructural, thermal, chemical, magnetic, dielectric, and shielding effectiveness properties is responsible for the overall improvement of these nanocomposite structures as EMI shielding material. Modulation of each of the mentioned physical properties of these nanocomposite structures and their significant contributions in shielding effectiveness has been discussed below in a very distinct manner.

6.1 Structural Property

Magnetic nanoparticle embedded PVDF materials are nanocomposite in nature and due to this reason, each of these structures will show the presence of both the crystallographic phases of respective component materials such as the crystallographic phases of magnetic nanoparticles and PVDF in their XRD pattern. The crystallographic phases of different magnetic nanoparticles such as ferrite and hexaferrite nanoparticles with different compositions and the polar and non-polar phases of semicrystalline PVDF (α-, β-, γ-, δ-, and ε-phase) signify the formation of the multi-phase structure of these nanocomposite materials. The detail of the crystallographic phases of magnetic nanoparticles (ferrite and hexaferrite nanoparticles) and semicrystalline PVDF with all its polar and non-polar phases have already been discussed in our published articles [28–30]. Also, the modulation of different polar (β-phase and γ-phase) and non-polar phases (α-phase, δ-phase, and ε-phase) of PVDF with the incorporation of magnetic nanoparticles as an effective nanofillers inside the matrix of PVDF has been discussed therein. This change in polymeric phases of PVDF towards the polar phases due to the incorporation of magnetic nanofillers inside the matrix of PVDF signifies the orientational modulation of PVDF molecular chain in form of all trans-planar zig-zag (TTTT) conformation. The improvement of the polarization property of magnetic nanoparticles embedded PVDF nanocomposite materials makes them one-piece to expect changes in the other physical properties in the direction of shielding effectiveness applications.

6.2 Microstructural Property

The microstructural analysis is one of the most interesting studies to understand the formation of the multi-phase nanocomposite structure of magnetic nanoparticles embedded in PVDF materials. The presence of magnetic nanofillers inside the matrix of PVDF, surface morphology of magnetic nanofillers, and the change in surface morphology of PVDF both in the presence and absence of magnetic nanofillers can be understood very clearly with the help of this technique. Moreover, this technique will be very helpful

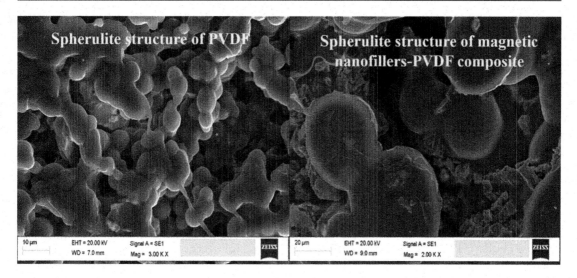

FIGURE 12.3 FESEM image of PVDF and magnetic nanofillers-PVDF nanocomposite structures.

to understand the underlying possibilities of these nanocomposite structures in the direction of shielding effectiveness applications. It has been observed in various research articles on magnetic nanoparticles embedded PVDF nanocomposite materials that the surface morphology of the nanocomposite structures takes the spherulitic shape when the polar phases get increased due to the incorporation of the magnetic nanoparticles inside the matrix of PVDF. The spherulite structure of PVDF and magnetic nanofillers-PVDF nanocomposite structures are displayed in Figure 12.3. These spherulitic patterns over the surface of nanocomposite PVDF structures at the cost of lamella patterns are the signature pattern of having a greater fraction of the polar phases over non-polar phases in the nanocomposite structures that can make them an important candidate for EMI shielding effectiveness applications.

6.3 Thermal Property

Understanding the thermal stability and the determination of the decomposition temperature of the materials is an important physical property in this direction. The thermal stability of magnetic nanoparticles embedded PVDF nanocomposite materials is also an important issue to understand the usefulness of these nanocomposite structures in different applications at different temperatures. The most important application of magnetic nanoparticles embedded PVDF nanocomposite structures is its coating over the surface of the fifth generation stealth fighter aircraft for stealth technology. The surface temperature of a flying aircraft at a speed of Mach 1 can be raised to 120 °C. So, it is very important to understand the thermal stability and to know the decomposition temperature of these nanocomposite structures for the better use of these materials in stealth applications. The DTA graph of these magnetic nanoparticles embedded PVDF nanocomposite structures show that the endothermic reaction is going on inside the matrix of PVDF nanocomposite structures and this endothermic reaction starts at nearly 150 °C and extends up to 500 °C or so, depending on the nature of the magnetic nanofillers and their weight percentages inside the matrix of PVDF. The TGA graph also shows that the complete decomposition of these magnetic nanofillers-PVDF nanocomposite materials appears at a quite high temperature. So, these magnetic nanoparticles embedded PVDF nanocomposite structures can retain their various physical properties such as chemical, mechanical, magnetic, dielectric/polarization, etc. at an ambient temperature that is significantly higher than the normal atmospheric temperature say 300 K. Therefore, the high degree of thermal stability of these magnetic nanoparticles embedded PVDF nanocomposite structures makes them useful for different applications in the area of EMI shielding and microwave absorption.

6.4 Chemical Property

The chemical property of magnetic nanofillers-PVDF nanocomposite materials related to the modulation of their polymeric chain inside the nanocomposite structure can be extracted by using the FTIR spectra. The presence of various individual peaks corresponding to α-phase, β-phase, and γ-phase in PVDF nanocomposite systems can be estimated herein. The individual peaks of magnetic nanofillers-PVDF nanocomposite materials have already been found at 494 cm^{-1}, 539 cm^{-1}, 620 cm^{-1} and 769 cm^{-1}, 894 cm^{-1} and 982 cm^{-1} corresponding to the infrared bands of non-polar α-phase of PVDF and four more peaks at 518 cm^{-1}, 606 cm^{-1}, 846 cm^{-1} and 1081 cm^{-1} have also been found for the polar β-phase of PVDF [28, 29]. It is to be mentioned here that the solution casting method, selection of nanofillers, the weight percentage of nanofillers inside the matrix of PVDF, etc., can modulate the β-phase fraction (F(β)%) of these magnetic nanofillers-PVDF nanocomposite materials and a very high value of β-phase fraction (F(β)%) (nearly ~ 43.2% or 45%) can be found for these nanocomposite materials. The occurrence of magnetic nanofillers inside the PVDF matrix can enhance the fractional contribution of accompanying chains with β phase-related all-trans (TTTT) conformation with respect to the α phase-related trans-gauche (TGTG) conformation and thereby it can lead to the effective enhancement of the β-phase fraction (F(β)%) of magnetic nanofillers-PVDF nanocomposite materials. The schematic diagram of the mechanism for the formation of this electrostatic interaction between the magnetic nanofillers and the CH_2 groups of the PVDF matrix has been displayed in Figure 12.4. This electrostatic interaction between magnetic nanofillers and the positively charged CH_2 groups is accountable for the generation of β-phase consistent with the all-trans (TTTT) conformation at the cost of non-polar α-phase consistent with the trans-gauche (TGTG) conformation of the magnetic nanofillers-PVDF nanocomposite materials.

6.5 Magnetic Property

Magnetic property is one of the important aspects of these magnetic nanoparticles embedded PVDF nanocomposite materials to expect superior shielding effectiveness behavior from them. The magnetic

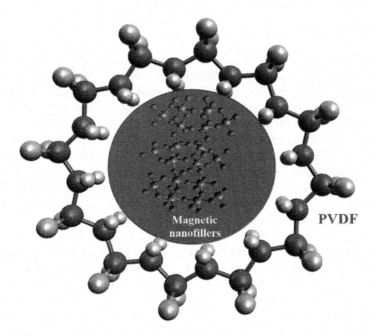

FIGURE 12.4 Schematic diagram of magnetic nanofillers embedded PVDF nanocomposite materials.

response is very much helpful to understand the presence of magnetic orderings of the resultant nanocomposite materials. The order of magnitude of various magnetic parameters such as magnetization, coercive field, and remanence is also important to comprehend the usefulness of the nanocomposite materials in shielding effectiveness applications. The presence of magnetic nanofillers such as ferrite and hexaferrite nanofillers inside the matrix of PVDF is responsible for the development of the magnetic responses in these magnetic nanofillers-PVDF nanocomposite laminated structures. Usually, PVDF is a non-magnetic material. So, the magnetic contribution of PVDF-based nanocomposite materials comes due to the presence of magnetic nanofillers inside the matrix of PVDF. Many articles are available on magnetic nanofillers embedded PVDF nanocomposite systems such as Ni-Zn-Cu-Ferrite-PVDF, Mn-Zn-Cu-ferrite-PVDF, Sr-hexaferrite-PVDF, U-type hexaferrite-PVDF nanocomposite systems. All these nanocomposite materials show significantly high magnetic responses such as high magnetization, coercive field, remanence. Moreover, the most interesting thing i.e., the tailoring of the magnetic responses by the variation of the weight percentage of the magnetic nanofillers inside the matrix of PVDF has also been observed in this type of nanocomposite structure. Usually, the magnetization of the nanocomposite materials increases due to the presence of a larger fraction of magnetic nanofillers inside the matrix of PVDF. However, this increase in magnetization of the nanocomposite structures improves at the cost of the structural deformities, when the loading percentage of the nanofillers is very large inside the matrix of PVDF. This deformation can lead to a higher degree of rigidity of the nanocomposite structures and reduce flexibility. So, it is very important to make the optimization of the magnetic response and the flexibility of the nanocomposite materials by making the right choice and weight percentage of the magnetic nanofillers inside the matrix of PVDF. The presence of magnetic responses and the modulation of the magnetic responses therein make these nanocomposite materials the important candidate for different high-frequency device applications where shielding effectiveness application is one of them.

6.6 Dielectric Property

Dielectric property study of magnetic nanofillers incorporated PVDF is an important observation that is helpful to understand the polarization phenomenon of the nanocomposite materials. This dielectric study of magnetic nanofillers incorporated PVDF nanocomposite materials is performed in the range of 40 Hz to 10^6 Hz and various dielectric parameters can be extracted therein [28]. Dielectric permittivity, tangent loss, dielectric modulus, dielectric impedance, etc. can be estimated from the dielectric measurement. Though the shielding effectiveness study is performed in the GHz frequency range and most of the published articles 8–18 GHz frequency range corresponding to the X-band (8–12 GHz) and K_u-band (12–18 GHz) have been mentioned. So, there is no direct correlation between dielectric study in the frequency range of 40 Hz to 10^6 Hz and the observations like complex permittivity, complex permeability, and shielding effectiveness study in the frequency range of 8–18 GHz. However, this dielectric study will help us to understand the possibility of these nanocomposite materials being evolved out as an important material for shielding effectiveness study. The presence of various forms of polarization effects such as ionic, molecular, or interfacial polarization effects can be found and their variation as a function of frequency can also be observed. All together, these indicate the overall ability of these magnetic nanofillers incorporated PVDF systems to interact with the electric field vector of external electromagnetic radiation to produce significant attenuation of microwave radiation.

6.7 EMI Shielding Effectiveness Property

Shielding's effectiveness study of magnetic nanofillers incorporated PVDF nanocomposite systems is the most important and key observation of the article. The process in which the intensity of the transmitted EM waves gets reduced due to the absorption or reflection of electromagnetic (EM) waves by the materials and produces attenuation of the EM waves is called EMI shielding. The overall improvement of the EMI shielding effectiveness/total shielding effectiveness (SE_T) of magnetic nanofillers incorporated PVDF nanocomposite materials depends on shielding effectiveness due to reflection (SE_R), shielding effectiveness due to absorption (SE_A), and shielding effectiveness due to multiple reflections (SE_{MR}). The detailed discussion on shielding effectiveness of various kind produced by these magnetic nanofillers incorporated PVDF nanocomposite materials are given below. This is to be mentioned here that the EMI shielding capacity of magnetic nanofillers incorporated PVDF nanocomposite materials can be estimated by extracting two scattering (S) parameters i.e., S_{11} (coefficient of reflection) and S_{21} (coefficient of transmission), respectively, within the frequency range of 8–12 GHz for X-band and 12–18 GHz for K_u-band by using PNA Series Network Analyzer.

6.7.1 Shielding Effectiveness Due to Reflection

Shielding effectiveness due to reflection (SE_R) of the nanocomposite material is an important phenomenon and it also contributes significantly to the development of a large amount of total shielding effectiveness of the nanocomposite material. Now, this reflection phenomenon would be maximum from the surface of the nanocomposite material if it contains electrons. The interaction between the electric field of the incident EM wave and the quantum particles (electrons) will produce a reflection of the incident EM wave from the surface of the materials and this interaction is accountable for the enhancement of the reflectivity of the materials as well. In this case, the poor conducting magnetic nanomaterial does not possess a high value of SE_R of the nanocomposite materials. The estimation of SE_R of magnetic nanofillers incorporated PVDF nanocomposite materials can be done by using equation (1)

$$SE_R (dB) = 10 \log_{10}\left(\frac{1}{1-|S_{11}|^2}\right) \quad (1)$$

6.7.2 Shielding Effectiveness Due to Absorption

Shielding effectiveness due to absorption (SE_A) of the nanocomposite material is also an important phenomenon in this direction and like the previous one, it can also contribute significantly to the development of a large amount of total shielding effectiveness. Now, the interaction among the magnetic field of the incident EM wave with magnetic dipoles and the electric field of the incident EM wave with electric dipoles convert this microwave radiation into heat energy and these interactions are accountable for the enhancement of the absorptivity of the materials. The presence of magnetic nanofillers such as ferrites and hexaferrites with significant magnetic moments inside the nanocomposite systems are mostly responsible for the improvement of this SE_A of these nanocomposite materials. The estimation of SE_A of magnetic nanofillers incorporated PVDF nanocomposite materials can be done by using equation (2)

$$\left| SE_A (dB) = 10 \log_{10}\left(\frac{1-|S_{11}|^2}{|S_{12}|^2}\right) \right. \quad (2)$$

6.7.3 Shielding Effectiveness Due to Multiple Reflections

Shielding effectiveness due to multiple reflections of these magnetic nanofillers incorporated PVDF nanocomposite materials is negligible because of the presence of a large difference between the skin depth of the EM wave corresponding to these laminated nanocomposite materials and the actual thicknesses of these laminated nanocomposite materials. Therefore, the net shielding effectiveness of these laminated nanocomposite materials can be exhibited in three dissimilar ways and these are (1) SE_A, (2) SE_R, and (3) SE_T due to both SE_A and SE_R.

6.7.4 Total Shielding Effectiveness

The magnetic nanofillers incorporated PVDF nanocomposite materials that consist substantial amount of SE_R and SE_A in the broadband frequency range of 8–18 GHz can possess a high value of total shielding effectiveness (SE_T) in the same frequency range. The estimation of SE_T from the estimated values of SE_R and SE_A of magnetic nanofillers incorporated PVDF nanocomposite materials can be done by using equation (3)

$$SE_r(dB) = 10\log_{10}\left(\frac{P_I}{P_0}\right) = SE_A + SE_R + SE_{MR} \tag{3}$$

Here, SE_{MR} stands for shielding effectiveness due to multiple reflections and this can be neglected for magnetic nanofillers incorporated PVDF nanocomposite materials, as already mentioned. Under this circumstance, the SE_T can be rewritten as

$$SE_r(dB) = 10\log_{10}\left(\frac{P_I}{P_0}\right) = SE_A + SE_R \tag{4}$$

7 EMI SHIELDING APPLICATIONS OF MAGNETIC NANOFILLERS-PVDF NANOCOMPOSITE MATERIALS

Magnetic nanofillers incorporated PVDF nanocomposite materials are multifunctional materials, and they have a wide range of applications. The most important application is the fabrication of an EMI shielding jacket to prevent the absorption of EM radiation by the living objects, coating material for the electrical cables, and the use of these nanocomposite materials as an important part of the stealth technology. Some details of these applications are listed below.

7.1 EM Pollution Reducer

The high EMI shielding effectiveness of magnetic nanofillers incorporated PVDF nanocomposite systems makes it an important material for the application of EM pollution reducers. If we look around us, then we will find that cell phone tower numbers are growing all around us very rapidly. These cell phone towers can emit high-frequency electromagnetic radiation to our environment. Now, the time has come when we need to realize by ourselves and be aware of our society about the adverse effects of high-frequency electromagnetic radiation on living objects. All the lives living within the range of 100–150 meters or so from these cell phone towers are affected badly because of the absorption of this high-frequency electromagnetic radiation. Since most living objects hold a large amount of water inside of their body, that is why the absorption of this high-frequency electromagnetic radiation makes adverse effects on our bodies. It is to be mentioned here that these adverse effects will be even more to some of our body parts such as eyes, brain, joints, heart, abdomen, etc. where the water has almost no movement. The absorption of this emitted electromagnetic radiation from the cell phone towers will cause localized heating inside of our bodies. It results in the evaporation of the water from various parts of our body such as the eyes, brain, joints, heart, abdomen, etc. The radiation can also affect various implants present inside of our body such as Pacemaker, Implantable Cardiovascular Defibrillators (ICDs), and Impulse Generators. The radiation that comes out from the cell phones causes electromagnetic interference (EMI) with those devices and disrupt their functioning. Radiations can also affect the immune system of living objects. So, keeping all these in mind, the state-of-art nanocomposite materials applicable for the reduction of electromagnetic pollution from various electromagnetic devices can be successfully made using magnetic nanofillers incorporated PVDF nanocomposite materials. As already mentioned, the fabrication of an EMI shielding jacket can prevent the absorption of microwave radiation by our body and keep us safe from these adversative effects of microwave radiation.

7.2 Coating Material for Electrical Cable

These magnetic nanofillers incorporated PVDF nanocomposite materials can also be used as coating material over the electrical cable. Usually, these EM radiation absorbers materials are good insulators with high breakdown strength. The insulating property of these nanocomposite systems makes it useful to prevent charge leakage from the electrical cables and the high breakdown strength of these materials is useful to withstand the significantly high electric field develop therein. Now, the presence of high EM radiation absorption capacity of these materials gives these magnetic nanofillers incorporated PVDF nanocomposite materials a degree of efficacy for the use as the coating material over electrical cables, where all the requirements such as prevention of charge leakage, prevention of EM radiation leakage and the capacity of holding large electric field can be fulfilled properly.

7.3 Stealth Technology

These magnetic nanofillers incorporated PVDF nanocomposite materials are getting rising attention in defense sectors of many countries due to their usefulness in the high-frequency region. Military systems are moving towards the higher frequency region and there is a growing need for microwave absorbers that work to 100 GHz. Laird Technologies has designs for specific resonant frequencies in the millimeter-wave band, as well as broadband designs. Laird Technologies is working on several military programs at these frequencies, as well as automotive radars and millimeter-wave communications programs. Also, the automotive electronics scientists of the US Air Force under the advanced tactical fighter (ATF) program are working in the field of radar absorbing materials (RAMs) for the reduction of radar cross-section (RCS) as a part of stealth technology along with the other technologies like Avionics, Cockpit, Armament, etc. in fighter aircraft. Radar absorption technology is of immense importance as the radar signal reflects from the surface of the aircraft and could be detected by radar detectors. Present-day radar detectors can detect several high frequencies in the GHz region. On the other hand, microwave absorber or radar absorbing systems and materials available are not only effective in such high-frequency GHz regions, but they are also effective in the ultra-high frequency GHz region. Microwave absorbers or radar absorbing materials offer highly effective radar absorption properties at such frequency ranges. This microwave absorption by such functionalized materials is effective for a wide band of the frequency range. Recently, the People's Republic Army (China) has shown interest in radar absorbing materials and they have officially revealed their fifth generation stealth fighter aircrafts Chengdu J-20 (Black Eagle) in the 2016 China International Aviation & Aerospace Exhibition. The aircraft was in their military service from March 2017. In India, Aeronautical Development Agency (ADA) and Hindustan Aeronautics Limited (HAL) jointly have taken up the project to develop the single-seat, twin-engine, stealth all-weather multirole fighter aircraft. It was reported that the first flight of a full-scale prototype is scheduled to occur in 2032. So, keeping the usefulness of these microwave absorber or radar absorbing materials (RAMs) for the reduction of electromagnetic pollution and the use of these materials in fifth generation stealth fighter aircraft in mind, this project will be very much interesting and useful for future applications in high-frequency electronic devices and defense applications.

8 CONCLUSION

In this chapter, I have discussed various aspects and the EMI shielding effectiveness property of magnetic nanofillers incorporated PVDF laminated nanocomposite structure. Magnetic nanofillers are also good for shielding effectiveness applications, but when they are considered inside the PVDF matrix, the applicability and efficacy of the resultant nanocomposite structure both get enhanced extensively as compared to the bare magnetic nanomaterials. Keeping this in mind and to make the resultant nanocomposite materials more useful in a different area of applications, this article has emphasized more the magnetic nanofillers (ferrites and hexaferrites) based PVDF nanocomposite structures for EMI shielding effectiveness study. In this article different possible applications have also been mentioned to make it more interesting to the reader and to make this field of research more popular to the researchers. This field of research needs more attention from the researchers and in the future, more and more information will be available in this direction.

REFERENCES

[1] H. Gargama, A.K. Thakur, S.K. Chaturvedi, Polyvinylidene Fluoride/nickel Nanocomposite Materials for Charge Storing, Electromagnetic Interference Absorption, and Shielding Applications, *J. Appl. Phys.* 117 (2015) 224903–224911.

[2] Q.M. Zhang, V. Bharti, X. Zhao, Giant Electrostriction and Relaxor Ferroelectric Behavior in Electron-Irradiated Poly(vinylidene fluoride-trifluoroethylene) Copolymer, *Science* 280 (1998) 2101–2104.

[3] S. Bauer, Poled Polymers for Sensors and Photonic Applications, *J. Appl. Phys.* 80(10) (1998) 5531–5558.

[4] X.J. Zhang, J.Q. Zhu, P.G. Yin, A.P. Guo, A.P. Huang, L. Guo, G.S. Wang, Tunable High-Performance Microwave Absorption of $Co_{1-x}S$ Hollow Spheres Constructed by Nanosheets within Ultralow Filler Loading, *Adv. Funct. Mater.* 28 (2018) 1800761–1800767.

[5] Y.F. Pan, G.S. Wang, L. Liu, L. Guo, S.H. Yu, Binary Synergistic Enhancement of Dielectric and Microwave Absorption Properties: A Nanocomposite of Arm Symmetrical PbS Dendrites and Polyvinylidene Fluoride, *Nano Res.* 10(1) (2016) 284–294.

[6] W. Xu, Y.F. Pan, W. Wei, G.S. Wang, Nanocomposites of Oriented Nickel Chains with Tunable Magnetic Properties for High-Performance Broadband Microwave Absorption, *ACS Appl. Nano Mater.* 1 (2018) 1116–1123.

[7] W. Xu, Y.F. Pan, W. Wei, G.S. Wang, P. Qu, Microwave Absorption Enhancement and Dual-nonlinear Magnetic Resonance of Ultra Small Nickel with Quasi-one-dimensional Nanostructure, *Appl. Surf. Sci.* 428 (2017) 54–60.

[8] W. Xu, G.S. Wang, P.G. Yin, Designed Fabrication of Reduced Graphene Oxides/Ni Hybrids for Effective Electromagnetic Absorption and Shielding, *Carbon* 139 (2018) 759–767.

[9] Y.H. Hao, L. Zhao, R.Y. Peng, Effects of Microwave Radiation on Brain Energy Metabolism and Related Mechanisms, *Mil. Med. Res.* 2 (2015) 4–11.

[10] W.J. Zhi, L.F. Wang, X.J. Hu, Recent Advances in the Effects of Microwave Radiation on Brains, *Mil. Med. Res.* 4 (2017) 29–42.

[11] C.D. Robinette, C. Silverman, S. Jablon, Effects upon Health of Occupational Exposure to Microwave Radiation (RADAR), *Am. J. Epidemiol.* 112 (1980) 39–53.

[12] E. Diem, C. Schwarz, F. Adlkofer, O. Jahn, H. Rüdiger, Non-Thermal DNA Breakage by Mobile-phone Radiation (1800 MHz) in Human Fibroblasts and in Transformed GFSH-R17 Rat Granulosa Cells in Vitro, *Mutat. Res.* 583 (2005) 178–183.

[13] V.G. Vrhovac, D. Horvat, Z. Koren, The Effect of Microwave Radiation on the Cell Genome, *Mutat. Res. Lett.* 243 (1990) 87–93.

[14] J.L. Sagripanti, M.L. Swicord, DNA Structural Changes Caused by Microwave Radiation, *Int. J. Radiat. Biol.* 50 (1986) 47–50.

[15] Z. Chen, C. Xu, C. Ma, W. Ren, H.M. Cheng, Lightweight and Flexible Graphene Foam Nanocomposites for High-Performance Electromagnetic Interference Shielding, *Adv. Mater.* 25 (2013) 1296–1300.
[17] A.J. Lovinger, Ferroelectric Polymers, *Science* 220 (1983) 1115–1121.
[18] E. Fukada, History and Recent Progress in Piezoelectric Polymers, *IEEE Trans. Ultrason. Ferroelectr. Freq. Control.* 47 (2000) 1277–1290.
[19] T. Prabhakaran, J. Hemalatha, Ferroelectric and Mmagnetic Studies on Unpoled Poly (Vinylidine Fluoride)/Fe3O4 Magnetoelectric Nanocomposite Structures, *J. Mater. Chem. Phys.* 137 (2013) 781–787.
[20] Y. Lu, J. Claude, B. Neese, Q.M. Zhang, Q. Wang, A Modular Approach to Ferroelectric Polymers with Chemically Tunable Curie Temperatures and Dielectric Constants, *J. Am. Chem. Soc.* 128 (2006) 8120–8121.
[21] V. Tomer, E. Manias, C.A. Randall, A High Field Properties and Energy Storage in Nanocomposite Dielectrics of Poly(Vinylidene Fluoride-Hexafluoropropylene), *J. Appl. Phys.* 110 (2011) 044107–044116.
[22] A. Salimi, A.A. Yousefi, Analysis Method: FTIR Studies of β-phase Crystal Formation in Stretched PVDF Films, *Polym. Test.* 22 (2003) 699–704.
[23] J. Gomes, J. Serrado Nunes, V. Sencadas, S. Lanceros-Mendez, Influence of the β-phase Content and Degree of Crystallinity on the Piezo- and Ferroelectric Properties of Poly(Vinylidene Fluoride), *Smart Mater. Struct.* 19 (2010) 065010–065016.
[24] V. Sencadas, V.M. Moreira, S. Lanceros-Mendez, A.S. Pouzada, R. Gregorio, α to β Phase Transformation and Microestructural Changes of PVDF Films Induced by Uniaxial StretchMater. *Sci. Forum.* 514 (2006) 872–876.
[25] P. Martins, A.C. Lopes, S. Lanceros-Mendez, Electroactive Phases of Poly(Vinylidene Fluoride): Determination, Processing and Applications, *Prog. Polym. Sci.* 39 (2014) 683–706.
[26] C.M. Kanamadi, B.K. Das, C.W. Kim, D.I. Kang, H.G. Cha, E.S. Ji, A.P. Jadhav, B.E. Jun, J.H. Jeong, B.C. Choi, B.K. Chougule, Y.S. Kang, Dielectric and Magnetic Properties of (x)CoFe$_2$O$_4$ + (1 − x)Ba$_{0.8}$Sr$_{0.2}$TiO$_3$ Magnetoelectric Nanocomposites, *Mater. Chem. Phys.* 116 (2009) 6–10.
[27] B.-W. Li, Y. Shen, Z.-X. Yue, C.-W. Nan, Enhanced Microwave Absorption in Nickel/Hexagonal-Ferrite/Polymer Nanocomposites, *Appl. Phys. Lett.* 89 (2006) 132504–132506.
[28] P. Saha, S. Das, S. Sutradhar, Influence of Ni-Zn-Cu-ferrite on Electroactive b-phase in Poly(Vinylidenefluoride)-Ni-Zn-Cu-ferrite Nanocomposite Film: Unique Metamaterial for Enhanced Microwave Absorption, *J. Appl. Phys.* 124 (2018) 045303–045312.
[29] P. Saha, T. Debnath, S. Das, S. Chatterjee, S. Sutradhar, β-Phase Improved Mn-Zn-Cu-ferrite-PVDF Nanocomposite Film: A Metamaterial for Enhanced Microwave Absorption, *Mater. Sci. Eng. B* 245 (2019) 17–29.
[30] S. Sutradhar, S. Saha, S. Javed, Shielding Effectiveness Study of Barium Hexaferrite-Incorporated, β-Phase-Improved Poly(vinylidene fluoride) Nanocomposite Film: A Metamaterial Useful for the Reduction of Electromagnetic Pollution, *ACS Appl. Mater. Interfaces* 11(26) (2019) 23701–23713.
[31] S. Sutradhar, K. Mukhopadhyay, S. Pati, S. Das, D. Das, P.K. Chakrabarti, Modulated Magnetic Property, Enhanced Microwave Absorption and Mössbauer Spectroscopy of Ni$_{0.40}$Zn$_{0.40}$Cu$_{0.20}$Fe$_2$O$_4$ Nanoparticles Embedded in Carbon Nanotubes, *J Alloys Compd.* 576 (2013) 126–133.

Iron-Based Materials to Remove Toxic Waste from the Environment

13

Srimathi Krishnaswamy, Puspamitra Panigrahi,
Ganapathi Subramaniam Nagarajan

Contents

1 Introduction	198
2 Synthesis Method of Magnetic Particles	199
2.1 Magnetic Performance	199
2.2 Types of Magnets	200
2.2.1 Zero-Valent Iron (ZVI) Nanoparticles (NP)	200
2.2.2 Iron Oxides (Fe_2O_3)	200
2.2.3 Hematite (α-Fe_2O_3)	200
2.2.4 Magnetite (Fe_3O_4)	200
2.2.5 Maghemite (γ-Fe_2O_3)	201
2.2.6 Spinel Ferrites	201
2.3 Analysis of Adsorbents	201
3 Industrial Dye and Its Hazard	201
3.1 Adsorption	201
3.1.1 Batch Adsorption Process	202
3.1.1.1 Effect of pH	202
3.1.1.2 Effect of Adsorbent Dosage	202
3.1.1.3 Effect of Concentration of Adsorbate	202
3.1.1.4 Effect of Contact Time	203
3.1.1.5 Effect of Temperature	203
3.1.1.6 Effect of Stirring Rate	203
3.1.1.7 Effect of Ionic Strength	203
3.1.1.8 Reusability	203
3.1.1.9 Isotherm Studies	204
3.1.1.10 Kinetic Studies	204
3.1.2 Fixed Bed Experiment	204
3.1.2.1 Fixed Bed Mode Studies	204
3.1.2.2 Effect of Bed Height	205

DOI: 10.1201/9781003196952-13

		3.1.2.3	Effect of Flow Rate	205
		3.1.2.4	Effect of Influent Concentration	205
		3.1.2.5	Mechanism of Adsorption	205
	3.1.3	Adsorbent for Removal of Inorganic Pollutants		206
	3.1.4	Adsorbent for Removal of Textile Dye		206
	3.1.5	Adsorbent for Removal of Pharmaceutical Products		206
	3.1.6	Adsorbent for Removal of Pathogens		207
	3.1.7	Adsorbent for Removal of Pesticides		207
	3.1.8	Bioadsorbents		208
3.2	Photocatalysis			208
	3.2.1	Mechanism		209
	3.2.2	Effect of Contact Time		209
	3.2.3	Effect of Intensity of Light		209
	3.2.4	Effect of Concentration of the Organic Molecule		209
	3.2.5	Effect of Photocatalyst Amount		210
	3.2.6	Effect of pH		210
	3.2.7	Photocatalysis of Removal of Organic Dyes		211
	3.2.8	Photocatalysis for Removal of Drug		211
4 Conclusion				211
References				211

1 INTRODUCTION

The main sources for drinking water are seawater, groundwater, rainwater harvesting, lakes, reservoirs, and canals. They could be contaminated by various sources of pollutants (Figure 13.1) such as industrial wastewater, domestic wastewater, and natural source water [1]. Water remediation can be done by adopting physical, chemical, and biological methods. Few physical methods are air stripping and incineration. In the case of chemical method, adsorption, oxidation, colloid treatment, and coagulation is adopted [2]. In the biological treatment method, organic compounds are degraded by bio-aided reactants. The main techniques in water purification are classified into adsorption, catalytic processes, biotechnology, ionizing radiation processes, membrane processes, and magnetically assisted processes. Recently, nanomaterials-based techniques emerged as improved versions for the water treatment process [3]. Industrially adopted

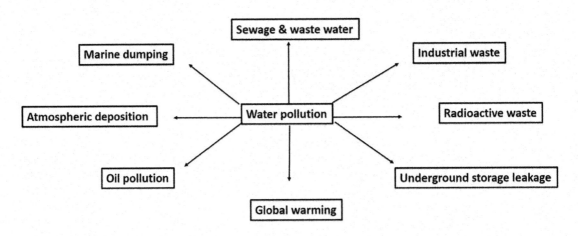

FIGURE 13.1 Sources of water pollution.

technique and frequently studied method is the adsorption process. The most used popular adsorbent is activated carbon derived from jute sticks, coconut coir, rice husk, etc., [4] The water treatment includes an adsorption technique using extractants or a combination of technique adsorption with redox or magnetic techniques. Organic and inorganic pollutants are reduced or oxidized by catalytic processes. Some of the catalytic processes for water purification are photo-assisted oxidation, electrocatalysis, solar-assisted oxidation, heterogeneous catalytic ozonation, and the photo-electro Fenton process [5]. Recently, magnetic nanomaterials have been efficiently used to remove water pollutants such as organic wastes (chlorinated hydrocarbons, aromatics, dyes, and pesticides) and heavy metals. In this chapter, different types of magnetic materials and their composites in eradicating the toxins are discussed.

2 SYNTHESIS METHOD OF MAGNETIC PARTICLES

Different types of methods are involved in the synthesis of magnetic nanoparticles which are sol-gel, polyol process, photolysis, microwave-assisted/sonochemical, chemical vapor deposition, thermal decomposition, co-precipitation, solvothermal, electrodeposition, laser pyrolysis, and electrochemical method has been explored in the literature [6]. Figure 13.2 depicts the classification of synthesis methods for the magnetic material. Control in particle size and distribution, crystal structure, crystallinity, and morphology are the factors in synthesizing magnetic materials.

2.1 Magnetic Performance

The property of magnetic particles is determined by applying an external magnetic field. Various forms of magnetism such as paramagnetism, diamagnetism, ferromagnetism, ferrimagnetism are observed based on the orientations of the magnetic moment [7]. Diamagnetism is very tiny and disparate from the applied magnetic field. As the orbit increases from zero paramagnetism arise when it is parallel to the applied magnetic field. In the case of ferromagnetism, without the need to utilize an external magnetic field it develops spontaneous magnetization which is inherently magnetically ordered. For ferrimagnetism, different atom possesses different moment strengths, but at below critical temperature, it is in an ordered state. Based on the magnetic susceptibility, the magnetic material can be divided.

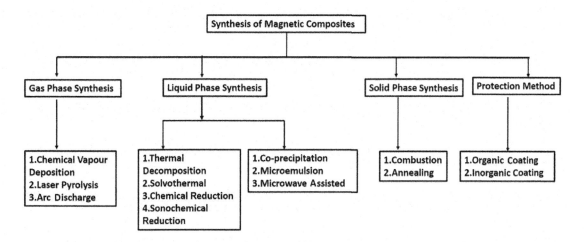

FIGURE 13.2 Classification of synthesis methods for magnetic material.

2.2 Types of Magnets

Variations of a type of magnet employed can enhance the intensity of magnetization of a separator. Various type of magnets is present which are permanent magnet, electromagnet, and superconducting magnet. Permanent magnets are from metals such as iron, nickel, and cobalt which have weak magnetic intensity owing to the generation of magnetic force less than 1 Tesla (T). But due to the advances in material development and shape, a high-power magnetic field could be created [8]. Owing to the production of magnetic intensity greater than 2T per unit length, multipole magnets are formed. In this case, electricity is not required for the creation of a magnetic field. In the case of electromagnets, the magnetic field is created within their cavity by the passage of electric current on the solenoid of electrical conducting wires. Depending on their application, the space of solenoid is varied from linear annular and space also varied. The maximum field generated for these electromagnets is 2.4T. The highest intensity magnetic field of 2 to 10T is generated by superconducting magnets.

2.2.1 Zero-Valent Iron (ZVI) Nanoparticles (NP)

Owing to the bioavailability, abundance, non-toxic nature, ease to synthesize, zerovalent iron particles have been used to eliminate the pollutants. In order to degrade the organic pollutants, hydrogen peroxide is generated from zero-valent iron particles and oxygen. The high value of magnetization is obtained by metallic iron NP and it is used to promote magnetic iron np and it is used to promote magnetic recovering step. ZVI exhibits high surface activity (E = −0.44V). By generating a high oxidation state, the toxic heavy metals are eradicated from wastewater.

2.2.2 Iron Oxides (Fe_2O_3)

Iron oxides nanoparticles have exceptional sorption capacity due to the porosity structure, robust magnetic response, and high surface area. Iron oxide nanoparticles exist in various structures. Owing to the polymorphism and temperature-induced phase transition, Hematite (α-Fe_2O_3), Magnetite (Fe_3O_4), and Maghemite (γ-Fe_2O_3) are proper candidates for water purification.

2.2.3 Hematite (α-Fe_2O_3)

Hematite is a stable material and it occurs in rocks and soils. Owing to its slow cost, high resistance to corrosion, small dimensions, and convenient operations it is utilized in photocatalysis and environmental treatment. Hematite is weakly ferromagnetic at room temperature in the bulk form and converts to the antiferromagnetic phase at morin temperature (Tm = −13.15 °C). Further, it exists as a paramagnetic phase above the curie temperature (TC = 682.85 °C). Several parameters such as size, crystallinity, and exchange interaction influence the magnetic properties of hematite. The adsorption properties of the hematite can be enhanced by substituting with numerous metal ions. Moreover, the bandgap of hematite is 2.0–2.2 eV and it is an n-type semiconductor. Many researchers have reported that can able to absorb around 43% of the visible light region.

2.2.4 Magnetite (Fe_3O_4)

Only magnetite two types of oxides such as Ferrous and ferric ions and further it has inverse phase structure. The structure has tetrahedral (A-site) and octahedral (B site) which can produce two magnetic sublattices that have an antiparallel spin on each other. For the inverse spinel structure, B sites have an equal number of ferric and ferrous ions. In the meantime, A sites have only ferrous ions. Magnetite possesses low bandgap energy and it shows p-type semiconductor and n-type semiconductor.

2.2.5 Maghemite (γ- Fe$_2$O$_3$)

Maghemite possesses a similar structure to magnetite and further, it exhibits a cubic structure with ferrous ions are shared over A sites and b sites. Moreover, due to the availability of ferrous ions, it has high chemical stability. Maghemite can be used as an adsorbent in wastewater purification due to the high value of magnetization saturation.

2.2.6 Spinel Ferrites

Recently spinel ferrites are broadly studied due to their exceptional properties, particularly their magnetic properties. Spinel ferrites have the formula MFe$_2$O$_4$ where M indicates divalent metal ions such as Cu, Ni, Co, and Zn. The properties and applications of spinel ferrite may depend upon the synthesis method and the nature and distribution of the cations. It possesses a low bandgap of 1.1–3 eV. Due to the low bandgap, magnetic behavior, and chemical stability, the material can be used in photocatalysis. The spinel ferrites have been employed to remove these nitrophenols because of this simplistic synthesis method. Also, it is highly resistant to acidic and basic conditions.

2.3 Analysis of Adsorbents

The powder X-ray diffraction studies are used to investigate the crystal structure of the samples. The morphology, particle size distribution, etc. of the nanocomposites are examined by Scanning electron microscopy (SEM) and Transmission electron microscopy (TEM). The elemental composition of the nanocomposite is examined by energy dispersive spectroscopy (EDS) analysis. The stability of the photocatalyst is evaluated by thermal gravimetric analysis (TGA). The optical properties of the nanocomposites are studied by a UV-visible spectrophotometer. The emission properties of the nanocomposites are studied using a spectrofluorometer. Raman spectroscopy is employed to study the chemical structure, molecular interactions, etc. The functional group of the nanocomposites is investigated by Fourier transform infrared spectroscopy (FTIR). The chemical status and elemental composition are analyzed by X-ray photoelectron spectroscopy (XPS). The surface area of the nanocomposites is examined by **Brunauer-Emmett-Teller** (BET) analysis. The particle size of the nanocomposites is evaluated by a particle size analyzer. Inductively coupled plasma mass spectroscopy (ICP-MS) is utilized to provide information about all constituent elements of the original sample. Atomic absorption spectroscopy (AAS) is utilized to determine the concentration of adsorbate and investigate the adsorption kinetics and isotherm. Further, AAS is used to determine the metals and metalloids.

3 INDUSTRIAL DYE AND ITS HAZARD

Industrial dye is applied in the textile industry to produce color for the textile. Further dyes are organic molecules and are classified into three types such as cationic, anionic, and non-ionic dyes. Figure 13.3 depicts the classification of dyes. The cationic dyes are basic. Direct, acid, and reactive dyes are classified as anionic dyes. The waste dyes which are released by textile industries combine with water and cause harmful effects to humans and the environment. The toxic effect of dyes exposure causes heart attack, vomiting, and jaundice [9]. The oral administration of dyes in animals produces testicular lesions [10].

3.1 Adsorption

Adsorption is one of the processes that is widely used for the removal of dyes, metal ions, and pesticides in wastewater treatment. It is a process of removal of containment such as liquid or gas which is

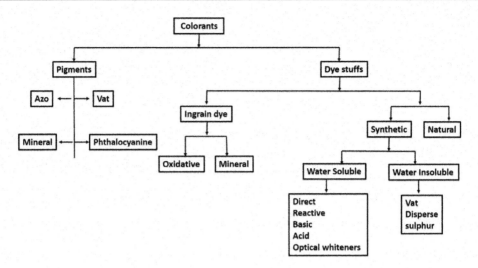

FIGURE 13.3 Classification of colorants.

an adsorbate by the adsorbent. Two types of adsorptions occur. Physisorption occurs due to the van der Waals force of attraction between adsorbate and adsorbent and it is reversible in nature. Chemisorption: A strong chemical bonding takes place between adsorbent and adsorbate and it is difficult to separate the adsorbate and the adsorbent. From the adsorption isotherm, the quantity of adsorbate adsorbed is calculated. The adsorbent should have a high surface area and less time to attain adsorption equilibrium. Modifying naturally available adsorbent as magnetic composite is an excellent method owing to the magnetic separation method. Basically, the adsorption of adsorbate onto an adsorbent surface is based on various physicochemical factors influencing the total process. Two types of methods are followed. One is a batch experiment and another one is fixed bed methodology. Several parameters can be calculated using the primitive batch experiment.

3.1.1 Batch Adsorption Process

3.1.1.1 Effect of pH
One of the important factors which have a significant role in the adsorption process is the pH of the solution. The pH has a major effect on the ionization of active site present and surface charge density of the adsorbent and elements of adsorbate in wastewater. At acidic conditions more H+ are present whereas in the alkaline domain more OH- are found which affects the adsorption process. For example, metal cations are stable at lower pH and can be adsorbed on the surface of the adsorbent. On the other hand, at higher pH values their solubility decreases and precipitation occurs which complicates the adsorption process.

3.1.1.2 Effect of Adsorbent Dosage
The adsorbent dose is another vital parameter to obtain higher adsorption performance by optimizing the amount of adsorbent dosage. The adsorption efficiency is increased with a higher amount of adsorbent due to the larger surface area with a higher number of vacant adsorption sites. For instance, an adsorbent dose from (0.1–0.4g) of ECCSB@Fe_3O_4 exhibited the removal percentage of Pb(II) to 97.64% [11]. Further increase in adsorbent dose does not increase the removal efficiency indicating that saturation of active sites has been attained.

3.1.1.3 Effect of Concentration of Adsorbate
The concentration of the contaminants has a vital role in determining the efficiency of the adsorption process. For a fixed amount of adsorbent, a higher pollutant concentration increases the adsorption

performance. It has been reported that with an increase in the initial concentration of both Cr(VI) and MB dye from 100 to 200 mg/L, the adsorption capacity of CQSs-Ac adsorbent increased from 231 and 391 to 282.1 and 462.6 mg/g [12] respectively for the pollutants.

3.1.1.4 Effect of Contact Time

Undoubtedly, contact time is another significant factor that is involved in the adsorption process. At the initial stage, the availability of more active sites on the surface of the adsorbent influences the adsorption rate. After attaining the equilibrium stage, the concentration gradient decreases due to the saturation of active sites of the adsorbent. Hence it is vital to study the optimum time to reach the equilibrium.

3.1.1.5 Effect of Temperature

Certainly, an important parameter to determine the adsorption performance is the temperature. On increasing the temperature, the adsorption efficiency increases by increasing the rate of diffusion of adsorbate molecules on the boundary layer of the adsorbent. At an elevated temperature, the viscosity of the liquid reduces which in turn increases the collision frequency between the adsorbent and adsorbate. This process increases the diffusion rate from the aqueous phase to the adsorbent phase and hence the adsorption performance increases. Moreover, the high affinity of the adsorbent affirms the adsorption method is endothermic in nature. The spontaneity feasibility predicts whether it is exothermic or endothermic. Further, it can be calculated by thermodynamic parameters value such as Gibbs free energy change (DG^0), enthalpy change (ΔH^0), and entropy change (ΔS^0). The free energy change (ΔG), standard free energy change (ΔG^0), and equilibrium constant (Kc) at constant temperature are related by can't Hoff reaction. If the ΔH^0 and ΔS^0 exhibit positive magnitude values, it refers to endothermic nature of adsorption. If ΔG^0 value is in the range −20–0 KJ/mol, it implies physisorption, while in the range −400–80 KJ/mol, it suggests chemisorption.

3.1.1.6 Effect of Stirring Rate

Usually, in the batch method, enough interaction between the adsorbent and adsorbate is greatly dependent on the agitation speed. The rate of stirring has a higher impact on the adsorption process. By increasing the degree of stirring, the external mass transfer has been overcome. Various parameters such as agitator type and agitator time also influence the adsorption process.

3.1.1.7 Effect of Ionic Strength

Estimation of adsorbent performance towards pollutants with the existence of other species such as organic, inorganic materials (acids, alkalis, and salts) in the form of dissolved and suspended necessary to be accomplished. When the effect of interfering species is well investigated, optimization of the adsorption process is established. These interfering species might hinder the adsorption process. When the ionic strength increases, the adsorption decreases which might be attributed to the limiting between the electrolyte and the adsorbate species.

3.1.1.8 Reusability

The other important aspect of the adsorption method is stability and recyclability from an economic perspective. The desorption process reduces the cost of the process and furthers the metals are recovered from an aqueous solution by easy approach. The extracted metal can be utilized as starting material to other industrial sectors. The desorption process not only reduces the overall cost of the process but also extends to include recovery of pollutants (metal) extracted from aqueous solutions by an effective approach. The extracted metal can be used as feedstock to other industrial sectors. For example, the recovery and reusability of chitosan/MWCNT-COOH nanocomposite to desorb Cr(VI) was successfully performed by using sodium hydroxide up to four cycles, the adsorption behavior was 100% [13]. Generally, adsorption data for the containment system are performed to kinetics isotherms and thermodynamics.

3.1.1.9 Isotherm Studies
At equilibrium conditions, at a constant temperature, the adsorption isotherm delivers the graphical representation of the amount of adsorbate adsorbed by the adsorbent. They inform the nature of adsorption such as monolayer form or multilayer form. Further, it notifies the nature of adsorbent whether it is a homogeneous or heterogeneous system. And also, they differentiate between physisorption and chemisorption. An evaluation of a suitable isotherm model is vital to be carried out for every pollutant. Various equilibrium isotherm models such as Langmuir, Freundlich, Temkin, Flory-Huggins, Dubinin, Fowler-Guggenheim, Jovanoic, and Frunking have been put together to study the adsorption behavior.

3.1.1.10 Kinetic Studies
Kinetic studies are one of the important factors to determine the mechanism and the rate-determining step of pollutant adsorbing from the wastewater onto the adsorbent. Basically, three steps are involved in the adsorption process.

1. The external mass transfer of adsorbate from the bulk solution to the surface of the adsorbent.
2. Diffusion from adsorbent surface to the inner surface.
3. Adsorption onto adsorbent surface binding sites.

When the adsorbent comes into contact with the adsorbate solution, transfer of adsorbate ions takes place from the bulk solution to the boundary layer of the adsorbent takes place. Due to the adsorbate diffusion on the surface of the adsorbent, the adsorbate concentration decreases in the bulk solution. Lastly, the interaction takes place between adsorbate and active sites of adsorbent. Generally, the adsorption data are assessed by various kinetic models such as pseudo-first-rate equation, pseudo-second-order rate equation, Boyd model, Weber and Morris model. Elovich model and Binghams model. Commonly, the linearized form of mentioned models is used for explaining the adsorption process.

3.1.2 Fixed Bed Experiment

The adsorption processes were performed by fixed-bed method or batch method through lab-scale. The data obtained from the batch method do not comply with water treatment plants. Hence in predicting large-scale operations fixed bed method was performed. In the fixed-bed column, two important concepts are appraised one is a breakthrough point and the exhaustion point is evaluated through effluent concentration with time. These data are essential to design for a continuous flow adsorption process. When the concentration of influent increases, total exhaustion occurs. Breakthrough shape, curve, time exhaustion time, and ion exchange properties depend on variable parameters such as flow rate, bed height, influent concentration, and chemical composition of adsorbent. By mathematical calculation of fixed-bed column performance, many parameters such as a number of bed volume (NBV) empty bed contact time (EBCT) adsorbent exhaustion rate, and mass transfer zone (MTZ) can be performed. The interaction between adsorbate and adsorbent is measured by empty bed contact time (EBCT). Further, the fixed bed length where adsorption takes place is represented by mass transfer (MTZ) which is the ratio of volume treated until the breakthrough point to the packed bed volume. Next, the adsorbent exhaustion rate (AER) is defined as the mass of adsorbent to the volume of water treated till the exhaustion point. When the AER has a lower value, it would depict better and more efficient adsorption.

3.1.2.1 Fixed Bed Mode Studies
At the lab scale, a continuous fixed-bed column consists of a column packed with adsorbent material with a fixed diameter and height. The column is packed with upper and lower supportive layers such as cotton to prevent adsorbent particles overflow. Different parameters such as the effect of initial concentration, flow rate, and bed height were studied in detail. Owing to the availability of numerous vacant adsorption sites, each molecule is rapidly adsorbed by the adsorbent. Subsequently, the target pollutants are eliminated by the adsorbent and the bed bottom discharges pure effluent. Over some time, when the pollutant

continuously flows through the bed column, the adsorption ability decreases and attains a saturation point owing to the occupation of the available adsorption area.

3.1.2.2 Effect of Bed Height

To study the effect of bed height on the adsorption processes, different bed height is chosen with fixed pollutant concentration and constant flow rate. A suitable adsorbent particle size is necessary to control the decrease in pressure in the column. By the enhancement of the bed height, the removal efficiency of the column increases. Owing to an increase in surface area larger number of vacant binding spots will be available for adhesion of the pollutant. Enhancement of bed height increases the broadening in the MTZ and later breakthrough time increases. Because the pollutant more time to penetrate the adsorbent sites and get adsorbed on the active surface.

3.1.2.3 Effect of Flow Rate

With a fixed pollutant concentration and constant bed height, various flow rates are propelled into the adsorbent in order to evaluate the effect of the rate of flow on contaminating elimination in the aqueous water. At a low flow rate, the adsorbate has a large time to interact with adsorbent active sites, which leads to high adsorption efficiency. On the other hand, at a higher flow rate, the adsorption efficiency is reduced due to their shorter retention time for adsorbate to interact with the adsorbent. Also, the adsorbate limits the penetration into the pore sites of the adsorbent. Before the attainment of equilibrium, the adsorbate leaves the column at a higher flow rate because residence time is reduced. Besides this, owing to the higher flow rate quicker breakthrough and exhaustion time is observed with an increase in MTZ. It is generally known that by the intraparticle mass transfer control the adsorption process to be performed at a slower rate.

3.1.2.4 Effect of Influent Concentration

With different initial pollutant concentrations, the adsorption studied was investigated while other factors such as flow rate and bed height were kept constant. On increasing the concentration of influent, the adsorption capacity is enhanced because the adsorption sites of adsorbent become saturated more rapidly as a result of a greater concentration gradient between the specific adsorbent active site and the adsorbent. On increasing the concentration of influent, the MTZ becomes smaller. Additionally, the breakthrough time and exhaustion are accelerated at a higher concentration of influent concentration due to the saturation of the fast bed. Increasing the bed height from 1 to 7cm exhibited higher adsorption efficiency of thallium by rice husk [14]. Enhancement of breakthrough and saturation point was noticed with increasing the bed height. Boeris et al. [15] employed P. Putida to adsorb aluminum from wastewater. On reducing the flow rate from 1ml/min to 0.5ml/min, higher breakthrough time and exhaustion time were attained.

3.1.2.5 Mechanism of Adsorption

Usually, the adsorption process is based on various physicochemical features such as size, chemical composition, solubility, surface charge, hydrophobicity, and reactivity of the adsorbent. The structural components depend upon various functional groups such as hydroxyl, amino, carboxyl, thiol, etc., on their surface. Generally, the interaction between the adsorbent surface and adsorbate can occur through ion exchange, oxidation, and reduction, electrostatic interaction, coordination or complexation, aggregation, and precipitation. The adsorption interaction happens in two opposite conditions: Surface adsorption and interstitial adsorption. During the first process, the adsorbate molecules shift from the aqueous solute to the adsorbent surface. When the pollutant surpassed the boundary layer surrounding the adsorbent, they get adsorbed on the surface of the adsorbent which is an active site and subsequently detached from the solutions. This type of adsorption is attained by hydrogen bonding, dipole interactions, and van der Waals forces. While during interstitial adsorption, adsorbate diffuses into adsorbent pores and is finally adsorbed to the interior surface of the adsorbent. The main adsorption mechanism between the adsorbate and the adsorbent is electrostatic interaction.

3.1.3 Adsorbent for Removal of Inorganic Pollutants

Due to industrial activities, heavy metals such as cadmium, lead, copper, vanadium, and titanium are released into the water. The toxic water is consumed by human beings, plants, animals. The metals can form complexes with biological substances and break the hydrogen bond in protein and can affect the liver, kidney, bone, and nervous system. Hence an eco-friendly reusable, and efficient method such as the adsorbent process is employed to eliminate the heavy metals from wastewater. The magnetic nanocomposite is employed to remove the toxins from polluted water. Table 13.1 depicts the Degradation efficiency for Metal ions using adsorbents.

3.1.4 Adsorbent for Removal of Textile Dye

The common pollutant in wastewater is organic material. The organic compounds are released from textile dye industries. These materials are non-biodegradable and have a higher resistance to aerobic digestion. They are colored, toxic, carcinogenic, and a potential threat to living organisms in the sea. It created various types of diseases and disorders in humans and living organisms. For example, methylene blue is an inhibitor of monoamine oxidase and causes serotonin syndrome. Recently, magnetic materials are used as adsorbents to remove organic material from wastewater. Table 13.2 exhibits the various type of magnetic materials employed to remove the toxic organic dyes from water.

3.1.5 Adsorbent for Removal of Pharmaceutical Products

The pharmaceutical companies let out waste such as antibiotics, anticonvulsants, antipyretics drugs which settle in surface and groundwater possesses a major challenge to the environment. Even a trace amount

TABLE 13.1 Degradation Efficiency for Metal Ions Using Adsorbents

SL.NO	MAGNETIC ADSORBENT	ADSORBATE	ADSORPTION CAPACITY OR REMOVAL EFFICIENCY	REF
1	ZVI np	Pb(II)	807.23mg/g, 90.11%	[16]
2	ZVI- coffee ground	Pb(II) Cd(II) As(II)	164.1mg/g-1hr 112.5mg/g(24h) 23.5mg/g(1h)	[17]
3	ZVI/Chitosan	U(VI)	591.72mg/g	[18]
4	ZVI/biochar	Cr(VI)	96%. 35.3mg/g	[19]
5	ZVI/activated carbon	As(V)	100%(2 hr)	[20]
6	Fe2O3-Al2O3	Hg2+	52%, 23.75mg/g	[21]
7	Chitosan/clay/Fe2O3	Cu(II)	17.2mg/g	[22]

TABLE 13.2 Degradation Efficiency for Adsorption of Textile Dyes Using Adsorbents

SL.NO	MAGNETIC ADSORBENT	POLLUTANT	DEGRADATION EFFICIENCY	REF
1	ZVI/Kaolinite	Acid Black 1	98% (120 min)	[23]
2	CuO-Fe2O3	RhB	98% (2 hr)	[24]
3	CdS/γ-Fe2O3	MB	90% (120 min)	[25]
4	Fe3O4@MnO2	CR	95%	[26]
5	Zn-Fe2O4.CeO2	MG	96% (180 min)	[27]
6	Activated carbon/γ-Fe2O3	Alizarin red S	108.69 mg/g	[28]

TABLE 13.3 Magnetic Nanomaterials Used for Removal of Pharmaceuticals

SL.NO	MAGNETIC MATERIALS USED	PHARMACEUTICALS PRESENT IN WATER	ADSORPTION EFFICIENCY	REF
1	nZVI and PEG and Zeolite supported nZVI	Amoxicillin Ampicillin	60.3 min 43.5 min for complete removal. First-order kinetics	[31]
2.	MFe2O4 (M = Fe, Mn, Co, Zn)	Tetracycline Oxytetracycline Chlortetracycline	96% removal in 5 min. Pseudo second order kinetics	[32]
3	MnFe2O4/activated carbon	Sulfamethoxazole	159mg/g, pH-7	[33]
4	MgFe2O4/γ-Fe2O3	Minocycline	200.8mg/g	[34]

of the contaminant is a serious concern to the aquatic species. In this aspect, Attia [29] prepared zeolite with magnetic nanoparticles for the adsorption of pharmaceutical and personal care products (PPCP) and found 95% removal efficiency in 10 min. Magnetic poly(styrene-2-acrylamido-2 methylpropane sulphonic acid) is used as an adsorbent for the removal of pharmaceutical waste [69]. Diclofenac (DF) is an anti-inflammatory drug found as a pollutant that is photo catalytically degraded by Fe-ZnO and the degradation follows first-order kinetics [30]. Table 13.3 exhibits the different Magnetic nanomaterials used for the removal of pharmaceuticals.

3.1.6 Adsorbent for Removal of Pathogens

The transmission of infectious diseases occurs mainly due to microbial pollution. these microbes can penetrate deep into the soil and migrate to groundwater due to the excellent solubility in water. When the polluted water containing micro-organisms is consumed by humans, it creates significant problems to the body and ultimately to death. Usually, chlorine gas and ozone gas are employed to remove these microbes. However, the unwanted properties create a risk to operation workers. Hence magnetic materials such as spinel ferrites are employed to remove these toxic micro-organisms. In this aspect, Fe_3O_4@C@MgO.Cu was synthesized and found as a disinfectant in water purification toward gram-negative Escherichia coli and gram-positive Staphylococcus aureus. Ferromagnetic Ni-doped ZnO nanoparticles were synthesized by Rana et al [35] and applied as an antibacterial agent to control the growth of pathogens. Bacteria such as E.Coli and V.Cholera have been tested for their antibacterial activity and found to be effective. Iron oxide nanoparticles coated with chitosan oligosaccharide were synthesized by Shukla [36] and tested for the removal of pathogenic protozoan cysts, entamoeba cyst from water. E.histolytica can be efficiently removed using the synthesized nanocomposites. Amine functionalized magnetic nanoparticle [Fe3O4-SiO2-NH2] was been prepared by Zhan et al. and used for the removal of pathogenic bacteria and viruses such as S.aureus, E.Coli, P.aeuginosa, Salmonella, and B.subtilis.

3.1.7 Adsorbent for Removal of Pesticides

Organophosphorus pesticides (OPP) are used in the agricultural field to protect crops and eradicate insects. Moreover, it is economical and has high efficiency in protecting crops. But their application has caused severe pollution to the environment. The residue has posed a great threat to public health by inhibiting the activity of acetylcholine stearase and causing damage to the organs. Hence effective technologies should be employed to remove toxic OPP. Various methods have been employed to remove OPP. One such method is adsorption which is a promising technique due to the simple operation and low cost. The adsorbent contains Zr-OH which has a high affinity for phosphate groups, hence it possesses selective recognition and a higher adsorption ability for glyphosphate. The incorporation of SiO_2 reduces the electron transfer between UiO-67 and Fe_2O_3 which leads to identifying and absorbing the organophosphorus pesticide [37]. Moreover, the Fe3O4 facilitates the separation and removal process through an external magnetic field. The adsorption capacity is 256.54mg/g for the smart adsorbent and a low detection limit (0.093mg/L) for glyphosphate.

Liu et al. [38] fabricated ZIF-8/magnetic multi-walled carbon nanotube (M-M-ZIF-8) and evaluated it as an adsorbent to remove organophosphorus pesticides. The adsorption data was well explained by the Freundlich adsorption model. Within 20 min, the pesticide was completely removed. Due to the sharing of electrons between the pesticide and adsorbent, higher adsorption efficiency was observed.

3.1.8 Bioadsorbents

Agricultural waste refers to organic-by products eliminated by humans such as plant waste, rural waste created from household activities from poultry manure and livestock. When compared to the inorganic adsorbent, these materials are from the vast source, eco-friendly, biodegradability and reproducibility make them worthful materials as adsorbents. The bio adsorbents are gaining importance in water purification due to their lower cost, readily available nature and it is easily decomposable. Low-cost materials such as agricultural waste materials and naturally found materials have been used as adsorbents. The consumption of these agricultural residues plays a substantial role in the decolorization process and the national economy. Further, agricultural waste would be turned into valuable products. Some of the agricultural adsorbents used are coconut shell, palm shell, and coffee residue are used in the literature.

A novel magnetic chitosan/polyvinyl alcohol hydrogel beads(m-CS/PVA.HB) were prepared by gelation method [39]. The adsorption capacity of Congo red on the adsorbent is 470.1mg/g. Thermodynamic studies revealed the reaction is endothermic and physical in nature. Zhu et al [40] successfully fabricated magnetic cellulose/Fe_3O_4/activated carbon composite and evaluated it as adsorbent for Congo red. The adsorption capacity of the adsorbent on Congo red is 66.09 mg/g. The isotherm was well explained by the Langmuir model and the kinetics was demonstrated by the pseudo-second-order model. Magnetic Fe_3O_4/C core-shell nanoparticle was synthesized by Zhang and Kang [41] and evaluated as adsorbent for removal of organic dyes from aqueous solution. The particle size of the adsorbent is 250 nm. The maximum adsorption capacity for methylene blue and cresol red dye is 44.38mg/g and 11.22 mg/g respectively.

Salgveiro et al. [42] synthesized K-Carrageenan coated magnetic iron oxide nanoparticles and tested them as adsorbents for removal of methylene blue from aqueous solution. The maximum adsorption was reached within 5 min. Aminoguanidine modified magnetic nanoparticles were produced by Li et al. [43] and tested as adsorbents for different acid dyes and maximum adsorption was achieved at pH 1.3–2.5. The adsorption behavior was well explained by the Langmuir isotherm and the kinetics was explained by the pseudo-second-order model. Adsorption performance remained constant for three cycles of reusability. Fan et al. [44] synthesized magnetic chitosan with graphene oxide and evaluated it for the removal of Methylene blue (MB) dye from wastewater. The adsorption efficiency was 179.6 mg/g for the adsorbent on MB dye (10mg/L) at pH 10.

3.2 Photocatalysis

Generally, catalysis is divided into two categories, one is heterogeneous and another homogeneous. In heterogeneous catalysis, the catalyst can be easily separated and can be reused. But it possesses a low reaction rate due to controlled surface area. But in the case of homogenous catalysis, it is difficult to separate the catalyst. Further, the photocatalyst is expensive and hence it should be reused. For large-scale applications, a high surface area is needed which is available in nanoparticles hence it is highly desirable to use nanoparticles as catalysts. Various semiconductors such as ZnO, ZnS, TiO_2 are non-magnetic materials that are utilized as heterogenous photocatalysis [45]. However, their separation is difficult after treatment, owing to their nano size. Then, insufficient recovery leads to loss of the photocatalyst. Further, the remaining catalyst behaves as an additional pollutant in wastewater. Due to these problems, the practical application of non-magnetic semiconductors is decreased in photocatalysis.

In order to improve the performance of non-magnetic semiconductor photocatalysis, magnetic materials are incorporated into it which improves the recovery and reuse of the photocatalyst. Further, the incorporation of magnetic materials creates the heterojunction between the magnetic material and non-magnetic material and reduces the recombination between the electron and holes. Additionally, in the acidic and basic medium, the materials are stable. Usually, the organic pollutants in wastewater are treated by photocatalysis method which is one of the economical techniques. Owing to their low bandgap energy, magnetic materials use visible light energy and easily convert on top chemical energy for oxidation and reduction reaction. The organic contaminants are degraded into CO_2, H_2O, and other gaseous chemicals without leaving any residue. The photocatalytic method is considered a simple, environmentally friendly approach to degrade organic pollutants. The magnetic materials are used as simple or composite or with an oxidant such as H_2O_2.

3.2.1 Mechanism

Figure 13.4a depicts the schematic representation of photocatalytic degradation of organic pollutants. When the light is illuminated on the wastewater in the presence of a magnetic semiconductor, an electron is shifted from the valence band to the conduction band of the photocatalyst, if the bandgap of the material is higher than the light energy. The hole remains in the valence band itself. The electrons in the conduction band react with oxygen to produce oxygen free-radical which in turn reacts with hydrogen to produce free hydroxy radicals (Figure 13.4b). On the other hand, the holes in the valence band react with water to form hydroxyl free radicals [46]. The produced hydroxyl radicals with close vicinity of contaminants are attacked and easily degraded to CO2 and water. Owing to their low bandgap energy, magnetic composites can utilize visible light and tremendously degrade wastewater.

3.2.2 Effect of Contact Time

In order to study the effect of illumination time on photocatalytic degradation of organic molecules are exposed at various intervals of time. With the increase in contact time, the degradation efficiency will be high because more photons will be available for the photocatalyst to effectively degrade the organic molecules.

3.2.3 Effect of Intensity of Light

The intensity of the light is another parameter that depicts the photocatalytic degradation efficiency of the photocatalyst. The absorption of radiation by the photocatalyst determines the reaction kinetics of the reaction. The rate of absorption of radiation is the quanta of light absorbed by the photocatalyst. Various forms of light such as UV light, visible light, and solar light are shined on the photocatalyst and the effect of light on the photocatalytic performance is investigated in detail. The thermal recombination between electron and hole reduces the photocatalytic degradation efficiency of organic molecules.

3.2.4 Effect of Concentration of the Organic Molecule

Another vital parameter to optimize in order to attain higher photocatalytic efficiency is the concentration of the organic molecule. The organic molecules adsorb on the photocatalyst effectively and are susceptible to oxidation. Depending upon the substituent group of aromatic compounds, the photocatalytic efficiency varies. In the case of nitrophenol degradation, the photocatalytic performance is higher compared to phenol due to the presence of the –NO2 substituent group. Similarly, dechlorinated phenol degrades faster than monochlorinated phenol. When compared to electron-donating groups, electron-withdrawing groups adsorb remarkably. The photocatalysis performance decreases as the concentration of organic molecules increases due to the saturation of the photocatalyst on the surface.

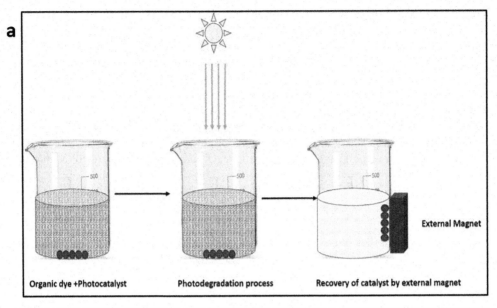
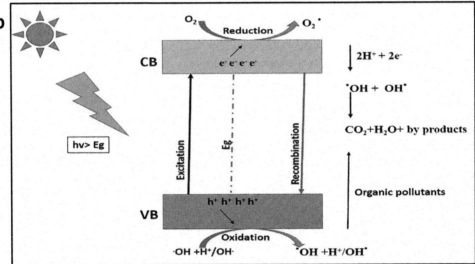

FIGURE 13.4 (a) Schematic representation of photocatalytic degradation of organic pollutants, (b) Mechanism of electron/hole formation in a photocatalyst in the presence of pollutant.

3.2.5 Effect of Photocatalyst Amount

Usually, with the increase in catalyst loading, the photocatalytic reactions rate increases. To dodge extra catalyst amount in the reaction, optimum catalyst concentration must be found out. Furthermore, the shined light should be absorbed by the photocatalyst completely. The excess photocatalyst increases the cost of the process, unfavorable light scattering which reduces the light diffusion on the solution.

3.2.6 Effect of pH

The pH of the aqueous solution is important for photocatalytic reactions. In an acidic medium, the surface is protonated and in the case of a basic medium, it is deprotonated. For example, the surface of titania is positively charged in an acidic medium and possesses a higher oxidizing ability. But in the alkaline medium, the surface remains negatively charged. Therefore, in order to have higher photocatalytic efficiency, optimization of pH is essential.

3.2.7 Photocatalysis of Removal of Organic Dyes

Sultan et al [47] synthesized iron oxide nanoparticles using pomegranate seed extract and evaluated the photocatalytic activity for the degradation of a textile dye such as reactive blue. The particle size of iron oxide nanoparticles is in the range of 25–55 m. The degradation efficiency for reactive blue is 95.08% within 56 min was attained in UV light. PPy coated on iron oxide was synthesized and tested for water purification. In this context, Mohammed et al. [48] fabricated PPy/PPy/NF/Zn-Fe LDH nanocomposite and evaluated the adsorption and photocatalytic performance. Falahian et al [49] synthesized PPy/Fe2O3 by polymerization of pyrrole, with Fe3O4 and FeCl3 (oxidizing agent) and employed as adsorbent for Hg^{2+}. The adsorbent capacity of the magnetic adsorbent is 173.16 mg/g. The kinetic studies depicted that it followed the pseudo-second-order kinetic model. The synthesized magnetic material indicated 90% regeneration efficiency.

3.2.8 Photocatalysis for Removal of Drug

The waste drugs released from pharmaceutical industries create problems for the environment. They are toxic, carcinogenic and a potential threat to living organisms. It creates various types of diseases and disorders in humans and living organisms. Hence it should be effectively removed by the photocatalysis approach. A nonsteroidal anti-inflammatory drug Meloxicam is degraded photo catalytically using gallium coated magnetic nanoparticle (GA-MNP) [50].

4 CONCLUSION

This chapter discusses the magnetic nanocomposite employed to treat the wastewater let out from various types of industries. The magnetic nanocomposites are economical, low-energy, effective, safe, environmentally friendly, and reusable. Various types of magnetic materials, synthesis methods, and their properties are discussed. Both techniques such as adsorption and photocatalysis are discussed to remove pollutants such as dyes, heavy metals, phenols, pesticides, pharmaceutical compounds, insecticides, and microorganisms from the wastewater streams. The incorporation of magnetic materials is not only used to separate the adsorbents, but it also reduces the rapid recombination of photo-induced e-/h+ which in turn increases the photocatalytic efficiency. The kinetics of the adsorption depends on various factors such as adsorbent concentration, contact time, pH, temperature, and initial concentration of the pollutant. Next, the bio-adsorbent-based magnetic materials are discussed. Agricultural waste such as chitosan is mixed with magnetic materials to enhance adsorption and photocatalytic performance. This chapter provides various magnetic materials employed to eliminate toxic chemicals from wastewater.

REFERENCES

1. George Tchobanoglous, Frank Burton QC, Metcalf and Eddy Inc, *Wastewater Engineering: Treatment Disposal Reuse*, McGraw Hill Inc., New York, 1991.
2. Imran Ali, New generation adsorbents for water treatment, *Chem. Rev.* 2012, 112, 5073.
3. S. Qadri, A. Ganoe, Y. Haik, Removal and recovery of acridine orange from solutions by use of magnetic nanoparticles, *J. Hazard. Mater.* 2009, 169, 318.

4. Carla F.S. Rombaldo, Antonio C.L. Lisboa, Manoel O.A. Mendez, Aparecido R. Coutinho, Brazilian natural fiber (jute) as raw material for activated carbon production, *An. Acad. Bras. Ciênc.* 2014, 86 (4)
5. Sergi Garcia, Segura Enric Brillas, Applied photoelectrocatalysis on the degradation of organic pollutants in wastewaters, *J. Photochem. Photobiol. C.* 2017, 31, 1.
6. R. Sivashankar, A.B. Sathya, K. Vasantharaj, V. Sivasubramanian, Magnetic composite an environmental super adsorbent for dye sequestration – A review, *Environ. Nanotechnol. Monit. M.* 2014, 1–2, 36.
7. A. Akbarzadeh, M. Samiei, S. Davaran, Magnetic nanoparticles: Preparation, physical properties, and applications in biomedicine, *Nanoscale Res. Lett.* 2012, 7, 144.
8. N.A. Frey, S. Peng, K. Cheng, S. Sun, Magnetic nanoparticles: Synthesis, functionalization, and applications in bioimaging and magnetic energy storage, *Chem. Soc. Rev.* 2009, 38, 2532.
9. V. Vadivelan, K.V. Kumar, Equilibrium, kinetics, mechanism, and process design for the sorption of methylene blue onto rice husk, *J. Colloids Interface Sci.* 2005, 286, 90.
10. P.C. Tiwari, Bhagwati Joshi, Environmental changes and sustainable development of water resources in the Himalayan Headwaters of India, *Water Res. Dev.* 2012, 2, 48.
11. Y. Yan, G. Yuvaraja, C. Liu, L. Kong, K. Guo, G.M. Reddy, G.V. Zyryanov, Removal of Pb(II) ions from aqueous media using epichlorohydrin crosslinked chitosan Schiff's base@Fe$_3$O$_4$ (ECCSB@Fe$_3$O$_4$), *Int. J. Biol. Macromol.* 2018, 117, 1305.
12. H. Guo, C. Bi, C. Zeng, W. Ma, L. Yan, K. Li, K. Wei, *Camellia oleifera* seed shell carbon as an efficient renewable bio-adsorbent for the adsorption removal of hexavalent chromium and methylene blue from aqueous solution, *J. Mol. Liq.* 2018, 249, 629.
13. Y. Huang, X. Lee, F.C. Macazo, M. Grattieri, R. Cai, S.D. Minteer, Fast and efficient removal of chromium (VI) anionic species by a reusable chitosan-modified multi-walled carbon nanotube composite, *Chem. Eng. J.* 2018, 339, 259.
14. H.A. Alalwan, M.N. Abbas, Z.N. Abudi, A.H. Alminshid, Adsorption of thallium ion (Tl+ 3) from aqueous solutions by rice husk in a fixed-bed column: Experiment and prediction of breakthrough curves, *Environ. Technol. Innovation.* 2018, 12, 1.
15. P.S. Boeris, A.S. Liffourrena, G.I. Lucchesi, Aluminum biosorption using non-viable biomass of *Pseudomonas putida* immobilized in agar-agar: Performance in batch and in fixed-bed column, *Environ. Technol. Innovation.* 2018, 11, 105.
16. Z. Dongsheng, G. Wenqiang, C. Guozhang, L. Shuai, J. Weizhou, L. Youzhi, Removal of heavy metal lead(II) using nanoscale zero-valent iron with different preservation methods, *Adv. Powder Technol.* 2019, 30, 581.
17. M.H. Park, S. Jeong, G. Lee, H. Park, J.Y. Kim, Removal of aqueous-phase Pb (II), Cd (II), as (III), and as (V) by nanoscale zero-valent iron supported on exhausted coffee grounds, Waste Manage. 2019, 92, 49.
18. Q. Zhang, D. Zhao, S. Feng, Y. Wang, J. Jin, A. Alsaedi, T. Hayat, C. Chen, Synthesis of nanoscale zero-valent iron loaded chitosan for synergistically enhanced removal of U(VI) based on adsorption and reduction, *J. Colloid Interface Sci.* 2019, 552, 735.
19. S. Li, T. You, Y. Guo, S. Yao, S. Zang, M. Xiao, Z. Zhang, Y. Shen, High dispersions of nano zero valent iron supported on biochar by one-step carbothermal synthesis and its application in chromate removal, *RSC Adv.* 2019, 9, 12428.
20. X. Dou, R. Li, B. Zhao, W. Liang, Arsenate removal from water by zero-valent iron/activated carbon galvanic couples, *J. Hazard. Mater.* 2010, 182, 108.
21. A. Mahapatra, B.G. Mishra, G. Hota, Electrospun Fe2O3-Al2O3 nanocomposite fibers as efficient adsorbent for removal of heavy metal ions from aqueous solution, *J. Hazard. Mater.* 2013, 258, 116.
22. D. Chauhan, J. Dwivedi, N. Sankararamakrishnan, Novel chitosan/PVA/zerovalent iron biopolymeric nanofibers with enhanced arsenic removal applications, *Environ. Sci. Pollut. Res.* 2014, 21, 9430.
23. B. Kakavandi, A. Takdastan, S. Pourfadakari, M. Ahmadmoazzam, S. Jorfi, Heterogeneous catalytic degradation of organic compounds using nanoscale zero-valent iron supported on kaolinite: Mechanism, kinetic and feasibility studies, *J. Taiwan Inst. Chem. Eng.* 2019, 96, 329.
24. R.C. Pawar, D.-H. Choi, C.S. Lee, Reduced graphene oxide composites with MWCNTs and single crystalline hematite nanorhombohedra for applications in water purification, *Int. J. Hydrogen Energy.* 2015, 40, 767.
25. Mohammad Saud Athar, Mohtaram Danish, Mohammad Muneer, Fabrication of visible light-responsive dual Z-Scheme (α-Fe$_2$O$_3$/CdS/g-C$_3$N$_4$) ternary nanocomposites for enhanced photocatalytic performance and adsorption study in aqueous suspension, *J. Environ. Chem. Eng.* 2021, 9(4), 105754.
26. Q. Yang, H. Song, Y. Li, Z. Pan, M. Dong, F. Chen, Z. Chen, Flower-like core-shell Fe$_3$O$_4$@MnO$_2$ microspheres: Synthesis and selective removal of Congo red dye from aqueous solution, *J. Mol. Liq.* 2017, 234, 18.
27. V. Ramasamy Raja, A. Karthika, S. Lok Kirubahar, A. Suganthi, M. Rajarajan, Sonochemical synthesis of novel ZnFe2O4/CeO2 heterojunction with highly enhanced visible light photocatalytic activity, *Solid State Ion.* 2019, 332, 55.

28. M. Fayazi, M. Ghanei-Motlagh, M.A. Taher, The adsorption of basic dye (Alizarin Red S) from aqueous solution onto activated carbon/γ-Fe2O3 nanocomposite: Kinetic and equilibrium studies mater. *Sci. Semicond. Process.* 2015, 40, 35.
29. T.M.S. Attia, X.L. Hu, D.Q. Yin, Synthesized magnetic nanoparticles coated zeolite for the adsorption of pharmaceutical compounds from aqueous solution using batch and column studies, *Chemosphere.* 2013, 93, 2076.
30. Sihui Zhan, Yang, Zhiqiang Shen, Junjun Shan, Yi Li, Shanshan Yang, Dandan Zhu, Efficient removal of pathogenic bacteria and viruses by multifunctional amine-modified magnetic nanoparticles, *J. Hazard. Mater.* 2014, 274, 115.
31. A. Ghauch, A. Tuqan, H.A. Assi, Antibiotic removal from water: Elimination of amoxicillin and ampicillin by microscale and nanoscale iron particles, *Environ. Pollut.* 2009, 157, 1626.
32. X. Bao, Z. Qiang, W. Ling, J.H. Chang, Sonohydrothermal synthesis of MFe2O4 magnetic nanoparticles for adsorptive removal of tetracyclines from water, *Sep. Purif. Technol.* 2013, 117, 104.
33. J. Wan, H.P. Deng, J. Shi, L. Zhou, T. Su, Synthesized magnetic manganese ferrite nanoparticles on activated carbon for sulfamethoxazole removal, *CLEAN – Soil, Air, Water.* 2014, 42, 1199.
34. L. Lu, J. Li, J. Yu, P. Song, D.H.L. Ng, A hierarchically porous MgFe$_2$O$_4$/γ-Fe$_2$O$_3$ magnetic microspheres for efficient removals of dye and pharmaceutical from water, *Chem. Eng. Sci.* 2016, 283, 524.
35. S.B. Rana, R.P. Singh, Investigation of structural, optical, magnetic properties and antibacterial activity of Ni-doped zinc oxide nanoparticles, *J. Mater. Sci. Electron.* 2016, 27, 9346.
36. S. Shukla, V. Arora, A. Jadaun, J. Kumar, N. Singh, V.K. Jain, Magnetic removal of Entamoeba cysts from water using chitosan oligosaccharide-coated iron oxide nanoparticles, *Int. J. Nanomedicine.* 2015, 10, 4901.
37. Qingfeng Yang, Jing Wang, Xinyu Chen, Weixia Yang, Hanna Pei, Na Hu, Zhonghong Li, Yourui Suo, Tao Li[c], Jianlong Wang, The simultaneous detection and removal of organophosphorus pesticides by a novel Zr-MOF based smart adsorbent, *J. Mater. Chem. A.* 2018, 6, 2184.
38. Guangyang Liu, Xiaodong Huang, Donghui Xu, Xiaomin Xu, Novel zeolitic imidazolate frameworks based on magnetic multiwalled carbon nanotubes for magnetic solid-phase extraction of organochlorine pesticides from agricultural irrigation water samples, *Appl. Sci.* 2018, 8, 959.
39. H.Y. Zhu, Y.Q. Fu, R. Jiang, J. Yao, L. Xiao, G.M. Zeng, Novel magnetic chitosan/poly(vinyl alcohol) hydrogel beads: Preparation, characterization and application for adsorption of dye from aqueous solution, *Bioresour. Technol.* 2012, 105, 24.
40. H.Y. Zhu, Y.Q. Fu, R. Jiang, J.H. Jiang, L. Xiao, G.M. Zeng, S.L. Zhao, Y. Wang, Adsorption removal of congo red onto magnetic cellulose/Fe$_3$O$_4$/activated carbon composite: Equilibrium, kinetic and thermodynamic studies, *Chem. Eng. J.* 2011, 173, 494.
41. Z. Zhang, J. Kong, Novel magnetic Fe3O4@C nanoparticles as adsorbents for removal of organic dyes from aqueous solution, *J. Hazard. Mater.* 2011, 193, 325.
42. A.M. Salgueiro, A.L. Daniel-da-Silva, A.V. Girao, P.C. Pinheiro, T. Trindade, Unusual dye adsorption behavior of K-carrageenan coated superparamagnetic nanoparticles, *Chem. Eng. J.* 2013, 229, 276.
43. D.-P. Li, Y.-R. Zhang, X.-X. Zhao, B.-X. Zhao, Magnetic nanoparticles coated by aminoguanidine for selective adsorption of acid dyes from aqueous solution, *Chem. Eng. J.* 2013, 232, 425–433.
44. L. Fan, C. Luo, M. Sun, X. Li, F. Lu, H. Qiu, Preparation of novel magnetic chitosan/graphene oxide composite as effective adsorbents toward methylene blue, *Bioresour. Technol.* 2012, 114, 703.
45. U.I. Gaya, A.H. Abdullah, Heterogeneous photocatalytic degradation of organic contaminants over titanium dioxide: A review of fundamentals, progress and problems, *J. Photochem. Photobiol. C. Rev.* 2008, 9, 1.
46. M.N. Chong, B. Jin, C.W.K. Chow, C. Saint, Recent developments in photocatalytic water treatment technology: A review, *Water Resour.* 2010, 44, 2997.
47. Ismat Bibi, Nosheen Nazar, Sadia Ata, Misbah Sultan, Abid Ali, Ansar Abbas, Kashif Jilani, Shagufta Kamal, Fazli Malik Sarim, M. Iftikhar Khang, Fatima Jalal, Munawar Iqbal, Green synthesis of iron oxide nanoparticles using pomegranate seeds extract and photocatalytic activity evaluation for the degradation of textile dye, *J. Mater. Res. Technol.* 2019, 8, 6115.
48. Fatma Mohamed, Mostafa R. Abukhadra, Mohamed Shaban, Removal of safranin dye from water using polypyrrole nanofiber/Zn-Fe layered double hydroxide nanocomposite (Ppy NF/Zn-Fe LDH) of enhanced adsorption and photocatalytic properties, *Sci. Total Environ.* 2018, 640, 352.
49. Zohreh Falahian, Firoozeh Torki, Hossein Faghihian, Synthesis and application of polypyrrole/Fe$_3$O$_4$ nanosize magnetic adsorbent for efficient separation of Hg^{2+} from aqueous solution, *Glob. Chall.* 2018, 22, 1700078.
50. J. Madhavan, P.S. Kumar, S. Anandan, M. Zhou, F. Grieser, M. Ashokkumar, Ultrasound assisted photocatalytic degradation of diclofenac in an aqueous environment, *Chemosphere.* 2010, 80, 747.

Nanoferrite-Based Structural Materials for Aerospace Vehicle Radomes

14

Manish Naagar, Sonia Chalia, Preeti Thakur, Atul Thakur

1 Introduction	215
2 Types of Radomes and Application Areas	216
2.1 Dielectric Space Frame (DSF) Radome	217
2.2 Inflatable Radome	218
2.3 Metal Space Frame (MSF) Radome	218
2.4 Application Areas	219
3 Basic Construction and Design	219
3.1 Design Considerations for Terrestrial-Based Radomes	220
4 Radome Operation	222
4.1 Basic Working Principles	222
4.2 Radome Operation with Nanoferrites as Frequency Selective Surfaces (FSS)	223
4.3 Key Performance Parameters	224
4.4 Effect of Radome Material on Antenna Effectiveness	226
5 Conventional Radome Materials	228
5.1 Limitations of Existing Materials	228
6 Nanoferrites as Structural Materials for Radomes	230
6.1 Suitability of Nanoferrites for Radome Materials	230
6.2 Configurations of Radome Geometries	231
7 Advances in the Use of Nanoferrites for Radome Materials	232
8 Outlook	234
References	238

1 INTRODUCTION

A radome is an essential element of the antenna system. It protects the antenna and other subsystems against adverse weather conditions like extreme winds, changes in temperature, icing, and rainy conditions. Its fundamental role is to create a sheltered enclosure between the antenna system and its environment with a

DOI: 10.1201/9781003196952-14

minimum effect on the antenna's performance [1]. Electromagnetic interference (EMI) is a critical disturbance in electronic devices, antennas, and radar systems. This disturbance interferes with the efficiency of the devices and equipment that may be a part of medical, industrial, commercial, and defense equipment [2]. Therefore, a paramount necessity exists to develop efficient and realistic EMI protection. In particular, there is a critical demand for lower density electromagnetic (EM) wave absorption materials that effectively adapt to the environment over a larger band [3]. It is essential to develop microwave absorbers that offer higher performance, and at the same time, have the wideband capability, least thickness, and minimum filler loading to curb electromagnetic pollution [4]. Stricter legislative regulations have motivated scientists and engineers to find effective microwave absorbers in different frequency bands of operation. Nanostructured materials have enhanced microwave absorption capacity and can shield them in the GHz frequency regime due to their size effects [5]. Also, there is a need to pay close attention to shielding materials and techniques to avoid the adverse effects of EM radiation on human health and electronic device interference [6]. This requirement has led to an urgent need for novel shielding technologies, wherein a slim material layer can replace a bulky physical protector. These materials shall possess properties like good absorption over a wideband and low weight, rational impedance matching, and ease of synthesis [7].

There has been comprehensive research on a particular class of ceramics, called ferrites, and their impact on microwave shielding. Various ferrites have been made in nanopowders, nanocrystalline, hybrid composites, and thin films. Ferrites are particularly preferred for microwave absorption in radio devices used for scientific, medical, and industrial frequency bands, as they are lightweight, compact, less expensive, and easily developed [8]. For example, a new type of nickel-coated strontium ferrite has been found to have effective absorption capabilities around 8 GHz [9]. The fabrication of epoxy composites by a careful blend of MWCNTs (multi-walled carbon nanotubes) with nickel-zinc (NiZn) ferrite nanopowder has shown good absorption properties in the X-band, usable for high-performance radar absorbing structures [10]. NiZn ferrites have a relatively low dielectric constant and magnetic coercivity. They also possess high corrosion resistance and higher permeability and resistivity in the higher frequency bands. The losses due to these eddy currents are also low [11]. However, different materials are required for use in applications that utilize various frequency bands, i.e., L band (1–2 GHz), S-band (2–4 GHz), C band (4–8 GHz), X band (8–12 GHz), Ku band (12–18 GHz), K band (18–26.5 GHz), and Ka-band (26.5–40 GHz).

Ferrite materials have a permeability tensor, which is controllable through the strength and direction of a magnetic DC bias. A range of frequencies results in an evanescent wave of the layer. A significant wave loss is transmitted through that material due to quasi-TEM mode generation as well as due to the generation of higher-order mode. The propagation of the magnetostatic surface wave is transverse to the quasi-TEM mode. Such symbiotic behavior yields the capability to tune the frequency of operation of a microstrip antenna [12].

Radars are built at higher frequency microwave bands. So, the radome's thickness gives rise to wavelength multiples that impact the antenna pattern. Therefore, the radome structure and the protective layers above it should be considered while designing the radar. If such optimization is not done, it results in gain loss, or BSE (bore vision errors), deteriorating the range of the radar, artifact detections, or detection of angular errors. Hence, optimizing and designing the radome layers, ensuring minimum reflections is essential. Safe, secure, and reliable operations are crucial components of telemetry and radar communication systems [13].

This chapter deals with various nanoferrites being used in different applications, their comparison, technological developments in the past 15 years, and how the contemporary advancements would shape the future applications of such materials. We will also be looking into the feasibility of magnetically switchable ferrite-based radomes where a thin ferrite layer is fabricated as a radome superstrate layer of a specific thickness.

2 TYPES OF RADOMES AND APPLICATION AREAS

There are multiple configurations of radomes deployed to minimize radio frequency reflections. These are typically A-sandwich, double and multiple A-sandwich, and half-wave configuration, as illustrated in

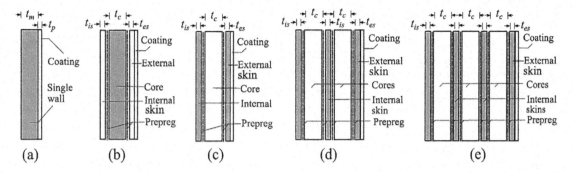

FIGURE 14.1 Radome Types; (a) Monolithic, (b) B-sandwich, (c) A-sandwich, (d) Double A-sandwich, and (e) Multiple A-sandwich (Here, t_m is material thickness, t_p is coating-ply thickness, t_c is core thickness, t_{is} is internal skin thickness, and t_{es} is external skin thickness.)

Source: (Adapted with permission from [15], under IEEE's Creative Commons License).

Figure 14.1 [14]. The radome configuration for any specific application depends on the radome's operating frequency and mechanical specifications. It has been found that monolithic (half-wave) is ideal for delivering uniformity. However, this is true only when the radome and antenna are joined together. If not, due to the significant incident angle variation during the antenna scan, an unstable degradation in the EM performance occurs. Multilayer radomes offer larger bandwidth and scan windows. Their key advantage is the high strength-to-weight factor, a critical requirement in specific practical cases. The A-sandwich configuration is widely deployed in multilayer radomes. It comprises three parts: one core of low density and two skins of higher density. The skin permittivity is higher than the core permittivity. B-sandwich is a consequence of the dual nature of A-sandwich, wherein the skin permittivity is lesser than the core permittivity. The double A-sandwich type is called C-sandwich (Figure 14.1d), and comprises five parts. This configuration helps further annulate residual reflections arising from the A-sandwich. Conventional double A-sandwich consists of two layers of the core, either foam or a honeycomb structure. They are partitioned centrally by a layer of skin, with two skin layers situated external to each core. These two skin layers constitute the radome exterior and interior.

Multiple A-sandwich radomes have more than two A-sandwich radome walls (Figure 14.1e). The radome electrical properties are a factor of critical specifications, i.e., incidence angle, permittivity, and thickness. The transmission performance gets impacted by the polarization angle derived by wave polarization, radome normal, and the incident direction [15].

2.1 Dielectric Space Frame (DSF) Radome

The construction of the wall leads to four categories of dielectric radomes. In each type of radome, the reinforcement of the panel edges to flanges helps assemble the adjacent panels. Based on the radome wall parameters, the adjacent flanges render the load-bearing capability. The constitution of each panel is a molded unitary portion that has no bond or seam lines. The panel arrays form a truncated spherical surface, especially when assembled with other panels. The panels are flat or doubly arched, resulting in a spherical appearance. A two- or three-layer sandwich architecture is produced when the radome wall gets foam-insulated. The DSF Radomes are of four types: thin membrane, solid laminate, two-layer, and three-layer. Radomes, where adjacent panels endure the wind loads, are called thin membrane wall DSF radomes. In this case, the wall thickness is mostly 1.02 mm or lesser. Where doubly curved fiberglass panels are used, radomes are known as solid laminate radomes. Radomes whose diameter is less than 1m and made from a single piece are called small radomes. Large-sized radomes comprise panels systematically arranged in horizontal and vertical rows. The wall thickness of solid laminate radomes is usually ~2.3 mm [16].

A typical doubly-curved multilayer design fabricated in a single unit is called a sandwich radome. The radome shell is constructed by highly engineered composites so that the strength and panel consistency are maximized. The skins are made of fiberglass, and they cover each core of the panels so that they are weather-proof. The sandwich radome performs well over narrowband or multiple discrete frequency spots. There are two types of sandwich radomes; 2-layer and 3-layer. The DSF radome becomes a two-layer sandwich by adding a foam layer to the interior wall of the membrane. The foam thickness is tuned to ensure thermal insulation and minimize costs. In the 3-layer sandwich radome, the core thickness is selected as 1/4th of the wavelength of the signal's highest frequency [17].

2.2 Inflatable Radome

An inflatable radome is specifically used to shield portable antenna systems. It is made from a flexible membrane and offers easy packing, shipping, and quick deployability. It gives resilience to stress caused by repetitive inflation and deflation. It also possesses excellent abrasion resistance. The electromagnetic properties of the membrane need to be considered so that it offers minimum interference [18]. It is a truncated balloon of spherical shape and is constructed by a resilient fabric material, which retains its shape due to high pressure. The inflatable radome is better than other radome types in EM performance because of the high value of transmittance cutting across lower to higher frequencies. The main disadvantage is its pressurizing system, which is susceptible to power outages. This drawback potentially leads to the collapse of the entire radome over the antenna in the event of a pressurization system malfunction. This issue may be mitigated by installing an uninterrupted power supply (UPS) system as a power backup. It is to be noted that this type of radome is vulnerable to extreme weather and windy conditions [17].

2.3 Metal Space Frame (MSF) Radome

MSF radomes are those where the radome panel flange framework is metal. The construction of this radome is done by using frames that are generally triangular, arranged in a quasi-random manner, and fused into a geodesic-structured dome. Such frames are usually built using extruded aluminum [17]. Strong metal members are joined together to offer a structurally capable shell to sustain large loads. The thin fiberglass skin sandwiches the windows between the elements. The skin receives membrane stresses while transmitting the load to the space frame [19]. The materials that form the thin membranes have low losses and low dielectric constant. These membranes are joined into the frames. To protect against snow and rain, a polytetrafluoroethylene (PTFE)-based thin laminate is attached to the membranes [17].

The transmission loss of the MSF radome frame is a seamless decreasing function of frequency. The transmission losses are higher at lower frequencies (<1 GHz). The frame losses decrease dramatically in the 2–20 GHz frequency band. This decrease is due to the frame reaching its pure optical blockage limit. The losses increase or decrease with changing metal beam cross-sections due to wind speed factors. In the 50–200 GHz band, the losses due to transmission are smooth, increasing with frequency. This increase is dependent on the membrane thickness [20].

Specific to the frequency band (1–12 GHz), the combined effect of a steady drop in transmission loss and the slow growth of membrane transmission loss results in a flat region in the transmission loss vs. frequency plot. Beyond 12 GHz, pure optical blockage of the frame is observed, wherein the losses remain flat. A slow ramp-up in the total transmission loss is attributed to the additional loss from the radome membrane [20]. Table 14.1 reports the features and drawbacks of various radome support configurations. From Table 1, we can conclude that while the self-supporting sandwich and solid laminate radome types pose minor drawbacks, inflatable, metal space, and dielectric space frame radomes offer the best feature among other types. The self-supporting sandwich radome outperforms the rest in more features and fewer drawbacks.

TABLE 14.1 Features and Drawbacks of Radome Support Configurations [21]

RADOME TYPE	FEATURES				DRAWBACKS	
	CAN WITHSTAND >150 MPH WINDS	ELECTRICALLY THIN BROADBAND PERFORMANCE	TUNED MULTIBAND PERFORMANCE	PROVIDES THERMAL INSULATIVE PROPERTIES	REQUIRES CONSTANT POSITIVE PRESSURE	SUPPORT FRAME ADDS SIGNIFICANT LOSS
Self Supporting Sandwich	✓	-	✓	✓	-	-
Inflatable	✓	✓	-	-	✓	-
Metal Space Frame (MSF)	✓	✓	✓	-	✓	✓
Dielectric Space Frame (DSF)	✓	✓	✓	-	✓	✓
Solid Laminate	✓	-	-	✓	-	-

2.4 Application Areas

Radomes are used for various purposes, including civilian and military radar, weather radar, surveillance, telecommunications, satellite communications, broadcast equipment, microwave radars, and civil flight simulation [22]. High specific mechanical properties coupled with good dielectric characteristics are desirable in materials for radome applications. Using RF transparent materials can achieve efficient performances [23]. Automotive applications include, but are not limited to, radio and microwave communications. The growing demand for real-time information has resulted in advancements in mobile satellite equipment for broadband communications. Satellite systems operating at Ku-band frequencies (11 to 14 GHz) find extensive applications in civilian and defense communications. Novel radome technologies are being developed and tested to build a solid foundation for future radome deployments in various fields [24]. The radio frequency (RF) band for satellite and radar communication systems extends to mm-wave to meet the enhanced bandwidth needs. The W-band has gained more importance due to its impressive data rate output capability. For example, the International Telecommunication Union (ITU) has allocated 71–76 GHz for transmission and 81–86 GHz for the reception [25]. Autonomous vehicles with speed control radars use the 76–81 GHz frequency band [26]. Atmospheric windows in the 94 GHz band are utilized for mm-wave applications for interstellar research, as well as for defense requirements [25]. All primary commercial radar and satellite operators are interested in allocating W-band for such communications. More commercial applications and projects in the W-bands are foreseen in the coming years [27].

3 BASIC CONSTRUCTION AND DESIGN

Using various construction materials, such as fiberglass and PTFE coated fabrics, radomes can be made into many shapes, such as geodesic, planar or spherical, based on the specific application. In addition to such protection, when used on uninhabited aerial vehicles (UAVs), the radome antenna also streamlines the system, thereby reducing traction. Radome architecture combines materials science, geodesic domes, structures, and EM science. Reinforcements such as polyester, epoxy, and cyanate esters and materials such as fiberglass, quartz, graphite, and Kevlar are used to make sophisticated composites and specialty products. Core materials such as honeycomb (aluminum, graphite, and fiberglass) and foams (Polyisocyanate and thermoform capable cores) are also used. Based on the required application, overcuring these components at temperatures touching ~200 °C is done in an autoclave, with high-pressure curing. Regardless of the application, a suitable reinforcement and matrix combination must be selected to meet the requirements.

The critical properties sought for radome material are the least transmission loss, low loss tangent, and the optimum dielectric constant. The mechanical properties of the material shall be such that it can be molded into a relevant structure, which implies the best properties of strength, durability, and density. In addition, it must withstand environmental degradation and pollution conditions, including thermal degradation, for which a good knowledge of thermal conductivity, emission, heat shock, and behavior with temperature change is necessary.

3.1 Design Considerations for Terrestrial-Based Radomes

Most radomes are individually designed to meet a specific set of needs. These requirements include cost, operating frequency, broadband performance, signal noise, signal distortions such as insertion loss, signal error, and signal scattering. The cost of the radome depends mainly on its size and surface area. The material and labor costs in a radome increase significantly with its diameter. The main reason is that doubling the diameter of a sphere results in a four-fold increase in its surface area. Thin-walled dielectric space frame radomes are less expensive than the 3-layer sandwich core radome by around 35%.

Most sandwich radomes work with wind loads up to 150 mph and snow, and ice loading of up to 2400 Pa. Structural support is commonly used for thin laminate radomes, while the flange structure of sandwich radomes provides the basis for support. Regarding the wall design, the fundamental parameters include radome style, material thickness, number of layers, material permittivity, panel sizes and shapes, dome geometry, truncation, and flange design. Ancillary inclusions comprise hatches, doors, hoists, lightning protection, lighting, hydrophobic coatings, and accessory buildings [28].

Fiberglass, Kevlar, quartz, and graphite are a few materials generally used to construct a radome. Polyester, vinyl ester, cyanate ester, and epoxies are resins used. Construction techniques consist of infusion, hand lamination, and prepreg fibers. An essential element in radome performance is laminate consistency. Therefore, prepreg materials shall be avoided. Honeycomb and foams (thermo-formable cores) are the core materials that are generally used. A clean room is required for high tolerance blueprints. There is a significant reduction in the system performance if carbon gets into the laminate. Therefore, this situation should be avoided [29].

The radome function protects the antenna from extreme environmental influences, with as little impact on electrical performance as possible. Detecting weak incoming signals that arrive after further signal attenuation has a lower impact on system sensitivity. The design considerations must include endurance to heavy loads created by the pressure gradient between the interior and exterior of the dome. It must also consider the impact of high temperatures in supersonic aviation applications [30]. Table 14.2 presents the various functional requirements of a suitable radome.

An idealized radome is represented in Figure 14.2. The two crucial material properties in determining radome effectiveness are the electric loss tangent (**tan**δ_ε), and the dielectric constant $\left(\varepsilon_r\right)$. Reflection and absorption are the two types of losses. The part of the signal reflected is attributed to ε_r. There is a

TABLE 14.2 Design Requirements for a Radome [30]

FUNCTION	REQUIREMENT
Constraints	• Support pressure difference, $\Delta \rho$ • Tolerate temperature up to t_{max}
Objective	• Minimize dielectric loss in transmission of microwaves
Free Variables	• Skin thickness, t_s • Choice of material

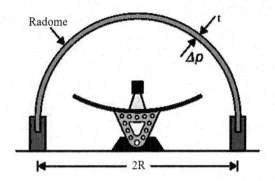

FIGURE 14.2 A Radome, transparent to microwaves.

direct proportion between the frequency and the reflected part of the signal [30]. Secondly, and importantly, are the absorption losses caused by the transmission of the signal across the skin.

The mathematical formula for the fractional power loss in the case of a wave (frequency f (Hz)) passing via a dielectric material having the loss tangent $tan\ \delta_\varepsilon$, with a thickness dt is given in Equation 1:

$$\left|\frac{dU}{U_0}\right| = \frac{fA^2\varepsilon_0}{2}(\varepsilon_r \tan\delta)dt \tag{1}$$

Here, A represents the electric amplitude, and ε_0 being the vacuum permittivity. The loss per unit area for a thin shell of thickness t is given by Equation 2.

$$\left|\frac{\Delta U}{U_0}\right| = \frac{fA^2\varepsilon_0 t}{2}(\varepsilon_r \tan\delta) \tag{2}$$

Now, this objective function needs to be minimized. To achieve this minimum, the skin must be made as thin as possible. However, the pressure difference given by Δp is a significant limitation. The pressure difference generates stress σ in the skin as per the expression given in Equation 3.

$$\sigma = \frac{\Delta pR}{2t} \tag{3}$$

Hence, if this stress has to support the pressure difference Δp, it must be less than the material's failure stress σ_f. This parameter limits the thickness, as expressed in Equation 4.

$$t \geq \frac{\Delta pR}{2\sigma_f} \tag{4}$$

Substituting Equation 4 in Equation 2 gives equation 5.

$$\left|\frac{\Delta U}{U}\right| = \frac{fA^2\varepsilon_0 \Delta pR}{4}\left(\frac{\varepsilon_r tan\delta}{\sigma_f}\right) \tag{5}$$

This index has to be maximized to achieve minimized power loss. Hence, Equation 6 represents the mathematical expression for M, an index which is the ratio of the elastic limit, σ_f, to the loss factor, $\varepsilon_r \tan\delta$.

$$M = \frac{\sigma_f}{\varepsilon_r \tan\delta} \tag{6}$$

It is evident that the design considerations for radome are related to thickness parameters, wall structure, mechanical (number of layers, material permittivity, panel sizes and shapes, dome geometry), and cost (mainly the diameter). Given a specific frequency f and its consequent optimization, the fractional power loss in a radome must be obtained to minimize power loss.

4 RADOME OPERATION

4.1 Basic Working Principles

Permittivity dispersion and permeability dispersion are the two fundamental mechanisms that drive microwave absorption. These parameters are expressed by the real components; μ', and ε' and by the imaginary components; μ'', and ε'' [31]. Three mechanisms help to shield electromagnetic radiation, viz. absorption, reflection, and multiple reflections. Reflection happens in structures that are highly conductive since they rely on mobile charge carriers. These primarily include metals and metallic structures. Dipoles in the shielding material, primarily electrical or magnetic, help wave absorption.

It is observed that attenuation due to absorption depends on the relative permeability (μ_r), relative permittivity (σ_r) and shield thickness. The absorption is directly proportional to the increase in frequency, while reflection is inversely proportional. The schematic in Figure 14.3 represents the EMI shielding mechanism. One more phenomenon is multiple reflections seen at interior interfaces and surfaces.

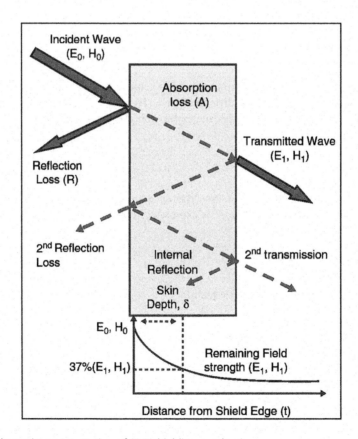

FIGURE 14.3 Schematic representation of EMI shielding mechanism.

The geometry of the material is impacted by larger surface areas, which leads to multiple reflections. The surfaces or interfaces inside the shield cause multiple reflections [32].

4.2 Radome Operation with Nanoferrites as Frequency Selective Surfaces (FSS)

Frequency selective surfaces (FSS) have multiple applications in various fields. The composition of the material is metallic patch elements or apertures that are systematically arranged. These materials are characterized by total reflection or transmission in the vicinity of the element resonance [33]. When a ferrimagnetic material is subject to a DC magnetic field bias and a perpendicular RF field, energy is absorbed under suitable conditions. This phenomenon is called ferromagnetic resonance (FMR). Such absorption occurs where the preceding frequency of the magnetization vector about the direction of the external DC field is equal to the RF field frequency [34]. The magnetic field bias results in the switching behavior of the printed antenna. This phenomenon is due to the superstrate layer utilized together with the antenna. Thereby, the magnetic bias can either turn the antenna on or off. This technique takes advantage of the characteristics of an extraordinary quasi-TEM plane wave's negative permeability state, which propagates in a ferrite region. Let us consider a situation where a plane wave is propagating in the direction of the z-axis. We also believe in an unbounded ferrite domain with an applied magnetic bias along the x-axis. If the wave undergoes polarization in the direction of the y-axis, this results in the magnetic field is parallel to the applied DC field. In this case, no interaction occurs with the magnetic characteristics of the ferrite. This particular scenario is termed an 'ordinary wave'. When the wave polarization occurs along the x-axis, the RF field becomes perpendicular to the applied bias. In this scenario, there is a significant interaction with the magnetic characteristics of the ferrite. This scenario is termed an 'extraordinary wave'. The extraordinary wave has a propagation constant which is defined by Equation 7.

$$\gamma_e = \alpha_e + j\beta_e = j\omega\sqrt{\mu_{eff}\epsilon} \tag{7}$$

where the effective permeability μ_{eff} is given by Equation 8.

$$\mu_{eff} = \left(\mu^2 - \kappa^2\right)/\mu \tag{8}$$

Here, μ and κ are the ferrite permeability tensor elements, given by Equations 9 and 10.

$$\mu = \mu_0\left[1 + \omega_0\omega_m / \left(\omega_0^2 - \omega^2\right)\right] \tag{9}$$

$$\kappa = \mu_0\omega\omega_m / \left(\omega_0^2 - \omega^2\right) \tag{10}$$

Where, $\omega_0 = \mu_0\gamma H_0$, $\omega_m = \mu_0\gamma M_s$, $\gamma = 1.759\times10^{11}$ C/Kg. M_S refers to the saturation magnetization, H_0 being the internal bias field. It is evident that when the loss is zero, γ_e will be purely imaginary if $\mu_{eff} > 0$ and purely real if $\mu_{eff} < 0$. The former case refers to a plane wave in propagation, while the latter refers to a non-propagating plane wave evanescent in the ferrite region. Using equations (9) and (10), we can prove that μ_{eff} will be negative, in the condition expressed in Equation 11:

$$\sqrt{\omega_0\left(\omega_0 + \omega_m\right)} < \omega < \omega_0 + \omega_m \tag{11}$$

Assuming the ferrite layer has a non-impacting effect near the antenna field and ignoring the multiple reflections between the ferrite and the antenna, an approximate value of attenuation (one-way) of a plane wave passing through the layer is given by the expression presented in Equation 12.

$$\exp(\alpha_e d) \tag{12}$$

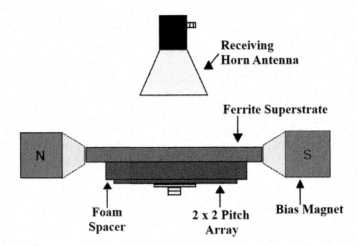

FIGURE 14.4 Measurement setup for transmission loss through a ferrite superstrate layer that is magnetically biased.

Where, d is the ferrite layer thickness [35].

Figure 14.4 demonstrates that a ferrite substrate mounted above a typical microstrip antenna or array can be set up as a switch. The antenna continues to perform transmission and reception when the ferrite is either unbiased or subject to a bias condition wherein $\mu_{eff} > 0$. When $\mu_{eff} < 0$, i.e., when the ferrite is biased to a level reaching the cutoff point, there is a significant reflection of the incident wave. However, due to the finiteness of α_e, and due to the finite thickness of the ferrite layer, there is not a total reflection. There is some power leakage through the layer, similar to a waveguide attenuator beyond cutoff. The expression given in Equation 12 demonstrates that it is possible to increase the amount of attenuation by operating a bias so that α_e is maximized. Alternatively, the extent of attenuation is increased by enhancing the thickness of the ferrite layer. To this effect, we could consider placing dielectric matching layers on both sides to mitigate reflection. In sharp contrast to the lossless state, the dielectric and magnetic losses directly impact the amount of attenuation. However, there may be seen a slight decrease in attenuation at the maximum cutoff point when the magnetic losses increase. Because scattering occurs in both directions across the layer, the radar cross-section of the antenna reduces with a correlation given by the square of the value in Equation 12.

When the ferrite radome is subject to a bias equal to cutoff, it exhibits a specular reflection of the incident wave. However, there is no contribution to the scattered field from the antenna system. It is to be noted that the above cutoff effect manifests only in the situation wherein the RF field is perpendicular to the applied bias. However, this is manageable for linearly polarized antennas but challenges circularly or dual-polarized antenna systems. A switchable polarizer can be made by leveraging the birefringence phenomenon of the ferrite. Alternatively, the cutoff effect can be utilized in a circularly polarized antenna by switching between linear and circular polarization. It is practical to apply the DC bias in the same plane as the ferrite layer due to the absence of demagnetization factors. The convenience is also because by not disturbing the aperture of the antenna, a closed magnetic circuit is set up [35].

4.3 Key Performance Parameters

A well-engineered radome provides solid climatic and environmental security along with the minimum impact on the antenna's RF performance. Insertion loss is the main parameter to be controlled from an electrical standpoint. It reduces the signal strength, leading to lesser adequate radiated power and G/T ratio, where G is the antenna gain in decibels, and T is the equivalent noise temperature in Kelvin.

Radomes have a characteristic of increasing the sidelobes of the antenna, resulting in interference with other wireless systems. This tendency increases the odds of the signal getting detected and intercepted by unwanted observers. Radomes also affect the polarization of the antennas. The depolarization of circularly polarized antennas is a typical example. Though depolarization is not significant for spherical radomes, it highly impacts large incident angles, especially in aircraft and military applications. Other electrical effects on the performance are beam width variations and boresight alteration.

Besides the above impacts of radome materials, a thin sheet of water significantly affects radome performance degradation. Water, whose loss tangent and dielectric constant are high at microwave frequencies, impacts the performance of a radome. Water tends to get attached to the radome surface, creating a film – especially on surfaces that are non-hydrophobic. This film acts as a shielding layer to radio signal transmission, leading to high signal attenuation [36]. Therefore, well-engineered radomes have a hydrophobic surface that helps water move away from the surface. In rainy weather, a hydrophobic surface has exhibited less incremental loss [37].

While the usefulness of radomes is relatively high, it is essential to consider that they degrade the electronic system connected to the antenna system. Radomes are also known to decrease the effective range of the radar due to their characteristics to reflect and attenuate incident plane waves. Caution must be exercised, primarily while determining the orientation of a reflecting body (or a radiation source), as errors may arise in such determination. Additionally, radar clutter caused by reflections due to multipath is also an effect. This characteristic also gets factored in the EM performance of a radome.

Four critical parameters measure a radome's EM performance: (a) far-field radiation pattern, (b) power transmittance, (c) boresight error, and (d) boresight error slope [38]. Far-field radiation pattern refers to the angular distribution of the radiated or intercepted intensity. Radomes impact side lobe levels, beamwidths, maxima or minima values, directions of pattern, and axial ratio. Such patterns are measured conventionally, except for special components to support the radome and the antenna. Power transmittance refers to the maxima peak intensity of the main beam of the pattern. This parameter is measured by dividing the received peak intensity (in the radome's presence) by the received intensity received (in the radome's absence). However, the above definition is in theory. More accurate measurements are made by holding the antenna stationary while receiving the signal from a remote source for all practical purposes. In this method, pivoting of the radome is done to the surrounding antenna system. Continuous measurements are made while the radome is in motion. Boresight error refers to the shift made by the radome in the direction of the deepest minimum of the difference mode pattern, specifically to monopulse antennas. Alternatively, in a pair of antennas, it represents the angular shift calculated by the relative change of phases. To measure the boresight error, there is a need to pivot the radome, as expressed earlier. Boresight error slope refers to the slope of the boresight error to the angle between the axes of the antenna and the radome. As a convention, the antenna's axis is considered in an assumed direction, and the revolution axis is the radome axis [38]. The specific application drives the significance of the radome-induced effects. Table 14.3 summarizes the most common degradations experienced in various electronics or sensor applications while radome design.

As per the transmission line theory, the reflection loss (RL) is given by Equations 13 and 14 for a single layer.

$$RL(dB) = 20\log\left|(Z_{in} - Z_0)/(Z_{in} + Z_0)\right| \qquad (13)$$

$$Z_{in} = Z_0\sqrt{\mu_r/\varepsilon_r}\tanh\left(-j2\pi f\sqrt{\mu_r\varepsilon_r}d/c\right) \qquad (14)$$

where f refers to the microwave frequency, d being the absorber thickness, c being the speed of light, ε_r and μ_r are, respectively, the relative complex permittivity and permeability, Z_0 is the free space impedance, and Z_{in} being the absorber input impedance [39].

TABLE 14.3 Most Important Radome Degradations in Various Applications [25]

APPLICATION	RADOME LOSS INCREASE	ANTENNA SIDELOBE INCREASE	RADOME DEPOLARIZATION	BSE INCREASE	BSES INCREASE	ANTENNA VSWR INCREASE
Ground/Sea	-	-	-	-	-	-
Weather Radar	✓	✓	✓	-	-	✓
Microwave Communication	✓	-	-	-	-	✓
Ground & Marine SATCOM	✓	✓	✓	-	-	✓
Satellite Navigation/GPS	✓	-	✓	-	-	-
Airborne	-	-	-	-	-	-
Weather Radar	✓	✓	✓	-	-	✓
Airborne SATCOM	✓	✓	✓	-	-	✓
Microwave air-to-ground communications	✓	-	-	-	-	-

BSE: Boreside Error | BSES: Boreside Error Slope | VSWR: Voltage Standing Wave Ratio

4.4 Effect of Radome Material on Antenna Effectiveness

Many characteristics of radome materials affect the antenna performance. Firstly, due to transmission losses, there is a reduction of gain of the primary beam antenna. Secondly, because of the radome resistive losses, there is an increase in the antenna noise temperature. Thirdly, as discussed before, there is a reduction in the G/T ratio due to the combined effect of the first two factors. These factors play a crucial role in satellite communications receiver systems as the end-to-end dynamic range decreases, impacting the system efficiency. Considering the impact of noise on the system performance, it is a general convention to assume a noiseless receiver and account for the noise parameter at the output, with the help of F, the noise factor. This disturbance is either of thermal origin or from other noise-generating sources. Alternative processes lead to noise generation, and those possess probability distribution functions and spectrum identical to thermal noise. The noise effects in a receiver system can be modeled by considering a single receiver subsystem. Using the principles of black body radiation, the total output noise of a single state subsystem (P_{no}) is given by Equation 15 [40].

$$P_{no} = k_b T_0 BGF \tag{15}$$

where G is the power gain of the subsystem, B is the noise bandwidth (Hz), k_b is the Boltzmann's constant (1.38×10^{-23} w-s/K), T_0 is the ambient temperature (290 K). $K_b T_0 B$ is the thermal noise power referenced to the input port. The noise figure is expressed in Equation 16.

$$NF = 10\log(F) \tag{16}$$

As an alternative, it is possible to take into account the subsystem generated noise at a temperature T_e through a noise-free termination. For a single-stage subsystem, the total noise at the output is evaluated from Equation 17.

$$P_{no} = k_b T_e BG \tag{17}$$

Here, it is evident that T_e correlates to the noise factor F, as per equation 18.

$$T_e = T_0 F \tag{18}$$

The effect of noise temperature on the overall performance of the antenna is also evaluated. Considering a radome-enclosed antenna, as shown in Figure 14.5, the essential contribution to the noise temperature of the system is given by Equation 19.

$$T_{sys} = T_a L_r + 290(1-L_r) + T_e \tag{19}$$

Where, T_{sys} is the noise temperature of the system, T_a is the noise temperature of the antenna, T_e is the electronic noise temperature, and the loss factor is connected to the radome loss in decibels (L_{db}) as per equation 20.

$$L_r = 10^{L_{db}/10} \tag{20}$$

Where, L_r equals 1 for a lossless radome. It is implied from Equation 20 that a power loss L_{db} drastically deteriorates the receiving noise temperature. Therefore, the noise temperature from the radome loss is the temperature reached when the radome has to be thermally heated, assuming it is lossless, to receive a matching contribution of the noise power. T_a is calculated using the antenna's radiation pattern and its surroundings. For instance, for a satellite antenna with a temperature of ~10 K (low cosmic sky temperature), the G/T can be given by Equation 21.

$$G/T = G_a - 10\log(T_a L_r + 290(1-L_r) + T_e) \tag{21}$$

The additive system noise temperature is due to both physical temperature and radome loss. The analysis shows that a significant contributor to the system noise temperature happens to be the radome loss at ambient temperatures. For example, every radome loss of the order of −0.1 dB degrades the noise temperature of the system by approximately 10 K [25]. Also, radomes impacted by rainwater exhibit enhanced degradation of system noise temperature [41].

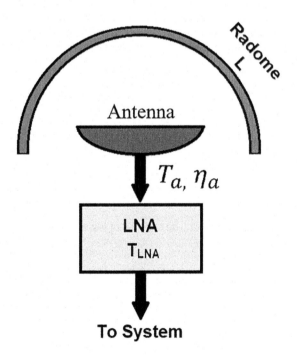

FIGURE 14.5 Receiving antenna encased inside a radome (loss L_r).

5 CONVENTIONAL RADOME MATERIALS

The typical materials used for radome construction are selected considering their electrical performance and higher strength. The usage domain of the radome also determines the material selection [42]. Figure 14.6 represents the strength-dielectric loss chart, and it is seen that the polymers have excellent M values, but they fall short when it comes to strength. We could also deduce that some ceramics have good M value but perform poorly in strength. An analysis of Figure 14.6 reflects that M is the maximum for materials in the first row, i.e., PTFE, polyethylene, and polypropylene. In applications requiring higher strength or impact resistance, the best choices are the second-row candidates, i.e., fiber-reinforced polymers. In high-temperature applications, the ceramics in the third row provide optimal performance. Overall, among polymers, polycarbonate and PTFE are the most prevalent, as they are very flexible. In situations requiring structural rigor, it is preferred to use glass fiber reinforced plastic) (GFRP) materials, which is either epoxy or woven-glass-reinforced polyester. Their selection, however, results in a marginal loss of performance. It shall also be noted that when performance overrides other objectives, a glass-reinforced PTFE material shall be used. Ceramics are the right choice in applications wherein the skin temperatures exceed 300 °C. For temperatures till 300 °C, it is optimal to use polyimides. Table 14.4 summarizes several materials used for radomes and their particular performance indicators.

5.1 Limitations of Existing Materials

The critical material properties to be taken into consideration for the design of radomes are low dielectric constant, low loss tangent, ability to sustain pressure differences, resistance to environmental impacts, impermeability to water, tenacity to tolerate aerodynamic and structural loads, high thermal resistance, and being lightweight as well [43]. There are various ceramics used for high-temperature applications. These comprise pyroceram, nitroxyceram, silicon nitride (Si_3N_4), alumina ceramics, celsian, fused silica, silicon oxynitride (Si_2N_2O), and β-SiAlON. Besides the higher dielectric constant, alumina demonstrates lower shock resistance due to heat at high temperatures. This property arises due to its more significant linear thermal expansion coefficient (5×10^{-6} to 9×10^{-6} C^{-1}). Such a radome gets

TABLE 14.4 Suitable Radome Materials for Particular Performances [31]

PERFORMANCE REQUIREMENT	SUITED MATERIAL
Minimum dielectric loss at near room temperature	Polytetrafluoroethylene (PTFE)
	Polyethylene
	Polypropylene
	Polystyrene
	Polyphenylene sulfide (PPS)
Greater strength and temperature resistance (at the expense of losses)	Glass-reinforced polyester
	Polytetrafluoroethylene (PTFE)
	Polyethylenes
	Polypropylenes
	Polyamide-imide
For re-entry vehicles and rockets to resist high aerodynamic heating	Silica
	Alumina
	Beryllia
	Silicon carbide

heated beyond 980 °C at supersonic speeds. It shall be noted that there are challenges during sintering such materials. This limitation exists because, at sintering temperatures, alumina tends to exhibit twisting and deformation. It is also accompanied by brittleness and stiffness, leading to difficulties during construction – presenting limitations where curvatures are required. Therefore, this has limitations for defense, missile, and aircraft radome applications [44]. Reaction-bonded silicon nitride (RBSN) demonstrates excellent electrical characteristics, rain performance, with a melt limit of ~1500 °C. It also shows thermal shock resistance and rain erosion at high aerodynamic temperatures to velocities up to Mach 7 [44].

Dense silicon nitride, which is realized by hot pressing of α-Si_3N_4 (HPSN) with suitable additives used for the high-temperature application, shows limitations due to its complex machining process and high production cost [45]. Nonetheless, silicon nitride has become a favorable material for such applications due to its optimum thermal shock resistance, rain erosion resistance, and mechanical characteristics. However, it is to be noted that its high dielectric constant and loss tangent at ambient and higher temperatures are its fundamental limitations [46,47]. Research has shown similar limitations in silicon oxynitride [48] and SiAlON composites [49,50]. One solution to reduce the dielectric constant is to make the material porous [51,52]. Some most commonly used radome materials and their properties have been presented in Table 14.5.

We see tremendous focus and research in developing better materials for radome applications. There are enormous tractions specific to the aerospace, supersonic, and military domains, primarily due to rapidly increasing speeds and operational altitudes. While sustaining high temperatures, the materials shall overcome the limitations of their unacceptable dielectric constant and loss tangent [44].

It has been observed that nanocrystalline materials have overcome lots of limitations, as they possess large surface areas and a higher amount of hanging bond atoms than conventional materials [53]. In the previous decade, there has been incisive research on developing ferrites and magnetic metal-based materials. A few of these include nickel-zinc ferrite [54], barium ferrite ($BaFe_{12}O_{19}$) [55,56], Fe [57], Ni [57], Iron (II,III) oxide (Fe_3O_4) [57, 58]. These materials suffer from certain deficiencies, such as loss of ferromagnetic characteristics beyond the Curie temperature and relatively higher densities. Nanosized ferromagnetic particle-based composites are ideal in applications such as microwave absorbers since nanoparticles demonstrate unique magnetic characteristics regarding bulk materials [59].

TABLE 14.5 Properties of Most Common Radome Materials [44]

PROPERTY	AT T °C	ALUMINA (99%)	PYROCERAM	SCFS	DENSE Si_3N_4	RBSN	β-SIALON	Si_2N_2O
Density (g/cm³)	-	3.9	2.6	2.2	3.2	2.4	2.92–3.07	2.81
Dielectric Constant (10 GHz)	25 °C	9.6	5.65	3.42	7.9	5.6	6.84–7.46	5.896
	1000 °C	11.4	6.1	3.8	8.5	5.8	7.34	-
Loss Tangent (10 GHz)	25 °C	0.0001	0.0002	0.0004	0.004	0.001	0.0013–0.002	0.002–0.003
	1000 °C	0.0014	0.0018	0.002	0.01	0.005	0.003–0.004	-
Flexural Strength (MPa)	25 °C	270	235	44	800	180	266	140
	1000 °C	220	75	66	500	180	-	-
Thermal Shock	-	Fair	Good	Very Good	Very Good	Very Good	-	-
Water Absorption	-	0%	0%	5%		20%	-	-
Rain Erosion	-	Excellent	Very Good	Poor	Very Good	Good	-	-

6 NANOFERRITES AS STRUCTURAL MATERIALS FOR RADOMES

Nanomaterials show outstanding properties vis-à-vis bulk counterparts. Changes in their physical, structural, chemical, magnetic, electrical, and optical properties are attributed to their nanoscale size [60]. Nanoferrites are ceramics that exhibit excellent ferrimagnetic characteristics due to ferric oxides being a key component [61]. Compared to their bulk counterparts and metals, they demonstrate exceptional magnetic properties [62]. Ferrites are classified into three types basis their chemical structure. These are spinel ferrites, garnet ferrites, and hexagonal ferrites. Spinel ferrites represent a class of ceramics, given by the general formula AB_2O_4. Here, A is a divalent ion (e.g., Ni^{2+}, Fe^{2+}, Zn^{2+}, Co^{2+}, Mg^{2+}, Cu^{2+}, Mn^{2+}, etc.), and B is trivalent (e.g., Al^{3+}, Fe^{3+}, etc.) [63]. The general formula $R_3Fe_5O_{12}$ represents garnet ferrites, wherein R is a trivalent ion. Hexagonal ferrites are generally represented as $MeFe_{12}O_{19}$. Here, Me represents a divalent ion having a higher ionic radius, like Sr^{2+}, Ba^{2+}, Pb^{2+} or a trivalent ion such as Al^{3+}, Cr^{3+}, La^{3+}, and Ga^{3+}.

Nanoparticles demonstrate magnetic characteristics, hence are susceptible to easy modifications by applying an external field [64]. Two categories of nanoferrites exist based on their coercivity. These are termed soft and hard nanoferrites. These categories depict their ability to bear external magnetic fields without getting demagnetized [65]. Soft nanoferrites are characterized by their lower coercivity and changeable magnetization properties. Hard nanoferrites are characterized by their higher coercivity and non-changeable magnetization properties [66,67]. Domains such as gas sensors [68], contaminated water treatment [69], electromagnetic devices [70,71], radar absorbing materials [72–74], supercapacitors [75,76], targeted drug delivery [77], catalysis [78,79], medical imaging [80,81], and storage devices [82] are some of the few notable areas where nanoferrites find their applications.

6.1 Suitability of Nanoferrites for Radome Materials

Ferrites are extraordinarily versatile and adaptive magnetic materials with tremendous applications in high-frequency microwave systems. The lithium ferrites are the aptest for microwave device applications because of their intrinsic characteristics like high dielectric constant, high saturation magnetization, high Curie temperature, and low dielectric losses [83]. Most recently, nanoferrites have emerged as novel materials for their extensive applications in aerospace, military, chemical, pharmaceutical, and other critical industrial fields [84,85].

The characteristics of large anisotropy, moderate saturation magnetization, high coercivity, and mechanical hardness render them excellent candidates for data storage, magnetic sensors, EM wave absorbers, information delivery devices, and magnetic composites. Despite their high thermal stability and impressive electrical and magnetic properties, there are limitations in using ferrite-based ceramic materials in complex devices, primarily due to their brittleness and lack of structural adjustability. Besides the above limitations, the requirements of very high temperatures (>1200 °C) for processing sintered ferrite bodies make it complicated to prepare complex ferrite structures using pure ferrite nanopowders. Therefore, a solution is to deploy hybridized ceramic materials with organic polymers. This solution helps in enhanced mechanical adaptation and efficient processing. The synthesis method is a crucial parameter that governs the nanoparticles' physical, morphological, structural, optical, chemical, magnetic, and dielectric properties [86]. Another aspect to note is that there is a substantial dependence of the magnetic properties (of the nanoferrites substituted) on the lattice site preferences of the ions. There are various reports available that have elaborated on the study of nanoparticles of Ni–Zn ferrite [87,88], Mn–Zn ferrite [89,90], $BaFe_{12}O_{19}$ [91], and $CoFe_2O_4$ [92,93] dispersed in a polymeric matrix.

The absorption of electromagnetic waves by $Ni_{0.5}Zn_{0.4}Cu_{0.1}Fe_2O_4$ spinel ferrite nanoparticles ingrained in the poly(methyl methacrylate) matrix has been reported. Investigations have been conducted specifically to determine the efficiency of magnetic polymer composites with multicomponent fillers [94]. Their applications as radio absorbents are worth mentioning. Researchers have studied the microwave absorption of polyaniline/BaFe$_{12}$O$_{19}$ and have noted that the absorbing properties are modified by controlling the polyaniline content for the specific frequency bands [95]. An immediate on-site polymerization process to get polyaniline/BaFe$_{12}$O$_{19}$ nanocomposites and the absorption characteristics of the nanocomposites have been reported [96].

In modern times, researchers are extensively exploring the potential of various ferrites for their microwave absorption properties. There is ongoing research in civil applications, such as reducing EM interference among electronic components and circuits in communication systems. There is also an increasing demand for radar-absorbing materials that should render characteristics of (i) lightweight, (ii) easy to fabricate, (iii) stable at high temperatures, (iv) resistant to corrosion, (v) flexible, and (vi) usable as a coating material. In these specific applications, nanostructured ferrites and polymer-ferrite composites are excellent candidates [97].

6.2 Configurations of Radome Geometries

The optimum arrangement for a specific domain depends on the operating frequency and mechanical prerequisites. A thin electrical radome (< 0.1 λ) is suited for optimum RF performance, as reflections at the dielectric/free space boundary and annulled by out-of-phase reflections on the opposite side of the dielectric. Research proves that the net transmission from such a laminate is maximum due to low signal losses. However, these radomes have significantly less heat insulation and are unsuitable for applications that require operation in extreme temperatures. Another critical application is a thick solid laminate configuration based on half-wavelength. This configuration is like the electrically thin arrangement as the reflections are annulled. The wave traverses 180° across the laminate and then reflects a −180° phase shift. This arrangement further traverses the wave by 180° on its return so that the net 180° phase shift essential for cancellation is achieved.

The operation of an A-sandwich radome is akin to the solid laminate with thickness equaling half-wavelength. However, the thickness here is 0.25 λ because the reflection coefficients have similar phases and amplitude. The roundtrip here from the second skin for the reflection is 0.5 λ. Therefore, it is to be noted that the reflections, precisely out of phase, annul each other. A C-sandwich typically comprises two foam layers and three skin layers. The thickness of both skin and foam layers are optimized for the best RF performance results in the frequency bands under consideration. Hence, many combinations can deliver good mechanical strength and RF performance characteristics. These provide improved performance vis-à-vis A-sandwich radomes. However, due to the nature of construction, the material and labor costs are relatively high.

There are three commonly used configurations, each of which is identified by the number of panels used in their construction. These are quasi-random, 3-panel, and 5-panel radomes. The chosen geometry causes scattering errors at certain frequency spots. Therefore, each geometry's design is done to minimize such scattering. The quasi-random geometry radomes generally have triangular, pentagonal, or hexagonal panel shapes. A typical example of a quasi-random radome is a geodesic radome that utilizes triangular panels as an option. Here, the shadowing of the framework is complex and depends on the geometry of the radome. The mathematical process of separating a truncated sphere into different panel shapes is called the tessellation of the sphere.

Other design considerations are also taken into account. For example, researchers have proposed a non-reciprocal ferrite radome that uses the Faraday rotation effect. This radome enables unidirectional transmission and effectively blocks the signal moving in the reverse direction. The composition has 2-layered gratings fused on both sides of the ferrite. They consist of conducting strips to enable effective reflection. In addition, thin lossy strips allow dissipation and three matching dielectric strips. This principle is identical to the principle used in waveguide Faraday isolators [98]. New structures to operate across a wide band of frequencies, using dielectric and ferrite layers, have also been investigated. Many layers are consecutively arranged in this novel structure to act as a matched layer [1].

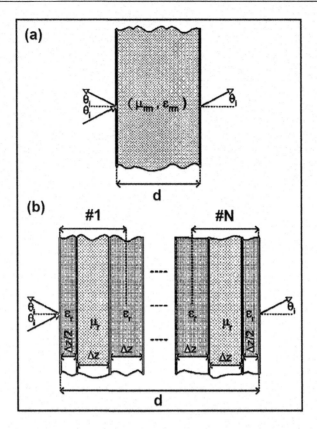

FIGURE 14.6 A cross-sectional view of (a) flat radome composed of mixed material, (b) flat radome using dielectric and ferrite layers (Adapted with permission from [1]. Under license number 5192600655825, *Source:* Copyright 2021).

Such a radome, as shown in Figure 14.6, consists of $2N+1$ consecutive layers of ferrite and dielectrics [1]. The thickness Δz is given by Equation 22.

$$\Delta z = \frac{d}{2N} \tag{22}$$

An extraordinary wave decaying in the ferrite slab prohibits the incident fields from arriving at the antenna. This tendency results in a considerable reduction in radar cross-section (RCS). Generally, as large amounts of power get translated into magnetostatic waves for antennas, a minimal amount is left to radiate into the air. Therefore, the antennas are OFF in this configuration as they do not act as scatterers or radiators. Though a considerable reduction in RCS is achieved with an outer ferrite covering, caution must be exercised for dipoles, as additional RCS resonances also happen at some other frequencies [99].

7 ADVANCES IN THE USE OF NANOFERRITES FOR RADOME MATERIALS

The core/shell/shell $ZnFe_2O_4@(SiO_2)_{1.0}@PPy$ nanocomposites can be fabricated using a scalable and straightforward procedure. The impedance matching can be precisely reached by changing the SiO_2

content in the nanocomposites [100]. E-glass fabric/epoxy composites containing multi-walled carbon nanotubes and Ni-Zn ferrites in radar absorbing structures can be developed for realizing enhanced absorption in the X-band. With the increase in the content of MWCNTs, the measured permeability and permittivity increase in direct proportion. At temperatures beyond 360 °C, all synthesized samples demonstrate good thermal stability. Perceptible losses, both magnetic and dielectric, are seen in Ni-Zn ferrite materials (NZFM 4 and NZFM 5), doped with 20 wt % Ni-Zn ferrite and 1.6, 2 wt % of MWCNTs, respectively. For double-layered radar absorbing structures (RAS), excellent microwave absorption properties are observed specific to the X-band. Their performance results from the perfect matching of the magnetic and dielectric losses. This tendency proves beneficial for applications requiring satisfactory thermal stability and high mechanical strengths.

A series of rGO/$Sm_3Fe_5O_{12}$/$CoFe_2O_4$ ternary composites can be fabricated with different $Sm_3Fe_5O_{12}$/ $CoFe_2O_4$ mass ratios by a facile hydrothermal technique. Studies show remarkable absorption performance of such composites. Four factors leading to improved dielectric loss include electron polarization, interfacial polarization, dipolar polarization, and the Debye relaxation. The enhancement in wave absorption is attributed to excellent impedance matching and the ability for microwave attenuation. The magnetic loss contribution is attributed to the natural ferromagnetic resonance loss and eddy losses [101].

The Co^{2+} and Al^{3+} ion substitution in synthesized $Ba_{0.5}Sr_{0.5}Co_xAl_xFe_{12-2x}O_{19}$ ferrite impacts enhanced microwave absorption. Composition at x = 0.2 performs well with a bandwidth of 3.27 GHz in the 9.12–12.4 GHz frequency range, with a thickness of 1.5 mm. The frequency of reflection loss peak microwave absorption and the extent of absorption range is tunable by controlling the thickness, the ion substitution, and impedance in $Ba_{0.5}Sr_{0.5}Co_xAl_xFe_{12-2x}O_{19}$ hexagonal ferrites [102].

Successful fabrication of 3D (Fe_3O_4/ZnO) @ C double-core @ shell nanocomposite is feasible by the self-propagating high-temperature synthesis method, followed by thermal treatment and phenolic polymerization. Subsequently, an in-situ carbothermal reduction is conducted. The absorption performance of the nanocomposite is tunable by altering the Zn content. The molar ratio of Zn to Fe being 1:2 provides adequate absorption bandwidth (RL < – 10 dB) of the nanocomposite – which is 6.4 GHz at 2 mm thickness and 7.11GHz at 1.9 mm [103].

The $Co_{0.6}Zn_{0.4}Fe_2O_4$ ferrite nanofibers can be successfully fabricated using the electrospinning technique. By controlling the Co^{2+} content, the coercivity and saturation magnetization of the nanofibers can be strengthened. Concerning the coating for microwave absorption, 15 wt.% of $Co_{0.6}Zn_{0.4}Fe_2O_4$ ferrite nanofibres demonstrate excellent absorbing properties in the entire X-band and more than 75% of the *Ku*-band. Researchers provide insights for solutions to resolve the key issues of heavy mass of conventional spinel ferrites so that a lightweight and more efficient microwave absorber is fabricated. This study can refer to a setup to render lightweight characteristics with surface densities as low as 2.4 kg/m^2 [104].

Fabrication of nanocrystallites using facile sol-gel synthesis procedures can be incorporated into E-Glass/epoxy-based radar absorbing structures (RASs). These synthesized structures exhibit good absorption characteristics, caused primarily due to excellent impedance matching. The thermal stability and load-bearing capabilities, the key advantages of RASs vis-à-vis conventional coatings, are observed to be comparatively better due to incorporating E-glass fibers with ceramic Ni-Zn nanoferrites. The only limitation of the solution is the high viscosity of the fluid used to synthesize the samples. This limitation arises due to higher weight percentages of the filler by over 25%. Better methods for overcoming this challenge and decreasing the viscosity are under development [105].

Fabrication of polyaniline/zinc (PANI/Zn) ferrite composite with fluffy structure has been successfully synthesized using a two-step method. Adding Zn ferrite increases the magnetic loss and enhances the impedance matching. An improvement in the attenuation efficiency and microwave transmission wave propagation can be obtained. The outstanding microwave attenuation performance can be attributed to the synergized operation of the structure with interfacial polarization, phase cancellation effect, optimum dielectric, and magnetic loss. The peak frequency is mainly due to the phase cancellation phenomenon. This composite is optimum for the Ku-band [106].

The coprecipitation method can synthesize samples of $SrFe_{12-2x}(Mn_{0.5}Cd_{0.5}Zr)xO_{19}$ hexaferrite. The coercive force decreases continuously by increasing the Zr–Mn–Cd concentration. At low doping content (x = 0.2), the saturation magnetization increases to an optimum level. Even with thin layers, good microwave absorption performance can be seen in the *Ku*-band with substantial RL at x = 1.4. Such hexaferrite samples hold vital importance in the GHz range as thin microwave absorbers [107].

The gel precursor transformation process can synthesize hard or soft bismuth ferrite/Ni-Zn ferrite (BFO/NZFO) nanocomposite microfibers with varied mass ratios and high aspect ratios. The thickness of the specimen and the overall composition contributes to their excellent absorption performance. With specimen thickness of 3 mm and a mass ratio of 7:3, the nanofibers show an RL_{min} of −35.5 dB at 12.4 GHz. The absorption bandwidth was found to be 6.6 GHz. For the same specimen, with the mass ratio being 9:1, the absorption bandwidth measured ~10.1 GHz. These substantial performance improvements result from the exchange-coupling interaction, shape anisotropy, interfacial polarization, and the small-size advantages of the nanocomposites [108].

Synthesis of plane hexagonal Y ferrite (Co_2Y) can be done successfully using a sol-gel method. There exists an inverse relationship between the saturation magnetization (M_s) and the sintering temperature. The real (ε') and imaginary (ε'') parts of permittivity increase with an increase in temperature, and the minimum absorption peak frequency shifts towards upper frequencies. The addition of Ni powder also enhances the absorption performance, with low thickness and low-frequency bands [109].

A novel absorbing material, reduced graphene oxide (rGO) functionalized with Co ferrite nanocomposites (CoFe@rGO), exhibits excellent absorption characteristics in the entire *Ku*-band when a 5 wt.% amount of paraffin is added. This remarkable performance becomes possible due to the polarization of rGO and the dielectric relaxation. Also, the scattering at the interfaces and the loss attributed to hysteresis of $CoFe_2O_4$ leads to the enhancement of microwave absorption [110].

The core-shell structured nanocomposites can be synthesized as a promising material for EM shielding through in-situ emulsion polymerization. The shielding efficiency of the composite is mainly achieved due to the presence of interconnected reduced graphene oxide/strontium fluoride (RGO/SrF) in the poly(3,4-ethylenedioxythiophene) (PEDOT) matrix. Also, the electric and magnetic loss contributes to the excellent shielding effectiveness. Moreover, the PEDOT@RGO@$SrFe_{12}O_{19}$ composites show good thermal performance. The performance further improves with the growing SrF nanoparticle concentration. The SiO_2-$MnFe_2O_4$ composite with significant microwave absorption performance in S-band can be synthesized through one-pot hydrothermal synthesis. The magnetic and dielectric losses contribute to the microwave absorption capability. Strong eddy current loss, good attenuation characteristics, optimal impedance matching, and multiple Debye relaxation lead to excellent absorbing capacity [111].

Higher annealing temperatures increase the nanosized barium ferrite's anisotropy and size, enhancing EM wave absorption. Studies using transmission line measurements (TLM) have shown that the thickness for the best performance in the X-band is ~2 mm. This $BaFe_{12}O_{19}$/NPR composite with 50 wt % demonstrates the highest RL of −37.06 dB at 9.5 GHz along with absorption bandwidths of 1.04 GHz and 1.01 GHz. Excellent thermal performance up to 400 °C of the material makes it ideal for thin absorbers in X-band applications [112]. The coercivity is easily tunable, and microwave attenuation can be achieved by increasing dielectric losses. When Co–Ti is substituted in barium ferrite, it yields satisfactory reflection loss and microwave absorption in the X-band [113]. Table 14.6 elucidates the potential nanoferrite materials and their detailed microwave absorption characteristics. The material classification according to the varying thickness with their respective frequency bands has been depicted in Figure 14.7.

8 OUTLOOK

Nanoferrites form an important emerging field of research and development, given their numerous applications in technology domains. This advancement is primarily due to their intrinsic properties and

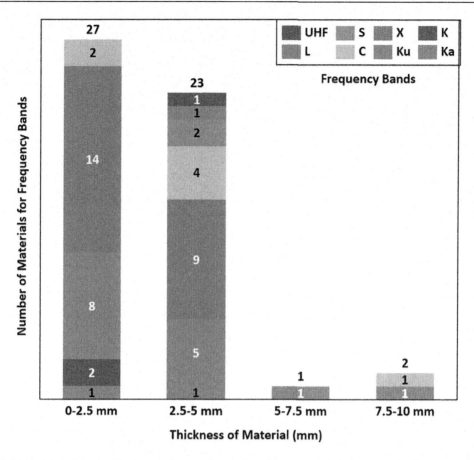

FIGURE 14.7 Materials classified according to the varying thickness and their respective frequency bands.

performance characteristics. While detailing specific radome applications, this chapter has also dealt with the comparison, technological advances, and specific future outlook of the nanoferrites. A summary of radome geometry, design considerations, and their suitability for varied frequency band applications have been elucidated. There is immense scope for further development of such materials in wideband applications.

Future nanomaterial properties should focus on the dielectric constant and the dielectric loss tangent as these two parameters determine radome performance. At the same time, two fundamental mechanisms, i.e., permittivity dispersion and permeability dispersion, are levers to assess and optimize microwave absorption. Insertion loss is one key parameter to be controlled from an electrical perspective, as it reduces signal strength, leading to lesser adequate radiated power and G/T ratio. While the effects of nanomaterials on radome performance have been analyzed, future research should take sensitive consideration of noise temperature, as it is a significant bottleneck to maximizing system efficiency. Critical reviews for future radar absorbing materials shall focus on lightweight characteristics, ease of fabrication, stability at high temperature, corrosion resistance, flexibility, and usability as coating materials. In this context, nanostructured ferrites and polymer-ferrite composites offer excellent potential. The present chapter suggests that if the above aspects are considered in the design, it shall open opportunities to develop novel metamaterials that can overcome existing limitations, leading to higher performance. There is a critical need to examine and evaluate the potential and suitability of nanoferrites and their derivatives in future research.

TABLE 14.6 Potential Nanoferrite Materials and Their Detailed Microwave Absorption Characteristics

S. NO	NAME	OPERATING FREQUENCY RANGE	SUPPORTED FREQUENCY BAND(S)	APPROX. SHIELDING EFFICIENCY AT FREQUENCY F	EFFECTIVE ABSORPTION BANDWIDTH	THICKNESS	REF.
1	$ZnFe_2O_4@SiO_2@PPy$ nanocomposite with varying SiO_2 content	19.5–26.5 GHz	K	−29.72 RL @ 24.96 GHz	7 GHz	1.5 mm	[100]
2	$ZnFe_2O_4@SiO_2@PPy$ nanocomposite with varying SiO_2 content	30.44–40 GHz	Ka	−36.75 RL @ 38.38 GHz	9.56 GHz	1 mm	[100]
3	E-glass fabric/epoxy composite containing MWCNTs and Ni-Zn ferrite (NZF) nanopowder ($Ni_{0.5}Zn_{0.5}Fe_2O_4$)	8.2–12.4 GHz	X	−22 RL @ 9.6 GHz	2.2 GHz	3 mm	[4]
4	Doped barium hexaferrite ($BaMg_{0.25}Mn_{0.25}Co_{0.5}Ti_{1.0}Fe_{10}O_{19}$)	12–18 GHz	Ku	−40 RL @ 16 GHz	4.5 GHz	2.7 mm	[114]
5	$rGO/Sm_3Fe_5O_{12}/CoFe_2O_4$ ternary nanocomposite	10.80–17.92 GHz	X	−73.71 RL @ 14.88 GHz	7.12 GHz	2.09 mm	[101]
6	Graphene/$BaFe_{12}O_{19}$/$CoFe_2O_4$ nanocomposite	4.5–7.5 GHz	C	−32.4 RL	3 GHz	5 mm	[115]
7	$BaFe_{12}O_{19}$-PVDF-RGO composite	9.6–12.8 GHz	X	−32 RL @ 11 GHz	3.2 GHz	2 mm	[116]
8	Double substituted M-type $Ba_{0.8}La_{0.15}Na_{0.15}Fe_{10}CoZrO_{19}$ Hexaferrites	26.5–40.0 GHz	Ka	−25.29 RL	6.75 GHz	2.6 mm	[117]
9	Co-Al substituted M-type hexagonal ferrite ($Ba_{0.5}Sr_{0.5}Co_{0.2}Al_{0.2}Fe_{11.6}O_{19}$)	11.14–12.23 GHz	X	−40.4 RL @ 11.72 GHz	1.09 GHz	1.4 mm	[102]
10	3D (Fe_3O_4/ZnO)@C double-core@shell nanocomposite (FZC3)	11.5–18 GHz	Ku	−40 RL @ 15.3 GHz	6.5 GHz	2 mm	[103]
11	Co-doped ZnNi ferrite/polyaniline composite ($Zn_{0.4}Ni_{0.4}Co_{0.2}Fe_2O_4$/PANI)	12–18 GHz	Ku	−54.3 RL	6.02 GHz	6.8 mm	[118]
12	Cobalt-Zinc ferrite nanofiber ($Co_{0.6}Zn_{0.4}Fe_2O_4$)	7.8–16.8 GHz	X, Ku	~−16 RL	9 GHz	2.9 mm	[104]
13	(RGO/SF/PANI) ternary nanocomposite	12.52–18 GHz	Ku	−45 RL @ 16.08 GHz	5.48 GHz	1.5 mm	[119]
14	$Ni_{0.5}Zn_{0.5}Fe_2O_4$-reinforced E-glass/epoxy nanocomposite	8.2–10.6 GHz	X	−33 RL @ 9.6 GHz	2.4 GHz	4 mm	[105]
15	multi-walled carbon nanotube/doped barium hexaferrite nanocomposite	~6.2–12.2 GHz	C, X	−23.1 RL @ 8.2 GHz	6 GHz	2.1 mm	[120]
16	EPDM ferrite composite containing manganese zinc ferrite (RFC-4)	~5–12 GHz	C, X	−26 RL @ 10.2 GHz	7.4 GHz	2.5 mm	[121]
17	Polyaniline (PANI)/Zn ferrite composites	13.1–17.9 GHz	Ku	−54.4 RL @ 17.6 GHz	4.8 GHz	1.4 mm	[106]
18	M-type Sr hexaferrite nanoparticles ($SrFe_{9.2}(Mn_{0.5}Cd_{0.5}Zr)_{1.4}O_{19}$)	13.6–16.5 GHz	Ku	−52 RL @ 14.5 GHz	2.9 GHz	1.9 mm	[107]

14 • Nanoferrite-Based Structural Materials   237

#	Material	Frequency range	Band	RL / Value	Frequency	Thickness	Ref.
19	Ce-doped barium hexaferrite ($Ba_{0.8}Ce_{0.2}Fe_{12}O_{19}$)	~15.5–16.8 GHz	Ku	−20.40 RL @ 16.1 GHz	1.3 GHz	10 mm	[122]
20	$BaFe_{12}O_{19}$ (BFO)/$Ni_{0.5}Zn_{0.5}Fe_2O_4$ (NZFO)-aligned nanocomposite microfibers	9.1–15.7 GHz	X, Ku	−35.5 RL @ 12.4 GHz	6.6 GHz	3 mm	[108]
21	Co_2Y ferrite	~10.5–13 GHz	X, Ku	−24.82 RL @ 11.68 GHz	2.56 GHz	2.5 mm	[109]
22	(CoFe@rGO) nanocomposite	10.87–18.04 GHz	X, Ku	−21.64 RL @ 13.41 GHz	7.17 GHz	2.7 mm	[110]
23	core-shell morphology-based PEDOT/RGO nanocomposite incorporated with Sr-Fe nanoparticles	8.2–12.4 GHz	X	42.29 dB @ 12.4 GHz	4.2 GHz	2.5 mm	[2]
24	SiO_2-$MnFe_2O_4$ composite	1.55–3.4 GHz	S	−14.87 RL @ 2.25 GHz	1.85 GHz	4 mm	[111]
25	RGO/$Ni_{0.4}Zn_{0.4}Co_{0.2}Fe_2O_4$ nanocomposite	11.8–18.0 GHz	Ku	−38.7 RL @ 15.2 GHz	6.2 GHz	1.9 mm	[123]
26	$Ni_{0.8}Co_{0.2}Fe_2O_4$ nanoferrite	9–12 GHz	X	−36.2 RL @ 11.52 GHz	3 GHz	2.5 mm	[32]
27	Nanosized $Ni/SrFe_{12}O_{19}$ magnetic powder	7–15 GHz	C, X, Ku	−41.3 RL @ 9.2 GHz	~8 GHz	3 mm	[8]
28	Ag/$SrFe_{12}O_{19}$/NanoG composite	9.4–12.4 GHz	X	−29 RL @ 10.9 GHz	3 GHz	2 mm	[124]
29	Spinel $Mn_{0.1}Ni_{0.45}Zn_{0.45}Fe_2O_4$ ferrite/thermoset PU nanocomposite	2–8 GHz	S, C	−29.7 dB @ 5.86 GHz	4.28 GHz	5 mm	[125]
30	$Ni_{0.5}Co_{0.5}Fe_2O_4$/graphene composites	0.58–1.19 GHz	UHF, L	−30.92 RL @ 0.84 GHz	0.61 GHz	4 mm	[126]
31	Nanosized $BaFe_{12}O_{19}$ in NPR matrix	~9.7–10.3 GHz	X	−37.06 RL @ 9.5 GHz	0.60 GHz	2 mm	[112]
32	Co-Ti substituted barium hexagonal ferrite ($BaFe_{10.6}Co_{0.7}Ti_{0.7}O_{19}$)	~9–10 GHz	X	−25.15 RL @ 9.54 GHz	0.94 GHz	3.3 mm	[113]
33	Ru-doped M-type strontium ferrite ($SrFe_{12-x}Ru_xO_{19}$; x = 1.0)	7.8–14.35 GHz	X, Ku	−31.16 RL @ ~9.8 GHz	6.55 GHz	2.3 mm	[127]
34	M-type hexaferrite $Ba_{0.8}Ca_{0.2}Fe_{12}O_{19}$ (BCFO) ceramics	7.6–9.8 GHz	X	−30.8 RL @ 8.5 GHz	2.2 GHz	2 mm	[128]
35	CQD@$BaTiO_3$ core shell/$BaFe_{12}O_{19}$ nanoparticle composite	8–12 GHz	X	−25.38 RL @ 10.21 GHz	2.7 GHz	2.75 mm	[129]
36	Paraffin-FeCo/$CoFe_2O_4$ composite powder	~6.6–7.5 GHz	C	−20 RL @ 7.1 GHz	~0.9 GHz	9 mm	[130]
37	Ho-doped W-type hexagonal ferrite $Ba_{0.85}Ho_{0.15}Co_2Fe_{16}O_{27}$-polyaniline composite	~6.9–11.3 GHz	C, X	−15.1 RL @ 9.4 GHz	~4.4 GHz	3.5 mm	[41]
38	La-Na co-substituted M-type Co-Ti-Mn barium hexaferrite ($Ba_{0.8}La_{0.1}Na_{0.1}Fe_{10}Co_{0.5}TiMn_{0.5}O_{19}$)	18–26.5 GHz	K	−45.94 RL @ 22.93 GHz	8.33 GHz	1.3 mm	[131]
39	Ce-Zn doped Ni ferrites ($Ni_{0.9}Ce_{0.1}Fe_{1.9}Zn_{0.1}O_4$)	11.56–18 GHz	X, Ku	−21.44 RL @ 15.52 GHz	6.44 GHz	2.5 mm	[132]
40	$Co_{0.2}Ni_{0.8}Fe_2O_4$/ RGO composite	~9–11.5 GHz	X	−19.56 RL @ 10.04 GHz	2.49 GHz	2 mm	[133]

REFERENCES

[1] M. Khalaj-Amirhosseini, S.M.J. Razavi, Wideband and wide-angle flat Radomes using dielectric and ferrite layers, *Int. J. Microw. Wirel. Technol.* 4 (2012) 529–535. https://doi.org/10.1017/S1759078712000281.

[2] P. Bhattacharya, S. Dhibar, G. Hatui, A. Mandal, T. Das, C.K. Das, Graphene decorated with hexagonal shaped M-type ferrite and polyaniline wrapper: A potential candidate for electromagnetic wave absorbing and energy storage device applications, *RSC Adv.* 4 (2014) 17039–17053. https://doi.org/10.1039/C4RA00448E.

[3] T. Zhang, D. Zhao, L. Wang, R. Meng, H. Zhao, P. Zhou, L. Xia, B. Zhong, H. Wang, G. Wen, A facile precursor pyrolysis route to bio-carbon/ferrite porous architecture with enhanced electromagnetic wave absorption in S-band, *J. Alloys Compd.* 819 (2020) 153269–153269. https://doi.org/10.1016/j.jallcom.2019.153269.

[4] R. Shu, J. Zhang, C. Guo, Y. Wu, Z. Wan, J. Shi, Y. Liu, M. Zheng, Facile synthesis of nitrogen-doped reduced graphene oxide/nickel-zinc ferrite composites as high-performance microwave absorbers in the X-band, *Chem. Eng. J.* 384 (2020) 123266–123266. https://doi.org/10.1016/j.cej.2019.123266.

[5] V.N. Archana, M. Mani, J. Johny, S. Vinayasree, P. Mohanan, M.A. Garza-Navarro, S. Shaji, M.R. Anantharaman, On the microwave absorption of magnetic nanofluids based on barium hexaferrite in the S and X bands prepared by pulsed laser ablation in liquid, *AIP Adv.* 9 (2019) 35035–35035. https://doi.org/10.1063/1.5088080.

[6] A.K. Mishra, P.A. Kokate, S.K. Lokhande, A. Middey, G.L. Bodhe, Time-dependent study of electromagnetic field and indoor meteorological parameters in individual working environment, in: V.P. Singh, S. Yadav, R.N. Yadava (Eds.), *Environmental Pollution*, Springer, Singapore, 2018: pp. 181–191. https://doi.org/10.1007/978-981-10-5792-2_15.

[7] J. Dalal, S. Lather, A. Gupta, R. Tripathi, A.S. Maan, K. Singh, A. Ohlan, Reduced graphene oxide functionalized strontium ferrite in poly(3,4-ethylenedioxythiophene) conducting network: A high-performance EMI shielding material, *Adv. Mater. Technol.* 4 (2019) 1900023–1900023. https://doi.org/10.1002/admt.201900023.

[8] A. Kakirde, B. Sinha, S. Sinha, Development and characterization of nickel-zinc spinel ferrite for microwave absorption at 2•4 GHz, *Bull. Mater. Sci.* 31 (2008) 767–770. https://doi.org/10.1007/s12034-008-0121-2.

[9] X. Pan, J. Qiu, M. Gu, Preparation and microwave absorption properties of nanosized Ni/SrFe12O19 magnetic powder, *J. Mater. Sci.* 42 (2007) 2086–2089. https://doi.org/10.1007/s10853-006-1210-5.

[10] P. Siva Nagasree, K. Ramji, M.K. Naidu, T.C. Shami, X-band radar-absorbing structures based on MWCNTs/NiZn ferrite nanocomposites, *Plast. Rubber Compos.* (2020) 1–12. https://doi.org/10.1080/14658011.2020.1836882.

[11] A. Thakur, P. Thakur, J.-H. Hsu, Smart magnetodielectric nanomaterials for the very high frequency applications, *J. Alloys Compd.* 509 (2011) 5315–5319. https://doi.org/10.1016/j.jallcom.2011.02.021.

[12] N.K. Saxena, N. Kumar, P.K.S. Pourush, Study of LiTiMg-ferrite radome for the application of satellite communication, *J. Magn. Magn. Mater.* 322 (2010) 2641–2646. https://doi.org/10.1016/j.jmmm.2010.03.032.

[13] J. Varghese, N. Joseph, H. Jantunen, S.K. Behera, H.T. Kim, M.T. Sebastian, Microwave materials for defense and aerospace applications BT, in: Y.R. Mahajan, R. Johnson (Eds.), *Handbook of Advanced Ceramics and Composites: Defense, Security, Aerospace and Energy Applications*, Springer International Publishing, Cham, 2020: pp. 165–213. https://doi.org/10.1007/978-3-030-16347-1_9.

[14] W.M. Cady, M.B. Karelitz, L.A. Turner, *Radar Scanners and Radomes*, McGraw-Hill Book Company, New York, 1948.

[15] Z. Qamar, N. Aboserwal, J.L. Salazar-Cerreno, An accurate method for designing, characterizing, and testing a multi-layer radome for mm-Wave applications, *IEEE Access.* 8 (2020) 23041–23053. https://doi.org/10.1109/ACCESS.2020.2970544.

[16] M. Wahab, Radar radome and its design considerations, in: *International Conference on Instrumentation, Communication, Information Technology, and Biomedical Engineering 2009*, IEEE, 2009: pp. 1–5. https://doi.org/10.1109/ICICI-BME.2009.5417229.

[17] Introduction, in: *Radome Electromagn. Theory Des.*, John Wiley & Sons, Ltd, Croydon, UK, 2018: pp. 1–13. https://doi.org/10.1002/9781119410850.ch1.

[18] D. Wienke, E.O. Van, Inflatable Radome, WO2012126885A1, 2012. https://patents.google.com/patent/WO2012126885A1/en (accessed May 12, 2021).

[19] A. Kay, Electrical design of metal space frame radomes, *IEEE Trans. Antennas Propag.* 13 (1965) 188–202. https://doi.org/10.1109/TAP.1965.1138397.

[20] Metal Space Frame Radomes, L3HarrisTM Fast Forw. (n.d.). www.l3harris.com/all-capabilities/metal-space-frame-radomes (accessed May 12, 2021).

[21] L. Griffiths, R.D. Engineer, A fundamental and technical review of radomes, *Microw Prod. Dig. Featur. Artic.* 2008 (2008) 1–4.

[22] V. Soumya, S. Navaneetha, A. Reddy, J.J. Kumar, Design considerations of Radomes: A review, *Int. J. Mech. Eng. Technol.* 8 (2017) 42–48.
[23] G. Pulvirenti, P. Tromboni, M. Marchetti, A. Delogu, A. Maccapani, R. Aricò, *Surveillance System Airborne Composite Radome Design*, Dip Ing Aerospaziale Astronaut. Univ. Roma "Sapienza"-Italy, 2005.
[24] K. Rana, P. Thakur, M. Tomar, V. Gupta, A. Thakur, Investigation of cobalt substituted M-type barium ferrite synthesized via co-precipitation method for radar absorbing material in Ku-band (12–18GHz), *Ceram. Int.* 44 (2018) 6370–6375. https://doi.org/10.1016/j.ceramint.2018.01.028.
[25] D. Kozakoff, *Analysis of Radome Enclosed Antennas*, Second Edition, Artech, Norwood, MA, 2009. http://ieeexplore.ieee.org/document/9100664.
[26] Z. Chen, Y.P. Zhang, A. Bisognin, D. Titz, F. Ferrero, C. Luxey, A 94-GHz dual-polarized microstrip mesh array antenna in LTCC technology, *IEEE Antennas Wirel. Propag. Lett.* 15 (2016) 634–637. https://doi.org/10.1109/LAWP.2015.2465842.
[27] J. Hasch, E. Topak, R. Schnabel, T. Zwick, R. Weigel, C. Waldschmidt, Millimeter-wave technology for automotive radar sensors in the 77 GHz frequency band, *IEEE Trans. Microw. Theory Tech.* 60 (2012) 845–860. https://doi.org/10.1109/TMTT.2011.2178427.
[28] H. Xin, M. Liang, 3-D-printed microwave and THz devices using polymer jetting techniques, *Proc. IEEE.* 105 (2017) 737–755. https://doi.org/10.1109/JPROC.2016.2621118.
[29] N. Khatavkar, B. K, Composite materials for supersonic aircraft radomes with ameliorated radio frequency transmission-a review, *RSC Adv.* 6 (2016) 6709–6718. https://doi.org/10.1039/C5RA18712E.
[30] S.J. Mumby, An overview of laminate materials with enhanced dielectric properties, *J. Electron. Mater.* 18 (1989) 241–250. https://doi.org/10.1007/BF02657415.
[31] M.F. Ashby, Chapter 6 – case studies: Materials selection, in: M.F. Ashby (Ed.), *Materials Selection Mechanical Design*. Fourth Ed., Butterworth-Heinemann, Oxford, 2011: pp. 125–195. https://doi.org/10.1016/B978-1-85617-663-7.00006-0.
[32] B. Chen, D. Chen, Z. Kang, Y. Zhang, Preparation and microwave absorption properties of Ni – Co nanoferrites, *J. Alloys Compd.* 618 (2015) 222–226. https://doi.org/10.1016/j.jallcom.2014.08.195.
[33] N. Joseph, S.K. Singh, R.K. Sirugudu, V.R.K. Murthy, Effect of silver incorporation into PVDF-barium titanate composites for EMI shielding applications, *Mater. Res. Bull.* 48 (2013) 1681–1687.
[34] Y. Liu, Analysis of frequency selective surfaces with ferrite substrates, in: *IEEE Antennas and Propagation Society International Symposium.* IEEE, 1996.
[35] G.P. Srivastava, B.K. Kuanr, Microwave ferrites, in: V.R.K. Murthy, S. Sundaram, B. Viswanathan (Eds.), *Microwave Materials*, Springer, Berlin, Heidelberg, 1994: pp. 112–140. https://doi.org/10.1007/978-3-662-08740-4_5.
[36] D.M. Pozar, A magnetically switchable ferrite radome for printed antennas, *IEEE Microw. Guid. Wave Lett.* 3 (1993) 67–69. https://doi.org/10.1109/75.205667.
[37] I. Anderson, Measurements of 20-GHz transmission through a radome in rain, *IEEE Trans. Antennas Propag.* 23 (1975) 619–622. https://doi.org/10.1109/TAP.1975.1141134.
[38] F.J. Dietrich, D.B. West, An experimental radome panel evaluation, *IEEE Trans. Antennas Propag.* 36 (1988) 1566–1570. https://doi.org/10.1109/8.9706.
[39] G.P. Tricoles, Radome electromagnetic design, in: Y.T. Lo, S.W. Lee (Eds.), *Antenna Handbook: Theory, Applications, and Design*, Springer, Boston, MA, 1988: pp. 2051–2081. https://doi.org/10.1007/978-1-4615-6459-1_31.
[40] F. Guo, R. Li, J. Xu, L. Zou, S. Gan, Electromagnetic properties and microwave absorption enhancement of Ba0.85RE0.15Co2Fe16O27-polyaniline composites: RE = Gd, Tb, Ho, *Colloid Polym. Sci.* 292 (2014) 2173–2183. https://doi.org/10.1007/s00396-014-3234-8.
[41] J. Dijk, A.C.A. Van der Vorst, Depolarization and noise properties of wet antenna radomes, *Electromagn. Noise Interf. Compat., Agard Conference Proceedings* No. 159, Springfield, VA, 1975.
[42] G.S. Mani, Radome materials, in: V.R.K. Murthy, S. Sundaram, B. Viswanathan (Eds.), *Microwave Materials*, Springer, Berlin, Heidelberg, 1994: pp. 200–239. https://doi.org/10.1007/978-3-662-08740-4_8.
[43] A. Nag, R.R. Rao, P.K. Panda, High temperature ceramic radomes (HTCR) – A review, *Ceram. Int.* 47 (2021) 20793–20806. https://doi.org/10.1016/j.ceramint.2021.04.203.
[44] S. Chalia, M.K. Bharti, P. Thakur, A. Thakur, S.N. Sridhara, An overview of ceramic materials and their composites in porous media burner applications, *Ceram. Int.* 47 (2021) 10426–10441. https://doi.org/10.1016/j.ceramint.2020.12.202.
[45] E.I. Suzdal'tsev, D.V. Kharitonov, A.A. Anashkina, Analysis of existing radioparent refractory materials, composites and technology for creating high-speed rocket radomes. part 1. analysis of the level of property indices and limiting possibilities of radioparent inorganic refractory materials, *Refract. Ind. Ceram.* 51 (2010) 202–205. https://doi.org/10.1007/s11148-010-9289-2.
[46] K.K. Kandi, N. Thallapalli, S.P.R. Chilakalapalli, Development of silicon nitride-based ceramic radomes – A review, *Int. J. Appl. Ceram. Technol.* 12 (2015) 909–920. https://doi.org/10.1111/ijac.12305.

[47] G. Gilde, P. Patel, C. Hubbard, B. Pothier, T. Hynes, W. Croft, J. Wells, Sion low dielectric constant ceramic nanocomposite, *US5677252A*, 1997. https://patents.google.com/patent/US5677252A/en (accessed June 4, 2021).

[48] D.K. Kim, H.N. Kim, Y.H. Seong, S.S. Baek, E.S. Kang, Y.G. Baek, Dielectric properties of SiAlON ceramics, *Key Eng. Mater.* 403 (2009) 125–128. https://doi.org/10.4028/www.scientific.net/KEM.403.125.

[49] I. Ganesh, Development of β-SiAlON based ceramics for radome applications, *Process. Appl. Ceram.* 5 (2011) 113–138.

[50] M. Saeedi, J. Ghezavati, M. Abbasgholipour, B. Alasti, Various types of ceramics used in radome: A review, *Sci. Iran.* 24 (2017) 1136–1147. https://doi.org/10.24200/sci.2017.4095.

[51] N.K. Georgiu, I.F. Georgiu, K.V. Klemazov, M.G. Lisachenko, A.O. Zabezhailov, M. Yu. Rusin, porous reaction-bonded silicon nitride ceramics: Fabrication using hollow polymer microspheres and properties, *Inorg. Mater.* 55 (2019) 1290–1296. https://doi.org/10.1134/S0020168519120057.

[52] K. Qian, Z. Yao, H. Lin, J. Zhou, A.A. Haidry, T. Qi, W. Chen, X. Guo, The influence of Nd substitution in Ni – Zn ferrites for the improved microwave absorption properties, *Ceram. Int.* 46 (2020) 227–235.

[53] A. Kumar, V. Agarwala, D. Singh, Effect of particle size of BaFe12O19 on the microwave absorption characteristics in X-band, *Prog. Electromagn. Res. M.* 29 (2013) 223–236. https://doi.org/10.2528/PIERM13011604.

[54] M. Jamalian, A. Ghasemi, M.J. Pourhosseini Asl, Magnetic and microwave properties of barium hexaferrite ceramics doped with Gd and Nd, *J. Electron. Mater.* 44 (2015) 2856–2861. https://doi.org/10.1007/s11664-015-3720-x.

[55] Z. Wang, P. Zhao, D. He, Y. Cheng, L. Liao, S. Li, Y. Luo, Z. Peng, P. Li, Cerium oxide immobilized reduced graphene oxide hybrids with excellent microwave absorbing performance, *Phys. Chem. Chem. Phys.* 20 (2018) 14155–14165. https://doi.org/10.1039/C8CP00160J.

[56] L.L. Adebayo, H. Soleimani, N. Yahya, Z. Abbas, F.A. Wahaab, R.T. Ayinla, H. Ali, Recent advances in the development OF Fe3O4-BASED microwave absorbing materials, *Ceram. Int.* 46 (2020) 1249–1268. https://doi.org/10.1016/j.ceramint.2019.09.209.

[57] P. Thakur, D. Chahar, S. Taneja, N. Bhalla, A. Thakur, A review on MnZn ferrites: Synthesis, characterization and applications, *Ceram. Int.* 46 (2020) 15740–15763. https://doi.org/10.1016/j.ceramint.2020.03.287.

[58] M.K. Bharti, S. Chalia, P. Thakur, S.N. Sridhara, A. Thakur, P.B. Sharma, Nanoferrites heterogeneous catalysts for biodiesel production from soybean and canola oil: A review, *Environ. Chem. Lett.* 19 (2021) 3727–3746. https://doi.org/10.1007/s10311-021-01247-2.

[59] M.K. Bharti, S. Gupta, S. Chalia, I. Garg, P. Thakur, A. Thakur, Potential of magnetic nanoferrites in removal of heavy metals from contaminated water: Mini review, *J. Supercond. Nov. Magn.* 33 (2020) 3651–3665. https://doi.org/10.1007/s10948-020-05657-1.

[60] M.K. Bharti, S. Chalia, P. Thakur, A. Thakur, Effect of lanthanum doping on microstructural, dielectric and magnetic properties of Mn0.4Zn0.6Cd0.2LaxFe1.8-xO4 (0.0 ≤ x ≤ 0.4), *J. Supercond. Nov. Magn.* 34 (2021) 2591–2600. https://doi.org/10.1007/s10948-021-05908-9.

[61] A. Thakur, P. Thakur, J.-H. Hsu, Magnetic behaviour of Ni0.4Zn0.6Co0.1Fe1.9O4 spinel nanoferrite, *J. Appl. Phys.* 111 (2012) 07A305. https://doi.org/10.1063/1.3670606.

[62] P. Mathur, A. Thakur, M. Singh, Effect of nanoparticles on the magnetic properties of Mn – Zn soft ferrite, *J. Magn. Magn. Mater.* 320 (2008) 1364–1369. https://doi.org/10.1016/j.jmmm.2007.11.008.

[63] A. Thakur, P. Thakur, J.-H. Hsu, Enhancement in dielectric and magnetic properties of $\rm In^{3+}$ substituted Ni-Zn nanoferrites by coprecipitation method, *IEEE Trans. Magn.* 47 (2011) 4336–4339. https://doi.org/10.1109/TMAG.2011.2156394.

[64] S. Taneja, D. Chahar, P. Thakur, A. Thakur, Influence of bismuth doping on structural, electrical and dielectric properties of Ni – Zn nanoferrites, *J. Alloys Compd.* 859 (2021) 157760. https://doi.org/10.1016/j.jallcom.2020.157760.

[65] T. Tatarchuk, M. Bououdina, W. Macyk, O. Shyichuk, N. Paliychuk, I. Yaremiy, B. Al-Najar, M. Pacia, Structural, optical, and magnetic properties of zn-doped CoFe2O4 nanoparticles, *Nanoscale Res. Lett.* 12 (2017) 141. https://doi.org/10.1186/s11671-017-1899-x.

[66] I. Malinowska, Z. Ryżyńska, E. Mrotek, T. Klimczuk, A. Zielińska-Jurek, Synthesis of CoFe2O4 nanoparticles: The effect of ionic strength, concentration, and precursor type on morphology and magnetic properties, *J. Nanomater.* 2020 (2020) e9046219. https://doi.org/10.1155/2020/9046219.

[67] P. Punia, M.K. Bharti, S. Chalia, R. Dhar, B. Ravelo, P. Thakur, A. Thakur, Recent advances in synthesis, characterization, and applications of nanoparticles for contaminated water treatment: A review, *Ceram. Int.* 47 (2021) 1526–1550. https://doi.org/10.1016/j.ceramint.2020.09.050.

[68] P. Mathur, A. Thakur, M. Singh, Low temperature synthesis of Mn0.4Zn0.6In0.5Fe1.5O4 nanoferrite for high-frequency applications, *J. Phys. Chem. Solids.* 69 (2008) 187–192. https://doi.org/10.1016/j.jpcs.2007.08.014.

[69] P. Mathur, A. Thakur, J.H. Lee, M. Singh, Sustained electromagnetic properties of Ni – Zn – Co nanoferrites for the high-frequency applications, *Mater. Lett.* 64 (2010) 2738–2741. https://doi.org/10.1016/j.matlet.2010.08.056.
[70] U.R. Lima, M.C. Nasar, R.S. Nasar, M.C. Rezende, J.H. Araújo, Ni – Zn nanoferrite for radar-absorbing material, *J. Magn. Magn. Mater.* 320 (2008) 1666–1670. https://doi.org/10.1016/j.jmmm.2008.01.022.
[71] P. Smitha, I. Singh, M. Najim, R. Panwar, D. Singh, V. Agarwala, G. Das Varma, Development of thin broad band radar absorbing materials using nanostructured spinel ferrites, *J. Mater. Sci. Mater. Electron.* 27 (2016) 7731–7737. https://doi.org/10.1007/s10854-016-4760-6.
[72] K. Khan, Microwave absorption properties of radar absorbing nanosized cobalt ferrites for high frequency applications, *J. Supercond. Nov. Magn.* 27 (2014) 453–461. https://doi.org/10.1007/s10948-013-2283-4.
[73] K. Malaie, M.R. Ganjali, Spinel nanoferrites for aqueous supercapacitors; linking abundant resources and low-cost processes for sustainable energy storage, *J. Energy Storage.* 33 (2021) 102097. https://doi.org/10.1016/j.est.2020.102097.
[74] G. Singh, S. Chandra, Copper doped manganese ferrites PANI for fabrication of binder-free nanohybrid symmetrical supercapacitors, *J. Electrochem. Soc.* 166 (2019) A1154. https://doi.org/10.1149/2.1081904jes.
[75] S. Rana, A. Gallo, R.S. Srivastava, R.D.K. Misra, On the suitability of nanocrystalline ferrites as a magnetic carrier for drug delivery: Functionalization, conjugation and drug release kinetics, *Acta Biomater.* 3 (2007) 233–242. https://doi.org/10.1016/j.actbio.2006.10.006.
[76] A. Goyal, S. Bansal, S. Singhal, Facile reduction of nitrophenols: Comparative catalytic efficiency of MFe_2O_4 (M = Ni, Cu, Zn) nano ferrites, *Int. J. Hydrog. Energy.* 39 (2014) 4895–4908. https://doi.org/10.1016/j.ijhydene.2014.01.050.
[77] J.A. Vara, P.N. Dave, S. Chaturvedi, Investigating catalytic properties of nanoferrites for both AP and NanoAP based composite solid propellant, *Combust. Sci. Technol.* (2020) 1–15. https://doi.org/10.1080/00102202.2020.1734582.
[78] Q.-Y. Wang, W. Huang, X.-L. Jiang, Y.-J. Kang, Breast cancer cells synchronous labeling and separation based on aptamer and fluorescence-magnetic silica nanoparticles, *Opt. Mater.* 75 (2018) 483–490. https://doi.org/10.1016/j.optmat.2017.11.003.
[79] Y. Hadadian, D.R.T. Sampaio, A.P. Ramos, A.A.O. Carneiro, M. Mozaffari, L.C. Cabrelli, T.Z. Pavan, Synthesis and characterization of zinc substituted magnetite nanoparticles and their application to magneto-motive ultrasound imaging, *J. Magn. Magn. Mater.* 465 (2018) 33–43. https://doi.org/10.1016/j.jmmm.2018.05.069.
[80] A. Hossain, M.S.I. Sarker, M.K.R. Khan, F.A. Khan, M. Kamruzzaman, M.M. Rahman, Structural, magnetic, and electrical properties of sol – gel derived cobalt ferrite nanoparticles, *Appl. Phys. A.* 124 (2018) 608. https://doi.org/10.1007/s00339-018-2042-2.
[81] N. Saxena, N. Kumar, P. Pourush, S. Khah, Study of magnetic properties of substituted LiTiZn-ferrite for microwave antenna applications, *Optoelectron. Adv. Mater. Rapid Commun.* 4 (2010) 328–331.
[82] G. Catalan, J.F. Scott, Physics and applications of bismuth ferrite, *Adv. Mater.* 21 (2009) 2463–2485. https://doi.org/10.1002/adma.200802849.
[83] J.D. Adam, L.E. Davis, G.F. Dionne, E.F. Schloemann, S.N. Stitzer, Ferrite devices and materials, *IEEE Trans. Microw. Theory Tech.* 50 (2002) 721–737. https://doi.org/10.1109/22.989957.
[84] A.G. Roca, R. Costo, A.F. Rebolledo, S. Veintemillas-Verdaguer, P. Tartaj, T. González-Carreño, M.P. Morales, C.J. Serna, Progress in the preparation of magnetic nanoparticles for applications in biomedicine, *J. Phys. Appl. Phys.* 42 (2009) 224002. https://doi.org/10.1088/0022-3727/42/22/224002.
[85] A. Pathania, P. Thakur, A.V. Trukhanov, S.V. Trukhanov, L.V. Panina, U. Lüders, A. Thakur, Development of tungsten doped Ni-Zn nanoferrites with fast response and recovery time for hydrogen gas sensing application, *Results Phys.* 15 (2019) 102531. https://doi.org/10.1016/j.rinp.2019.102531.
[86] A. Saini, A. Thakur, P. Thakur, Matching permeability and permittivity of NiZnCoInFeO ferrite for substrate of large bandwidth miniaturized antenna, *J. Mater. Sci. Mater. Electron.* 27 (2016).
[87] M.K. Bharti, S. Chalia, P. Thakur, G.C. Hermosa, A.-C. Aidan Sun, A. Thakur, Low-loss characteristics and sustained magneto-dielectric behaviour of cobalt ferrite nanoparticles over 1–6 GHz frequency range, *Ceram. Int.* 47 (15) (2021) 22164–22171. https://doi.org/10.1016/j.ceramint.2021.04.239.
[88] Gh.R. Amiri, M.H. Yousefi, M.R. Abolhassani, S. Manouchehri, M.H. Keshavarz, S. Fatahian, Magnetic properties and microwave absorption in Ni – Zn and Mn – Zn ferrite nanoparticles synthesized by low-temperature solid-state reaction, *J. Magn. Magn. Mater.* 323 (2011) 730–734. https://doi.org/10.1016/j.jmmm.2010.10.034.
[89] X. Zhang, W. Sun, Microwave absorbing properties of double-layer cementitious composites containing Mn – Zn ferrite, *Cem. Concr. Compos.* 32 (2010) 726–730. https://doi.org/10.1016/j.cemconcomp.2010.07.013.
[90] A. Kumar, V. Agarwala, D. Singh, Effect of Mg substitution on microwave absorption of $BaFe_{12}O_{19}$, *Adv. Mater. Res.* 585 (2012) 62–66. https://doi.org/10.4028/www.scientific.net/AMR.585.62.

[91] Y. Liu, X. Liu, X. Wang, Double-layer microwave absorber based on CoFe2O4 ferrite and carbonyl iron composites, *J. Alloys Compd.* 584 (2014) 249–253. https://doi.org/10.1016/j.jallcom.2013.09.049.

[92] F.E. Carvalho, L.V. Lemos, A.C.C. Migliano, J.P.B. Machado, R.C. Pullar, Structural and complex electromagnetic properties of cobalt ferrite (CoFe2O4) with an addition of niobium pentoxide, *Ceram. Int.* 44 (2018) 915–921. https://doi.org/10.1016/j.ceramint.2017.10.023.

[93] D.-K. Kim, M.S. Toprak, M. Mikhaylova, Y.S. Jo, S.J. Savage, H.B. Lee, T. Tsakalakos, M. Muhammed, Polymeric nanocomposites of complex ferrite, *Solid State Phenom.* 99–100 (2004) 165–168. https://doi.org/10.4028/www.scientific.net/SSP.99-100.165.

[94] A.V. Lopatin, N.E. Kazantseva, Yu.N. Kazantsev, O.A. D'yakonova, J. Vilčáková, P. Sáha, The efficiency of application of magnetic polymer composites as radio-absorbing materials, *J. Commun. Technol. Electron.* 53 (2008) 487–496. https://doi.org/10.1134/S106422690805001X.

[95] T.-H. Ting, K.-H. Wu, Synthesis, characterization of polyaniline/BaFe12O19 composites with microwave-absorbing properties, *J. Magn. Magn. Mater.* 322 (2010) 2160–2166. https://doi.org/10.1016/j.jmmm.2010.02.002.

[96] P. Xu, X. Han, J. Jiang, X. Wang, X. Li, A. Wen, Synthesis and characterization of novel coralloid polyaniline/BaFe12O19 nanocomposites, *J. Phys. Chem. C.* 111 (2007) 12603–12608. https://doi.org/10.1021/jp073872x.

[97] S. Hazra, N.N. Ghosh, Preparation of nanoferrites and their applications, *J. Nanosci. Nanotechnol.* 14 (2014) 1983–2000. https://doi.org/10.1166/jnn.2014.8745.

[98] A. Parsa, T. Kodera, C. Caloz, Ferrite based non-reciprocal radome, generalized scattering matrix analysis and experimental demonstration, *IEEE Trans. Antennas Propag.* 59 (2011) 810–817. https://doi.org/10.1109/TAP.2010.2103016.

[99] Hung Y. David Yang, Characteristics of switchable ferrite microstrip antennas, *IEEE Trans. Antennas Propag.* 44 (1996) 1127–1132. https://doi.org/10.1109/8.511821.

[100] Y. Ge, C. Li, G.I.N. Waterhouse, Z. Zhang, L. Yu, ZnFe2O4@SiO2@Polypyrrole nanocomposites with efficient electromagnetic wave absorption properties in the K and Ka band regions, *Ceram. Int.* 47 (2021) 1728–1739. https://doi.org/10.1016/j.ceramint.2020.08.290.

[101] W. Shen, B. Ren, S. Wu, W. Wang, X. Zhou, Facile synthesis of rGO/SmFe5O12/CoFe2O4 ternary nanocomposites: Composition control for superior broadband microwave absorption performance, *Appl. Surf. Sci.* 453 (2018) 464–476. https://doi.org/10.1016/j.apsusc.2018.05.150.

[102] J. Singh, C. Singh, D. Kaur, S.B. Narang, R. Joshi, S.R. Mishra, R. Jotania, M. Ghimire, C.C. Chauhan, Tunable microwave absorption in CoAl substituted M-type BaSr hexagonal ferrite, *Mater. Des.* 110 (2016) 749–761. https://doi.org/10.1016/j.matdes.2016.08.049.

[103] X. Meng, Y. Liu, G. Han, W. Yang, Y. Yu, Three-dimensional (Fe3O4/ZnO)@C Double-core@shell porous nanocomposites with enhanced broadband microwave absorption, *Carbon.* 162 (2020) 356–364. https://doi.org/10.1016/j.carbon.2020.02.035.

[104] X. Huang, J. Zhang, S. Xiao, G. Chen, The cobalt zinc spinel ferrite nanofiber: Lightweight and efficient microwave absorber, *J. Am. Ceram. Soc.* 97 (2014) 1363–1366. https://doi.org/10.1111/jace.12909.

[105] P. Siva Nagasree, K. Ramji, Ch. Subramanyam, K. Krushnamurthy, T. Haritha, Synthesis of Ni0.5Zn0.5Fe2O4-reinforced E-glass/epoxy nanocomposites for radar-absorbing structures, *Plast. Rubber Compos.* 49 (2020) 434–442. https://doi.org/10.1080/14658011.2020.1793080.

[106] H. Xing, Y. Liu, Z. Liu, H. Wang, H. Jia, Structure and microwave absorption properties of polyaniline/Zn ferrite composites, *Nano.* 13 (2018) 1850105–1850105. https://doi.org/10.1142/S1793292018501059.

[107] E. Kiani, A.S.H. Rozatian, M.H. Yousefi, Structural, magnetic and microwave absorption properties of SrFe12−2x(Mn0.5Cd0.5Zr)xO19 ferrite, *J. Magn. Magn. Mater.* 361 (2014) 25–29. https://doi.org/10.1016/j.jmmm.2014.02.042.

[108] X. Shen, F. Song, J. Xiang, M. Liu, Y. Zhu, Y. Wang, Shape anisotropy, exchange-coupling interaction and microwave absorption of hard/soft nanocomposite ferrite microfibers, *J. Am. Ceram. Soc.* 95 (2012) 3863–3870. https://doi.org/10.1111/j.1551-2916.2012.05375.x.

[109] Y. He, S. Pan, J. Yu, Research on magnetic and microwave absorbing properties of Co2Y ferrite fabricated by sol – gel process, *J. Sol-Gel Sci. Technol.* 96 (2020) 521–528. https://doi.org/10.1007/s10971-020-05235-w.

[110] Y. Ding, Q. Liao, S. Liu, H. Guo, Y. Sun, G. Zhang, Y. Zhang, Reduced graphene oxide functionalized with cobalt ferrite nanocomposites for enhanced efficient and lightweight electromagnetic wave absorption, *Sci. Rep.* 6 (2016) 32381–32381. https://doi.org/10.1038/srep32381.

[111] P. Yin, L. Zhang, J. Wang, X. Feng, L. Zhao, H. Rao, Y. Wang, J. Dai, Preparation of SiO2-MnFe2O4 composites via one-pot hydrothermal synthesis method and microwave absorption investigation in S-Band, *Molecules.* 24 (2019). https://doi.org/10.3390/molecules24142605.

[112] S. Ozah, N.S. Bhattacharyya, Nanosized barium hexaferrite in novolac phenolic resin as microwave absorber for X-band application, *J. Magn. Magn. Mater.* 342 (2013) 92–99. https://doi.org/10.1016/j.jmmm.2013.04.050.

[113] S. Bindra Narang, P. Kaur, S. Bahel, C. Singh, Microwave characterization of Co – Ti substituted barium hexagonal ferrites in X- band, *J. Magn. Magn. Mater.* 405 (2016) 17–21. https://doi.org/10.1016/j.jmmm.2015.12.044.

[114] M.K. Tehrani, A. Ghasemi, M. Moradi, R.S. Alam, Wideband electromagnetic wave absorber using doped barium hexaferrite in Ku-band, *J. Alloys Compd.* 509 (2011) 8398–8400. https://doi.org/10.1016/j.jallcom.2011.05.091.

[115] H. Yang, T. Ye, Y. Lin, M. Liu, Preparation and microwave absorption property of graphene/BaFe12O19/CoFe2O4 nanocomposite, *Appl. Surf. Sci.* 357 (2015) 1289–1293. https://doi.org/10.1016/j.apsusc.2015.09.147.

[116] H. He, F. Luo, N. Qian, N. Wang, Improved microwave absorption and electromagnetic properties of BaFe 12 O 19 -poly(vinylidene fluoride) composites by incorporating reduced graphene oxides, *J. Appl. Phys.* 117 (2015) 085502. https://doi.org/10.1063/1.4913396.

[117] A. Arora, S.B. Narang, Tuning of microwave absorptive behavior of double substituted barium hexaferrites with change in thickness in 26.5–40.0 GHz band, *Appl. Phys. A.* 123 (2017) 520–520. https://doi.org/10.1007/s00339-017-1138-4.

[118] Y. Lei, Z. Yao, H. Lin, J. Zhou, A. Haidry, P. Liu, The effect of polymerization temperature and reaction time on microwave absorption properties of Co-doped ZnNi ferrite/polyaniline composites, *RSC Adv.* 8 (2018) 29344–29355. https://doi.org/10.1039/C8RA05500A.

[119] J. Luo, P. Shen, W. Yao, C. Jiang, J. Xu, Synthesis, characterization, and microwave absorption properties of reduced graphene oxide/strontium ferrite/polyaniline nanocomposites, *Nanoscale Res. Lett.* 11 (2016) 141–141. https://doi.org/10.1186/s11671-016-1340-x.

[120] H. Nikmanesh, M. Moradi, G.H. Bordbar, R.S. Alam, Synthesis of multi-walled carbon nanotube/doped barium hexaferrite nanocomposites: An investigation of structural, magnetic and microwave absorption properties, *Ceram. Int.* 42 (2016) 14342–14349. https://doi.org/10.1016/j.ceramint.2016.05.089.

[121] M. Katiyar, M. Prasad, K. Agarwal, R.K. Singh, A. Kumar, N. Prasad, Study and characterization of E.M. absorbing properties of EPDM ferrite composite containing manganese zinc ferrite, *J. Reinf. Plast. Compos.* 36 (2017) 073168441769081–073168441769081. https://doi.org/10.1177/0731684417690816.

[122] Z. Mosleh, P. Kameli, A. Poorbaferani, M. Ranjbar, H. Salamati, Structural, magnetic and microwave absorption properties of Ce-doped barium hexaferrite, *J. Magn. Magn. Mater.* 397 (2016) 101–107. https://doi.org/10.1016/j.jmmm.2015.08.078.

[123] P. Liu, Z. Yao, J. Zhou, Preparation of reduced graphene oxide/Ni0.4Zn0.4Co0.2Fe2O4 nanocomposites and their excellent microwave absorption properties, *Ceram. Int.* 41 (2015) 13409–13416. https://doi.org/10.1016/j.ceramint.2015.07.129.

[124] X. Chen, X. Wang, L. Li, S. Qi, Preparation and excellent microwave absorption properties of silver/strontium ferrite/graphite nanosheet composites via sol – gel method, *J. Mater. Sci. Mater. Electron.* 27 (2016) 10045–10051. https://doi.org/10.1007/s10854-016-5076-2.

[125] M. Khadour, Y. Atassi, M. Abdallah, Preparation and characterization of a flexible microwave absorber based on MnNiZn ferrite (Mn0.1Ni0.45Zn0.45Fe2O4) in a thermoset polyurethane matrix, *SN Appl. Sci.* 2 (2020) 236–236. https://doi.org/10.1007/s42452-020-2004-0.

[126] P. Yin, Y. Deng, L. Zhang, W. Wu, J. Wang, X. Feng, X. Sun, H. Li, Y. Tao, One-step hydrothermal synthesis and enhanced microwave absorption properties of Ni0.5Co0.5Fe2O4/graphene composites in low frequency band, *Ceram. Int.* 44 (2018) 20896–20905. https://doi.org/10.1016/j.ceramint.2018.08.096.

[127] Y. Chang, Y. Zhang, L. Li, S. Liu, Z. Liu, H. Chang, X. Wang, Microwave absorption in 0.1–18 GHz, magnetic and structural properties of SrFe12-xRuxO19 and BaFe12-xRuxO19, *J. Alloys Compd.* 818 (2020) 152930–152930. https://doi.org/10.1016/j.jallcom.2019.152930.

[128] G. Feng, W. Zhou, C.-H. Wang, Y. Qing, D. Chen, L. Gao, F. Luo, D. Zhu, Microwave absorption of M-type hexaferrite Ba1-xCaxFe12O19 (x ≤ 0.4) ceramics in 2.6–18 GHz, *Ceram. Int.* 45 (2019) 7102–7107. https://doi.org/10.1016/j.ceramint.2018.12.214.

[129] S. Goel, A. Tyagi, A. Garg, S. Kumar, H.B. Baskey, R.K. Gupta, S. Tyagi, Microwave absorption study of composites based on CQD@BaTiO3 core shell and BaFe12O19 nanoparticles, *J. Alloys Compd.* 855 (2021) 157411–157411. https://doi.org/10.1016/j.jallcom.2020.157411.

[130] S. Golchinvafa, S.M. Masoudpanah, M. Jazirehpour, Magnetic and microwave absorption properties of FeCo/CoFe2O4 composite powders, *J. Alloys Compd.* 809 (2019) 151746–151746. https://doi.org/10.1016/j.jallcom.2019.151746.

[131] S.B. Narang, A. Arora, Broad-band microwave absorption and magnetic properties of M-type Ba(1−2x)LaxNaxFe10Co0.5TiMn0.5O19 hexagonal ferrite in 18.0–26.5 GHz frequency range, *J. Magn. Magn. Mater.* 473 (2019) 272–277. https://doi.org/10.1016/j.jmmm.2018.10.042.

[132] Z. Yan, J. Luo, Effects of CeZn co-substitution on structure, magnetic and microwave absorption properties of nickel ferrite nanoparticles, *J. Alloys Compd.* 695 (2017) 1185–1195. https://doi.org/10.1016/j.jallcom.2016.08.333.

[133] A. Das, P. Negi, S.K. Joshi, A. Kumar, Enhanced microwave absorption properties of Co and Ni co-doped iron (II,III)/reduced graphene oxide composites at X-band frequency, *J. Mater. Sci. Mater. Electron.* 30 (2019) 19325–19334. https://doi.org/10.1007/s10854-019-02293-x.

Miniaturization Techniques for Microstrip Patch Antenna for Telecommunication Applications

Preeti Thakur, Shilpa Taneja, Atul Thakur

Contents

1	Introduction	245
2	Microstrip Antennas	246
	2.1 Microstrip Patch Antenna	246
3	Miniaturization of Patch Antenna	248
	3.1 Material Loading	249
	3.2 Shorting and Folding	250
	3.3 Reshaping or Introducing Slots	251
	3.4 Modifications of the Ground Plane	251
	3.5 Use of Meta-Materials	252
4	Feeding Techniques of Antenna Miniaturization	253
5	Applications of MPA	256
6	Medicinal Applications of the Patch	257
7	Outlook	257
References		257

1 INTRODUCTION

The widespread use of wireless access devices like smartphones, organizers, computers, and navigation devices [1–6] sparked a lot of interest in compact antenna research and development. An antenna is an essential component of all such wireless devices from a technical standpoint. An antenna on the gadget, on the other hand, is something inelegant that should be avoided or at the very least rendered inconspicuous for designers and users. The antennas are mounted on buildings and masts both outside and inside. Wind load and weight become key factors for outdoor antennas. Indoor antennas must be built inside structures and made aesthetically pleasing to the general public.

For nearly seven decades, many investigations on antenna shrinking have been conducted [7–10]. Early research found that bandwidth and efficiency depend on the size of an antenna [11–14]. Because of the size restriction, the maximum feasible radiation quality factor and, as a result, the maximum achievable impedance bandwidth are both reduced. Many new studies are being undertaken today to minimize the total size of various types of antennas. The physical and electrical properties of antennas are frequently changed using these shrinking techniques. Antenna miniaturization has a plethora of published works. Many of these specified comparable principles to attain their goal of developing a methodology using simulation and analytic methodologies. These analyses frequently provide insight into the promise of a certain technique that has been proved experimentally utilizing built prototypes, they also provide information into the limitations of that technique. Microstrip patch antennas [15–17], fractal antennas [18], PCS (personal communication system) antennas [12], and slot antennas [19] are examples of antenna miniaturization approaches. Miniaturization of patch antennas is one of the following approaches that have a lot of attention in both defense and commercial applications. The wavelength of electromagnetic waves in the antenna substrate material determines the size of these fields [20]. Microstrip patch antennas have grown in popularity in recent years due to their advantages of planar construction, small size, low production cost, and simple fabrication technique [21,22]. The substrate of a patch antenna is a dielectric material. Because of the low permittivity of the substrate materials, antennae must be large to fit into the given space. A size reduction can be achieved by utilizing a high permittivity substrate, however, this affects bandwidth and radiation pattern performance. The use of magneto-dielectric materials [9,17,23,24] can overcome these difficulties. Magneto-dielectric materials are not naturally occurring materials but they have both dielectric and magnetic properties. These materials are ideally suited for specific applications because they have two basic properties: relative permeability (μ_r) and relative permittivity (ε_r). They produce strong dielectric strength which makes them ideal for high-frequency applications. As a result, using suitable values of μ_r and ε_r, antennas can be miniaturized [9,25–29]. Kong et al. [30] created a magnesium cobalt-based soft ferrite with permittivity and permeability value 10 and a maximum frequency of 30 MHz. For magnetic substrate materials, Rialet et al. calculated μ_{reff} which is equal to 2 [31]. Thakur et al. demonstrated high and matching μ_r and ε_r, but only up to 30 MHz [27]. Up to 200 MHz, Mathur et al. synthesized $Ni_{0.49}Zn_{0.49}Co_{0.02}Fe_2O_4$ with ε_r and μ_r up to 9.1 and 4.4, respectively [24].

2 MICROSTRIP ANTENNAS

Microstrip antennas have huge benefits over other antennas and are thus frequently used in a variety of applications. Figure 15.1 depicts microstrip antennas in their most basic arrangement.

On one face of the dielectric substrate ($\varepsilon_r < = 10$), there is a radiating patch, and on the other side, there is a ground plane. Microstrip antennas have a broader range of physical characteristics than traditional microwave antennas and have a variety of geometrical shapes and sizes [32]. As illustrated in Figure 15.2, microstrip antennas are classified into four fundamental kinds.

2.1 Microstrip Patch Antenna

A microstrip patch antenna is a device consisting of a ground plane on one side and a conducting patch on the other side of a dielectric substrate. It is a printed resonant antenna for microwave wireless communications which provides semi-hemispheric coverage in limited bands. It is frequently used in array

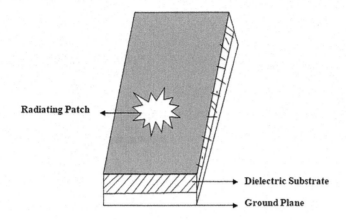

FIGURE 15.1 Microstrip antenna configuration.

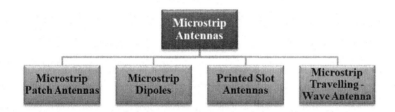

FIGURE 15.2 Basic categories of microstrip antennas.

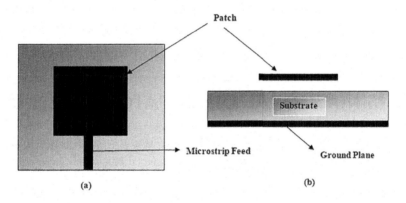

FIGURE 15.3 Structure of rectangular microstrip patch antenna (a) top view (b) side view.

elements because of its configuration and ease of integration with microstrip technology. There have been a lot of microstrip patch antennas investigated. The most basic and commonly used microstrip antennas are rectangular and circular patches. These patches are suitable for both simple and complex applications. Rectangular patch antennas are straightforward to analyze (Figure 15.3) and the circular patch antenna has the benefit of having a symmetrical radiation pattern (as presented in Figure 15.4).

The characteristics of microstrip patch antennas, microstrip slot antennas, and printed dipole antennas are compared in Table 15.1.

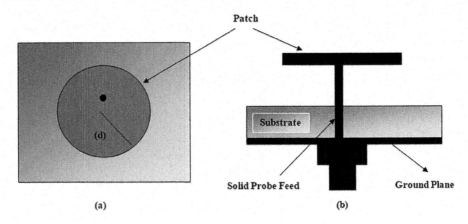

FIGURE 15.4 Circular microstrip patch antenna (a) top view (b) side view.

TABLE 15.1 The Characteristics of Microstrip Patch Antennas, Microstrip Slot Antennas, and Printed Dipole Antennas

S.NO.	CHARACTERISTICS	MICROSTRIP PATCH ANTENNAS	MICROSTRIP SLOT ANTENNAS	PRINTED DIPOLE ANTENNAS
1.	Profile	Thin	Thin	Thin
2.	Polarization	Both Linear and Circular	Both Linear and Circular	Linear only
3.	Shape Flexibility	Any Shape	Mostly Rectangular and Circular Shapes	Rectangular and Triangular Shapes
4.	Bandwidth	2–50%	5–30%	–30%
5.	Fabrication	Very Easy	Easy	Easy

3 MINIATURIZATION OF PATCH ANTENNA

Patch antenna miniaturization is a hot topic in recent trends. Wheeler introduced the fundamental constraints, concluding that shrinking an antenna resulted in reduced bandwidth and gain [7,33]. Chu and Harrington, as well as many others, came to the same results [34,35]. Equation 1 gives the theoretical lower constraint on the antenna Q for a tiny antenna that may be encompassed in a sphere of radius a:

$$Q = \frac{1}{ka} + \frac{1}{ka^3} \tag{1}$$

where "k" is the wavenumber. As can be shown from equation 1, shrinking the size of an antenna increases its Q. The ratio of quality factor (Q) to the antenna's efficiency gives the value of antenna performance. As a result, antenna Q can be reduced at the cost of efficiency and gain. The microstrip patch antenna can be miniaturized in two ways (MPA). The first way is to alter the substrate's material qualities to reduce the effective wavelength in the substrate region. The second strategy is to alter its geometry in such a way that the electrical size is increased (current path). Various strategies for MPA miniaturization based on these two major methodologies have been documented in the literature [36,37]. As illustrated in Figure 15.5, these two major strategies are divided into five categories.

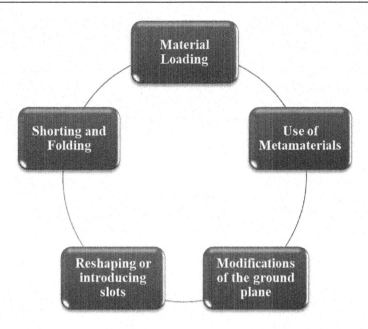

FIGURE 15.5 Various methods for Microstrip Patch Antenna (MPA) miniaturization.

3.1 Material Loading

To reduce the size of an MPA by using a high value of dielectric substrate, the length and width of the patch are inversely proportional to the square root of ε_r, as shown in equation 2:

$$n_i = \sqrt{\epsilon_r \mu_r} \qquad (2)$$

However, such a miniaturization strategy causes a rise in surface wave excitation within the substrate, which results in a smaller bandwidth and lesser efficiency. The ground plane's truncation not only reduces polarization purity but also alters the MPA's radiation properties. Several studies have looked into various materials and configurations to efficiently employ the above approach to miniaturize an MPA:

- **Antennas with Engineered Substrate:** Several downsizing strategies have been prepared using substrates with a high value of dielectric constant based on this relationship between size and electromagnetic properties of the substrate [38–41]. Surface wave excitation, energy storage, and dielectric loss all contribute to this reduction. A substrate with a high dielectric constant might cause antenna impedance matching problems [42–44]. High-dielectric-constant materials can be employed in the latter [45]. Magnetodielectric substrates having a value higher than one is an alternative to high dielectric constant substrates [29,41,46]. The intrinsic impedance of the substrate can be coordinated to the intrinsic impedance of the air by engineering it to have $\mu_r = \varepsilon_r$ while maintaining a high refractive index. Miniaturization is aided by better radiation efficiency, matching, and bandwidth.
- **Antennas with Meta-Material:** Veselago proposed the notion of negative permittivity and permeability composite materials, often known as double negative (DNG) or left-handed materials, about 50 years ago. A metallic array having magnetic or dielectric substrates is used to create such materials artificially [47,48]. In a restricted frequency range, they also reveal remarkable scattering and propagation features. As a result, they're known as meta-materials. The propagation vector is antiparallel to the path of usual power flow because the Poynting

vector shows the direction of usual power flow. This feature can be exploited to give a wave moving through such a medium a negative time delay. Many studies have been published on the use of meta-materials to enhance the efficiency of tiny antennas [49–54]. A meta-material shell was employed around an electrically tiny dipole antenna in [49]. A lossless matching circuit can aid in bettering the matching and efficiency of the system. The antenna with the corresponding network, on the other hand, will no longer be small. As a result, the notion of encasing the small antenna in a DNG meta-material shell was considered.

3.2 Shorting and Folding

MPAs have been folded and shorting posts have been employed to minimize their size and make them electrically tiny [20]. The E-field distribution under a half-wavelength rectangular MPA displays a sine wave pattern, with the greatest E-field at the edges. The patch would still resonate at the same frequency when the electric wall was erected in the center and the other half was removed. A quarter-wavelength MPA is a name for such a patch. A quarter-wavelength patch has the same quality factor as its half-wavelength counterpart, according to a theoretical study [20]. However, because the antenna aperture has shrunk, the antenna directivity has shrunk as well, affecting the antenna gain directly. For a quarter-wavelength MPA, it is difficult to implement a continuous conducting sheet at the edge between the patch and the ground plane. As shown in Figure 15.6, construction becomes easier for quarter-wavelength MPA when an array of shorting pins along the border of the patch is added. Many papers [55–63] have been published that describe the analysis of miniature MPAs employing folding techniques and to produce a 1/8 MPA in [64]. The antenna's radiation efficiency was determined to be 90%, with a bandwidth of 4%. Figure 15.7 shows a schematic diagram of the shorted-folded patch. The findings of parametric analysis

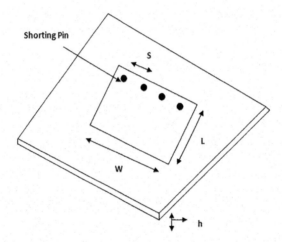

FIGURE 15.6 Quarter-wavelength MPA fabricated using shorting.

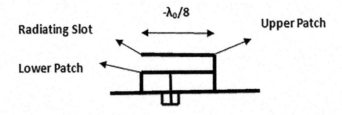

FIGURE 15.7 $\lambda/8$ shorted/folded patch antenna.

of employing single, double, or multiple shorting posts with a circular MPA are provided in [63]. When the shorting posts were placed optimally, MPA was found to be lowered by more than a factor of three.

3.3 Reshaping or Introducing Slots

Modification in the design of slots in the MPA is also used to miniaturize. The genetic algorithm [65], which runs on high-performance computing platforms, can be used to shape and tune miniaturized patches to produce a great electrical length in a small region. Fractal geometries have also been successfully employed to generate miniature MPAs [18,66]. Fractals are shapes that allow electrically big characteristics to be successfully packed into a compact space while lowering bandwidth. The ohmic losses in a downsized MPA are larger, resulting in reduced radiation efficiency. An engineered conductor, consisting of layers of conductors separated by laminations, was presented in [67] to alleviate this difficulty. The conductor's total thickness was the same as the conductor in a typical MPA. The designed conductor, separated by laminations, is depicted in Figure 15.6, and it aids in increasing the small antenna's gain and efficiency. Different miniature MPAs have also been employed with such a conductor. It has been demonstrated that as the number of conducting layers in an antenna increases, the gain and efficiency of the antenna improve. When compared to a normal single-layer conductor, a 5-layer conductor enhances the efficiency of a miniaturized antenna by 30%. In the literature, several works [58,68–76] have been published in which the size of the MPA has been lowered by introducing various types of slots into the MPA. When this approach is used, the polarization purity is usually low. However, such purity can be kept by arranging the slots on the patch in a symmetrical pattern [77]. Inserting various sorts of slots into the MPA resulted in a miniaturization of 40–75%. Despite the fact that this technology is extensively employed in a variety of designs and allows for various degrees of miniaturization, it lacks a common design methodology. The majority of the reported devices based on this technique had a poor radiation efficiency of approximately 25%, but when using slots, they gave larger operating bandwidths. Some had up to 5.5% fractional bandwidth [69].

3.4 Modifications of the Ground Plane

Modifying the ground plane of MPAs can also help to miniaturize them. An unlimited ground plane is assumed in most MPA models. The ground plane, on the other hand, is finite in any practical MPA design. For even more miniaturization, the ground plane is shrunk to the point where it is only slightly larger than the patch dimensions at times. Various works [78,79] have investigated MPAs with truncated ground planes analytically. It was discovered that these antennas had poor polarization purity, and that shrinking the ground plane had an impact on the input impedance. Furthermore, there was strong rear lobe radiation due to edge diffraction, lowering the front-to-back ratio. Many different ground plane modifications, in addition to lowering the ground plane, are conceivable to miniaturize an MPA. The inclusion of various sorts of slots in the ground plane is one of these adjustments. When appropriately planned, these slots aid in increasing the current path within the patch region. As a result, the MPA's resonance frequency is reduced, and the MPA's size is reduced. There are many papers on miniaturizing the MPA using the slots in the ground plane [80–83].

A single slot of 1 mm width was used to minimize the size of an MPA [80]. The position and length of the slot beneath the patch were changed in parametric research. Without the slot, the MPA primarily resonated at 2.87 GHz. The resonant frequency was reduced to 1.38 GHz using the best slot placement and length, which is a 52% reduction. The MPA's size was reduced by 90% in terms of area. To reduce the 50% MPA size, three slots were carved beneath the patch to change the shape [81]. In [82,83], a similar utilization of slots resulted in a 56–83% reduction in patch size. It's worth noting that, while all of the foregoing concentrated on MPA miniaturization. As a result, the impact of miniaturization on several factors is discussed in most of these publications. Furthermore, the research provided no physical insight into the miniaturization method's basic principles. Furthermore, they did not go into detail about the method's general application. They didn't offer any design suggestions for various frequency bands depending on the slot in the ground plane approach. Another way for antenna miniaturization that used an uneven ground construction was proposed in [62]. The MPA was created on a two-layer substrate in which tiny metallic cylinders were arranged in the lower substrate. The proposed construction concludes 75.6% reduction in the size of the MPA. The MPA's bandwidth was also reduced as a result of the size reduction. With a regular ground structure, 8.3% bandwidth was obtained with 5.32 GHz. Using the proposed approach, the antenna's resonance frequency was dropped to 2.635 GHz, while the bandwidth was reduced to 1.9%. A circularly polarized MPA was successfully demonstrated using the same size-reduction technique.

3.5 Use of Meta-Materials

Meta-materials (MTMs) are materials that have been artificially manufactured to have qualities that aren't easily available commercially. Using MTMs, materials with zero permittivity, negative permittivity, or permeability can all be realized. Epsilon negative (ENG) refers to a material with negative permittivity, whereas −negative (MNG) refers to a material with only negative permeability. Double-negative materials have both negative permittivity and permeability (DNG). Several structures have been developed over the last decade that, when placed periodically, display MTM features across a specific frequency range. Many people have been interested in these structures, and they have been extensively examined and developed. They've also been employed to achieve fascinating features in a variety of RF, microwave, and photonics devices.

MTM structures have high gain and efficiency. Antenna miniaturization has also been accomplished using them. Antennas that use MTM are referred to by two different designations in the literature such as MTM-based and MTM-inspired antennas. The former are primarily fictitious antennas that are explored in theory utilizing ENG, MNG, or DNG media. It is impossible to realize a proper MTM-based antenna. The latter isn't true MTM antennas because they don't utilize an MTM's ENG, MNG, or DNG features.

The effect of an MTM substrate may be determined using the cavity model and the expression of its resonant frequency. The inverse square root link between the MPA's resonant frequency and the permittivity and permeability of its substrate may be seen. As a result, an MPA sited on an ENG or MNG substrate will not resonate, and a homogeneous DNG substrate will not allow for any particular features.

MPAs on MTM substrates were theoretically studied in [84]. Figure 15.8 depicts the antenna's geometry. The patch's substrate was a mix of DNG medium and normal medium with both positive

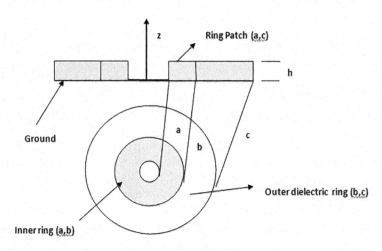

FIGURE 15.8 Geometry of the annular ring patch on an MTM substrate.

TABLE 15.2 Summary of MPA Miniaturization Techniques with Features

MINIATURIZATION TECHNIQUE	CHARACTERISTICS	MERITS	DEMERITS
Material loading	• Large dielectric substrates • Ceramic substrates • Magnetodielectric substrates	• High degree of miniaturization • Easy to design	• Expensive Materials • Limited bandwidth
Shorting and folding	• Shorting and folding pins and walls	• Cost-effective solution	• No standard design • Makes antenna geometry harder • Non-planar • Very low gain and directivity
Reshaping a patch or introducing slots	• Fractal antenna • Engineered conductors • Slots in the patch	• Provide wider bandwidth	• Make antenna geometry complex • No standard design • Poor polarization purity
Modifications in a ground plane	• Slots in a ground plane	• Antenna geometry is planar and simple	• Low efficiency • No standard design
Use of MTMs	• Use of ENG, MNG, or DNG substrates for inspired techniques	• High degree of miniaturization	• Limited bandwidth • Low efficiency • Complex antenna geometry • No standard design

and negative values. The inner-circle was produced by the DNG, which carefully determined the ratios of the area filled by the DNG and DPS media. The MPA on MTM substrate was further investigated analytically in [85], and numerical results were offered to support the analytical findings. The cavity model was used to investigate a rectangular patch with 1 and 2 permittivity and permeability values. The filling ratio was defined as the volumes of material media to the patch. Table 15.2 emphasizes the characteristics of each technique and outlines its primary benefits and drawbacks. As can be seen from the table, nearly all of the strategies listed result in significant miniaturization. However, some approaches, such as folding or MTM-inspired antennas, can result in a more complicated and non-planar antenna structure.

4 FEEDING TECHNIQUES OF ANTENNA MINIATURIZATION

A feed line is used to enthuse the radiations directly or indirectly. There are several methods of feeding which are as follows:

1) Coaxial probe feed patch antenna
2) Microstrip line feed patch antenna
3) Aperture coupling feed patch antenna
4) Proximity coupling feed patch antenna

Coaxial probe feeding: As illustrated in Figure 15.9, is a feeding technique in which the coaxial's inner conductor is connected to the antenna's radiating patch while the outside is connected to the ground surface. Coaxial feeding has the advantages of being simple to fabricate, easy to match, and minimal spurious radiation, but it has the disadvantages of having a narrow bandwidth and being difficult to predict, especially for thick substrates.

Microstrip line feed is the easiest technique to construct a conducting strip connected to the patch and can thus be considered a patch extension. Controlling the inset position, as shown in Figure 15.10, makes it straightforward to model and match. The problem of this approach is that as the thickness of the substrate grows, so does the amount of surface wave limiting the bandwidth.

Aperture coupled feed: In this method, two distinct substrates are separated by a ground plane as illustrated in Figure 15.11. The top substrate is made of a low dielectric constant substrate, whereas the bottom is made of a high dielectric constant substrate. The middle ground plane separates the feed from the radiation to reduce interference pattern formation. Advantages are allowed independent optimization of feed mechanism element.

Proximity coupling: It has the broad bandwidth and the least amount of spurious radiation fabrication. The pattern of coupling is managed by the length of the feeding stub and the width of the patch by using a capacitive coupling technique. Because of the two dielectric layers that must be aligned properly

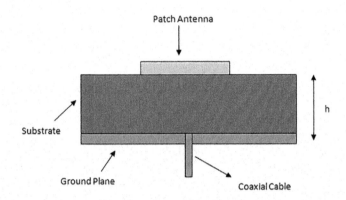

FIGURE 15.9 Coaxial probe feed patch antenna.

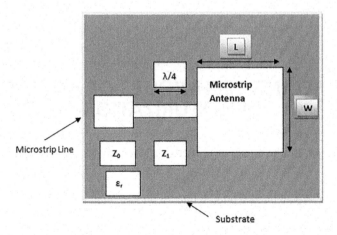

FIGURE 15.10 Microstrip line feed patch antenna.

15 • Miniaturization Techniques   255

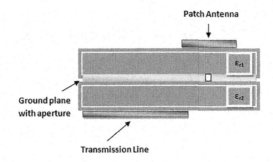

FIGURE 15.11 Aperture coupling feed patch antenna.

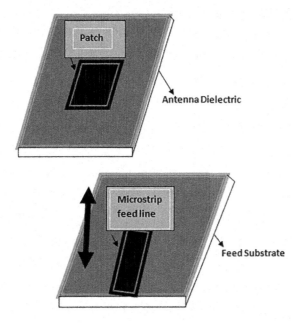

FIGURE 15.12 Proximity coupling feed patch antenna.

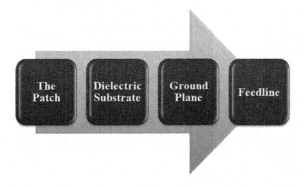

FIGURE 15.13 Components of patch antenna.

as illustrated in Figure 15.12, this feeding technique has a significant disadvantage in terms of fabrication. The antenna's overall thickness has also increased.

In the extensive range of antenna models, different structures of Microstrip antennas were defined with four basic parts which are shown in Figure 15.13.

5 APPLICATIONS OF MPA

MPA are well-known for their high performance, as well as their durable design, manufacture, and wide use. They overcome the limitations such as ease of design, lightweight, etc. Applications include medical applications, satellites, and, of course, military systems such as rockets, aircraft, missiles, and so on. Microstrip antennas are being used in a wide range of sectors and applications, and they are presently blooming in the commercial sector [86–91]. It is also projected that, as patch antennas become more widely used in a wide range of applications, they will eventually supplant traditional antennas for the majority of uses. As demonstrated in Figure 15.14, the microstrip patch antenna offers a variety of uses.

A. **Mobile and satellite communication application:** Basic requirements for mobile communication are low-cost and small-size antennas. To fulfill such requirements, circularly polarized radiation patterns are prepared for satellite communication and can be achieved using either a square or circular patch with one or two feed points.
B. **Global positioning system applications:** MPA with high permittivity value are now used for global positioning systems. Due to their placement, these antennas are circularly polarized, very tiny, and quite expensive.
C. **Radio frequency identification (RFID):** RFID has applications in a variety of fields, including logistics, manufacturing, mobile communication, transportation, and health care. Depending on the requirements, RFID systems use frequencies ranging from 30 Hz to 5.8 GHz. A tag or transponder and a transceiver or reader are the basic components of an RFID system.
D. **Worldwide interoperability for microwave access (WiMax):** WiMax refers to the IEEE 802.16 standard. It has a theoretical range of 30 miles and a data throughput of 70 Mbps. MPA attains the resonant modes at 2.7, 3.3, and 5.3 GHz, allowing it to be used in WiMax-compliant communication devices.
E. **Radar application:** Moving targets, such as people and automobiles, can be detected using radar. Microstrip antennas are an excellent alternative for low-cost, lightweight antenna systems. In comparison to conventional antennas, photolithography-based manufacturing technology allows for the mass manufacture of microstrip antennas with a reproducible performance at a lower cost and in a shorter time frame.

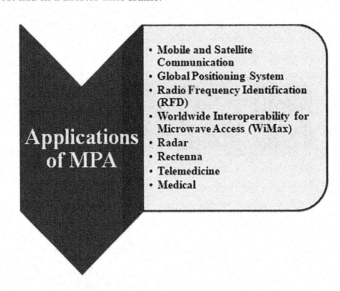

FIGURE 15.14 Applications of Microstrip patch Antenna (MPA).

F. **Rectenna application:** Rectenna is a rectifying antenna, which is a form of antenna that converts microwave energy directly into DC electricity. Antenna, ore rectification filter, rectifier, and post rectification filter are the four subsystems that make up a rectenna. To satisfy the demands of long-distance communications, it is required to develop antennas with extremely high directional qualities in rectenna applications to attain the goal to transfer DC power over great distances via wireless networks.
G. **Telemedicine application:** The antenna used in telemedicine applications operates at 2.45 GHz. A microstrip antenna that may be worn is suited for a Wireless Body Area Network (WBAN). In comparison to other antennas, the proposed antenna has a higher gain and front-to-back ratio, as well as a semi-directional radiation pattern, to avoid unnecessary radiation to the user's body and meets the requirements for on-body and off-body applications. Telemedicine applications can benefit from an antenna with a gain of 6.7 dB and an F/B ratio of 11.7 dB that resonates at 2.45GHz.

6 MEDICINAL APPLICATIONS OF THE PATCH

Microwave energy is shown to be the most effective means of producing hyperthermia in the treatment of malignant tumors. The design of the specific radiator that will be utilized for this function should be low weight, easy to handle, and durable. The printed dipoles and annular rings used in the earliest designs for the Microstrip radiator for producing hyperthermia were designed on the S-band. Later, in L-band, the design was based on a circular microstrip disc. The gadget performs a simple operation in which two connected Microstrip lines are separated by a flexible separator and the temperature inside the human body is measured. In the illustration below, a 430 MHz flexible patch applicator may be shown.

7 OUTLOOK

The development of compact antennas with multi-band operation remains a major problem in the advancement of communication devices and systems. Mirroring techniques have been used to reduce the size of both planar and non-planar antenna designs. PIFAs, stacked patches, and folded monopoles were all considered as well as shorted microstrip patches with notches and slots. To accomplish dual-band functioning, two band frequencies were included. There are still many studies to be done. New and growing wireless services require small antennas. There are no ways to get around the fundamental restrictions that relate to antenna size and electrical properties. Antenna engineers' innovation and breakthroughs in small antenna technology, on the other hand, open up new applications for better antenna designs.

REFERENCES

[1] A.M. Abbosh, S. Member, Miniaturization of Planar Ultrawideband Antenna via Corrugation, *IEEE Antennas Wireless Propag. Lett.* 7 (2009) 685–688.

[2] J.J. Adams, E.B. Duoss, T.F. Malkowski, M.J. Motala, B.Y. Ahn, R.G. Nuzzo, J.T. Bernhard, J.A. Lewis, Conformal Printing of Electrically Small Antennas on Three-Dimensional Surfaces, *Adv. Mater.* 23 (2011) 1335–1340.

[3] T.M. Factors, K. Buell, S. Member, H. Mosallaei, S. Member, K. Sarabandi, A Substrate for Small Patch Antennas Providing, *IEEE Antennas Wireless Propag. Lett.* 54 (2006) 135–146.

[4] M. Fallahpour, R. Zoughi, Antenna Miniaturization Techniques, *IEEE Antennas Wireless Propag. Lett.* (2018).
[5] A. Saini, A. Thakur, P. Thakur, Matching Permeability and Permittivity Bandwidth Miniaturized Antenna, *J. Mater. Sci. Mater. Electron.*, 27 (2016) 2816–2823.
[6] A. Saini, A. Thakur, P. Thakur, Effective Permeability and Miniaturization Estimation of Ferrite-loaded Microstrip Patch Antenna, *J. Electron. Mater.* 45 (2016) 4162–4170.
[7] H.A. Wheeler, Fundamental Limitations of Small Antennas, *Proc. IRE* 35 (12) (1947) 1479–1484.
[8] L.J. Chu, Physical Limitations of Omnidirectional Antennas, *J. Appl. Phys.* 19 (12) (1948) 1163.
[9] G.B. Gentili, P. Piazzesi, C. Salvador, K.L. Wong, W.S. Chen, K.L. Wong, M. Opt, R.C. Hansen, M. Burke, Antennas with Magneto-dielectrics, *Microw. Opt. Technol. Lett.* 26 (2000) 75–78.
[10] J.S. McLean, A Re-examination of the Fundamental Limits on the Radiation Q of Electrically Small Antennas, *IEEE Trans. Antennas Propag.* 44 (5) (1996) 672.
[11] I. Bibyk, R. Romanofsky, E. Wintucky, RF Technologies for Advancing Space Communication Infrastructure, *IEEE Trans. Antennas Propag.* (2006) 1–9.
[12] J. Trajkovikj, J. Zürcher, A.K. Skrivervik, Soft and Flexible Antennas on Permittivity Adjustable PDMS Substrates, *Loughborough Antennas & Propagation Conference* (2012).
[13] S.E. Lyshevski, I. Puchades, L.F. Fuller, Emerging MEMS and Nano Technologies: Fostering Scholarship, STEM Learning, Discoveries and Innovations in Microsystems, *IEEE Trans. Antennas Propag.* (2012).
[14] S. Bhukal, T. Namgyal, S. Mor, S. Bansal, S. Singhal, Structural, Electrical, Optical and Magnetic Properties of Chromium Substituted Co-Zn Nanoferrites Co 0.6Zn 0.4Cr xFe 2-xO 4 (0 ≤ x ≤ 1.0) Prepared via Sol-gel Auto-combustion Method, *J. Mol. Struct.*, 1012 (2012) 162–167.
[15] Y. Horii, M. Tsutsumi, Harmonic Control by Photonic Bandgap on Microstrip Patch Antenna, *IEEE Microw. Guided Wave Lett.* 9 (1) (1999) 13–15.
[16] R.B. Waterhouse, Small Microstrip Patch Antenna, *Electron. Lett.* 31 (1995) 604–605.
[17] P.M.T. Ikonen, S. Member, K.N. Rozanov, A. V Osipov, P. Alitalo, S.A. Tretyakov, S. Member, Magnetodielectric Substrates in Antenna Miniaturization: Potential and Limitations, *IEEE Trans. Antennas Propag.* 54 (2006) 3391–3399.
[18] J.P. Gianviffwb, Y. Rahmat-samii, A Novel Antenna Miniaturization Technique, D Applications, *Int. J. Recent Trends Eng.* 44 (2002).
[19] R. Azadegan, S. Member, K. Sarabandi, A Novel Approach for Miniaturization of Slot Antennas, *IEEE Trans. Antennas Propag.* 51 (2003) 421–429.
[20] M. Antenna, D. Handbook, Microstrip Antenna Design Handbook Ramesh Garg Abstract – Recently, Microstrip Patch Antenna is Widely Used in Such Applications Like (6) Ramesh Garg, Prakash Bhartia, Inder Bahl, Apisak Ittipiboon. "*Microstrip Antenna Design Handbook*", Artech Ho (2005) 3–6.
[21] C.A. Palacio Gómez, J.J. McCoy, M.H. Weber, K.G. Lynn, Effect of Zn for Ni Substitution on the Properties of Nickel-Zinc Ferrites as Studied by Low-energy Implanted Positrons, *J. Magn. Magn. Mater.* 481 (2019) 93–99.
[22] R. Bancroft, Radiation Properties of an Omnidirectional Planar, 50 (2008) 55–58.
[23] S.W. Lee, C.S. Kim, Superparamagnetic Properties Ni-Zn Ferrite for Nanobio Fusion Applications, *J. Magn. Magn. Mater.* 304 (2006) 418–420.
[24] A. Thakur, J. Hsu, Novel Magnetodielectric Nanomaterials with Matching Permeability and Permittivity for the Very-High-Frequency Applications, *Scr. Mater.* 64 (2011) 205–208.
[25] A.O. Karilainen, C.R. Simovski, S.A. Tretyakov, A.N. Lagarkov, S.A. Maklakov, K.N. Rozanov, S.N. Starostenko, Experimental Studies on Antenna Miniaturisation Using Magneto-dielectric and Dielectric Materials (2011) 495–502.
[26] S. Sharma, K.S. Daya, S. Sharma, K.M. Batoo, M. Singh, Sol – Gel Auto Combustion Processed Soft Z-type Hexa Nanoferrites for Microwave Antenna Miniaturization, *Ceram. Int.* 41 (2015) 7109–7114.
[27] P. Mathur, A. Thakur, M. Singh, S. Hill, Impact of Processing and Polarization on, *Int. J. Mod. Phys. B* 23 (2009) 2523–2533.
[28] P.M. Ã, A. Thakur, M. Singh, Nanoferrite for High-frequency Applications, *J. Appl. Phys.* 69 (2008) 187–192.
[29] H. Mosallaei, K. Sarabandi, Antenna Miniaturization and Bandwidth Enhancement Using a Reactive Impedance Substrate, *IEEE Trans. Antennas Propag.* 52 (9) (2004) 2403–2414.
[30] V. Voronkov, Microwave Ferrites: The Present and the Future, *J de Physique IV Proc. EDP Sci.* 07 (C1) (1997) C1-35–C1-38.
[31] K. Rana, P. Thakur, A. Thakur, M. Tomar, V. Gupta, J. Luc, P. Queffelec, In Fluence of Samarium Doping on Magnetic and Structural Properties of M Type Ba – Co Hexaferrite, *Ceram. Int.* (2016) 1–6.
[32] H.E. Zhang, B.F. Zhang, G.F. Wang, X.H. Dong, Y. Gao, The Structure and Magnetic Properties of Zn 1-x Ni x Fe 2 O 4 Ferrite Nanoparticles Prepared by Sol-gel auto-Combustion, *J. Magn. Magn. Mater.* 312 (2007) 126–130.

[33] J.H. Nam, Y.H. Joo, J.H. Lee, J.H. Chang, J.H. Cho, M.P. Chun, B.I. Kim, Preparation of NiZn-ferrite Nanofibers by Electrospinning for DNA Separation, *J. Magn. Magn. Mater.* 321 (2009) 1389–1392.

[34] D.F. Sievenpiper, D.C. Dawson, M.M. Jacob, S. Member, T. Kanar, S. Member, S. Kim, J. Long, R.G. Quarfoth, S. Member, Experimental Validation of Performance Limits and Design Guidelines for Small Antennas, *Electron. Lett.* 60 (2012) 8–19.

[35] A.R. Raslan, S. Member, A.M.E. Safwat, N-Internal Port Design for Wide Band Electrically Small Antennas with Application for UHF Band, *IET Microw. Antennas Propag.* 61 (2013) 4431–4437.

[36] A.A. Salih, A Miniaturized Dual-Band Meander Line Antenna for RF Energy Harvesting Applications (2015).

[37] E.A.M. Souza, P.S. Oliveira, A.G.D. Assunção, L.M. Mendonça, C. Peixeiro, Miniaturization of a Microstrip Patch Antenna with a Koch Fractal Contour Using a Social Spider Algorithm to Optimize Shorting Post Position and Inset Feeding, *IEEE Jordan Conference on Applied Electrical Engineering and Computing Technologies* (2019).

[38] A. Morisako, T. Naka, K. Ito, A. Takizawa, M. Matsumoto, Y.K. Hong, Properties of Ba-ferrite/AlN Double Layered Films for Perpendicular Magnetic Recording Media, *J. Magn. Magn. Mater.* 242–245 (2002) 304–310.

[39] A. Natarajan, S. Member, A. Komijani, S. Member, X. Guan, A 77-GHz Phased-Array Transceiver with On-Chip Antennas in Silicon: Transmitter and Local LO-Path Phase Shifting, *IEEE J. Solid-State Circuits* 41 (2006) 2807–2819.

[40] P. Momenroodaki, Z. Popovic, R. Scheeler, A 1. 4-GHz Radiometer for Internal Body Temperature Measurements, *European Microwave Conference* (2015) 694–697.

[41] P. Momenroodaki, R.D. Fernandes, Z. Popovi, Air-substrate Compact High Gain Rectennas for Low RF Power Harvesting, *European Conference on Antennas & Propagation* (2016).

[42] J.S. Kula, D. Psychoudakis, W. Liao, C. Chen, J.L. Volakis, J.W. Halloran, Patch-antenna Miniaturization Using Recently Available Ceramic Substrates, *IEEE Antennas Propag. Mag.* 48 (6) (2006) 13–20.

[43] J.S. Kula, D. Psychoudakis, W. Liao, C. Chen, J.L. Volakis, J.W. Halloran, Patch-Antenna Miniaturization Using Recently Available Ceramic Substrates, *IEEE Antennas Propag. Mag.* 48 (6) (2006) 13–20.

[44] P. Mookiah, S. Member, K.R. Dandekar, S. Member, Metamaterial-Substrate Antenna Array for MIMO Communication System, *IEEE Antennas Propag. Mag.* 57 (2009) 3283–3292.

[45] A. Thakur, M. Singh, Preparation and characterization of nanosize $Mn_{0.4}Zn_{0.6}Fe_2O_4$ Ferrite by Citrate Precursor Method, *Ceramics Int.* 29 (2003) 505–511.

[46] N. Altunyurt, M. Swaminathan, P.M. Raj, V. Nair, Antenna Miniaturization using Magneto-Dielectric Substrates, *59th Electronic Components and Technology Conference* (2009) 801–808.

[47] J.B. Pendry, A.J. Holden, D.J. Robbins, W.J. Stewart, Magnetism from Conductors and Enhanced Nonlinear Phenomena, *IEEE Trans. Microw. Theory Techn.* 47 (11) (1999) 2075–2084.

[48] N. Engheta, R.W. Ziolkowski, A Positive Future for Double-Negative Metamaterials, *IEEE Trans. Microw Theory Techn.* 53 (2005) 1535–1556.

[49] R.W. Ziolkowski, A.D. Kipple, S. Member, Application of Double Negative Materials to Increase the Power Radiated by Electrically Small Antennas, *IEEE Trans. Antennas Propag.* 51 (2003).

[50] F. Qureshi, M.A. Antoniades, G.V. Eleftheriades, A Compact and Low-Profile Metamaterial Ring Antenna with Vertical Polarization, *IEEE Antennas Wireless Propag. Lett.* 4 (2005) 333–336.

[51] R.W. Ziolkowski, A. Erentok, Metamaterial-Based Efficient Electrically Small Antennas, *IEEE Trans. Antennas Propag.* 54 (2006) 2113–2130.

[52] R.O. Ouedraogo, E.J. Rothwell, A.R. Diaz, K. Fuchi, A. Temme, Miniaturization of Patch Antennas Using a Metamaterial-Inspired Technique, *IEEE Trans Antennas Propag.* 60 (5) (2012) 2175–2182.

[53] F. Farzami, S. Member, K. Forooraghi, Miniaturization of a Microstrip Antenna Using a Compact and Thin Magneto-Dielectric Substrate, *IEEE Trans. Antennas Propag.* 10 (2012) 1540–1542.

[54] D. Psychoudakis, J.L. Volakis, J.H. Halloran, Design Method for Aperture-Coupled Microstrip Patch Antennas on Textured Dielectric Substrates, *IEEE Trans Antennas Propag.* 52 (2004) 70–72.

[55] R. Li, S. Member, G. Dejean, S. Member, M.M. Tentzeris, S. Member, J. Laskar, S. Member, Development and Analysis of a Folded Shorted-Patch Antenna with Reduced Size, *IEEE Trans. Antennas Propag.* 52 (2004) 555–562.

[56] C.Y. Chiu, C.H. Chan, K.M. Luk, Study of a Small Wide-band Patch Antenna with Double Shorting Walls, 3 (2004) 230–231.

[57] A. Holub, M. Polivka, A Novel Microstrip Patch Antenna Miniaturization Technique: A Meanderly Folded Shorted-Patch Antenna, *2008 14th Conference on Microwave Techniques* (2008) 1–4.

[58] A.K. Shackelford, K. Lee, K.M. Luk, Design of Small-Size Wide-Bandwidth Microstri p-Patch Antennas, *IEEE Antennas Propag. Soc.* 75–83.

[59] S. Moon, H. Ryu, J. Woo, H. Ling, Miniaturisation of 1/4 Microstrip Antenna Using Perturbation Effect and Plate Loading for Low-VHF-Band Applications, *Electron. Lett.* 47 (2011) 0–1.

[60] R. Porath, Theory of Miniaturized Shorting-post Microstrip Antennas, *IEEE Trans. Antennas Propag.* 48 (1) (2000) 41–47.
[61] R.R. Nayak, N. Pradhan, D. Behera, K.M. Pradhan, S. Mishra, L.B. Sukla, B.K. Mishra, Green Synthesis of Silver Nanoparticle by Penicillium Purpurogenum NPMF: The Process and Optimization, *J. Nanoparticle Res.* 13 (2011) 3129–3137.
[62] S. Wang, H.W. Lai, K.K. So, K.B. Ng, Wideband Shorted Patch Antenna With a Modified Half U-Slot, *Prog. Electromag. Res.* 11 (2012) 689–692.
[63] R.B. Waterhouse, S.D. Targonski, D.M. Kokotoff, Design and Performance of Small Printed Antennas, *IEEE Trans. Antennas Propag.* 46 (11) (1998) 1629–1633.
[64] W. Li, L. Fa-Shen, Structural and Magnetic Properties of Co 1- x Zn x Fe 2 O 4 Nanoparticles, *Chinese Phys. B.* 17 (2008) 1858–1862.
[65] N. Herscovici, M.F. Osorio, C. Peixeiro, Miniaturization of Rectangular Microstrip Patches Using Genetic Algorithms, *IEEE Antennas Wireless Propag. Lett.* 1 (2002) 94–97.
[66] H. Oraizi, S. Hedayati, Miniaturization of Microstrip Antennas by the Novel Application of the Giuseppe Peano, *IEEE Trans. Antennas Propag.* 60 (2012) 3559–3567.
[67] S.I. Latif, L. Shafai, C. Shafai, An Engineered Conductor for Gain and Efficiency Improvement of Miniaturized Microstrip Antennas, *IEEE Antennas Propag. Mag.* (2013).
[68] C.G. Kakoyiannis, P. Constantinou, A Compact Microstrip Antenna with Tapered Peripheral Slits for CubeSat RF Payloads at 436MHz: Miniaturization Techniques, Design & Numerical Results, *IEEE International Workshop on Satellite and Space Communications* (2008) 1–5.
[69] J. Anguera, L. Boada, C. Puente, C. Borja, J. Soler, Stacked H-shaped Microstrip Patch Antenna, *IEEE Trans. Antennas Propag.* 52 (4) (2004) 983–993.
[70] S.A. Bokhari, J.-F. Zurcher, J.R. Mosig, F.E. Gardiol, A Small Microstrip Patch Antenna with a Convenient Tuning Option, *IEEE Trans. Antennas Propag.* 44 (11) (1996) 1521–1528.
[71] A. Faramarzi, K.L. Wong, A. Imani, J. Nourinia, M. Ojaroudi, R.A. Sadeghzadeh, M. Ojaroudi, A. Corporation, S. Chatterjee, K. Ghosh, J. Paul, S.K. Chowdhury, P.P. Sarkar, W. Bengal, W. Bengal, W. Bengal, Compact Microstrip Antenna for Mobile Communication, *IEEE Trans Antennas Propag.* 55 (2013) 954–957.
[72] W. Chen, C. Wu, K. Wong, S. Member, Square-Ring Microstrip Antenna with a Cross Strip for Compact Circular Polarization Operation, *ETRI J.* 47 (1999) 1566–1568.
[73] H. Iwasaki, A Circularly Polarized Small-Size Microstrip Antenna with a Cross Slot, *IEEE Trans. Antennas Propag.* 44 (10) (1996) 1399–1401.
[74] O.T. Letters, J. Wiley, S. Dey, R. Mittra, C. Engineermg, Compact Microstrip Patch Antenna, *Microw. Opt. Technol. Lett.* 13 (1997) 12–14.
[75] X. Zhang, F. Yang, Study of a Slit Cut on a Microstrip Antenna and Its Applications, *Microw. Opt. Technol. Lett.* 18 (1998) 297–300.
[76] H.T. Nguyen, S. Noghanian, L. Shafai, Microstrip Patch Miniaturization by Slots Loading, *IEEE Antennas and Propagation Society International Symposium* (2005) 215–218.
[77] A.R. Conductance, Rectangular Microstrip Patch Antennas with Infinite and Finite Ground Plane Dimensions (1983), *IEEE Trans. Antennas Propag.*
[78] A.K. Bhattacharyya, Effects of Finite Ground Plane on the Radiation Characteristics of a Circular Patch Antenna, *IEEE Trans. Antennas Propag.*, 38 (2) (1990) 152–159.
[79] I. Introduction, The Finite Ground Plane Effect on the Microstrip Antenna Radiation Patterns, *IEEE Trans. Antennas Propag.* (1983) 649–653.
[80] J.H. Cho, S.W. Yun, K.J. Chen, W. Menzel, M.J. Lancaster, H.S. Kim, T.S. Hyun, S.S. Kwoun, H.G. Kim, L.E. Davis, Y.K. Yoon, B. Lee, J.J. Choi, J.Y. Kim, J.C. Lee, S. Sarkar, A. Das Majumdar, S. Mondal, S. Biswas, D. Sarkar, P.P. Sarkar, W. Bengal, Miniaturization of Rectangular Microstrip Patch Antenna Using Optimized Single-Slotted Ground, *World J. Rhetumatol.* 53 (2011) 111–115.
[81] S.O. Kundukulam, M. Paulson, C.K. Anandan, P. Mohanan, K. Vasudevan, A Circular-Sided Compact Microstrip Antenna, *IEEE Antennas and Propagation Society International Symposium (IEEE Cat. No.02CH37313)*, vol. 1 (2002) 38–41.
[82] H.V. Prabhakar, U.K. Kummuri, R.M. Yadahalli, V. Munnappa, Effect of Various Meandering Slots in Rectangular Microstrip Antenna Ground Plane for Compact Broadband Operation, *Electron. Lett.* 43 (2007) 16–17.
[83] L. Chen, J. Lu, C.Y. Tan, R. Forse, M.S. Dig, B. Avenhaus, M.J. Lancaster, P. Inst, E. Eng, C.H. Loh, C.H. Poh, J. Kuo, K. Wong, A Compact Microstrip Antenna with Meandering Slots in the, *Microw. Opt. Technol. Lett.* 29 (2001) 95–97.
[84] S.F. Mahmoud, A New Miniaturized Annular Ring Patch Resonator Partially Loaded by a Metamaterial Ring With Negative Permeability and Permittivity, *IEEE Antennas Wireless Propag. Lett.* 3 (2004) 19–22.

[85] A. Alù, S. Member, F. Bilotti, S. Member, N. Engheta, Subwavelength, Compact, Resonant Patch Antennas Loaded With Metamaterials, *IEEE Trans. Antennas Propag.* 55 (2007) 13–25.
[86] M.S. Sharawi, S. Member, M.U. Khan, S. Member, A.B. Numan, D.N. Aloi, S. Member, A CSRR Loaded MIMO Antenna System for ISM Band Operation, *IEEE Trans. Antennas Propag.* 61 (2013) 4265–4274.
[87] M.M. Bait-Suwailam, Size Reduction of Microstrip Patch Antennas Using Slotted Complementary Split-Ring Resonators, *The International Conference on Technological Advances in Electrical, Electronics and Computer Engineering, TAEECE* 2013 (2013) 528–531.
[88] Y.-L. Ban, C. Li, C.-Y.-D. Sim, G. Wu, K.-L. Wong, 4G/5G Multiple Antennas for Future Multi-Mode Smartphone Applications, *IEEE Access* 4 (2016) 2981–2988.
[89] K. Goodwil, V.N. Saxena, M.V. Kartikeyan, Dual Band CSSRR Inspired Microstrip Patch Antenna for Enhancing Antenna Performance and Size Reduction, *Prog. Electromagn. Res.* 495–497.
[90] X. Cheng, S. Member, D.E. Senior, S. Member, C. Kim, A Compact Omnidirectional Self-Packaged Patch Antenna With Complementary Split-Ring Resonator Loading for Wireless Endoscope Applications, *IEEE Antennas Wireless Propag. Lett.* 10 (2012) 1532–1535.
[91] Y. Liu, F. Min, T. Qiu, M. Zhang, C. Xu, Effect of Zn^{2+} Content on the Microstructure and Magnetic Properties of Nanocrystalline $Ni_{1-x}Zn_xFe_2O_4$ Ferrite by a Spraying-Coprecipitation Method, *J. Wuhan Univ. Technol. Mater. Sci. Ed.* 25 (2010) 429–431.

Applications of Magnetic Materials in Batteries

16

Shiva Bhardwaj, Felipe M. de Souza, Ram K. Gupta

Contents

1 Introduction   263
2 Synthesis of Nanostructured Magnetic Materials   264
3 Working Principle of Batteries   266
4 Nanostructured Magnetic Materials for Batteries   269
    4.1 Nanostructured Magnetic Materials for Metal-Ion Batteries   269
    4.2 Nanostructured Magnetic Materials for Metal-Sulfur Batteries   273
    4.3 Nanostructured Magnetic Materials for Metal-Air Batteries   276
5 Conclusion   279
References   280

1 INTRODUCTION

The constantly increasing demand for smaller and lighter portable electronic devices led to the establishment of efficient energy storage systems with high-power, high-energy-density, and long-life span which became important in our daily life for comfort and practicality. The idea of electrochemical energy storage began with a scientific investigation into electricity during 1789 from where the electrochemical cell and then battery consisting of two or more electrochemical cells were developed [1]. These devices are termed electrochemical energy storage devices which are of different varieties of range like batteries, supercapacitor, and fuel cells among which batteries are the popular source for the storage of energy from the generations as they have good reversibility and pollution-free operations. Scientists are also focusing on rechargeable batteries as means to improve their cyclability which is a desired property for both small devices as well as heavy storage systems. During the early days of batteries development, undercharging or overcharging problems generally occurred along with fast-fading capacity, dissolution of cathode, and anode materials in the electrolytic solution, and formation of dendrites during recharging. The latter causes an increase in battery size leading to a deterioration in performance as well as leakage of electrolyte [2].

    One of the ways to address such issues came through nanotechnology which includes a plethora of nanostructures i.e., nanowires, nanobelts, nanotubes, nanorods, among others which can potentially present superior properties compared to their bulky counterparts. On top of that, the use of magnetic materials can also lead to an enhancement of energy storage properties. These nanostructured magnetic materials offer direct current pathways due to an extra number of electrons compared to non-magnetic materials. Also, ion-diffusion length is greatly shortened due to which rate performance increases as the amount of time required by the ions to diffuse through the electrode material depends on the diffusion length and diffusion coefficient [3]. Their high surface area enables a large electrolyte-electrode contact area which reduces the charge timing. A long-life cycle was enabled by these materials as they can accommodate the volume expansion and reduce mechanical degradation. With the progress in energy storage technology, many next-generation rechargeable batteries such as metal-ion batteries (MIBs), metal-sulfur batteries (MSBs), and metal-air batteries (MABs) were developed. The properties of electrode materials and thus the performance of these batteries can be tuned by tailoring the chemical, electronic, and morphological properties of electrode materials. In the following sections, various synthesis routes and energy applications of magnetic materials are discussed.

2 SYNTHESIS OF NANOSTRUCTURED MAGNETIC MATERIALS

Nanostructured magnetic materials arrived as a novel technology for the development of batteries. These materials can be synthesized through several well-known techniques such as physical (mechanical, physical vapor deposition) and chemical (vapor condensation, micellular, chemical reduction, sonochemical, sol-gel) methods. Owing to the simple setup and facile operation, the hydrothermal and solvothermal methods have been widely used in the synthesis of various nanomaterials. Typically, the synthesis is performed above ambient temperature and pressure in aqueous (hydrothermal method) [4] or non-aqueous (solvothermal method) [5] media sealed in Teflon-lined stainless-steel autoclaves. Importantly, the reaction can be carried out at a temperature beyond the normal boiling point of solvents, which is inaccessible in other synthesis methods. At relatively high temperatures and high pressure, the reactivity and solubility of reactants can be greatly increased, resulting in variations of physicochemical properties of solvents. Accompanied with the optimum ratio of reacting species, such as precursors, reductants, and ligands, the nucleation and growth processes of metal nanocrystals could be regulated to form uniform products with fine-tuned morphology and size. The shortage of this method lies in its difficulty in real-time observation for the shape evolution process of metal nanocrystals.

Moreover, compared to other wet-chemical methods, hydrothermal and solvothermal methods are relatively faster. The hydrothermal method, which uses water as a solvent, has been successfully applied to synthesize various nanomaterials including different dimensions like Ag nanoplates, Co nanosheets, Ni nanoplatelets and can be applied to other magnetic materials by simply changing the reaction time and temperature. The solvothermal method adopts non-aqueous solvents like benzyl-alcohol, formamide, ethylene glycol, and di-methyl fluoride (DMF). In composite metallic systems, desired energy properties can be attained through readily varied ratios of oxidizer and fuels. A complete balance between the oxidizer and fuel may be reached to maximize energy density. However, due to the granular nature of composite energetic materials, reaction kinetics are typically controlled by the mass transport rates between reactants. In composite energetic materials, decreasing reactant sizes effectively increase the interfacial surface area which is important for the storage of energy.

Sol-gel chemical methodology has been investigated over years. Sol-gel chemistry is a solution-based synthetic route to highly pure organic or inorganic materials that have homogeneous particle and pore sizes as well as densities [6]. The method is commonly used to prepare metal-oxide-based materials. However, sol-gel methods do exist for the preparation of organic fuel-based materials so they can also be used to prepare nanomaterials of both oxidizers and fuels. Its benefits include the convenience of low-temperature preparation using general and inexpensive laboratory equipment. From a chemical point of view, the method affords easy control over stoichiometry and homogeneity than conventional methods lack. In addition, one of the integral features of the method is its ability to produce materials with special shapes such as monoliths, fibers, films, and powders of uniform and very small particle sizes. Sol can be formed through the hydrolysis and condensation of dissolved molecular precursors. This produces nanometer-sized particles, which aggregate to form clusters, with very uniform size, morphology, and composition. The pH of the solution, solvent, temperature, and die concentrations of reactants used dictate the die size of the clusters, which can be from 1 to 1000 nm in diameter. By controlling the solution conditions, the sol can be condensed into a robust gel. This process can lead to sol clusters into either aggregates or linear chains which results in the formation of the stiff monolith. The gel can be dried by evaporation of the solvent to produce a monolithic xerogel or it can be removed under supercritical conditions to produce an aerogel. Ambient drying results in the exertion of large capillary forces on the gel framework and causes a significant amount of shrinkage of the material to produce a medium-density material. Later, with supercritical drying high-density material is proposed. Fe and Al-based compounds can also be obtained through the sol-gel technique.

Electrospun is another technique commonly used for the preparation of nanofibers of magnetic materials. Electrospray process is another variation of this procedure. In both procedures, liquid jets are ejected by using the high-voltage-induced electrostatic force [7]. The major differences between these two electrostatic techniques rest with the liquid involved and the products obtained. High-viscosity non-Newtonian fluids, such as concentrated polymer solutions, are often employed in electrospinning to generate continuous nanofibers, instead of forming mono-dispersed particles as with electrospray that relies on low-viscosity liquids. A syringe pump is usually involved to ensure a stable and controllable infusion rate of polymeric-based solution. The forming process of electrospun nanofibers undergoes the following four consecutive procedures: (i) the polymer droplet is charged and deformed into a conical jet, (ii) the charged jet is ejected along a straight line at the tip of the "Taylor cone", (iii) the jet is further stretched and refined with bending/whipping instability in the external electric field, and (iv) the polymer nanofibers are cured and formed on the collector. Generally, a single syringe or a hollow needle is employed as the spinneret for conventional electrospinning, where polymer-based nanofibers with solid interiors are often produced. A processing image is shown in Figure 16.1 for the understanding of the electrospinning

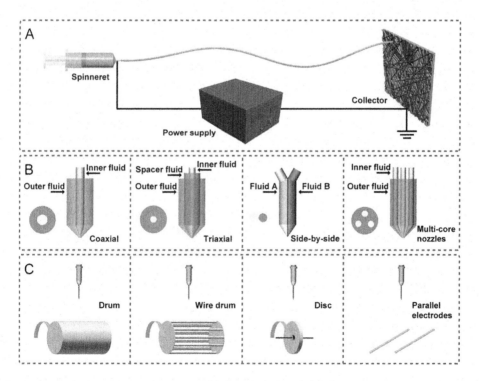

FIGURE 16.1 (a) The detailed processing unit of electrospinning device setup. (b) Spinneret type. (c) Collector on which fiber gets deposited. Adapted with permission [7].

Source: Copyright (2021), American Chemical Society.

process. Depending on the material, different parameters decide the morphology of the produced fibers. Polymers with high molecular weight produce fibers with larger diameters along with the increase in solution viscosity and surface tension of polymer solution. Low viscous solutions can lead to the formation of beads. Also, fiber's diameter tends to decrease as the temperature and humidity increase. Apart from the discussed methods, there is a variety of other techniques that can be used for the synthesis of nanostructured magnetic materials.

3 WORKING PRINCIPLE OF BATTERIES

The working principle of a battery is based on the conversion of chemical energy into electricity based on the reduction and oxidation reactions that occur simultaneously at the cathode and anode, respectively. That process can be used to power an electronic device due to the release of electrons. The flow of ions occurs at the electrolyte which can be composed of ionic species dissolved aqueous media, organic solvents, or dispersed in solid-state components that can transport charged species through their structural framework. On the other hand, the electrons flow through a metal wire connected at the anode and cathode as shown in Figure 16.2.

Depending on the type of battery's electrodes they can be classified into different types like MIBs e.g., Li-ion, Zn-ion, and Na-ion batteries. MSBs e.g., Li-S, Mg-S, and Al-S battery. MABs e.g., Li-air, Zn-air, Na-air, and battery, among others. MIB consists of two electrodes (anode and cathode). The cathode of MIBs is mainly composed of metallic compounds, and generally, the anode material is composed of graphite. MIBs include redox reactions that result in the movement of metallic ions back and forth between the electrodes through the electrolyte during the charging and discharging. This transportation of ions is a deciding feature for the kinetic response of a battery. It is described with the flux expressions

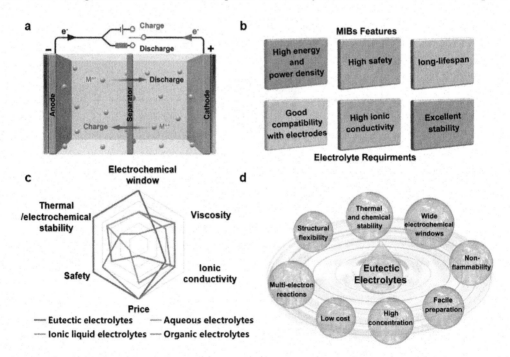

FIGURE 16.2 Schematic representation of the working principle of typical MIBs. Adapted with permission [10].
Source: Copyright (2022), American Chemical Society.

that relate ion flux to the gradient in chemical potential. During the intercalation of a metallic compound in MIB, metal flux emerges in the presence of a gradient in chemical potential according to Equation 1.

$$J_M = -L\nabla\mu_M \tag{1}$$

Where μ is the chemical potential and L is the kinetic coefficient, which measures the non-equilibrium process, and describes the fluctuations at equilibrium conditions. It also measures the metallic mobility in the crystal structure of the intercalation compound. Most electrodes for MIBs are intercalation compounds that host guest cations in the interstitial sites of their crystal. In general, the phase that emerges during the conversion reaction upon metal insertion has little or no relation with the initial crystallographic phase. The reconstructive nature of phase transformation provides the knowledge for the growing and shrinking phases that are incoherent or semi-coherent and occur due to the presence of oxygen ions in their stack [8,9].

MIBs are important for portable electronic devices as they offer higher energy density than other rechargeable systems. Still, there is a further requirement in increasing the energy density, which is currently limited by the liquid electrolyte in the MIBs. To overcome the charge storage limitations of insertion-compound electrodes, materials that undergo conversion reactions while accommodating more ions and electrons are becoming a promising option. In that sense, Li-S batteries have drawn the attention of the scientific community as Li-S-based electroactive materials can offer a high theoretical capacity of around 1672 mAh/g, leading towards the development of MSBs. Electrical energy is stored in the sulfur electrodes of the MSB. A metal-sulfur cell consists of a metallic anode, an organic electrolyte, and a sulfur composite cathode. During the initial phase, sulfur is in the charged state, due to which the cell operation starts with the discharge. Metal gets oxidized at the negative electrode during the discharge to produce metallic ions and electrons. These metal ions produced the move to the positive electrode through the electrolyte internally. At the same time, electrons travel towards the positive electrode externally, resulting in the generation of the electric current at which the schematic of this process is presented in Figure 16.3. Therefore, sulfur is reduced to produce metal-sulfide by

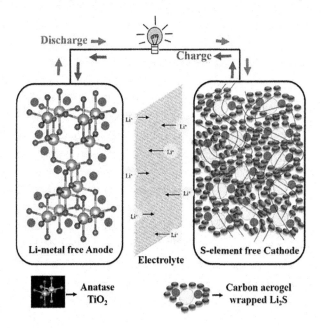

FIGURE 16.3 The Schematic representation of working of MSBs with an example of LSBs. Adapted with permission [11].

Source: Copyright (2021), American Chemical Society.

accommodating the metal ions and electrons at the positive electrode. However, one of the challenges lies in the strong tendency of sulfur to catenate leading to the formation of long homoatomic chains or homocyclic rings of various sizes. Within that line, S_8 (octa-sulfur) is the most stable allotrope at room temperature. During the ideal discharge process, S_8 is reduced, and the ring opens, resulting in high-order metallic polysulfides.

MSBs meet the requirement of specific energy, specific power, and low-temperature performance. However, their rate capability and recharge time are barely met and far below the minimum requirements, limiting their commercial applications. Hence, further research is still required. Alongside that line, another viable technology for energy storage is the MABs which can potentially present higher specific energy than state of art MIBs. In general, MABs present a hybrid architecture that includes the combined feature of both batteries and fuel cells with its components presented in Figure 16.4. They are composed of a metallic anode and air (O_2) as the active cathode mass and an M^+ ion-containing electrolyte solution. The cathode in this system is a composite electronically conducting porous matrix that enables the electrochemical contact between O_2 and the metal-ion in the electrolyte solution phase. O_2 electrochemistry is dependent on the nature of solvent and electrolyte salt, and it shows reversible redox behavior in some cases.

One of the main parameters that influence the mechanism of O_2 reduction to metal-oxide is the donor number (DN) of solvent and salt as they determine the solubility and dissociation of metal-oxide. Low DN indicates the growth of metal-oxide on the electrode whereas, the high value of DN indicates its dissolution in the solution. Contaminants and additives in the electrolyte solution. A fundamental hysteresis during the oxidation mechanisms in charging reveals the continuous phase changes from metal-oxide to original metal in the reverse process. The ideal number of electrons per O_2 is determined by the given Equation 2.

$$O_2 + xe^- + xM^+ \leftrightarrow M_xO_2 \qquad (2)$$

A significant amount of side products is also formed during the charging and discharging process in MABs and are considered the parasitic reaction inside the batteries [13].

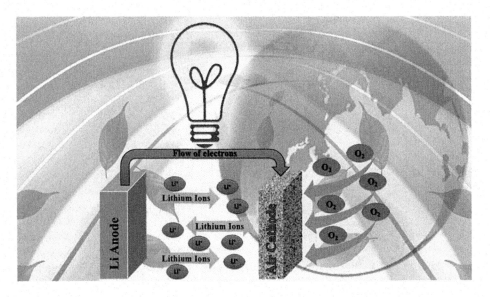

FIGURE 16.4 Schematic representation of working of MABs with an example of LABs. Reproduced with permission [12].

Source: Copyright (2019), American Chemical Society.

4 NANOSTRUCTURED MAGNETIC MATERIALS FOR BATTERIES

4.1 Nanostructured Magnetic Materials for Metal-Ion Batteries

Developing new methods for preparing nanomaterials and modifying their size, morphology, and porosity has been pursued as a fundamental scientific interest and many technological applications. Nanoparticles and nanowires/nanorods with controlled size and shape are important because most materials' electrical, optical, and magnetic properties strongly depend on their size and shape. Materials play an important role in developing different batteries or energy storage devices depending on their working principle. All MIBs such as LIBs, ZIBs, SIBs, among others work on the same principle, leading to the importance of the materials used. In that sense, LIB is the current technology that dominates the market. Yet, the scientific community is working on another alternative to diversify the use of resources and introduce new technologies within the field of energy storage. In that sense, Wang et al. [14] developed porous nanoparticles made from iron in the form of Fe_2O_3 referred to as α-Fe_2O_3, the most stable form of the Iron-oxide family, and was synthesized using two different approaches one in the microwave (MW) assisted, and the other is hydrothermal (HT), resulting in Fe-MIL-88A-MW, and Fe-MIL-88A-HT. After calcination Fe-MIL-88A-MW was converted into Fe_2O_3-MW-4h and Fe-MIL-88A-HT into Fe_2O_3-HT as described in Figure 16.5.

The nanoparticle's small size along with its surface area allow proper interaction with the electrolyte, leading to an enhancement in the LIB's capacity. After the complete testing, the whole process of lithiation/delithiation is summarized in Equation 3.

$$Fe_2O_3 + 6Li^+ + 6e^- \leftrightarrow Fe^0 + 3Li_2O \tag{3}$$

The cyclic voltammetry (CV) curves at the scanning rate of 0.1 mV/s from 0.01 to 3 V showed a sharp reduction peak at 0.63 V during the first discharge which appeared due to the reduction process from Fe^{3+}

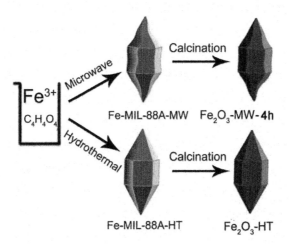

FIGURE 16.5 Schematic fabrication of Fe_2O_3 using both microwave and hydrothermal assisted methods. Reproduced with permission [14].
Source: Copyright (2021), American Chemical Society.

to Fe^0 by Li, and the formation of solid electrolyte interphase film. Also, two oxidation peaks at 1.62 V 1.88 V appeared during the oxidation process which indicated the formation of Fe^0 → Fe^{2+} → Fe^{3+}. Almost all the peaks are superimposable to each other except the first cycle. Based on these redox processes it was observed that Fe$_2$O$_3$-MW-4h and Fe$_2$O$_3$-HT presented a capacity of 1010 and 933 mAh/g, respectively after 300 cycles. The improvement in performance observed for the α-Fe$_2$O$_3$ obtained through a microwave-assisted method in comparison to hydrothermal was likely attributed to the smaller size of nanoparticles along with a multicavity in metal oxide's structure which eased the contact with the electrolyte allowing more ions to diffuse within its pores. Sn-based nanostructured materials are also towards the candidacy of the LIBs because, in principle, they can store more than twice the amount of the charge as that of Li$^+$ and are derived from oxides of tin (SnO$_2$). Sn-based anodes were prepared using a sol-gel synthesis route with a diameter of 100 nm while presenting a brush-like nanofiber structure [15]. These obtained nanostructured electrodes had far better-improved capabilities, indicating the low discharge capacities. They also exhibited high cyclability due to the slight change in absolute volume of nanofibers, and the brush-like structure that created space. At the same time, there was an expansion in the volume of each fiber and it reached 1400 charge/discharge cycles without any loss of capacity.

A large amount of nanostructured magnetic composite of ternary nickel–cobalt–manganese (in the ratio of 6:2:2) was prepared by Chang et al. [16] using the solvothermal synthesis route, which led to the formation of Ni$_{0.6}$Co$_{0.6}$Mn$_{0.2}$O$_x$ composite oxide magnetic material, which showed uniform particle distribution and regular spherical morphology. Also, nanomagnetic metal oxide endowed high capacity, excellent rate performance, and long cycle stability. The CV curves almost overlapped over the repeated cycles, which indicated high strength and the initial high coulombic efficiency of 71% with the experimental reversible capacity of 770.3 mAh/g with its theoretical capacity of 816 mAh/g. The obtained output capacity after 500 cycles was 559.8 mAh/g indicating the capacity retention of 70.3% with superior cycling stability in lithium charge storage performance compared to other magnetic metal-based anode materials leading towards high-energy-density LIBs. When combined with oxygen, a paramagnetic element, Manganese formed manganese oxide (MnO$_2$), which showed remarkable attraction towards the magnet. When prepared in nanodimension, this magnetic compound showed outstanding characteristics for the LIB. Meng et al. [17] developed a Ni-based MnO$_2$/reduced graphene oxide (RGO) composite whose schematic synthesis steps are depicted in Figure 16.6.

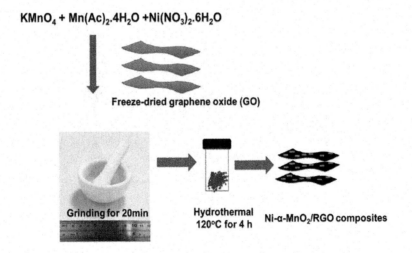

FIGURE 16.6 Schematic representation of the synthesis steps for the preparation of Ni-based MnO$_2$/RGO. Reproduced with permission [17].

Source: Copyright (2021), American Chemical Society.

Ni-doped-α-MnO$_2$ loaded on RGO functioned as a new type of nanocomposite for LIB anode applications. Also, the nanoparticles could be synthesized through a solvent-free synthesis which showed potential large-scale production. The introduction of carbon-based nanomaterials with the metal oxide enhanced its cyclability and suppressed the aggregation of non-conductive Li$_2$O, which would result in the increase of internal resistance over time, causing severe battery failure. This synthesis of magnetic composite showed good cyclic performance for LIB anode and retained the reversible capacity of 610 mAh/g over 200 cycles with high coulombic efficiency [18]. Aside from the global use of LIB, there is the requirement to develop other energy storage technologies that can decrease the strain on the use of Li, which is a non-renewable source. Based on this challenge researchers are also focusing on SIB as a viable and more abundant energy storage technology. In that sense, Daweisu et al. [19] developed both α-MnO$_2$ and β-MnO$_2$ nanorods, synthesized through the hydrothermal method. The electrodes were prepared using the standard parameters, resulting in good electrochemical stability. For α-MnO$_2$ nanorods, the initial specific discharge and charge capacities are 278 mAh/g. At the second cycle, they dropped to 204 mAh/g. On the other hand, β-MnO$_2$ nanorods showed the initial discharge capacities of 298 mAh/g, which dropped to 240 mAh/g displaying a better result when compared to the α-MnO$_2$. Also, the charge transfer resistance of β-MnO$_2$ is much lower than that of α-MnO$_2$ nanorods suggesting better suitability as a magnetic nanostructures material to be used for the SIB.

When combined with oxygen, another magnetic element, vanadium, forms vanadium pentoxide (V$_2$O$_5$), showing favorable characteristics for their bi-layered structure with larger d-spacing than orthorhombic V$_2$O$_5$ and achieved a good electrochemical performance as cathode material for SIB. Wang et al. [20] prepared the bi-layered V$_2$O$_5$ using the solvothermal facile synthesis method having a unique crystal structure. The discharge capacity during the first cycle was 206.3 mAh/g with the voltage plateau at 2.4 V. This discharge capacity increased for the second (231.4 mAh/g) and third (222.9 mAh/g) cycles due to the gradual diffusion of electrolyte into the large space between the bilayer stacks in the first cycle. After repeated cycles, the stack facilitated the electrochemical reaction, which increases its discharge capacities. The highest capacity at 40 mA/g was 236 mAh/g. Each unit of V$_2$O$_5$ could accommodate approximately 2 Na atoms, leading to an open crystal structure providing open channels for the facile Na-ion insertion and extraction. Another approach within SIB has been explored by using titanium dioxide or titania (TiO$_2$). TiO$_2$ itself has low ion diffusion coefficients and relatively poor electrical conductivity (~ 10^{-12} S/cm), making it difficult for the requirement of high power/energy density devices [21]. Doping TiO$_2$ with foreign atoms and designing their nanocomposite with CNTs enhances electron transport with low-dimension structures, which shorten the ion insertion/extraction pathways [22,23].

Chen et al. [24] synthesized a nitrogen-doped TiO$_2$ (N-TiO$_2$) using the one-step hydrothermal facile synthesis. Standard parameters have been used to prepare the coin cell, then tested with an Arbin battery cycler (Galvanostatic charge/discharge, GCD) between 3.0–0.01 V vs. Na$^+$/Na. The capacities reached 258 mAh/g at 0.5 C. The obtained capacity retention after 300 cycles was 93.6% at 5C. This magnetic nanocomposite showed excellent cycling stability and good rate capabilities. With the concern of higher safety, better performance, and low cost compared to the prevailing MIBs. Taking from that line, considerable research work has been devoted to the aqueous rechargeable Zn-ion batteries (ZIBs) because of their outstanding operating voltage range (~1.75 V) [25,26]. The widespread application of ZIB with Ni is severely affected by their unsatisfactory cyclic performance, self-corrosion, and dendrite growth of Zn-based anodes [27]. Various ways have been implemented to solve the above issues. Yet, most efforts have been directed to the development of novel Ni-based magnetic nanostructured cathodes that can provide better durability and high capacity. Lu et al. [28] provided the study of a porous activated Ni nanoparticle/N-doped carbon matrix (A-Ni/NC) composite that functioned as an ultra-stable cathode for the advanced ZIB. A-Ni/NC was prepared using the solvothermal synthesis route. After several tests both, the equivalent series resistance (R$_s$) and the charge transfer resistance (R$_{ct}$) decreased continuously, which showed the improved electrode/electrolyte contact originated from the enhanced hydrophilicity of the electrode. It is observed from the curves that the aqueous device possesses a wide potential range of 1.40–1.90 V, which exhibited more evident plateaus and substantially larger capacities. The capacity of the A-Ni/NC//Zn battery achieved 381.2 μAh/cm³ with an excellent reversible capacity of 381.2 and

250 μAh/cm³ at 6 and 16 mA/cm². The structure of the prepared composite was well preserved after the 36000th GCD cycle which indicated that the robust porous N-doped carbon matrix prevented the degradation and aggregation of Ni nanoparticles that would occur at long-term charge/discharge cycles. The improved cycle stability in this work indicated great potential for materials used in energy storage and conversion devices.

At present, the main research directions of ZIBs include seeking suitable positive electrode materials, exploring the modification of Zn anode [29]. Different magnetic materials have been investigated for MIBs like Mn-based oxides despite having a complicated reaction mechanism that proves themselves the suitable candidate for the ZIBs. Wang et al. [30] tested the Mn-based oxide magnetic material for the cathode of ZIB in the mechanism of the battery depending on the transformation reaction of Zn^{2+}. For that, the hydrothermal synthesis route was used for the fabrication of $NaMnO_2$. Through that, it was noticed that during the electrochemical testing, there was an inhibition of the dissolution of manganese in the electrode material. The reversible capacity of Mn-based nanoparticles was around 260 mAh/g after 50 and 100 mA/g cycles which indicated the activation process during the initial cycles that further improved the penetration of electrolytes into the cathodes. This led towards the enhancement of the structural stability during the charge and discharge. Within that line, some of the results were a specific capacity of 301.3 mAh/g at the current density of 100 mA/g. Also, there was a 69.3% of capacity retention after 800 cycles performed at a rate of 1 A/g. The obtained properties could be partially explicated through the proposed mechanism of the Zn^{2+} (de)intercalation process, which appeared to occur at the c axis which was the only one that could properly accommodate the metallic cations. In this sense, a grinding process was performed to increase the presence of available c axis to improve the intercalation of Zn^{2+} ions. This process is described in Figure 16.7.

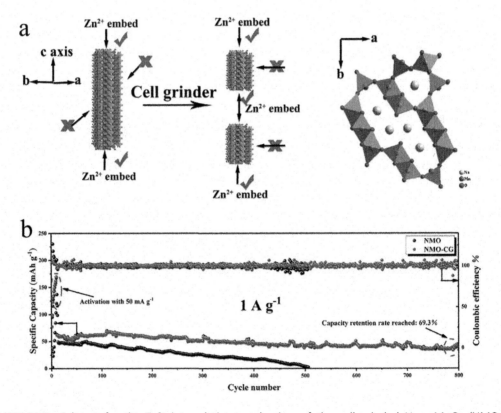

FIGURE 16.7 Scheme for the Zn^{2+} intercalation mechanism of the cell grinded $Na_{0.44}MnO_2$ (NMO-CG) Adapted with permission [30].

Source: Copyright (2020), American Chemical Society.

4.2 Nanostructured Magnetic Materials for Metal-Sulfur Batteries

Rechargeable MSBs have colossal success in electrochemical energy storage devices. Because of that, there is a strong incentive to develop batteries with higher energy, power density, and low cost leading to the development of MSBs. These devices utilize the active metal anodes to increase the specific energy due to the high capacity of metal anodes compared to intercalation compounds. While sulfur as cathode exhibits high theoretical specific capacity along with lower cost and toxicity components used for developing the MSBs [31]. MSBs can be of various types depending on the electrode and materials used. Nanostructured magnetic materials show promising candidature towards the MSBs. Vanadium, a soft magnetic metal, attracts the researchers, among which Zhang et al. [32] studied the fourth group elements that have d electrons from the V. Small doping of Co in compounds of V and synthesizing it as nanostructured micro-flowers that became magnetized under the influence of a magnetic field. Density functional theory (DFT) calculation showed a minimal amount of magnetic cobalt (Co) doped into vanadium nitride (VN) which led to an enhancement in the polysulfide adsorption. Moreover, it exhibited the adsorption energy of −5.04 eV, proving its suitability for accumulating high-order polysulfides. Such factors contribute to the decrease of the shuttling effect which is one of the main drawbacks of MSBs. The synthesis was performed through a hydrothermal approach that yielded Co-VN nanostructured micro-flowers. CV tests are conducted to explore the contribution of Co doping to the reaction kinetic of polysulfide conversion, especially liquid–liquid interface conversion, with a small number of additives in an electrolyte to contribute the minor current [33]. A higher current response indicated lithiation/delithiation reactions conversion between intermediates. The discharge capacity of Co-VN/S cathodes at 0.2, 0.5, and 1.0 C were 1123, 986, and 843 mAh/g, respectively. The excellent rate capability showed the fading rate of 0.0034% after 500 cycles. The utilization of magnetic material to modulate electronic structure for accelerated redox reaction in LSB increased rapidly as LSBs have been considered potential candidates to overcome the low-energy-density problem of LIBs. In another approach Chen et al. [34] took advantage of the ferromagnetic properties of MnO at room temperature and used the electrospinning technique to overlap the transition metal oxide structure with carbon nanofibers (CNF) as the process is described in Figure 16.8.

Despite possessing magnetic properties, its oxide nature restricts its electronic properties in terms of the conversion reactions performed by the multielectron redox processes. Also, MnO particles acted as capture sites able to immobilize LiPSs, which prohibited the diffusion of polysulfides across the separator. The unique multichannel morphology provided more ion-binding areas and more electron-transport pathways, which led to an improvement of the internal reaction kinetics of the Li–S cells. Meanwhile, the multichannel structure enabled the fibers to encounter the electrolyte more sufficiently. The intermediate layer of MnO/CNF presented typical mesoporous designs with central pores ranging from 2 to 16 nm. Meso-pores design facilitated the storage and electron transfer of poly-sulfides, improving the electrochemical conversion rates of polysulfides. It exhibited a higher initial capacity of 1336 mAh/g at 0.1 C,

FIGURE 16.8 The fabrication process of MnO/CNF interlayer. Reproduced with permission [34].
Source: Copyright (2020), American Chemical Society.

274 Applications of Low Dimensional Magnets

showing the relatively overlapped charge/discharge curve in the density range of 0.1–5 C. As per the results exhibited, the high sulfur-loaded battery still presented good cycle repetitions (515 mAh/g until 140 cycles at 0.5 C). Hence, the developed MnO/CNF interlayer for LSBs using the electrospinning strategy suppressed the shuttle effect and improved the cycle and rate performance. The intercalated carbon structure inhibited the formation of polysulfides and provides sufficient electron transport pathways for insulated S cathode. Aside from the use of Mn-based composite, cobalt phosphide (CoP_3) can provide another viable option for the development of LSB as this metal phosphide has a thermal conductivity of 258 mW/cm. K and theoretical electrical resistivity of 0.47 mΩ/cm. Along with that, it shows super magnetic properties at room temperature along with ferromagnetic behavior at 2 K, which allows LSBs to operate at lower temperatures. Based on that, the properties of CoP_3 were explored by Wen et al. [35] synthesized CoP_3 nanoflakes-induced flower-like structures using the solvothermal method (Figure 16.9).

These assembled hollow spheres are more attractive to the uniform distribution of sulfur. During the heat treatment, the spherical structure collapsed but still maintained sulfur uniformity, and flake growth of sulfur provided a more efficient conversion ability of sulfur which explained the enhanced redox kinetics. Electrochemical performances of CoP@S cathode were tested in LSBs with voltage range 1.8–2.8 V, delivering a high discharge capacity of 1020 at 0.2 C. Apart from this, long cycling behavior was the crucial parameter for the commercialization of LSB, for which CoP@S showed an initial high discharge capacity of 1280 mAh/g and 520 mAh/g after 1000 cycles despite sulfur's insulating characteristics and poor conversion reaction mechanism [36]. Still, the above discussed nanostructured magnetic materials proved their conversion ability of sulfur in LSB and enhanced lithium's conductivity. Apart from LSB, sodium-sulfur batteries (Na-SB) are attracting scientists across the globe due to their low-cost, and natural abundance of both cathode and anode materials. The theoretical energy density for Na-SB is reported as

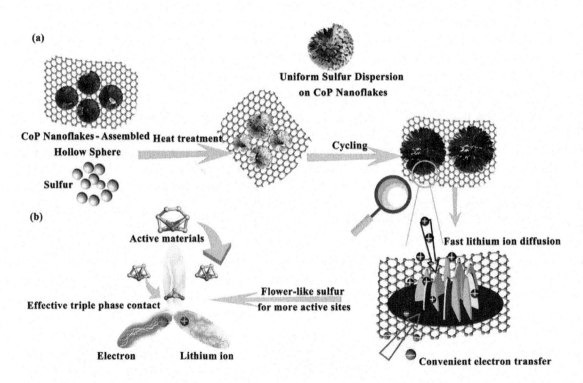

FIGURE 16.9 Schematic process for the fabrication of CoP nanoflower-like structure. (a) Assembly of CoP with sulfur loading using cycling method. (b) Formation of a flower-like structure presenting three perianths in which each perianth shows different characteristics. Adapted with permission [35].

Source: Copyright (2020), American Chemical Society.

1274 Wh/kg [37,38]. To date, the practical application of Na-SB is not possible due to sluggish reaction kinetics between S and Na from Na_2S_4 to Na_2S_2 or Na_2S_2 to Na_2S causing significant volumetric expansion during the cycling stability. This leads to low sulfur utilization and poor cyclic capacity. In another scenario Ni with C layer has plenty of surface catalytic active center to react on its surface, however, the appropriate amount of some magnetic materials like Fe, Mn, Co, and V creates a vast possibility of charge carriers to withdraw themselves from the electrolytic solution. Due to the electron affinity, these magnetic metals in their nanophase allow better conductivity. Herein, Yan et al. [39] synthesized a nanocomposite based on Fe@Ni having magnetic properties as Fe modulates the electron structure to enhance the sodium polysulfides (NaPSs) conversion reactions to achieve superior performances. The Gibbs free-energy diagram of NaPSs is shown in Figure 16.10. In that sense, from S_8 to Na_2S there are exothermic and spontaneous transformations that faced a Gibbs free energy decrease, which was followed by an increase in the last two conversion steps by displaying a higher energy barrier. The rate-determining step of the reaction for the $FeNi_3$@C nanostructured magnetic material was slightly lower when compared to Ni@C in the last step.

The rate performance of the developed cathode tested between the current density ranged from 0.2 to 5 A/g. An outstanding rate capability of 383 mAh/g at 5 A/g and an excellent long-term cyclic stability

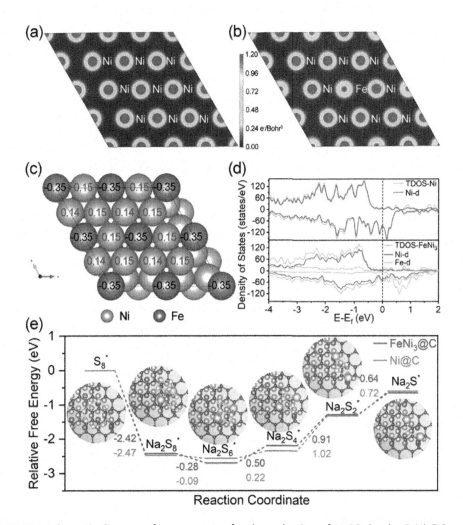

FIGURE 16.10 Schematic diagram of energy states for the reduction of NaPSs by the $FeNi_3$@C and Ni@C. Adapted with permission [39].

Source: Copyright (2021), American Chemical Society.

of 591 mAh/g after 500 cycles at 2 A/g with shallow decay capacity of 0.070% per cycle was observed. The Na-SB measurement was performed at room temperature displaying a viable use for this magnetic nanomaterial. In another line of research, it has been noted that cobalt-based sulfur composite may not show the most promising characteristics for the sulfur storage mechanism. Still, Co-MOF materials offer advantageous features for the multisulfiphillic host and induce the short-chain sulfur molecule [40]. In that sense, Rogach et al. [41] synthesized a MOF-based CoS_2/N-Doped composite using the hydrothermal approach. It showed that even though the composite is not magnetic, the battery's performance was raised to 518 mAh/g at 0.1 A/g and 380 mAh/g at 1 A/g. These superior properties were due to the Co-based MOF, a magnetic element whose nanocrystal possesses high adsorption energy to intermediate sodium polysulfides, inhibiting their dissolution into the electrolyte. The chemical interaction between the CoS_2, S_8, and the N-doping allowed the apparent chemical adsorption of S_8 by breaking the S-S bonds and opening its ring structure to shorten its length.

To further improve the electrochemical performance of Na-SB, more electroconductive and higher sulfiphilic attractive elements are required. These properties are found in the nanostructured magnetic part Mn [42]. The MnS_x composite @N-doped carbon is potentially applicable to the Na-SB, enabling the physical confinement and catalytic conversion of NaPS. Li et al. [43] synthesized the MnS@N-C using the carbonate precipitation method followed by etching with HCl. When employed as the cathode of Na-SB at room temperature, it delivered a high initial capacity of 893.9 mAh/g at 0.1 C with excellent cyclic stability at 0.5 C for 300 cycles leading to the capacity decay rate of 0.16% per cycle. However, not all these classes of cathode composite possess sufficient elasticity to resist the severe stress induced in the electrode during repeated Na^+ insertion/extraction processes. Therefore, all the approaches made above for Na-SB could successfully mitigate the conductivity and dissolution problems [44].

4.3 Nanostructured Magnetic Materials for Metal-Air Batteries

Over the last decades, the intensive use of fossil fuels and electricity has dramatically increased the standard of living. Electrical energy, the core of modern society, reduces our dependence on fossil fuels. Serious issues like global warming and air pollution make it extremely clear that it is imperative to reduce our dependence on fossil fuels to meet our energy requirements. As means to provide an alternative for that metal-air batteries show a promising feature with a practical, specific energy of around 1000 Wh/kg, which is much higher than state-of-the-art MIBs and MSBs. While looking at the basic structure design, MAB represents the metal anode and an oxygen-air electrode similar to the hybrid architecture of both fuel cells and batteries. MAB is different depending on the metals (Li, Zn, and Al) used in their anodes [45]. Lithium-oxygen batteries (Li-OBs) are attracting attention due to their high specific energy density. Still, they face critical problems like clogging of oxygen cathode by insoluble reaction products.

Another major factor is the decomposition of organic electrolytes upon cycling due to attack by reduced O_2 species [46,47]. Lee et al. [48] studied the detailed application of MnO_x based carbon composite nanofiber using electrospinning technique followed by a two-step heat process for catalytic performances in evolution and reduction of oxygen. MnO_x being magnetic and its nanostructured composite during the reaction, emitted a high number of electrons. There are two reactions: the four-electron transfer and the two-electron transfer reactions. The four-electron transfer reaction is more oxygen-efficient for current production, representing the higher current in the kinetics-diffusion controlled region. The oxygen reduction reaction (ORR) depends on the number of exchanged electrons for characterizing the catalyst performance. During ORR, the kinetic current produced by nanostructured magnetic composite was calculated to be 0.7 mA, larger in its segment. During the anodic potential scan, the polarization curve measured for composite resulted in 0.15 mA @ 2 V vs RHE. The powdered compound formed by the reaction between Mn and O did not show main OER activity as it was likely dependent on its morphological and chemical composition, including its crystallographic structure. The composite of MnO_x/C nanofiber showed improved bi-functional activity due to a more active site for oxygen to be reduced or oxidized and allowed the enhanced catalyst utilization. The Galvanostatic charge-discharge reported the difference between the open-circuit voltage and the steady-state voltage, represented as ΔE_{disc} (0.08 V) and ΔE_{ch} (0.25 V), respectively. This seems to allow effective mass transport toward the active sites suggesting that electrospun nanofibers can be used as an efficient bi-catalyst in rechargeable Li-OB. The activity of catalyst or the available surface area for the reaction of species played an important role in determining the kinetics of the reaction.

The atomic layer deposition (ALD) technique is an efficient and powerful technology for designing highly durable catalysts for batteries. Based on these concepts, Wang et al. [49] synthesized perovskite LaNiO$_3$ (LNO) by using ALD to deposit the magnetic iron oxide (Fe$_2$O$_3$). The ORR and OER kinetics studied before and after the deposition showed the remarkable ΔE (0.77 V) superior to other perovskite catalysts for Li-OBs. After modification with ALD-Fe$_2$O$_3$, there was an increase in active surface sites for electrochemical reactions. Also, there was a gradual increase in the relative concentration of not only surface O$_2$ but also the OH- species. This increment in O$_2$ refers to an enhancement on the OER and ORR activities as well as OH- species, indicating the hydroxide group's adsorption, which favors the synergistic effect for promoting other catalytic activities. The first discharge capacity at a current density of 100 mA/g was 10419 mAh/g and was almost double that of pristine LNO (6026 mAh/g), with the first Coulumbic efficiency as high as 99.4% for Li-OBs. Based on these results it was notable that this technology can open new insights into the development of advanced bifunctional catalysts and can be extended to the synthesis of flexible hybrid core/shell catalysts for batteries. The higher the number of donors in the solvent in Li-OBs, the better the ability to dissolve superoxide intermediates at higher rates, resulting in higher capacities and longer life cycles.

Ma et al. [50] developed an efficient soluble catalyst to address the above problems. The group synthesized vanadium (III) acetylacetonate (V(acac)$_3$). When mixed with acetylacetonate, the magnetic V worked as a bifunctional soluble catalyst that tuned the ORR activity by reducing the charge voltage after the electrolyte's transportation while controlling the superoxide intermediates formation which diminishes the side reaction occurring in Li-OB. It allowed discharge and charge overpotential and current density of 200 mA/g with a capacity of 500 mAh/g after 100 cycles. This work facilitates the practical use of Li-OBs with high energy density. Alongside the research trend to develop novel approaches for the generation of more sustainable energy, there has been an effort towards the development of Zn-air batteries (ZAB) as a viable alternative. Earlier, Pt-based materials used for ORR catalytic reactions showed weak stability and created problems for ZAB commercialization. The magnetic and transition element Co-based CoO nanoparticles have been applied for ORR electrocatalysts. Despite having high conductivity, nanoparticles of CoO suffer from small specific surface areas and poor durability [51,52]. CoO nanoparticles were confined in nitrogen-doped graphene (CoO/NG) for excellent durability and promote adsorption, transfer, and diffusion of O^{2-}, OH-, OOH-, and O$_2$ on the catalyst surface. The synthesis processes of CoO/NG are provided in the schematic diagram shown in Figure 16.11.

The number of electrons (n) transferred was found using the Koutecky-Levich equation and is given as 4.1, revealing that the transfer of four electrons process allowed the best conductivity during the ORR. The kinetic current density at 0.4 V was 6.46 mA/cm^2. This excellent performance was due to the presence of NG as the carbon substrate that achieved homogeneous distribution of CoO nanoparticles alloying catalyst to expose more accessible active sites. Along with that, N-doping changed the electronic structure, which facilitated the absorption of O$_2$ and intermediates. Also, in the presence of mesoporous graphene, there was an activation of the reactants during the reaction, which led to the synergistic effect that facilitated the ORR [53].

Aside from Co as a magnetic metal, there is also Mn which can potentially display better performances for the ORR and solve the issue of slow kinetics reaction. Within this line, Wang et al. [53] synthesized Mn-assisted carbon aerogel doped with nitrogen. When assisted towards the C-aerogel structure, the nanostructured magnetic Mn diminished the sluggish kinetics of the reactions due to the high specific surface area of 7.54 m^2/g. The output voltage was 1.124 V at 100 mA/cm^2 which was 97 mV higher than the traditional Pt-based ZAB and was only due to the high density of active catalytic sites with unimpeded mass transfer through its abundant macropores. When an original battery pack was tested, a slight drop of voltage from 1.30 to 1.28 V was seen after 55 h of operation due to the Zn-foil electrode exhaustion.

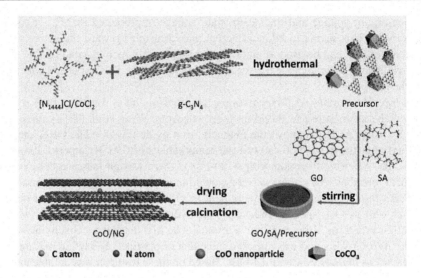

FIGURE 16.11 A Schematic illustration synthesis process of CoO/NG. Reproduced with permission [53].

Source: Copyright (2020), American Chemical Society.

Yet, the system showed excellent stability. Despite its advantages, ZAB development is plagued by low conversion efficiency and short lifespan, mainly caused by bifunctional air electrodes.

To overcome the above issues, Kang et al. [54] synthesized nanorods of cobalt oxide (Co_3O_4) doped with vanadium to develop a mesoporous structure (V-Co_3O_4). Bulk Co_3O_4 has a spinel structure and showed different configurations after doping with vanadium. It allowed the increment in occupancy due to abundant vacant d-orbitals leading to binary and ternary spinel oxides formation, which effectively enhanced the oxygen catalytic activity for high-performance ZAB. The OER was investigated by using CV which afforded a decent potential of 1.581 V vs. reversible hydrogen electrode (RHE) of 1.610 V at current densities of 10 and 20 mA/cm^2, respectively. The ΔE for V-Co_3O_4 was 0.760 V which was lower than standard ZAB-Pt. The ZAB exhibited a high maximum power density of 120.3 mW/cm^2 with a specific capacity of 814 mAh/g, which was high in its segment with a long lifespan. Aluminum-air batteries (AAB) are among viable candidates within the field of MAB due to their high theoretical energy densities of 2800 Wh/kg along with lower cost. However, the self-corrosion effect of the aluminum anode and slow dynamics of ORR at air cathode limits its application. Hence, intending to address such issues Fe-based composite found itself as a fit for the electrocatalyst in AAB as explored by Li et al. [55]. Within this line, iron-carbide doped with titanium nitrogen (FCTN@CNT) quantum dots heterostructure interface were synthesized through one-pot pyrolysis. In their work, the Fe source was embedded in CNTs whereas, Ti was responsible for the uniform distribution over the surface of the carbon matrix. The positive onset potential of 1.06 V demonstrated superior catalytic activity with the lowest Tafel slope of 53 mV/dec compared to commercial Pt/C based electrodes (70.3 mV/dec). This included the synergistic effect of heterostructure quantum dots as a reason for allowing them to absorb more oxygen by decreasing the energy barrier which resulted in good feasibility for practical application in ORR mechanisms. Still, Fe-based carbon electrocatalyst faces specific issues of low Fe-loading, which hinders the practical application and requires further studies. Another strategy that has been explored to improve the performance of AAB has been made through doping with carbon-based nanomaterials as they can act as electron donors which can improve the sluggish kinetics related to the ORR process. In that sense, Shao et al. [56] developed magnetic cobalt-based material Co_3O_4-CeO_2 doped with carbon (Co_3O_4-CeO_2/C) using a one-pot hydrothermal synthesis as schematized in Figure 16.12.

The CeO_2 acted as oxygen buffer leading to several redox couple reactions and storing extra oxygen. When there was low oxygen concentration in the high-current density region, the absorbed O_2 in

16 • Magnetic Materials in Batteries   279

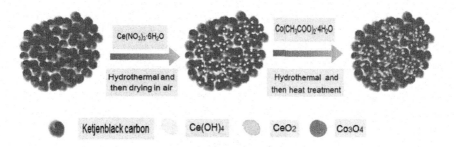

FIGURE 16.12 Schematic diagram for the preparation of Co_3O_4-CeO_2/C. Reproduced with permission [56].
Source: Copyright (2016), American Chemical Society.

CeO_2 sites received electrons to form HO_2- which was transferred to Co_3O_4 active sites and enhanced the catalytic activity of the composite fabricated for AAB. The four-electron pathway was favored in ORR because of its synergistic effect. It promotes itself as a suitable catalyst for AAB and can potentially replace traditional and costly Pt/C. MnO_2 proves itself as a well-suited candidate for different MIBs, MSBs, and MABs. Alongside that, MnO_2 based carbon-rich nanofibers can be also employed for the AAB. In that sense, Yang et al. [57] used a plasma-assisted synthesis route for the development of MnO_2 based carbon nanofiber doped with nitrogen. In their work, the plasma surface etching was implemented as an efficient strategy for constructing defects and creating vacancies for heteroatoms to attack graphitic carbon surfaces. The plasma surface etching process created vacancies that can function as active sites for ORR as heteroatoms incorporated in these defects can diminish the kinetic barrier for catalytic activity [58]. The positive effect of etching was visible when a practical battery was assembled and tested with a higher voltage range than that of Pt/C based on results in a power density of 129.7 mW/cm^2 comparable with Pt/C with excellent stability at a current density of 20 and 100 mA/cm^2 after 4 h. Throughout these studies, it has been noted that the scientific community is developing novel approaches to establish practical uses for MABs.

5 CONCLUSION

Throughout the discussion on this chapter, it was notable that energy storage devices specifically batteries are synthesized using magnetic materials and stepping towards sustainable energy development. Yet, the process for the development of nanostructured magnetic materials for application in batteries requires modern-day technology as the main challenge faced in the fabrication of nanostructured magnetic materials is their long-term performance and stability. To address that, scientists are emphasizing the formulation of different magnetic materials at the nanoscale that can potentially enhance the energy densities of their derived devices. Through that, various magnetic metallic compounds are diagnosed to use in batteries along with their structural morphology. At present, only selected materials are used due to cost-effectiveness and health factors as they release toxic gases or contaminate the environment when discarded. There is a development of MIBs with several examples discussed for different materials along with their composite and synthesis route which results in higher electrochemical performance. After this MSBs which are the current research perspective and a potential future technology that can compete with LIB are making progress as it attracts interest due to the abundance of S along with the MSB's high theoretical capacitance. Also, MABs propose an interesting and sustainable technology as their theoretical energy densities can be two times larger than MSBs, which also shows great potential to compose the future market. MABs are the only ones whose waste is minimum as compared to the other ones and can be recycled easily for the sustainable goal in the future. Apart from this, efforts to widen the range of applications of these nanostructure magnetic materials are being adapted for fuel cells, sensors, or biological applications even though those are still in the infant phase. Through that, it is likely that a considerable part of research in the field of energy storage application will be devoted to nanostructured magnetic materials.

REFERENCES

[1] Y. Wang, F. Chu, J. Zeng, Q. Wang, T. Naren, Y. Li, Y. Cheng, Y. Lei, F. Wu, Single Atom Catalysts for Fuel Cells and Rechargeable Batteries: Principles, Advances, and Opportunities, *ACS Nano.* 15 (2021) 210–239.

[2] A.J. da Costa, J.F. Matos, A.M. Bernardes, I.L. Müller, Beneficiation of Cobalt, Copper and Aluminum from Wasted Lithium-ion Batteries by Mechanical Processing, *Int. J. Miner. Process.* 145 (2015) 77–82.

[3] H. Wang, X. Yang, W. Shao, S. Chen, J. Xie, X. Zhang, J. Wang, Y. Xie, Ultrathin Black Phosphorus Nanosheets for Efficient Singlet Oxygen Generation, *J. Am. Chem. Soc.* 137 (2015) 11376–11382.

[4] S. Feng, R. Xu, New Materials in Hydrothermal Synthesis, *Acc. Chem. Res.* 34 (2001) 239–247.

[5] J. Lai, W. Niu, R. Luque, G. Xu, Solvothermal Synthesis of Metal Nanocrystals and Their Applications, *Nano Today.* 10 (2015) 240–267.

[6] A.E. Gash, R.L. Simpson, J.H. Satcher, Direct Preparation of Nanostructured Energetic Materials Using Sol-Gel Methods, in: *Defense Applications of Nanomaterials*, American Chemical Society, 2005: pp. 14–198.

[7] G. Nie, Z. Zhang, T. Wang, C. Wang, Z. Kou, Electrospun One-Dimensional Electrocatalysts for Oxygen Reduction Reaction: Insights into Structure – Activity Relationship, *ACS Appl. Mater. Interfaces.* 13 (2021) 37961–37978.

[8] A. Van der Ven, Z. Deng, S. Banerjee, S. Ping Ong, Rechargeable Alkali-Ion Battery Materials: Theory and Computation, *Chem. Rev.* 120 (2020) 6977–7019.

[9] A. Van der Ven, Z. Deng, S. Banerjee, S.P. Ong, Rechargeable Alkali-Ion Battery Materials: Theory and Computation, *Chem. Rev.* 120 (2020) 6977–7019.

[10] L. Geng, X. Wang, K. Han, P. Hu, L. Zhou, Y. Zhao, W. Luo, L. Mai, Eutectic Electrolytes in Advanced Metal-Ion Batteries, *ACS Energy Lett.* 7 (2021) 247–260.

[11] R.K. Bhardwaj, H. Lahan, V. Sekkar, B. John, A.J. Bhattacharyya, High-Performance Li-Metal-Free Sulfur Battery Employing a Lithiated Anatase TiO2 Anode and a Freestanding Li2S-Carbon Aerogel Cathode, *ACS Sustain. Chem. Eng.* 10 (2022) 410–420.

[12] A. Zahoor, Z.K. Ghouri, S. Hashmi, F. Raza, S. Ishtiaque, S. Nadeem, I. Ullah, K.S. Nahm, Electrocatalysts for Lithium-Air Batteries: Current Status and Challenges, *ACS Sustain. Chem. Eng.* 7 (2019) 14288–14320.

[13] W.J. Kwak, Rosy, D. Sharon, C. Xia, H. Kim, L.R. Johnson, P.G. Bruce, L.F. Nazar, Y.K. Sun, A.A. Frimer, M. Noked, S.A. Freunberger, D. Aurbach, Lithium-Oxygen Batteries and Related Systems: Potential, Status, and Future, *ACS Appl. Mater. Interfaces.* 120 (2020) 6626–6683.

[14] C. Zhang, Z. Chen, H. Wang, Y. Nie, J. Yan, Porous Fe2O3Nanoparticles as Lithium-Ion Battery Anode Materials, *ACS Appl. Nano Mater.* 4 (2021) 8744–8752.

[15] C.R. Sides, N. Li, C.J. Patrissi, B. Scrosati, C.R. Martin, Nanoscale Materials for Lithium-ion Batteries, *MRS Bull.* 27 (2002) 604–607.

[16] C. Chu, L. Chang, D. Yin, D. Zhang, Y. Cheng, L. Wang, Large-Sized Nickel – Cobalt – Manganese Composite Oxide Agglomerate Anode Material for Long-Life-Span Lithium-Ion Batteries, *ACS Appl. Energy Mater.* 4 (2021) 13811–13818.

[17] Y. Meng, Y. Liu, J. He, X. Sun, A. Palmieri, Y. Gu, X. Zheng, Y. Dang, X. Huang, W. Mustain, S.L. Suib, Large Scale Synthesis of Manganese Oxide/Reduced Graphene Oxide Composites as Anode Materials for Long Cycle Lithium Ion Batteries, *ACS Appl. Energy Mater.* 4 (2021) 5424–5433.

[18] A.M. Abakumov, S.S. Fedotov, E.V. Antipov, J.M. Tarascon, Solid State Chemistry for Developing Better Metal-ion Batteries, *Nat. Commun.* 11 (2020) 1–14.

[19] D. Su, H.J. Ahn, G. Wang, Hydrothermal Synthesis of α-MnO2 and β-MnO 2 Nanorods As High Capacity Cathode Materials for Sodium Ion Batteries, *J. Mater. Chem. A.* 1 (2013) 4845–4850.

[20] D.A. Semenenko, D.M. Itkis, E.A. Pomerantseva, E.A. Goodilin, T.L. Kulova, A.M. Skundin, Y.D. Tretyakov, Electrochemistry Communications Li x V 2 O 5 Nanobelts for High Capacity Lithium-ion Battery Cathodes, *Electrochem. Commun.* 12 (2010) 1154–1157.

[21] Graphical Abstract: Chem. Eur. J. 5/2014, *Chem. – A Eur. J.* 20 (n.d.) 1193–1202.

[22] C. Deng, C. Ma, M.L. Lau, P. Skinner, Y. Liu, W. Xu, H. Zhou, Y. Ren, Y. Yin, B. Williford, M. Dahl, H. (Claire) Xiong, Amorphous and Crystalline TiO2 Nanoparticle Negative Electrodes for Sodium-ion Batteries, *Electrochim. Acta.* 321 (2019) 134723.
[23] T. Lan, J. Tu, Q. Zou, X. Zeng, J. Zou, H. Huang, M. Wei, Synthesis of Anatase TiO2 Mesocrystals with Highly Exposed Low-index Facets for Enhanced Electrochemical Performance, *Electrochim. Acta.* 319 (2019) 101–109.
[24] Y. Yang, X. Ji, M. Jing, H. Hou, Y. Zhu, L. Fang, X. Yang, Q. Chen, C.E. Banks, Carbon Dots Supported Upon N-doped TiO2 Nanorods Applied into Sodium and Lithium Ion Batteries, *J. Mater. Chem. A.* 3 (2015) 5648–5655.
[25] J. Wang, J.G. Wang, H. Liu, Z. You, C. Wei, F. Kang, Electrochemical Activation of Commercial MnO Microsized Particles for High-performance Aqueous Zinc-ion Batteries, *J. Power Sources.* 438 (2019) 226951.
[26] J. Jindra, Sealed Ni – Zn Cells, 1996–1998, *J. Power Sources.* 88 (2000) 202–205.
[27] S. Wang, S. Lai, P. Li, T. Gao, K. Sun, X. Ding, T. Xie, C. Wu, X. Li, Y. Kuang, W. Liu, W. Yang, X. Sun, Hierarchical Cobalt Oxide@Nickel-vanadium Layer Double Hydroxide Core/shell Nanowire Arrays with Enhanced Areal Specific Capacity for Nickel – Zinc Batteries, *J. Power Sources.* 436 (2019) 226867.
[28] L. Meng, D. Lin, J. Wang, Y. Zeng, Y. Liu, X. Lu, Electrochemically Activated Nickel-Carbon Composite as Ultrastable Cathodes for Rechargeable Nickel-Zinc Batteries, *ACS Appl. Mater. Interfaces.* 11 (2019) 14854–14861.
[29] X. Guo, J. Zhou, C. Bai, X. Li, G. Fang, S. Liang, Zn/MnO2 Battery Chemistry with Dissolution-deposition Mechanism, *Mater. Today Energy.* 16 (2020) 100396.
[30] J. Li, L. Li, H. Shi, Z. Zhong, X. Niu, P. Zeng, Z. Long, X. Chen, J. Peng, Z. Luo, X. Wang, S. Liang, Electrochemical Energy Storage Behavior of Na0.44MnO2in Aqueous Zinc-Ion Battery, *ACS Sustain. Chem. Eng.* 8 (2020) 10673–10681.
[31] M. Salama, Rosy, R. Attias, R. Yemini, Y. Gofer, D. Aurbach, M. Noked, Metal – Sulfur Batteries: Overview and Research Methods, *ACS Energy Lett.* 4 (2019) 436–446.
[32] Z. Cheng, Y. Wang, W. Zhang, M. Xu, Boosting Polysulfide Conversion in Lithium-Sulfur Batteries by Cobalt-Doped Vanadium Nitride Microflowers, *ACS Appl. Energy Mater.* 3 (2020) 4523–4530.
[33] H. Lin, L. Yang, X. Jiang, G. Li, T. Zhang, Q. Yao, G.W. Zheng, J.Y. Lee, Electrocatalysis of Polysulfide Conversion by Sulfur-deficient MoS 2 Nanoflakes for Lithium-sulfur Batteries †, *1476 | Energy Environ. Sci.* 10 (2017) 1476.
[34] M. Chen, T. Li, Y. Li, X. Liang, W. Sun, Q. Chen, Rational Design of a MnO Nanoparticle-Embedded Carbon Nanofiber Interlayer for Advanced Lithium – Sulfur Batteries, *ACS Appl. Energy Mater.* 3 (2020) 10793–10801.
[35] C. Qi, Z. Li, C. Sun, C. Chen, J. Jin, Z. Wen, Cobalt Phosphide Nanoflake-Induced Flower-like Sulfur for High Redox Kinetics and Fast Ion Transfer in Lithium-Sulfur Batteries, *ACS Appl. Mater. & Interfaces.* 12 (2020) 49626–49635.
[36] X. Yang, X. Gao, Q. Sun, S.P. Jand, Y. Yu, Y. Zhao, X. Li, K. Adair, L.-Y. Kuo, J. Rohrer, J. Liang, X. Lin, M.N. Banis, Y. Hu, H. Zhang, X. Li, R. Li, H. Zhang, P. Kaghazchi, T.-K. Sham, X. Sun, Promoting the Transformation of Li2S2 to Li2S: Significantly Increasing Utilization of Active Materials for High-Sulfur-Loading Li – S Batteries, *Adv. Mater.* 31 (2019) 1901220.
[37] S. Wei, S. Xu, A. Agrawral, S. Choudhury, Y. Lu, Z. Tu, L. Ma, L.A. Archer, A Stable Room-temperature Sodium – Sulfur Battery, *Nat. Commun.* 7 (2016) 11722.
[38] S. Xin, Y.-X. Yin, Y.-G. Guo, L.-J. Wan, A High-Energy Room-Temperature Sodium-Sulfur Battery, *Adv. Mater.* 26 (2014) 1261–1265.
[39] L. Wang, H. Wang, S. Zhang, N. Ren, Y. Wu, L. Wu, X. Zhou, Y. Yao, X. Wu, Y. Yu, Manipulating the Electronic Structure of Nickel via Alloying with Iron: Toward High-Kinetics Sulfur Cathode for Na-S Batteries, *ACS Nano.* 15 (2021) 15218–15228.
[40] H. Liu, W. Pei, W.-H. Lai, Z. Yan, H. Yang, Y. Lei, Y.-X. Wang, Q. Gu, S. Zhou, S. Chou, H.K. Liu, S.X. Dou, Electrocatalyzing S Cathodes via Multisulfiphilic Sites for Superior Room-Temperature Sodium – Sulfur Batteries, *ACS Nano.* 14 (2020) 7259–7268.
[41] F. Xiao, H. Wang, T. Yao, X. Zhao, X. Yang, D.Y.W. Yu, A.L. Rogach, MOF-Derived CoS2/N-Doped Carbon Composite to Induce Short-Chain Sulfur Molecule Generation for Enhanced Sodium-Sulfur Battery Performance, *ACS Appl. Mater. Interfaces.* 13 (2021) 18010–18020.
[42] X. Liu, J.-Q. Huang, Q. Zhang, L. Mai, Nanostructured Metal Oxides and Sulfides for Lithium – Sulfur Batteries, *Adv. Mater.* 29 (2017) 1601759.
[43] F. Ma, P. Hu, T. Wang, J. Liang, R. Han, J. Han, Q. Li, Yolk@Shell Structured MnS@Nitrogen-Doped Carbon as a Sulfur Host and Polysulfide Conversion Booster for Lithium/Sodium Sulfur Batteries, *ACS Appl. Energy Mater.* 4 (2021) 3487–3494.

[44] A. Ghosh, A. Kumar, A. Roy, M. Ranjan Panda, M. Kar, D. R. MacFarlane, S. Mitra, Three-Dimensionally Reinforced Freestanding Cathode for High-Energy Room-Temperature Sodium – Sulfur Batteries, *ACS Appl. Mater. & Interfaces.* 11 (2019) 14101–14109.

[45] W.J. Kwak, Rosy, D. Sharon, C. Xia, H. Kim, L.R. Johnson, P.G. Bruce, L.F. Nazar, Y.K. Sun, A.A. Frimer, M. Noked, S.A. Freunberger, D. Aurbach, Lithium-Oxygen Batteries and Related Systems: Potential, Status, and Future, *ACS Appl. Mater. Interfaces.* 14 (2020) 6626–6683.

[46] J. Read, Characterization of the Lithium/Oxygen Organic Electrolyte Battery, *J. Electrochem. Soc.* 149 (2002) A1190.

[47] T. Kuboki, T. Okuyama, T. Ohsaki, N. Takami, Lithium-air Batteries Using Hydrophobic Room Temperature Ionic Liquid Electrolyte, *J. Power Sources.* 146 (2005) 766–769.

[48] K.N. Jung, J.I. Lee, S. Yoon, S.H. Yeon, W. Chang, K.H. Shin, J.W. Lee, Manganese Oxide/Carbon Composite Nanofibers: Electrospinning Preparation and Application as a Bi-functional Cathode for Rechargeable Lithium-oxygen Batteries, *J. Mater. Chem.* 22 (2012) 21845–21848.

[49] C. Gong, L. Zhao, S. Li, H. Wang, Y. Gong, R. Wang, B. He, Atomic Layered Deposition Iron Oxide on Perovskite LaNiO3 as an Efficient and Robust Bi-functional Catalyst for Lithium Oxygen Batteries, *Electrochim. Acta.* 281 (2018) 338–347.

[50] Q. Zhao, N. Katyal, I.D. Seymour, G. Henkelman, T. Ma, Vanadium(III) Acetylacetonate as an Efficient Soluble Catalyst for Lithium – Oxygen Batteries, *Angew. Chemie.* 131 (2019) 12683–12687.

[51] Z. Wang, B. Li, X. Ge, F.W.T. Goh, X. Zhang, G. Du, D. Wuu, Z. Liu, T.S. Andy Hor, H. Zhang, Y. Zong, Co@Co3O4@PPD Core@bishell Nanoparticle-Based Composite as an Efficient Electrocatalyst for Oxygen Reduction Reaction, *Small.* 12 (2016) 2580–2587.

[52] L. Tan, Q.-R. Pan, X.-T. Wu, N. Li, J.-H. Song, Z.-Q. Liu, Core@Shelled Co/CoO Embedded Nitrogen-Doped Carbon Nanosheets Coupled Graphene as Efficient Cathode Catalysts for Enhanced Oxygen Reduction Reaction in Microbial Fuel Cells, *ACS Sustain. Chem. & Eng.* 7 (2019) 6335–6344.

[53] L. Xu, C. Wang, D. Deng, Y. Tian, X. He, G. Lu, J. Qian, S. Yuan, H. Li, Cobalt Oxide Nanoparticles/Nitrogen-Doped Graphene as the Highly Efficient Oxygen Reduction Electrocatalyst for Rechargeable Zinc-Air Batteries, *ACS Sustain. Chem. Eng.* 8 (2020) 343–350.

[54] Y. Rao, S. Chen, Q. Yue, Y. Kang, Optimizing the Spin States of Mesoporous Co3O4 Nanorods through Vanadium Doping for Long-Lasting and Flexible Rechargeable Zn-Air Batteries, *ACS Catal.* 11 (2021) 8097–8103.

[55] K. Li, C. Wang, H. Li, Y. Wen, F. Wang, Q. Xue, Z. Huang, C. Fu, Heterostructural Interface in Fe3C-TiN Quantum Dots Boosts Oxygen Reduction Reaction for Al – Air Batteries, *ACS Appl. Mater. Interfaces.* 13 (2021) 47440–47448.

[56] K. Liu, X. Huang, H. Wang, F. Li, Y. Tang, J. Li, M. Shao, Co3O4-CeO2/C as a Highly Active Electrocatalyst for Oxygen Reduction Reaction in Al-Air Batteries, *ACS Appl. Mater. Interfaces.* 8 (2016) 34422–34430.

[57] R. Cheng, F. Wang, M. Jiang, K. Li, T. Zhao, P. Meng, J. Yang, C. Fu, Plasma-Assisted Synthesis of Defect-Rich O and N Codoped Carbon Nanofibers Loaded with Manganese Oxides as an Efficient Oxygen Reduction Electrocatalyst for Aluminum-Air Batteries, *ACS Appl. Mater. Interfaces.* 13 (2021) 37123–37132.

[58] Z. Liu, Z. Zhao, Y. Wang, S. Dou, D. Yan, D. Liu, Z. Xia, S. Wang, In Situ Exfoliated, Edge-Rich, Oxygen-Functionalized Graphene from Carbon Fibers for Oxygen Electrocatalysis, *Adv. Mater.* 29 (2017) 1606207.

Recent Advancement in Magnetic Materials for Supercapacitor Applications

17

Magdalene Asare, Felipe M. de Souza, Ram K. Gupta

Contents

1 Introduction   283
2 Synthesis of Magnetic Materials   285
    2.1 Solvothermal/Hydrothermal   285
    2.2 Sol-Gel   286
    2.3 Electrochemical Deposition   286
    2.4 Other Techniques   286
3 Mechanism of Charge Storage   287
    3.1 Electrochemical Double-Layer Capacitors   287
    3.2 Pseudocapacitors   288
    3.3 Hybrid Supercapacitors   289
4 Magnetic Materials for Supercapacitors   290
5 Conclusion   298
References   299

1 INTRODUCTION

With the continuous advancement in science and technology amidst the rapid increase in the world's population, there is a heavy demand for the constant supply of energy. Energy generated from non-renewable sources like fossil fuels comes with the issue of environmental pollution which has inspired a durative shift to alternative sources such as solar, tidal, and hydro. These renewable sources however have some drawbacks and require an efficient means of production and storage, hence, expanding the need for energy storage devices [1]. Typical electrochemical storage devices include batteries, supercapacitors, and fuel cells [2]. Currently, lithium-ion batteries (LIBs) and supercapacitors have attracted lots of attention due to their sustainability and unique properties [3]. Supercapacitors can be used in several applications such as memory backup, wearable devices, electric vehicles, forklifts, energy recovery, and grid storage among others [4]. They can be classified into pseudocapacitors and electrochemical double-layer capacitors (EDLC) concerning how they store energy. The former stores energy through a redox reaction and typically uses metal oxides as their active materials while the latter does it by forming an electrical double layer with the use of carbon-based materials [5,6]. A combination of both types of forms, another which is known as hybrid supercapacitors. Despite some drawbacks of supercapacitors such as low energy density, they have unique characteristics that make them stand out such as excellent power density, easy handling, prolonged cycle life, and fast charge/discharge rate leading to an outstanding performance [7–9].

A typical supercapacitor is composed of an electrolyte, two electrodes isolated by a separator, and a current collector as shown in Figure 17.1. The different parts of the supercapacitor affect the electrochemical behavior, the electrodes are the crucial component and one of the main determinants for performance [10,11]. Emerging materials used as electrodes include metal oxides, carbon-based materials, conducting polymers, and more recently, natural macromolecules and bioinspired materials [8]. Due to the low energy density of supercapacitors, the need for improved supercapacitor material composition is explored. In EDLCs, carbon-based electrodes can be functionalized or composited such as activated carbon and nanotubes to improve the performance of the supercapacitor. Due to the cheap, non-toxic, and easy processing with varying sizes of carbon, a high surface area can be manipulated for greater results such that large pore sizes can be channeled in the generation of a high-power density and low pore sizes for a high energy density [12,13]. Conducting polymers used for electrodes have a low-cost and environmental impact, high voltage window, high conductivity in the dope state, and high charge density with examples like polyaniline, polypyrrole, and polythiophene [14]. However, as compared to other carbon electrodes, they have a lower cycle life since their polymer backbone has a limited ability to sustain repeated redox cycles [15].

Metal oxides can generate higher energy density than carbon materials and better electrochemical stability as compared to polymeric materials [17]. Transition metal oxides for instance, which have several oxidation states are observed to have a higher theoretical capacitance as compared to carbon-based

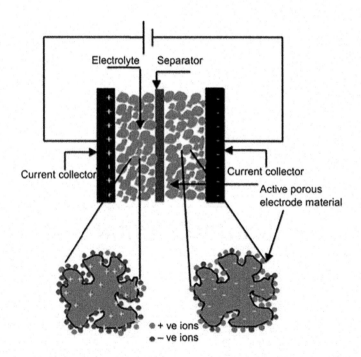

FIGURE 17.1 Illustration of the basic structure of supercapacitors. Adapted with permission [16].
Source: Copyright (2009), Royal Society of Chemistry.

electrodes. In addition, they have a fast reversible faradaic redox reaction and better electrochemical performance and stability. A classic example that has been studied is RuO_2, which has excellent conductivity and high specific capacitance. However, it is faced with an extremely high cost and toxic nature [18,19]. Other researched materials are MnO_2, V_2O_5/VO_4, and WO_3 which are enticing due to their easy availability, good chemical, and mechanical stability as well as environmentally friendly nature [5]. However, metal oxides have low conductivity and limited cycling ability [20]. Magnetic materials are another rising category in supercapacitors due to their affordability and easy production in large quantities. Broad examples explored include magnetic metal oxide nanoparticles and ferrites [21]. Ferrites, which are magnetic materials, have gained scientists' interest in their use as materials for supercapacitors due to their various redox states, good optical, catalytic, magnetic, and electrochemical stability. They can also be synthesized in different shapes and forms using a diverse range of techniques [5]. Quintessential electrodes made from magnetic materials that have been tested in the application of supercapacitors include Fe_2O_3 [22], Co_3O_4 nanoparticles [23], $MnCoFeO_4$ [5], $MnZnFe_2O_4$ [24], in addition to other ferrite-based compounds. It is also important to note that scientists have observed an effect of external magnetic fields on the performance of supercapacitors that are made of magnetic materials. However, in-depth research will be needed for a more robust finding on how magnetic-based supercapacitors are affected by an outside magnetic source.

2 SYNTHESIS OF MAGNETIC MATERIALS

The method of synthesis for materials in supercapacitor applications is very important [25]. Recently, the production of magnetic materials, particularly nanostructured metal oxides or ferrites for different applications such as drug delivery, catalysis, medical diagnosis, energy storage, and other uses, is a crucial part of research that has gained lots of attention. In addition, the surface area and morphology of the materials based on the synthetic route are very crucial. In the synthesis of magnetic nanoparticles for energy storage applications, the sizes and distribution of the materials can greatly influence the energy densities, specific capacitance, and thus the overall performance of a supercapacitor [26]. And as important as the specific size of the magnetic material is for the overall performance of the supercapacitor, the cost of production and the synthetic route is salient [27]. Currently, there are a lot of methods for the synthesis of magnetic materials which come with their advantages and disadvantages. Common examples include solvothermal, hydrothermal, sol-gel, spray pyrolysis, electrochemical deposition, physical vapor deposition, and others as discussed below:

2.1 Solvothermal/Hydrothermal

The solvothermal method of synthesis is one of the advantageous ways of generating materials, especially in nanosized forms. It is also typically known for the preparation of open-framework inorganic solids and currently for metal-organic framework structures [28]. Comparatively, they present a simple way of synthesis, they generate highly crystalline products at low temperatures, they can be manipulated to control the growth and size of the nanoparticles as well as being suitable for the preparation of samples in larger quantities [29]. In the solvothermal process, the reactions are done in tightly sealed containers with a mixture of the solid and liquid reagents or precursors present. Usually, the reaction is heated at elevated or close enough temperatures to the boiling point of the main solvent by the increased pressure in the vessels. This is particularly done to facilitate the reactivity and solubility of the reactants at high temperatures. However, some starting reactants could potentially undergo unforeseen reactions such that some metals can decompose to form nanomaterials on their own and others could be carbonized. In effect, the reaction and temperature can be adjusted throughout to produce the desired final material [30]. Additionally, the solvent plays a huge role in the synthesis of the materials. Due to the high polarity, low cost, and non-toxic nature of water, it is usually patronized for a solvent. Other solvents used in the synthesis of nanoparticles include dimethylformamide (DMF), ethanol, polyols, etc. due to their compatibility with other reagents and capability to shape the material into different nano morphologies [30]. Concerning the advantageous nature of solvothermal/hydrothermal synthesis, researchers have utilized it in the synthesis of magnetic materials for supercapacitors with promising results. Synthesized materials include $CoFe_2O_4$ on carbon composites, $NiCoO_4$ on reduced graphene oxide, and many more with results indicating that the electrochemical performance and overall supercapacitor behaviors using this method for magnetic materials improved [31,32].

2.2 Sol-Gel

Sol-gel is another convenient technique for the synthesis of catalytic and nanostructured materials due to some advantages such as the short processing time under low temperatures and the synthesis of fine particles with narrow size distribution [33]. In addition, the rheological characteristics of the sols or gels facilitate the formation of fibers, films, and composites using different techniques. It can also be used for glasses and ceramics, and it is even more advantageous for the synthesis of magnetic materials because it is cost-effective and simple to use through a gentle route in the production of fine porous materials [34]. However, it involves different steps such as hydrolysis, polymerization, gelation, condensation, drying, and densification. Sol-gel is derived from the idea that molecules in a solution (sol) will gather under controlled conditions to form an ordered network (gel) which can be latter treated to create nanomaterials [35]. Several scientists have adopted the sol-gel technique for the synthesis of magnetic materials for energy storage with examples such as $NiCo_2O_4$ and $NiFe_2O_4$ among others [33,36]. In a practical example, a research team synthesized Co_3O_4 nanoparticles for electrodes in supercapacitors using the sol-gel method as a quick inexpensive synthesis with urea. The experiment was successful and when compared to other methods for the production of Co_3O_4 nanoparticles, sol-gel proved the most effective in aiding the generation of a high specific capacitance value of 761.25 F/g at 11 mA/cm^2 [37].

2.3 Electrochemical Deposition

Electrochemical deposition utilizes electric current and potential as a means of depositing dissolved metal cations and further coating them on electrodes for energy storage applications. It has been adopted in different metal oxides, hydroxides, and magnetic materials with two broad categories which are the anodic and cathodic electrodeposition. This method is flexible and can be manipulated for specific structure properties, however, it is not practical for large-scale applications. Also, this procedure is faced with some fluctuations that can lead to an uneven distribution of substances on the electrodes. Researchers like Lui et al. [35] found a solution to this problem with the use of a strong magnetic field (SMF). As shown in Figure 17.2, SMF was adopted in an electrochemical deposition for the production of Ni-Co-Zn-P for supercapacitor applications. This modified method was successful and the electrochemical performance of the system, as well as the morphology, was improved [38]. The electrochemical deposition has also been used to produce magnetic materials such as NiO nanoparticles and films which overall had improved electrochemical properties.

2.4 Other Techniques

Several other methods such as chemical vapor deposition (CVD), physical vapor deposition (PVD), spray pyrolysis, and microwave are used for the synthesis of magnetic materials. Usually, CVD and PVD are costly, complex, and require specific equipment in their usage for synthesizing magnetic materials. Spray pyrolysis, on the other hand, has been used to make ferrites like Ni-Zn and different types of films with improved homogeneity and reproducibility. In addition, a simple microwave method was used to generate Co_3O_4 nanoparticles for supercapacitors. This fast and low energy budget procedure resulted in the establishment of excellent cyclic stability and improved electrochemical performance of the supercapacitor [39,40].

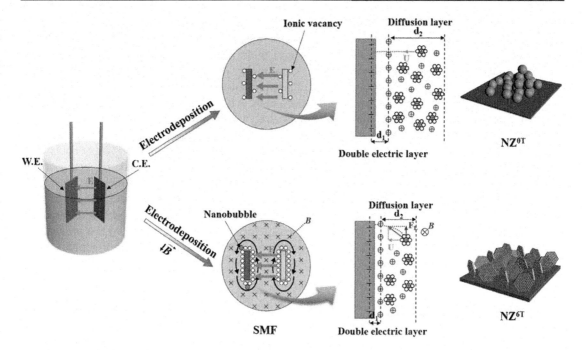

FIGURE 17.2 Modified electrodeposition with SMF for the synthesis of electrodes. Adapted with permission [38]. *Source:* Copyright (2022), American Chemical Society.

3 MECHANISM OF CHARGE STORAGE

Supercapacitors can be classified into three main categories based on their energy storage mechanism and flow of ions from the electrolytes to the surface of the electrodes. In general, supercapacitors have two electrodes that are separated using a semipermeable membrane referred to as a separator. The electrodes and separator are then filled with electrolytes that facilitate the flow of ionic currents between the electrodes. Based on charge storage mechanism, they can be divided into EDLC, pseudocapacitors and hybrid supercapacitors. EDLCs typically use carbon-based materials as electrodes, whereas pseudocapacitors use metal-based or conducting materials. This category of supercapacitors falls between conventional electrostatic capacitors and batteries concerning their power and energy densities [41,42]. Their unique charge storage and mechanism is described in more detail below:

3.1 Electrochemical Double-Layer Capacitors

EDLC represents the most common and highly used supercapacitors in the market. Their basic setup consists of two electrodes, an electrolyte, and a separator. In the case of EDLC, carbon-based materials are used for the electrodes. They can be compared to conventional capacitors that use dielectric mediums, solvents such as propylene carbonate and KOH are used for electrolytes in EDLC. The inception of EDLC was proposed in the 19th century by Helmholtz while researching the presence of opposite charges on the surface of colloidal particles. An overall idea on the model and observation showed that different charges are layered on the electrode/electrolyte surface which is separated by the atomic distance between them. This theory has been reviewed and modified by other researchers like Gouy, Chapman, and Stern [43]. In

EDLC, the separation of electric charges is attained by the movement of ions and electrons on the surface between the electrodes and electrolyte where the extra charges either accumulate on the positive or negative electrodes. To maintain a neutral network, the ions in the electrolyte are regulated in the solution. With the application of a load to the system, anions move toward the positive electrode, and cations in the electrolyte move to the negative electrode which collectively causes the formation of a double layer at the interface of the electrodes. When the load is removed, anions in the electrolyte attract the positive electric charges on the electrode, while the cations from the electrolyte and negative charges on the electrode get attracted to each other which in effect, causes the stability of the double layers on the electrodes. The stabilization and mechanism signify that the actual concentration of the electrolyte is constant or unchanged irrespective of the addition or removal of an external load and the EDLC accumulates energy on the double layer interface and stores energy electrostatically. This method of charge storage does not involve any electron or charge transfer, it does not occur through redox reactions and it can be classified as a non-faradaic process. Since there is only physical charge transfer, this type of supercapacitor has a relatively long cycle life. And to prevent the recombination of ions at the electrodes, the double layer is utilized. By increasing the specific surface area of the electrodes as well as decreasing the distance between them, EDLC can have a higher energy density [10,44].

3.2 Pseudocapacitors

This category can also be referred to as faradaic supercapacitors because they rely on faradaic reactions (Figure 17.3). The general mechanism is similar to that of batteries rather than capacitors and

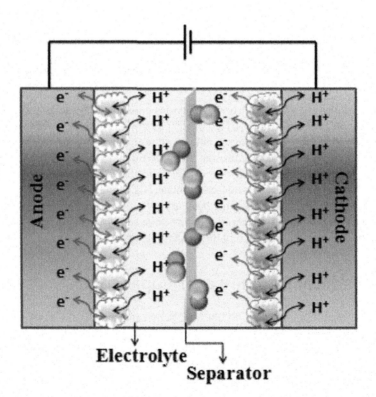

FIGURE 17.3 Working mechanism of pseudocapacitor. Adapted with permission [45].

Source: Copyright (2021), American Chemical Society.

they are not popularly used on the commercial scale as compared to EDLCs. On the application of a potential, reduction, and oxidation occur on the electrode and this results in the movement of charges across the double layer leading to the faradaic current passing through the supercapacitor. In the case of pseudocapacitors, the faradaic reaction facilitates the production of a higher specific capacitance and energy density as compared to EDLC. However, over long cycles, pseudocapacitors are faced with low stability and power density because during charging and discharging, the electrodes are overused and degrade faster. Typical electrodes used in pseudocapacitors include RuO_2 and MnO_2 as well as conducting polymers such as polyaniline, polypyrrole, and polythiophene [10,43,44].

3.3 Hybrid Supercapacitors

Hybrid supercapacitors have gained a lot of attention and hence a broadened research area in the quest of finding an optimum means for energy storage. Hybrid supercapacitors as inferred from the name describe a combination of two things which in this case: two different types of electrode materials. However, it still has the other basic parts of supercapacitors which are the electrolyte, separator, and collector as shown in Figure 17.4. The distinguishing properties of hybrid supercapacitors as compared to EDLCs and pseudocapacitors are the presence of both a redox reaction electrode and a non-faradaic electrode. It is made up of a part of an electrode from EDLC (double layer type) which is usually the positive electrode and from a pseudocapacitor (faradaic), usually the negative electrode. This adjustment is to improve the overall charge storage ability of supercapacitors in the journey of solving the energy crises we face. Concerning this combination, hybrid supercapacitors utilize both mechanisms simultaneously to provide a high energy density from the faradaic electrodes and a high-power density in addition to a longer cycle life with the non-faradaic electrode. In comparison, the construction and mechanism of hybrid supercapacitors are similar to LIBs to some extent [44,46].

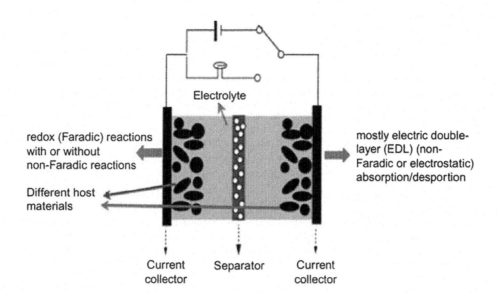

FIGURE 17.4 Image showing the basic make-up of hybrid supercapacitors. Adapted with permission [47].
Source: Copyright (2013), Royal Society of Chemistry.

4 MAGNETIC MATERIALS FOR SUPERCAPACITORS

The field of energy storage has been continuously evolving as there is an ever-growing need to develop materials with better performance to enable a technology powered by a more sustainable energy grid without great dependence on non-renewable sources as well as to further optimize electronic devices that have become a necessity. Within this line of development, supercapacitors have gained more attention as their electrochemical properties have been optimized based on the use of nanoparticles of several electroactive materials such as transition metal hydroxides, hydrides, sulfides, bimetallic compounds, composites, and others. Along with that, several synthesis approaches can yield a myriad of nanostructured morphologies that play a major influence on the properties. Yet, it has been noted that magnetic nanoparticles can potentially deliver an improvement in electrochemical performance when exposed to a magnetic field, leading to an expansion within the field of energy storage devices that has not been widely explored yet.

Even though there has been a modest number of studies in this regard it has been observed that the presence of a magnetic field near a magnetic electrode can lead to an increase in current, which can be related to the convection flow close to the electrode [48,49]. This process is likely to promote a better contact between the electrode-electrolyte interface resulting in a decrease in resistance as well as Nernst's layer dimension suppression. Through that, there is an increase in the concentration of ions near the electrode which facilitates the (de)intercalation process. The influence of magnetic field over a supercapacitor may be related to the Lorentz force, represented in equation 1, that can describe the effect of the movement of charged species, electronic states, charge density, among other parameters [50].

$$F_L = qE + q(v \times B) \tag{1}$$

Where q is the particle's charge, E is the electric field applied, v is the charge's velocity and B is the magnetic field. Based on that, when no magnetic field is applied then only influence comes from the applied electric field (qE) which contributes to the charging process that leads to an electrode's capacitance dominated by the redox activity. On the other hand, when a magnetic field is applied there is the appearance of a driving force that can accelerate the active species into the electrode's surface leading to a higher degree of intercalation process and therefore higher capacitance. In addition, increasing the magnetic field leads to an alignment of previously disorganized magnetic domains. However, higher values of B can cause the system to saturate and a decrease in F_L as ($v \times B$) tends to zero, resulting in saturation of the specific capacitance.

Based on the understanding of these phenomena for improvement in capacitance based on variations of a magnetic field over nanosized magnetic nanoparticles, Sharma et al. [51] studied nanoparticles based on ferromagnetic elements such as Ni, Mn, Co, and Fe. Within that line nano leaf-like, Fe_2O_3 were synthesized through forced hydrolysis and reflux condensation of Fe^{3+} mixed with $(NH_2)_2CO$ that acted as mineralizer agent followed by a calcination process. The electrochemical measurements were performed through cyclic voltammetry (CV) in the presence of a magnetic field as two magnetic coils connected to an energy supply were used and a three-electrode system was placed within them as seen in Figure 17.5a. The CV curves with different magnetic fields from 0, 1, 3, and 5 mT are presented in Figure 17.5b-e. Without a magnetic field, the CV profile of the Fe_2O_3 nano leaves displayed a commonly known behavior as with the increase of scan rate from 5 to 200 mV/s there was a decrease in specific capacitance from 88 to 29 F/g. This behavior was observed due to the shorter time that ions had to intercalate within the electrode's structure and perform the redox process. Hence, at higher scan rates the electrochemical behavior was predominantly based on the formation of an electric double layer as the pseudocapacitance process was suppressed. However, there was an increase in specific capacitance of around 61% when a magnetic field maximum of 5 mT was applied (Figure 17.5e). For values higher than that there was no further increment in the capacitance, which was referred to as saturation. Correlated to that, a similar behavior was also observed for galvanostatic charge-discharge (GCD) analysis as at a current density of 1 A/g there was

17 • Supercapacitor Applications 291

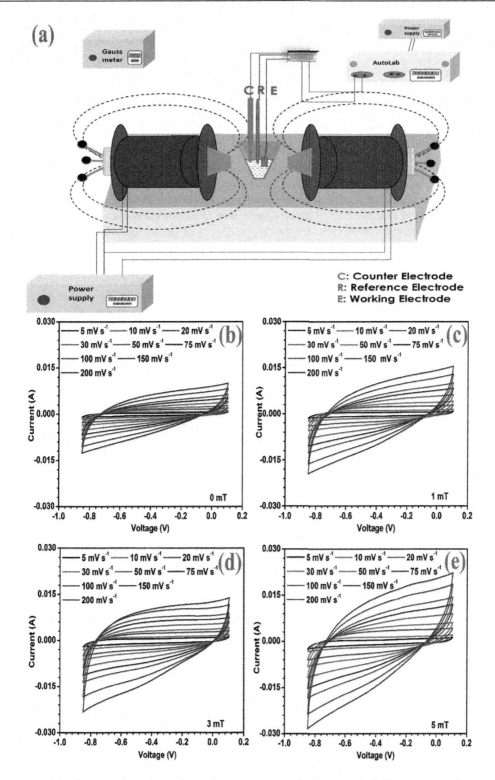

FIGURE 17.5 (a) Schematics for a three-electrode system exposed to a magnetic field. CV over applied magnetic fields of (b) 0, (c) 1, (d) 3, and (e) 5 mT, respectively. Adapted with permission [51].

Source: Copyright (2018), John Wiley and Sons.

an increase in specific capacitance from 86 to 134 F/g after exposure to a magnetic field that went from 0 to 5 mT whereas coulombic efficiencies remained nearly the same regardless of magnetic field variation. Similar effects have been observed for MnO_2 used as an electrode material for supercapacitor as when it was exposed to a magnetic field from 0 to 1.34 mT it had an increase of 119 to 142 F/g [50]. On the other hand, there has not been a focus on this effect for the case of batteries. However, a previous study demonstrated that Fe_2O_3-based electrodes that were induced to magnetization led to an improvement in cycling stability and energy stored [50]. Also, another work has demonstrated a considerable improvement in a Zn-air fuel cell after exposure to a magnetic field as there was an increase of transfer and diffusion coefficient by around 52 and 102%, respectively [49].

Based on this observation from several studies a mechanism for the improvement of electrochemical properties after exposure to a magnetic field has been proposed which can be described on four factors. (i) The presence of a magnetic field influences the movement of ionic species in solution. Furthermore, there is an increase in the ion's kinetic energy which results in an enhancement of the ion's population near the electrode. Also, the magnetic field can vary the charge density of anionic species due to the magnetohydrodynamic effect (MHDE) [52]. On top of that, MHDE can increase the limiting current as it leads to a diminishment of the Nernst layer as there is the emergence of a convection flow generated through the magnetic field. Within this line, the MHDE influences the electrochemical properties by decreasing the resistance of charge transfer and by improving the formation of an electric double layer at the electrode's surface. (ii) When the magnetic field is applied and aligned with the current lines ($F_L = 0$) there is an increase in the limiting current which is related to the variation in the magnetic field that appears between the neighboring domains within the material. This effect creates a paramagnetic force that changes the electrolyte's magnetic susceptibility (χ) [50]. Because of this magnetic susceptibility, which is a stimulated gradient, the solvated ions tend to move in the direction of the magnetic field which leads to an increment in ion (de)intercalation. Thus, increasing the overall specific capacitance. (iii) The presence of a magnetic field led to a stronger interaction between the electrode and electrolyte and therefore decreasing the charge transfer resistance. Also, the microscopic structure of water may change when exposed to a magnetic field leading to an increase in the dielectric constant of the electrolyte over the electrode's interface [49]. (iv) Alongside that, there is a distinguishability of the magnetic dipole moment of charged species as the magnetic field leads to the appearance of degenerated energy levels such as for electrons which were $\pm 1/2$ [49,53]. Because of that, electrons acquire a higher level of energy which is likely to play a role in the redox reactions that occur at the electrode. On top of that, there is an optimized EDLC behavior, which synergistically improves the overall system's capacitance. The steps of this proposed mechanism can be explicated in Figure 17.6.

This type of study enlightens novel possibilities for a drastic improvement in the electrochemical performance of magnetic nanomaterials. Based on this process Zhu et al. [53] proposed the synthesis of magnetic graphene nanocomposites (MGNCs) through a thermal decomposition approach which consisted of dissolution of graphene in dimethyl sulfoxide (DMF) followed by addition of $Fe(CO)_5$ and reflux of the mixture. Through that a composite of Fe_2O_3–graphene with relatively evenly distributed Fe_2O_3 was obtained with a preferential growth located at the defect of the edges. The authors compared the MGNC's electrochemical properties with those of pristine graphene. In that sense, the authors observed that without a magnetic field applied, MGNC presented a lower capacitance when compared to graphene which was likely due to the large loading of particles (52.5 wt.%) which increased the internal resistance and therefore hindered the electron transfer step in between the electrode and electrolyte. However, when a magnetic field was applied both MGNC as well as graphene presented a great enhancement in capacitance values. When graphene was under a scan rate of 2 and 10 mV/s its capacitance went from 67.1 and 26.8%, respectively. Yet, for the case of MGNC, also under 2 and 10 mV/s of scan rate, there was an increment of the capacitance of 154.6 and 98.2%, respectively. Based on these observations, it is notable that a magnetic field can function as an effective driving force to further enhance the supercapacitance. Also, for this case, there was an improvement even when the exposed material does not present magnetic properties. Hence, is likely to infer that the presence of a magnetic field prevented the relaxation at the electrode-electrolyte interface leading to an increase in

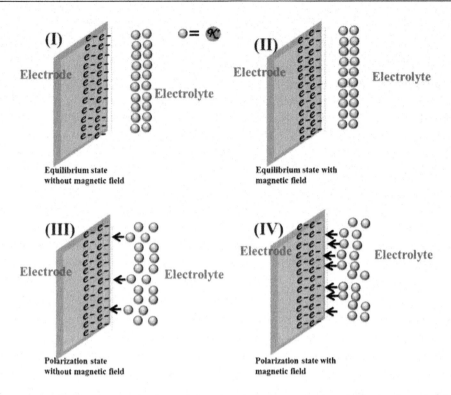

FIGURE 17.6 Scheme for the mechanism of the charging process at the electrode-electrolyte's interface when influenced by a magnetic field. Adapted with permission [51].

Source: Copyright (2018), John Wiley and Sons.

capacitance. However, further studies are still required to optimize the overall capacitance as during GCD testing the MGNC yielded values of specific capacitance that were as low as 6.5 F/g which were likely due to the high loading of Fe_2O_3 over graphene's surface which created a considerable transfer resistance. Hence, even though there was an advantageous effect on applying a magnetic field, there is also the requirement to optimize the morphology and proper distribution of electroactive materials over the matrix's surface.

Magnetic properties of Fe_2O_3 can also be employed in innovative ways other than inherently improving the electrochemical properties. Within this idea, Huang et al. [54] developed a yarn-based supercapacitor with self-healing properties based on the magnetism of Fe_2O_3. The fabrication of the supercapacitor composite consisted of a microwave-assisted hydrothermal method performed multiple times over a stainless-steel yarn which led to the formation of magnetic Fe_2O_3 particles over its surface. This process changed the yarn's aspect from silver metallic-like to black. An annealing process was performed to further increase the adhesion of Fe_2O_3 and magnetic properties. After that, a 2 μm layer of polypyrrole (PPy) was electrodeposited over the Fe_2O_3 functionalized yarn. The PPy functioned as electroactive material as well as a protective coating. Then, two electrode sets were intertwined with polyvinyl alcohol (PVA)-H_3PO_4 gel that functioned as a solid electrolyte. Finally, an outer layer of carboxylated polyurethane (PU) was introduced to promote self-healing properties. The advantage of this polymer in the composite lies in the spontaneous self-healing process due to a large number of hydrogen-bonding interactions throughout the polymeric chain that can heal a tear or a scratch caused to the device spontaneously. In addition, the PVA gel can reattach itself to some level due to adhesive interactions. On the other hand, PPy which is one of the active materials does not present self-healing properties. To counter that, the magnetic Fe_2O_3 film can function as a self-healing agent by performing magnetic interactions in between the yarn's fibers, leading to the reestablishment of conductivity and capacitive properties. The schematics for the fabrication and composite's self-healing properties are described in Figure 17.7(a-b). The composite's electrochemical behavior has been studied through CV which presented the EDLC mechanism of PPy (rectangular shape) and the redox behavior of Fe_2O_3 (non-rectangular shape). Based on the synergy of both components the self-healing supercapacitor device reached a specific capacitance of 61.4 mF/cm² at a scan rate of 10 mV/s. In addition, after the second and fourth healing cycles the specific capacitance dropped to 50.7 and 44.1 mF/cm², respectively, along with a 71.8% of capacity retention after the fourth healing process as shown in Figure 17.7c. The work presented in this study showed a simultaneous use of Fe_2O_3 both as a magnetic component to aid on self-healing properties as well as an electroactive material for supercapacitance, along with a relatively facile and well-known fabrication process. Hence, variations of such ideas can enlighten their use for similar approaches.

One of the reasons for the efficient synthesis of nanomaterials is due to the enhancement of properties that they acquire when compared to their bulky counterparts due to quantum confinement which is related to the transition between behavior that follows the classic mechanics to quantum mechanics that is reached when the material is within the nanoscale. Based on these concepts, it deems important to maintain the size of metal oxides in between 1 to 100 nm to further enhance their properties. However, there is a common tendency for these materials to agglomerate, which often leads to a decrease in properties. In that sense, one of the strategies lies in coating the metal oxide with carbon-based materials as it can provide a protective layer against agglomeration, improvement of chemical stability, and increase of the active area of nanomaterials. Through this, Vermisoglou et al. [55] proposed the synthesis of a Fe_3C/reduced graphene oxide core/shell nanostructure synthesized by solution intercalation of a graphene oxide dispersion with p-aminobenzoic acid and $FeCl_3 \cdot 6H_2O$. In addition, the synthesis was also performed under alkaline and acid pH which led to different mechanisms. In

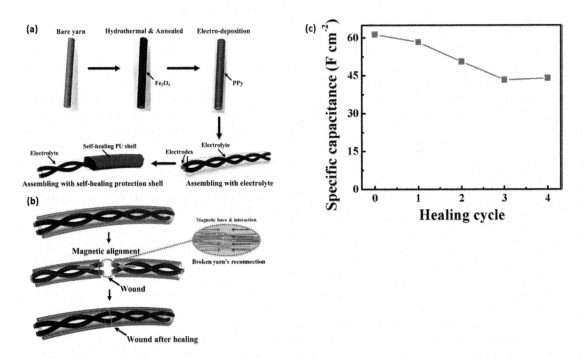

FIGURE 17.7 (a) Self-healing supercapacitor fabrication process. (b) Composite's self-healing mechanism. (c) Specific capacitance in the function of the healing cycles. Adapted with permission [54].

Source: Copyright (2015), American Chemical Society.

this line, acid media promoted nucleophilic substitution named IGO. On the other hand, when the synthesis was performed at basic media (IGO/b) the mechanism formation occurred through cation exchange. After that both samples, IGO and IGO/b, went through thermal annealing to promote the thermal reduction of the oxygenated groups over graphene's surface leading to r-IGO and r-IGO/b, respectively containing Fe_3C nanoparticles of around 35 to 50 nm distributed over reduced graphene's surface. The nanocomposites were first characterized through a CV. It was observed that, when the synthesis was performed at basic media, r-IGO/b, the nanocomposite presented a porosity of 163 m²/g which was almost double of the one obtained in acid media, r-IGO, which presented 87 m²/g. However, r-IGO/b displayed a slightly higher internal resistance of 3.7 Ω compared to r-IGO with 3.0 Ω, which could be attributed to the ionic entrapment within the structure of r-IGO/b. Yet, despite that the higher surface area of r-IGO/b was likely to counter that drawback as it displayed a specific capacitance of 17 F/g when compared to r-IGO with 6 F/g, which suggested that higher surface area played a stronger influence over the specific capacitance without causing a considerable disadvantageous ionic entrapment during ionic permeation process. On the other hand, the synthetical approach seemed to have played a role in the magnetic properties of r-IGO/b and r-IGO since they presented saturation magnetization at 300 K of 43 and 70 J/T/kg, respectively. This difference was attributed to the higher concentration of magnetic α-Fe in r-IGO.

Aside from Fe_2O_3 and its derivatives, other transition metal-based materials also display great applicability within the energy field such as NiO, MnO_2, Co_3O_4, $NiCo_2O_4$, among others. Among them, NiO is a viable option as it presents satisfactory electrochemical as well as magnetic properties when obtained in the nanoscale. More specifically, it has been noted that NiO can yield several morphologies such as nanoparticles, nanorods, and nanoflakes, which are 0D, 1D, and 2D, respectively. Such morphologies can lead to an overall increment in properties as these induce a structure with high porosity and conductivity, which are core factors for proper electrochemical and magnetic activity. Anandha et al. [56] synthesized NiO nanoflakes through a surfactant-free microwave-assisted method. The authors observed that higher microwave power of 600 and 900 W yielded nanomaterials with different properties which were further evaluated. It was observed that when the synthesis was performed at higher microwave wattage there was a higher magnetization that went from 0.0052 to 0.23–0.3 emu/g. Such variance was likely related to the morphology since higher microwave power led to morphologies with better uniformity and reduced agglomeration. Based on that, nanoflowers and nanoflakes were obtained when 450 and 900 W were applied, respectively. The schematic for this process is displayed in Figure 17.8. Nanoflakes are usually desired for NiO as they have demonstrated a higher specific capacitance based on previous studies, mostly due to higher conductivity and exposed surface area for better interaction with electrolytes [57,58]. It was observed that the NiO nanoflakes displayed satisfactory electrochemical behavior as they presented a relatively similar CV profile even with varying scan rates from 5 to 50 mV/s which suggested an optimized mass transport and electron transfer steps within the nanoflakes. In that regard, a pseudocapacitive behavior was observed which was determined by the Ni^{2+}/Ni^{3+} redox pair (equation 2). The highest specific capacitance obtained was 252 F/g at a scan rate of 5 mV/s when the microwave-assisted synthesis was performed at 900 W. Also, the NiO nanoflakes presented high cyclic stability with capacitance retention of 87% after 1200 suggesting a long-term use. Herein, this approach demonstrates a facile way to obtain highly organized and stable NiO nanoflakes with satisfactory magnetic and electrochemical properties.

$$NiO + OH^- \leftrightarrow NiOOH + e^- \qquad (2)$$

Another class of electroactive materials that are rapidly gaining attention within the scientific community is the transition metal sulfides, which can potentially surpass the current technology based on LIBs. Within this branch of research, there have been some efforts into using CoS and its stoichiometric variants such as Co_9S_8, Co_3S_4, and CoS_2 for example due to several factors such as relatively low cost, high redox activity as well as electrochemical stability. Hence, to further explore the properties of CoS, Ashok et al. [59] synthesized a 3D nanoflower structure of this metal sulfide through the facile hydrothermal

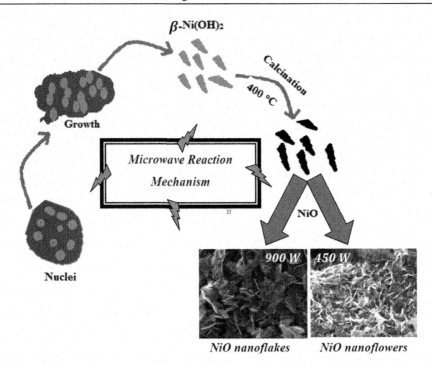

FIGURE 17.8 Scheme for the formation of NiO nanostructures based on microwave power. Adapted with permission [56].

Source: Copyright (2015), Royal Society of Chemistry.

method Figure 17.9a. Through that, satisfactory properties in terms of power density and cycling stability were obtained. On top of that, the CoS nanoflower-like structure (Figure 17.9b) presented temperature-dependent magnetic properties which can enlarge its field of application. Based on these properties, the electrochemical analysis of CoS was performed. In that sense, during the CV analysis, a profile influenced by redox activity of Co^{2+}/Co^{3+} was formed. The suggested mechanism is presented in equation 3 and 4. The CV plot in Figure 17.10c showed relatively similar anodic and cathodic peaks at lower scan rates which suggested a good rate capability for the redox process. Alongside that, the CoS nanoflowers also presented excellent cycling stability with capacitance retention of 97.2% after 1000 cycles. Such results infer that the CoS nanoflower could perform a fast redox process without compromising its structure or detaching from the electrode which are important factors for large-scale applications.

$$CoS + OH^- \leftrightarrow CoSOH + e^- \qquad (3)$$
$$CoSOH + OH^- \leftrightarrow CoSO + H_2O + e^- \qquad (4)$$

The effect of temperature in the magnetic analysis for the CoS was also performed. A metallic behavior was noticed as there was an increase in conductivity at lower temperatures. Yet, these factors also seemed to be influenced by the nanoflower structure which facilitated electron hopping due to the organized and bounded grains. In addition, the further decrease of temperature led to a change from metallic to semiconducting behavior which could be explained through the Peierls transition. In that sense, the analysis was made based on the CoS nanoflower with dimensions of 46 nm which can be considered as a 1D system along with the electronic configuration of Co as [Ar] $3d^7$, $4s^{2,}$ and S as [Ne] $3s^2\,3p^4$. Since there are 2 electrons provided by Co to occupy the S atom's valency band there is a hole in the d state of Co. However, the authors identified a Co_3S_4 defective structure which caused a charge imbalance that led to a metallic

17 • Supercapacitor Applications 297

FIGURE 17.9 (a) Hydrothermal approach for the synthesis of CoS nanoflowers. (b) High-resolution scanning electron microscopy (HRSEM) displays the nanoflower-like structure of CoS. (c) CV plot for the CoS at varying scan rates. (d) Curie-Weiss plot for the CoS in ZFC. Adapted from [59].

Source: Copyright (2019), The authors, Publisher Nature. This article is licensed under a Creative Commons Attribution 4.0 International License.

behavior of Co until there was a thermal disturbance that was large enough to promote fluctuations in this state. When the system was exposed to a lower temperature of around 20 K the atoms tended to get closer leading to a periodicity in the system causing it to behave as a 2D nanostructure. Through that, the electrons were unlikely to stay at their lattices causing them to only vibrate. Because of that, the vibrational distance of electrons can differ being shorter or longer due to the dimensional change. Yet, energy input is necessary to change the electron's state. Hence, because of this transition, the system behaved as a semiconductor as more energy was required allowing electrons to move more freely at higher temperatures. Through this discussion, it was observed that the CoS nanoflowers presented a lower magnetization when exposed to lower temperatures of around 120 K. However, above 120 K there was a quick increase in the magnetic properties which pointed to a ferromagnetic or ferrimagnetic transition which is seen in the Curie-Weiss plot for the zero-field cooled (ZFC) in Figure 17.9d.

Molybdate oxides and derivates represent an important type of materials that show great potential for energy storage application due to their applicable redox properties and versatility in synthetical approaches such as hydrothermal, hard template, ultrasonic-assisted, hydrothermal, precipitation, solution combustion, among others. Seevakan et al. [60] performed the synthesis of $Bi_2Mo_2O_9$ through microwave combustion followed by a study of its electrochemical and magnetic properties. This approach led to the formation of agglomerated spherical particles with dimensions between 80 to 100 nm, which could be attributed to the ion-by-ion addition along with the inherent magnetism of $Bi_2Mo_2O_9$. Following that, the magnetic properties were also measured through a vibrating sample magnetometer (VSM) that demonstrated a paramagnetic behavior observed from the magnetic hysteresis (M–H) loop. $Bi_2Mo_2O_9$ presented a coercivity (Hc) of 570.54 kOe, saturation magnetization (Ms) of 4.49×10^{-4} emu/g, and remnant (Mr) of 0.45×10^{-4} emu/g (Figure 17.10a). Such results were likely influenced by the nanoparticle's morphology,

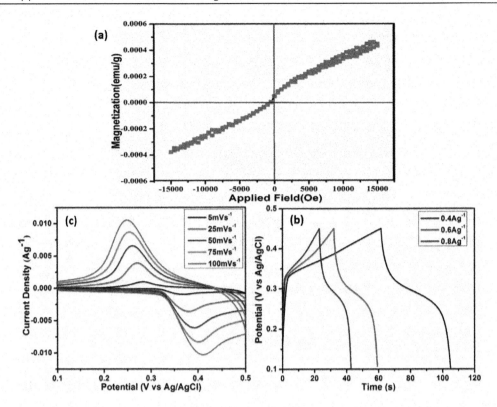

FIGURE 17.10 Magnetic and electrochemical properties of $Bi_2Mo_2O_9$ in terms of (a) VSM, (b) CV, and (c) GCD. Adapted with permission [60].

Source: Copyright (2019), Elsevier.

crystal structure, size, and synthetical approach. The electrochemical properties were analyzed first through CV which demonstrated a dominant pseudocapacitance behavior that could be originated by the redox process of Bi^{3+} to Bi^0 [61,62]. The CV profile of $Bi_2Mo_2O_9$ is provided in Figure 17.10b. Similarly, to that, the GCD curves also displayed a profile related to pseudocapacitive nature due to the asymmetric triangular shape Figure 17.10c. In addition, the $Bi_2Mo_2O_9$ presented good electrochemical stability after 1000 cycles presenting a 93% capacitance retention suggesting this nanomaterial as a potential candidate for long-term use as a supercapacitor.

5 CONCLUSION

Throughout the discussion in this chapter, it was notable that there is a vast and unexplored branch within the field of energy storage applications that focuses on the influence of magnetic properties over the electrochemical ones, which can provide novel insight and guidelines for research in the near future. The electrochemical properties can be studied under different lenses that can provide greater insight and areas of applications for nanomaterials such as transition metal oxides, hydrides, sulfides, bimetallic, and its derivate composites which were almost exclusively used in energy storage and electrocatalysis-related applications. As observed from the discussions in this chapter, there is a considerable improvement in electrochemical properties for some nanomaterials that are exposed to a magnetic field due to several proposed mechanisms such as better interaction between electrode and electrolyte that shorter the distance between them and therefore facilitates the diffusion of ions, among other factors. Also, further exploring the magnetic properties of nanomaterials such as Fe_2O_3, NiO, CoS, along with its derivatives and composites can provide novel ideas for applications that can make simultaneous use of both electrochemical and magnetic properties since there is well-established knowledge regarding these nanomaterials in terms of synthetical routes, characterization, and electrochemical properties in the current literature. On the other hand, a small amount of research work has been devoted to correlating and explaining the actual effects of magnetism and energy storage. Since the results demonstrated so far pointed to an advantageous and synergetic process, it can be seen as a promising research niche to be further explored. Within this line and as observed in the discussions provided by this chapter, the magnetic properties can also be influenced by the morphology, level of crystallinity, nanoparticle size, synthetical approach among other factors and parameters. There is a plethora of experimental designs that can be performed to tune the properties of these nanomaterials according to the desired application. Thus, based on the current literature, it is notable that nanomagnetic materials can be an interesting research path to provide an overall optimization to the performance of supercapacitor and electronic devices.

REFERENCES

[1] H. Chen, M. Ling, L. Hencz, H.Y. Ling, G. Li, Z. Lin, G. Liu, S. Zhang, Exploring chemical, mechanical, and electrical functionalities of binders for advanced energy-storage devices, *Chem. Rev.* 118 (2018) 8936–8982.
[2] S. Yang, R.E. Bachman, X. Feng, K. Müllen, Use of organic precursors and graphenes in the controlled synthesis of carbon-containing nanomaterials for energy storage and conversion, *Acc. Chem. Res.* 46 (2013) 116–128.
[3] M. Vijayakumar, R. Santhosh, J. Adduru, T.N. Rao, M. Karthik, Activated carbon fibres as high performance supercapacitor electrodes with commercial level mass loading, *Carbon N. Y.* 140 (2018) 465–476.
[4] M. Vijayakumar, A. Bharathi Sankar, D. Sri Rohita, T.N. Rao, M. Karthik, Conversion of biomass waste into high performance supercapacitor electrodes for real-time supercapacitor applications, *ACS Sustain. Chem. Eng.* 7 (2019) 17175–17185.
[5] A.E. Elkholy, F. El-Taib Heakal, N.K. Allam, Nanostructured spinel manganese cobalt ferrite for high-performance supercapacitors, *RSC Adv.* 7 (2017) 51888–51895.
[6] N.O. Laschuk, I.I. Ebralidze, E.B. Easton, O.V. Zenkina, Systematic design of electrochromic energy storage devices based on metal-organic monolayers, *ACS Appl. Energy Mater.* 4 (2021) 3469–3479.
[7] P. Nakhanivej, Q. Dou, P. Xiong, H.S. Park, Two-dimensional pseudocapacitive nanomaterials for high-energy- and high-power-oriented applications of supercapacitors, *Accounts Mater. Res.* 2 (2021) 86–96.
[8] L. Yang, B. Gu, Z. Chen, Y. Yue, W. Wang, H. Zhang, X. Liu, S. Ren, W. Yang, Y. Li, Synthetic biopigment supercapacitors, *ACS Appl. Mater. Interfaces.* 11 (2019) 30360–30367.
[9] W. Zhao, S.J.B. Rubio, Y. Dang, S.L. Suib, Green electrochemical energy storage devices based on sustainable manganese dioxides, *ACS ES&T Eng.* 2 (2022) 20–42.
[10] Z.S. Iro, C. Subramani, S.S. Dash, A brief review on electrode materials for supercapacitor, *Int. J. Electrochem. Sci.* 11 (2016) 10628–10643.
[11] M.A. Pope, S. Korkut, C. Punckt, I.A. Aksay, Supercapacitor electrodes produced through evaporative consolidation of graphene oxide-water-ionic liquid gels, *J. Electrochem. Soc.* 160 (2013) A1653–A1660.
[12] Y. Zhang, H. Feng, X. Wu, L. Wang, A. Zhang, T. Xia, H. Dong, X. Li, L. Zhang, Progress of electrochemical capacitor electrode materials: A review, *Int. J. Hydrogen Energy.* 34 (2009) 4889–4899.
[13] V.K. Yadav, N. Bhardwaj, Introduction to supercapacitors and supercapacitor assisted engine starting system, *Int. J. Sci. Eng. Res.* 4 (2013) 583–588.
[14] R. Ramya, R. Sivasubramanian, M. V. Sangaranarayanan, Conducting polymers-based electrochemical supercapacitors – Progress and prospects, *Electrochim. Acta.* 101 (2013) 109–129.
[15] L.Z. Fan, J. Maier, High-performance polypyrrole electrode materials for redox supercapacitors, *Electrochem. Commun.* 8 (2006) 937–940.
[16] L.L. Zhang, X.S. Zhao, Carbon-based materials as supercapacitor electrodes, *Chem. Soc. Rev.* 38 (2009) 2520–2531.

[17] G. Wang, L. Zhang, J. Zhang, A review of electrode materials for electrochemical supercapacitors, *Chem. Soc. Rev.* 41 (2012) 797–828.

[18] L. Liu, H. Zhang, Y. Mu, Y. Bai, Y. Wang, Binary cobalt ferrite nanomesh arrays as the advanced binder-free electrode for applications in oxygen evolution reaction and supercapacitors, *J. Power Sources*. 327 (2016) 599–609.

[19] M.K. Zate, S.M.F. Shaikh, V. V. Jadhav, K.K. Tehare, S.S. Kolekar, R.S. Mane, M. Naushad, B.N. Pawar, K.N. Hui, Synthesis and electrochemical supercapacitive performance of nickel-manganese ferrite composite films, *J. Anal. Appl. Pyrolysis*. 116 (2015) 177–182.

[20] C. Guan, J. Liu, Y. Wang, L. Mao, Z. Fan, Z. Shen, H. Zhang, J. Wang, Iron oxide-decorated carbon for supercapacitor anodes with ultrahigh energy density and outstanding cycling stability, *ACS Nano*. 9 (2015) 5198–5207.

[21] M.I.A. Abdel Maksoud, R.A. Fahim, A.E. Shalan, M. Abd Elkodous, S.O. Olojede, A.I. Osman, C. Farrell, A.H. Al-Muhtaseb, A.S. Awed, A.H. Ashour, D.W. Rooney, *Advanced Materials and Technologies for Supercapacitors Used in Energy Conversion and Storage: A Review*, Springer International Publishing, 2021.

[22] A. Chowdhury, A. Dhar, S. Biswas, V. Sharma, P.S. Burada, A. Chandra, Theoretical model for magnetic supercapacitors-from the electrode material to electrolyte ion dependence, *J. Phys. Chem. C*. 124 (2020) 26613–26624.

[23] R. Packiaraj, P. Devendran, K.S. Venkatesh, S. Asath Bahadur, A. Manikandan, N. Nallamuthu, Electrochemical investigations of magnetic Co3O4 nanoparticles as an active electrode for supercapacitor applications, *J. Supercond. Nov. Magn.* 32 (2019) 2427–2436.

[24] F.M. Ismail, M. Ramadan, A.M. Abdellah, I. Ismail, N.K. Allam, Mesoporous spinel manganese zinc ferrite for high-performance supercapacitors, *J. Electroanal. Chem.* 817 (2018) 111–117.

[25] T. Hyeon, Chemical synthesis of magnetic nanoparticles, *Chem. Commun.* 3 (2003) 927–934.

[26] C.R. Vestal, Z.J. Zhang, Effects of surface coordination chemistry on the magnetic properties of MnFe2O4 spinel ferrite nanoparticles, *J. Am. Chem. Soc.* 125 (2003) 9828–9833.

[27] C. Hu, Z. Gao, X. Yang, One-pot low temperature synthesis of MFe2O4 (M = Co, Ni, Zn) superparamagnetic nanocrystals, *J. Magn. Magn. Mater.* 320 (2008) 70–73.

[28] D.R. Modeshia, R.I. Walton, Solvothermal synthesis of perovskites and pyrochlores: Crystallisation of functional oxides under mild conditions, *Chem. Soc. Rev.* 39 (2010) 4303–4325.

[29] S. Yáñez-Vilar, M. Sánchez-Andújar, C. Gómez-Aguirre, J. Mira, M.A. Señarís-Rodríguez, S. Castro-García, A simple solvothermal synthesis of MFe2O4 (M = Mn, Co and Ni) nanoparticles, *J. Solid State Chem.* 182 (2009) 2685–2690.

[30] J. Lai, W. Niu, R. Luque, G. Xu, Solvothermal synthesis of metal nanocrystals and their applications, *Nano Today*. 10 (2015) 240–267.

[31] A.M. Elseman, M.G. Fayed, S.G. Mohamed, D.A. Rayan, N.K. Allam, M.M. Rashad, Q.L. Song, A novel composite CoFe2O4@CSs as electrode by easy one-step solvothermal for enhancing the electrochemical performance of hybrid supercapacitors, *ChemElectroChem*. 7 (2020) 526–534.

[32] E. Umeshbabu, G. Rajeshkhanna, P. Justin, G. Ranga Rao, Synthesis of mesoporous NiCo2O4-rGO by a solvothermal method for charge storage applications, *RSC Adv.* 5 (2015) 66657–66666.

[33] D.H. Chen, X.R. He, Synthesis of nickel ferrite nanoparticles by sol-gel method, *Mater. Res. Bull.* 36 (2001) 1369–1377.

[34] D.H.K. Reddy, S.M. Lee, Three-dimensional porous spinel ferrite as an adsorbent for Pb(II) removal from aqueous solutions, *Ind. Eng. Chem. Res.* 52 (2013) 15789–15800.

[35] R.S. Kate, S.A. Khalate, R.J. Deokate, Overview of nanostructured metal oxides and pure nickel oxide (NiO) electrodes for supercapacitors: A review, *J. Alloys Compd.* 734 (2018) 89–111.

[36] Y.Q. Wu, X.Y. Chen, P.T. Ji, Q.Q. Zhou, Sol-gel approach for controllable synthesis and electrochemical properties of NiCo2O4 crystals as electrode materials for application in supercapacitors, *Electrochim. Acta*. 56 (2011) 7517–7522.

[37] C.I. Priyadharsini, G. Marimuthu, T. Pazhanivel, P.M. Anbarasan, V. Aroulmoji, V. Siva, L. Mohana, Sol – Gel synthesis of Co3O4 nanoparticles as an electrode material for supercapacitor applications, *J. Sol-Gel Sci. Technol.* 96 (2020) 416–422.

[38] L. Liu, X. Yu, W. Zhang, Q. Lv, L. Hou, Y. Fautrelle, Z. Ren, G. Cao, X. Lu, X. Li, Strong magnetic-field-engineered porous template for fabricating hierarchical porous Ni – Co – Zn – P nanoplate arrays as battery-type electrodes of advanced all-solid-state supercapacitors, *ACS Appl. Mater. Interfaces*. 14 (2022) 2782–2793.

[39] A. Sutka, G. Strikis, G. Mezinskis, A. Lusis, J. Zavickis, J. Kleperis, D. Jakovlevs, Properties of Ni-Zn ferrite thin films deposited using spray pyrolysis, *Thin Solid Films*. 526 (2012) 65–69.

[40] S. Vijayakumar, A. Kiruthika Ponnalagi, S. Nagamuthu, G. Muralidharan, Microwave assisted synthesis of Co3O4 nanoparticles for high-performance supercapacitors, *Electrochim. Acta*. 106 (2013) 500–505.

[41] M. Vangari, T. Pryor, L. Jiang, Supercapacitors: Review of materials and fabrication methods, *J. Energy Eng.* 139 (2013) 72–79.
[42] J.T. Mefford, W.G. Hardin, S. Dai, K.P. Johnston, K.J. Stevenson, Anion charge storage through oxygen intercalation in LaMnO 3 perovskite pseudocapacitor electrodes, *Nat. Mater.* 13 (2014) 726–732.
[43] W. Raza, F. Ali, N. Raza, Y. Luo, K.H. Kim, J. Yang, S. Kumar, A. Mehmood, E.E. Kwon, Recent advancements in supercapacitor technology, *Nano Energy.* 52 (2018) 441–473.
[44] J. Libich, J. Máca, J. Vondrák, O. Čech, M. Sedlaříková, Supercapacitors: Properties and applications, *J. Energy Storage.* 17 (2018) 224–227.
[45] M. Tomy, A. Ambika Rajappan, V. VM, X. Thankappan suryabai, emergence of novel 2D materials for high-performance supercapacitor electrode applications: A brief review, *Energy and Fuels.* 35 (2021) 19881–19900.
[46] L. Zhou, C. Li, X. Liu, Y. Zhu, Y. Wu, T. van Ree, *Metal Oxides in Supercapacitors*, Elsevier Inc., 2018.
[47] F. Wang, S. Xiao, Y. Hou, C. Hu, L. Liu, Y. Wu, Electrode materials for aqueous asymmetric supercapacitors, *RSC Adv.* 3 (2013) 13059–13084.
[48] W. Kiciński, J.P. Sęk, E. Matysiak-Brynda, K. Miecznikowski, M. Donten, B. Budner, A.M. Nowicka, Enhancement of PGM-free oxygen reduction electrocatalyst performance for conventional and enzymatic fuel cells: The influence of an external magnetic field, *Appl. Catal. B Environ.* 258 (2019) 117955.
[49] J. Shi, H. Xu, L. Lu, X. Sun, Study of magnetic field to promote oxygen transfer and its application in zinc – air fuel cells, *Electrochim. Acta.* 90 (2013) 44–52.
[50] A. Bund, S. Koehler, H.H. Kuehnlein, W. Plieth, Magnetic field effects in electrochemical reactions, *Electrochim. Acta.* 49 (2003) 147–152.
[51] V. Sharma, S. Biswas, A. Chandra, Need for revisiting the use of magnetic oxides as electrode materials in supercapacitors: Unequivocal evidence of significant variation in specific capacitance under variable magnetic field, *Adv. Energy Mater.* 8 (2018) 1800573.
[52] R.N. O'Brien, The magnetohydrodynamic effect In electrochemical systems including batteries, *ECS Trans.* 3 (2007) 23–29.
[53] J. Zhu, M. Chen, H. Qu, Z. Luo, S. Wu, H.A. Colorado, S. Wei, Z. Guo, Magnetic field induced capacitance enhancement in graphene and magnetic graphene nanocomposites, *Energy Environ. Sci.* 6 (2013) 194–204.
[54] Y. Huang, Y. Huang, M. Zhu, W. Meng, Z. Pei, C. Liu, H. Hu, C. Zhi, Magnetic-assisted, self-healable, yarn-based supercapacitor, *ACS Nano.* 9 (2015) 6242–6251.
[55] E.C. Vermisoglou, E. Devlin, T. Giannakopoulou, G. Romanos, N. Boukos, V. Psycharis, C. Lei, C. Lekakou, D. Petridis, C. Trapalis, Reduced graphene oxide/iron carbide nanocomposites for magnetic and supercapacitor applications, *J. Alloys Compd.* 590 (2014) 102–109.
[56] G. Anandha Babu, G. Ravi, T. Mahalingam, M. Kumaresavanji, Y. Hayakawa, Influence of microwave power on the preparation of NiO nanoflakes for enhanced magnetic and supercapacitor applications, *Dalt. Trans.* 44 (2015) 4485–4497.
[57] P. Justin, S.K. Meher, G.R. Rao, Tuning of capacitance behavior of NiO using anionic, cationic, and nonionic surfactants by hydrothermal synthesis, *J. Phys. Chem. C.* 114 (2010) 5203–5210.
[58] Y.-G. Zhu, G.-S. Cao, C.-Y. Sun, J. Xie, S.-Y. Liu, T.-J. Zhu, X.B. Zhao, H.Y. Yang, Design and synthesis of NiO nanoflakes/graphene nanocomposite as high performance electrodes of pseudocapacitor, *RSC Adv.* 3 (2013) 19409–19415.
[59] K. Ashok Kumar, A. Pandurangan, S. Arumugam, M. Sathiskumar, Effect of bi-functional hierarchical flower-like CoS nanostructure on its interfacial charge transport kinetics, magnetic and electrochemical behaviors for supercapacitor and DSSC applications, *Sci. Rep.* 9 (2019) 1228.
[60] K. Seevakan, A. Manikandan, P. Devendran, Y. Slimani, A. Baykal, T. Alagesan, Structural, magnetic and electrochemical characterizations of Bi2Mo2O9 nanoparticle for supercapacitor application, *J. Magn. Magn. Mater.* 486 (2019) 165254.
[61] V. Vivier, A. Régis, G. Sagon, J.-Y. Nedelec, L.T. Yu, C. Cachet-Vivier, Cyclic voltammetry study of bismuth oxide Bi2O3 powder by means of a cavity microelectrode coupled with Raman microspectrometry, *Electrochim. Acta.* 46 (2001) 907–914.
[62] B. Sarma, A.L. Jurovitzki, Y.R. Smith, S.K. Mohanty, M. Misra, Redox-induced enhancement in interfacial capacitance of the titania nanotube/bismuth oxide composite electrode, *ACS Appl. Mater. Interfaces.* 5 (2013) 1688–1697.

Magnetic Nanomaterials for Flexible Spintronics

18

Felipe M. de Souza, Ram K. Gupta

Contents

1 Introduction 303
2 Fundamentals of Spintronic 305
3 Flexible Spintronic 307
   3.1 Graphene-Based Flexible Spintronic 307
   3.2 2D Materials beyond Graphene for Flexible Spintronic 310
   3.3 Metal Oxide-Based Flexible Spintronic 312
   3.4 Chalcogenides-Based Flexible Spintronic 314
4 Conclusion 316
References 316

1 INTRODUCTION

The latest technological advancements using nanomaterials are reaching several important sectors such as energy storage, biomedical, and electronic devices which enabled the advent of optimized applications. In this sense, the development in these areas is branching out to flexible electronics and wearable devices which can be conveniently used in several types of products such as prosthesis or epidermal implants, soft robotic instruments, transient and imperceptible electronics, solar cells, radio-frequency identification, among many other electronic devices [1,2]. The use of materials with magnetic properties can play a major role in such applications since they can be used as magnetic sensors to enable the mechanical movement of robotics as well as in vivo medical implants. Hence, materials with such properties can be employed as monitors of heart valves or joints, for instance. The proper synthesis of magnetic materials plays an important role in the fabrication of such devices. One of the aims for the satisfactory applications of such materials is related to the combination of ferroelectricity along with ferromagnetism, known as the magnetoelectric effect. Materials with magnetoelectric properties can induce magnetization when exposed to an electric field along with being able to polarize when exposed to a magnetic field [3].

However, compounds based on magnetic metals such as Fe, Ni, Co, among other rare metals, that can be obtained as oxides, bimetallic or other structures, are usually rigid and brittle which imposes an initial challenge for the fabrication of flexible devices. One of the main approaches adopted to counter such issues lies in the use of flexible substrates such as thin Si wafers or polymers, for instance, that allow the magnetic materials to be incorporated into its structure while maintaining its mechanical and flexible properties. In addition, it is also deemed important to control the morphology of the magnetic materials to simultaneously provide a higher surface area, proper distribution, and attractive interactions along the matrix to avoid detachment of active material. With this concept, it is usually preferable to use 1D or 0D nanomaterials such as nanowires or quantum dots, respectively, as these would be more likely to disperse within the substrate while promoting anisotropic magnetic properties. Yet, achieving those conditions can be often challenging, hence other types of polymeric nanocomposites can be designed that can consist of particulate, laminate, or polymers used as a binder [4].

Aside from these requirements to fabricate flexible devices that consist of polymeric substrates embedded with nanoparticles there is also the requirement of developing materials that present satisfactory magnetic and electroactive properties which have pointed to the study of spintronics. The field of spintronics explores the use of three features of the electron that can function as an information carrier: (a) electron's spin, (b) charge, and (c) photons (Figure 18.1). These electron's properties are the core factors for the information technology field as the data processing, storage, and transfer are based on the electron's flow (e⁻), spin, and optical connection (hv), respectively. The further exploration of these phenomena opens great potential for the optimization of current electronic devices. The improvement in this area can be taken into perspective when cassette tapes were used as magnetic media which consisted of granular ferromagnetic particles. The binary information (0 or 1) was coded based on the magnetic domains which could be magnetized to one direction or the opposite concerning the axis of anisotropy. However, such domains consisted of micrometer-sized grains which influenced the device's size as well. Nowadays, the size of these ferromagnetic materials has reached the nanoscale which in practice allows them to store more information. Yet, higher magnetic anisotropy is required to maintain the thermal stability for the grain's magnetization. Hence, nanotechnology played a major role in information technology. Another breakthrough came on the revelation of two phenomena: (a) tunneling magnetoresistance (TMR) and (b) giant magnetoresistance (GMR). TMR has attracted attention as it can now be performed at room temperature. Also, GMR was noted in metallic multilayer based on electron transport that depends on the spin, which is the current technology employed in hard disk drive [5]. It is worth mentioning that

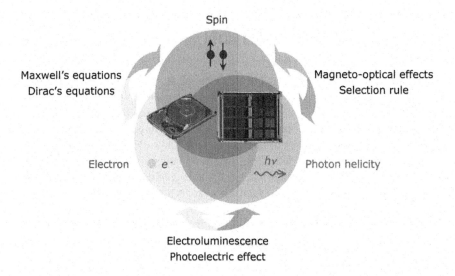

FIGURE 18.1 Scheme for the correlation between electron's characteristics with the application on spintronic devices. Adapted with permission [8].

Source: Copyright (2014), IOP Publishing.

semiconductor materials greatly contribute to this technological advance due to their properties such as carrier- and photo-induced ferromagnetisms [6,7].

2 FUNDAMENTALS OF SPINTRONIC

Spintronics is the core for the development of cutting-edge hardware technology to further improve their performance, speed, data transfer, storage, and processing, which makes it a field with superb relevance and investment. Based on that, it is deemed important to elucidate the fundamentals of spintronics. GMR and TMR are important phenomena that are based on the interaction between s–d orbitals with a local magnetic moment along with the conducting electron that is spin-polarized. Hence, there is a combination of both electronic and magnetism that is based on the transport phenomenon through spin-polarized electrons. Through that, a spin-transfer torque (STT) can be performed onto a neighbor magnetic moment on a conductor. Another way consists of a spin-wave propagation through local magnetic moments on an insulator.

One of the current aims within the scientific community lies in optimizing this process by developing 3D hierarchically organized structures along with 0D spintronics, which can allow patterning within the nanoscale. Hence, achieving the ideal size for the spintronic material is an important aspect to obtain appreciable properties. Reducing the ferromagnet material's size leads to the rise of a dipole field from the edges that can create a splitting of a single-domain state. Also, if the particle's size is further reduced then the exchange interaction becomes more relevant which leads to the predominance of the single-domain state. The spins of the electrons become aligned with the global easy axis which is determined by the magnetic anisotropy. There is a certain number of spins for each unit of magnetization at which the spins have their orientation preserved. This is named spin diffusion length (d_{sd}). Below d_{sd}, the exchange length (d_{ex}) of the spin-polarized moments or electrons can control the change in spin orientation. For conducting electrons, a further decrease of length from d_{ex} allows the electron's spins to move without scattering within the mean free path (λ). Through that, selecting a specific lateral size can promote specific magnetic properties. These concepts are applied on hard disk drives which consist of a system that the length between the magnetic materials is shorter than λ [8,9]. This condition allows the spin to be preserved during transport. Based on these concepts, the current goal of research in spintronics is to build a quantum computer by making use of the long spin coherence length of semiconductors [10]. The continuous study on that can lead to the development of single-electron transistors (SET) which can potentially process data through a single electron [11].

The generation of spins is the core aspect for spintronics as they can be generated in non-magnetic materials through spin injection methods from a magnetic or electric field, ferromagnet materials, thermal gradient, circularly polarized photoexcitation, and Zeeman splitting. These methodologies are presented in Figure 18.2. A widely used methodology consists of spin injection from a ferromagnetic material such as Ni, Co, Fe, Gd, half-metallic ferromagnetic, and dense media separation (DMS) which is placed in contact with a semiconductor or non-magnetic metal by a tunnel barrier or an ohmic contact. Also, spin-polarized carriers can be induced through an electric field in a non-magnetic material based on the principles of the Hall effect. The use of circularly polarized light functions as an energy source to excite spin-polarized electrons in a semiconductor material. Oppositely, a current of a spin-polarized electron can generate a circularly polarized light. This process can be extended by electromagnetic waves that include high-frequency spin induction and spin pumping [12]. Another way to induce spin-polarized electrons is through a temperature difference between semiconductors or conductors which leads to the generation of a voltage among the two materials (known as the spin Seebeck effect). Lastly, in a DMS, the Zeeman effect, which is the splitting of spectral lines into other components when in the presence of a static magnetic field, can promote imbalance at the Fermi level.

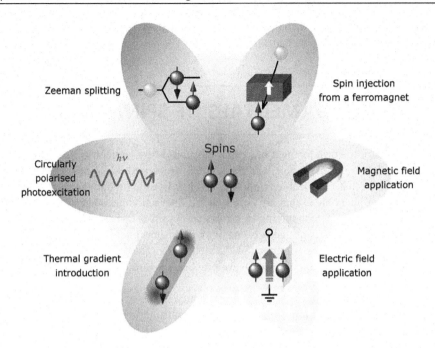

FIGURE 18.2 Methodologies that can be employed for the generation of spin-polarized electrons from a non-magnetic environment. Adapted with permission [8].

Source: Copyright (2014), IOP Publishing.

Even though spintronics is a relatively recent area, several phenomena are being unfolded and applied by scientists which are bringing these materials to the spotlight [9]. The phenomena such as TMR and GMR, which have been studied before spintronics, have been largely employed in the information technology industry. The TMR describes a process at which two ferromagnetic materials are separated by a thin insulator film. If the latter has a thickness that allows electrons to pass through it, then a tunneling effect from the electrons occurs allowing them to flow from one ferromagnet to the other. The resistance in such materials can vary depending on their magnetic configuration as it has been demonstrated in a magnetic tunnel junction (MTJ) at low temperatures [13]. However, recent works have demonstrated TMR at room temperature in MTJs based on Al, Co, Fe, and B thin films [14]. Based on that, one of the aims lies in increasing the TMR ratio which can be expressed in equation 1. Where P_1 and P_2 represent the spin polarization of each ferromagnetic material. Previous studies have shown an increase of 1,000% of the TMR ratio of an MTJ that consisted of an epitaxial MgO and Fe [15,16]. Such improvement was attributed to the coherent tunneling between Fe and MgO at their interface. Following that, CoFeB/MgO MTJs have also demonstrated an efficient TMR due to the smooth connection of their respective Δ_1 bands. The advantage of this aspect is that Δ_1 can be spin-polarized in CoFe alloys and Fe which allows the electrons to easily tunnel through MgO.

$$TMR\ ratio = \frac{2P_1 P_2}{(1 - P_1 P_2)} \tag{1}$$

Another important property explored in spintronics is the GMR which is observed in multilamellar materials composed of intercalated non-magnetic and ferromagnetic conducting layers. In that sense, the magnetization process of the adjacent ferromagnetic layers can take place in a parallel or antiparallel arrangement. The resistance decreased when there is a parallel alignment, and it increases when in an antiparallel alignment. In addition, applying an external magnetic field allows the direction of the magnetization to be controlled. Such effect is influenced by the electron's spin as well as scattering. Based on these concepts, previous studies were made on a thin film of Fe/Cr alloys that presented a decrease in their resistance by half when a magnetic field was applied at temperatures as low as 4.2 K [5]. In addition, measuring the magnetoresistance ratio provides important information regarding the magnetic transport efficiency which can be described in equation 2. Where R_{AP} and R_P are the antiparallel and parallel resistance of the ferromagnetic process, respectively.

$$MR\,ratio = \frac{\Delta R}{R} = \frac{(R_{AP} - R_p)}{R_p} \tag{2}$$

3 FLEXIBLE SPINTRONIC

The development of electronic devices that can be designed in various shapes opens great possibilities for novel applications that were not available due to the rigid materials that have been used in electronics. Hence, by targeting this type of technology scientists were able to develop a flexible solar cell that consisted of a 100 μm thin single layer of silicon, which was one of the first flexible electronic devices [17]. After that, several flexible devices were developed which included thin films based on polycrystalline silicon transistors and later a diversity of sensors, flexible displays, among others [18,19]. One of the most explored methods to fabricate flexible electronics is based on the deposition of magnetic films over a flexible polymer such as polyethylene terephthalate (PET), polyimide (PI), polyethylene naphthalate (PEN), polydimethylsiloxane (PDMS), and polyethersulphone (PES), which present a relatively lower cost, well-established manufacturing, and present satisfactory thermal stability to withstand post-annealing processes that can reach around 400 °C. Several materials can be incorporated into these polymers. The followings are some examples of flexible spintronics based on different materials.

3.1 Graphene-Based Flexible Spintronic

Understanding the physics of electron spin is enabling a promising path for the development of electronic devices for logic and memory components that can enhance computer performance. Among various materials, graphene has been studied as a novel material with several desired properties such as high electrical and thermal conductibility, flexibility, and the capability to transport spin-polarized electrons through tens of μm while maintaining the electron's spin orientation at room temperature [20,21]. Because of these factors, exploring the 2D nanostructure of graphene shows great viability as it can function as a flexible matrix for the deposition of magnetic materials. Such conditions have been met by employing graphene with heterostructures such as boron nitride, which diminished surface roughness to prevent carrier scattering [22,23]. Yet, the spin transport over graphene's surface is highly influenced by the substrate's roughness. Because of that, determining the spin signals lifetime in that type of topography is important to analyze the material's applicability.

Among the materials applicable for this matter, graphene shows a promising response as it presents satisfactory properties in terms of electron and energy transport which makes it a promising component for spintronics. With the idea to reinforce the uses of graphene within this field, Panda et al. [24] proposed the fabrication of a composite based on graphene grown through the commercial chemical vapor deposition (CVD) method over a SiO_2/Si at room temperature. Through this considerably long spin channels were obtained which consequently enabled long spin diffusion lengths. Because of that, there was diminishment in spin relaxation that improved the overall transport efficiency of the device as relatively long spin lifetimes of around 2.5 to 3.5 ns were obtained. The satisfactory performance was also attributed to the spin relaxation mechanism based on D'yakonov-Perel's over the 2D channels of graphene along with the contact regions. Throughout this initial analysis, it was also observed that graphene should present high-quality channels as well as contact to enable appreciable spin transport. When such conditions are achieved it was noted that electrostatic doping can considerably increase spin transport in graphene [20,22]. As a result, it can promote surface charge transfer doping (SCTD) that can cause a considerable shift in the Dirac point (V_D) [25,26]. Such effect can be studied when devices present a long channel (L ≥ 10 μm) which can provide a better understanding of the effects of SCTD and contact separation. On the other hand, it appears that is more difficult to elucidate spin transport in devices with long channels due to an exponential spin diffusion decay in relation to the channel length.

In addition, the spin signal can be obscured by electric noise which demands high-quality devices to enable proper measurement. This analysis can be performed through spin current injection into graphene's structure by using a current that passes through a ferromagnetic injector (i) connected to the graphene circuit as an isolated detector (d) with graphene circuit is used to measure the graphene's spin accumulation. The scheme for the graphene-based device measurement is provided in Figure 18.3. Spin transport can be measured through graphene by analyzing the difference in the voltage between parallel (↓↓ or ↑↑) and antiparallel (↓↑ or ↑↓). Through that, the signal is obtained through a spin-valve measurement technique, at which an in-plane magnetic field ($B_∥$) is applied in the direction of the easy axis concerning the ferromagnetic electrodes. Because of the difference of the injector and detector electrodes coercive fields, there is a switching from parallel to antiparallel electronic configuration. The CVD-grown graphene-based device presented satisfactory results such as long spin communication of 45 μm at room temperature, 13.6 μm for spin diffusion length along 3.5 ns of a spin lifetime. This study demonstrated that channel length plays an important role in SCTD which led to dependence between channel length over spin metrics and relaxation.

With the idea to provide insight into graphene-based flexible spintronics, Serrano et al. [27] fabricated a device composed of graphene over PEN through a large-scale CVD approach. It was observed that the composite was able to perform up to 15 μm spin transport at room temperature. Another aspect was its relatively large spin diffusion coefficients that were around 0.2 m²/s, which can be comparable to graphene devices over Si/SiO₂ substrate. One of the reasons for such improvements can be related to the large-scale CVD synthesis of graphene which led to a thin film with high uniformity. In this sense, the graphene was first grown over a Cu substrate to be transferred later onto a PEN substrate through a wet chemistry process. The fabricated device is displayed in Figure 18.4a. An individual device is presented in Figure 18.4b which is based on channels of patterned graphene with Ti, Co, Au based electrodes. The substrate's topography, displayed in Figure 18.4c, also played a major role to increase the spin-polarized electrons transport distance as a smoother surface aid on that process. The scheme for the detection of pure spin states on the device is shown in Figure 18.4d.

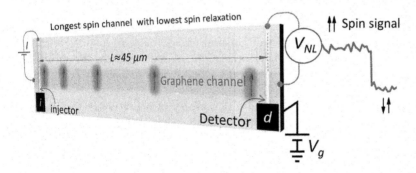

FIGURE 18.3 Scheme of the spintronic device based on CVD presenting a channel length (L) of around 45 μm presenting spin signal at room temperature. Adapted from reference [24].

Source: Copyright (2020), American Chemical Society. This is an open access article published under a Creative Commons Attribution (CC-BY) License.

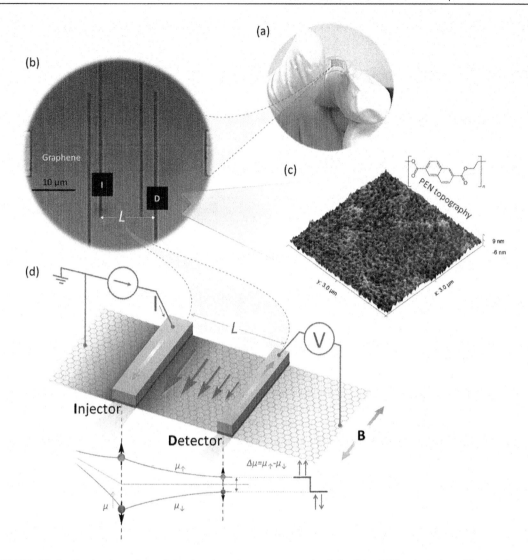

FIGURE 18.4 Device-based on lateral spin valves composed of flexible PEN substrate with graphene. (a) Photocopy of the flexible array over PEN substrate. (b) Micrograph for the flexible graphene spin circuit (FGSC) for the characterization of spin transport. (c) Atomic force microscopy (AFM) shows PEN's topography. (d) Scheme for the detection of pure spin signals which are represented by the arrows at the graphene channel. Below is represented the variation in the profile for the spin-down ($\mu_\downarrow$) and up ($\mu_\uparrow$), respectively. Adapted with permission [27].

Source: Copyright (2019), American Chemical Society.

During the AFM testing of PEN's surface, a relatively rough surface with the distance between the peaks and valley being around 50 to 100 nm was observed, which was around 5 times higher than the surface of Si/SiO_2. However, there was an advantageous interaction between PEN and graphene which decreased the negative effect of a rough surface on the spin-polarized electron transport as a Kurtosis smaller than 3 was obtained, since if Kurtosis is higher than 3 it suggests sharper peaks on the topology. The quantification of the spin transport was also performed utilizing the Hanle measurement geometry which consisted of a magnetic field out of plane ($B_\perp$) while the parallel configuration of magnetization of the injector and detector is maintained. In addition, this measurement includes causing the spin-polarized electrons in graphene to go through Larmor precession with a frequency (ω) defined by equation 3. Where the gyromagnetic ratio (g) equals 2.

$$\omega = \frac{g\mu_B}{\hbar} B \qquad (3)$$

The precession along with spin relaxation and diffusion contributions can be used in the Bloch diffusion expressed in equation 4 where spin polarization vector (Δμ) is defined as the net with the direction towards the axis of polarization, spin relaxation time (τ), spin diffusion (D). With those two parameters, the spin diffusion length (λ) can be obtained as it can be calculated in equation 5.

$$D\nabla^2 \Delta\mu - \frac{\Delta\mu}{\tau} + \omega \times \mu_s = 0 \qquad (4)$$

$$\lambda = \sqrt{D\tau} \qquad (5)$$

3.2 2D Materials beyond Graphene for Flexible Spintronic

Even though graphene holds great importance for the scientific community in many areas several materials within the field of spintronics are being explored which can provide interesting insights as electrodes suitable for the transportation of spin-polarized electrons. One of the much-desired possibilities that this process can provide is the fabrication of quantum computers. The foundation of this technology is based on the qubits (quantum bits) which can, theoretically, use a single electron's spin to store, transfer, and process information. Based on that, Drögeler et al. [22] proposed the fabrication of a single-layer graphene composite with an enhanced spin lifetime that went from 1.2 to 12.6 ns along with 30.5 μs for a diffusion length. Device's high performance was attributed to the presence of hexagonal boron nitride flakes (hBN), which prevented the contact of solvents over graphene that could cause a diminishment in the spin lifetimes to below 1 ns. The scheme for the spintronic device based on graphene and hBN flakes is presented in Figure 18.5.

Another important type of material that can be used for spintronics is the 2D transition metal-based compounds which are known for their magnetic properties due to the presence of electrons in their d and f electronic layers. However, most of the compounds that present a 2D monolayer structure do not present magnetic properties. This drawback can be countered as there are mainly two ways to introduce magnetic properties in a non-magnetic material, for this case. The first consists of the synthesis of organometallic 2D materials at which different types of organic moieties can be bonded with a transition metallic center. Through that, the flexibility of these materials can be improved depending on the organic ligand's structure whereas the metallic center of the structure provides the magnetic and metallic character that reaches 100% of spin-polarized current. The second approach lies in embedding a metal over a non-magnetic 2D layer. Through that, the nanoparticles of a magnetic material can adhere to the matrix's surface ideally without compromising its mechanical properties. The synthesis of organometallic materials has been studied by Zhou et al. [28], where several synthesized transition metal-based organic metallics derived from Zn, Ni, Co, Cu, Cr, and Mn were bonded with phthalocyanine (Pc) as an organic linker. It was observed that only 2D structured Mn-Pc presented ferromagnetic properties as it displayed 100% spin polarization around the Fermi level and half-metallic nature. Also, the presence of benzenic and pyrrolic rings allowed the spin-down electrons to be transported freely through their structure which was possibly related to the hybridization of *p* orbitals from the rings and *d* orbitals from the metallic center. Another aspect that can be tuned for these materials is the magnetic moment which can increase when the system is exposed to biaxial strain. However, it can lead to a transition from low to high spin. This phenomenon can be observed through a scanning tunneling microscope (STM). In another study, Zhou et al. [29] synthesized polymerized Fe-Pc (poly-Fe-Pc) along with polymerized Cr-Pc which presented an

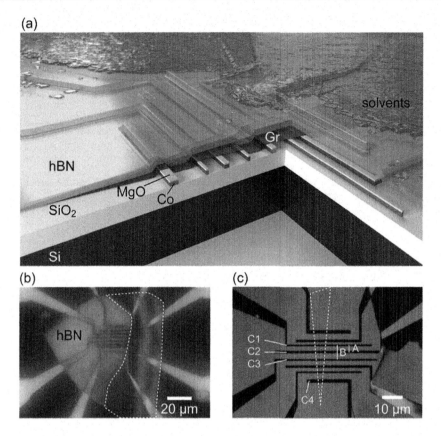

FIGURE 18.5 Representation of a spintronic device based on hBN flakes and graphene. Adapted with permission [22].

Source: Copyright (2016), American Chemical Society.

antimagnetic behavior. The authors were able to induce stable ferromagnetic states by promoting electron doping with Curie temperatures in the range between 130–140 K whereas hole doping promoted an enhancement of the stability of antimagnetic states. Through that, it was inferred that these changes in coupling were related to the balance between antiferromagnetic superexchange and ferromagnetic p–d exchange.

Other types of 2D materials have also been explored for spintronics applications such as metal carbide sulfides. For this group of materials, Zhao et al. [30] synthesized a half-metallic $Mn_3C_{12}S_{12}$ kagome spin-lattice which presented a spin-lattice with S = 3/2. This observation was based on d electrons' spin in Mn that were able to couple through ferromagnetism due to the formation of long ferromagnetic order. In that sense, by using Monte Carlo simulations the Curie temperature was around 212 K. Also, satisfactory carrier mobility along with a bandgap (E_g) 1.54 eV suggested a viable application for spintronics. On top of that, a small bandgap was observed at the Dirac point at the kagome bands which was likely attributed to the effects of spin-orbital. Through that, a quantum anomalous Hall effect may be achieved. In another related study Liu et al. [31] proposed the trade of S by N in the form of –NH, yielding $Mn_3C_{12}N_{12}H_{12}$, which led to an improvement in the ferromagnetic properties such as a Curie temperature of 450 K. The reason for such improvement was suggested due to the reduced lattice constant along with the presence of N which can promote mediation for magnetic coupling through p–d orbitals. The 2D structure of $Mn_3C_{12}S_{12}$ is provided in Figure 18.6.

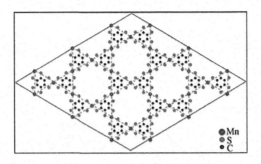

FIGURE 18.6 Scheme for the $Mn_3C_{12}S_{12}$ 2D monolayered structure where the Kagome lattice is represented by the dashed lines on the Mn ions. Adapted with permission [30].
Source: Copyright (2013), Royal Society of Chemistry.

3.3 Metal Oxide-Based Flexible Spintronic

The factors such as chemical stability, synthetical versatility, and facile fabrication are some of the reasons for the research on metal oxide-based materials applicable for flexible spintronics. Some material examples include $NiFe_2O_4$, CrO_2, $La_{0.7}Sr_{0.3}MnO_3$, and Fe_3O_4. Within those, Fe_3O_4 is an interesting candidate for spintronic applications as it presents a relatively high Curie temperature of 858 K compared to $La_{0.7}Sr_{0.3}MnO_3$ and CrO_2 which present 369 and 390 K, respectively. Curie temperature is an important characteristic that defines spintronic devices' thermal stability [32,33]. Alongside that, Fe_3O_4 can present Fe with varying valence states i.e., Fe^{2+} (d^6) and Fe^{3+} (d^5). It has been proposed that electrons can hop between Fe^{2+} and Fe^{3+} ions. That effect leads to a better conductivity compared to other ferromagnetic oxides at room temperature. Also, nearly 100% spin polarizability at the Fermi level allows Fe_3O_4 to be a useful material for applications related to magnetic random memory, which are the core components for the fabrication of computers and other electronic devices. However, one of the main challenges lies in growing Fe_3O_4 over flexible substrates which could allow different designs for the spintronics along with more robusticity. Yet, there has been some challenge to adhere Fe_3O_4 over flexible polymers. Despite that, some progress has been made for the epitaxial growth of Fe_3O_4 over substrates of Si (001) which required a heterostructure of double-buffer-layer composed of MgO and TiN [34]. Muscovite has been employed as a potential candidate for a flexible substrate that allows heteroepitaxy of oxide films [35]. Based on this idea, Wu et al. [36] proposed the use of muscovite as a flexible substrate for the growth of Fe_3O_4 film. The authors carried out an analysis of the magnetic properties along with bending of the device in which it was noted a retaining of properties in terms of coercivity, remanence, and saturation as seen in Figure 18.7a. On the other hand, the magnetic anisotropy varied with the bending (Figure 18.7b), yet it was restored after the composite went back to the neutral position (Figure 18.7c) which could be a potentially controllable magnetic anisotropy property based on mechanical flexibility. This effect may be related to the variation in magnetostriction as the shape anisotropy leans towards an easy axis in the in-plane direction. A microscopic analysis of the magnetic properties was also performed through magnetic force microscopy (MFM). In that sense, a 5 T magnetic field was applied along the in-plane in order to aligned with the magnetic domains. By maintaining one side fixed and the other movable, different bending radius could be applied. Through that, it was observed an increase in magnetic domains when a bending radius of 90 mm was applied which can be observed in Figure 18.7c-d. In addition to that, after the muscovite/Fe_3O_4 film returned to the neutral position its magnetic domains also returned to the initial state as demonstrated in Figure 18.7f.

In another approach, Li et al. [37] also used mica as a flexible substrate to develop flexible spintronics that contained epitaxial nanocolumns based on $La_{0.67}Sr_{0.33}MnO_3/SrTiO_3$. The $La_{0.67}Sr_{0.33}MnO_3$ is a material that presents appreciable ferromagnetic properties way above room temperature, strong lattice coupling, and magnetic anisotropy which makes it an interesting use as a spintronic and spin injection semiconductor.

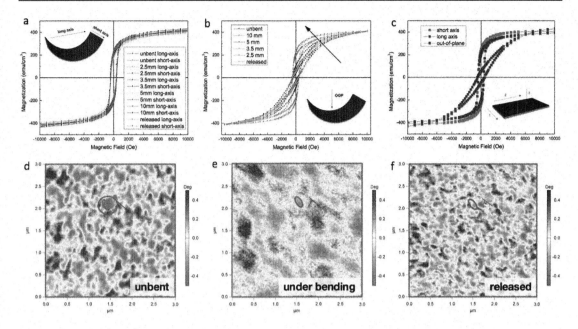

FIGURE 18.7 Magnetic properties of the muscovite/Fe$_3$O$_4$ when exposed to bending. (a) In-plane and (b) out-of-plane magnetizations with hysteresis loops under several bending radii. (c) Hysteresis loops after the release of the muscovite/Fe$_3$O$_4$ films and three-direction magnetizations. MFM images of the film (d) unbent, (e) under bending down to 50 μm, and (f) neutral state. Adapted with permission [36].

Source: Copyright (2016), American Chemical Society.

Also, SrTiO$_3$ is a viable buffer as it has a similar structure when compared to La$_{0.67}$Sr$_{0.33}$MnO$_3$. Based on that, the authors noted a variation in properties concerning the thickness of La$_{0.67}$Sr$_{0.33}$MnO$_3$ as the thinnest film (around 16 nm) had the highest saturation magnetization (M$_S$), coercivity, and remanence ratio. Such influence of film's thickness could be explained by the dependence of magnetization towards temperature from the analysis of zero-field cooled (ZFC) and field cooled (FC) at 500 Oe. The authors noted a separation between FC and ZFC curves that suggested a blocking process for the spin-glass phase. Also, for the lowest thickness of 16, 50, and 100 nm there were peaks on the ZFC that suggested spin freezing at the blocking temperature (T$_B$). On the other hand, when the film had a thickness of 300 nm it became apparent under 100 Oe magnetic field. Because of that, the suggestion of a spin-glass layer that occurs at either the grain surfaces or its boundaries, along the film's domains might be the reason for the difference in ZFC and FC curves. Is worth mentioning that, this effect occurs usually due to the broken bonds at the surface along with the translational symmetry breaking of the lattice of the La$_{0.67}$Sr$_{0.33}$MnO$_3$ [38]. Also, the film has a core that is ferromagnetic along with a shell that is spin disordered. To understand the magnetic properties, the magnetic-hysteresis analysis of the composite film was performed. Through it was notable that the morphology of nanocolumns is an important factor for the magnetic properties as the ratio of surface/volume is higher than the planar film. In addition, the surface spin-glass layer can lead to disorder spins and randomness in the double exchange, which culminates in a lower magnetization for the film with 50 nm of thickness. Following the same trend, the film with 100 nm of thickness presented larger values for the surface/volume ratio, which consequently led to a further decrease in magnetization.

The introduced flexibility and resonance field (H$_r$) tunability in electronic devices is highly influenced by the effective field (H$_{eff}$) of the thin film [38]. Based on that, field angular θ$_H$-dependent H$_r$ analysis can be performed to understand the effect of bending the thin films in relation to the magnetic properties. Through that, the authors observed that a film with 16 nm of thickness was more influenced by upward bending as it presented the highest H$_{eff}$ when the θ$_H$ > 60°. In addition, the 50 nm thick film presented a nearly stable magnetic behavior along with the θ_H range, which showed little influence of bending in the film's magnetic properties. On the other hand, for the case of a film with 300 nm thick it was observed that H_r had an increase for $\theta_H < 55°$ and a decrease for $\theta_H > 55°$ which could allow some degree of tuneability of properties according to the θ_H. Through that, the authors could infer that H_{eff} can aid in determining the bending tunability. Hence, through this analysis, it was demonstrated that controlling the magnetic film's thickness played a major role in the properties which open a way to developed magnetic flexible materials for spintronics that are influenced mostly by the thickness.

3.4 Chalcogenides-Based Flexible Spintronic

Transition metal chalcogenides have been explored for several technologies such as supercapacitors, batteries, while also finding space within the field of spintronics. These compounds are based on a semi-metal (M) such as Sn, Si, or Ge along with chalcogens (X) such as Te, Se, S. Through that, several MX-based structures can be obtained which have presented a considerable spin hall effect, spin polarization along with functioning as topological insulators [39,40]. These properties arise from the presence of several valleys, that are located on the axis of the Brillouin zone, which is a unique and defined primitive cell in reciprocal space, within the field of solid-state physics in mathematics. Such conditions can prompt strong spin-orbit coupling because of the presence of heavy elements which can allow these materials to be applicable for valleytronic, spintronics, and optical components for devices. Within this line, metal chalcogenides can be possibly employed due to their multiferroicity and appreciable piezoelectric properties. In addition, these materials go through a doping and strain process which provides a viable way to optimize these nanomaterial's performance [40].

The transition metal dichalcogenides are another vastly researched class of materials that also holds great importance within technology fields since they can present properties that are similar or relatable to those of graphene. Their magnetic properties are being investigated which led to findings based on their multiferroic and optical properties which allow control of valley polarization, coherence as well as considerably large spin-valley coupling [41,42]. The fabrication of composites based on 2D transition metal dichalcogenides grown over graphene has provided a considerable enhancement in spin-orbit coupling without compromising the inherently high electron mobility of graphene. Also, graphene can provide more flexibility for the transition metal dichalcogenides that are grown over its structure, leading to an advantageous combination. Wakamura et al. [43] synthesized a composite based on graphene with a monolayer of WS_2 grown over it (graphene/monolayer-WS_2) which was compared to the same composite with a bulkier counterpart of WS_2 (graphene/bulk-WS_2). Through that, it was observed an increase of spin-orbit coupling of one order of magnitude for the graphene/monolayer-WS_2 when compared to the bulky one which showed the importance of monolayered materials due to their capability of preserving magnetic states and improving other parameters such as spin lifetime, spin hall effect, among others. There is a considerable number of 2D transition metal dichalcogenides materials that can be employed for spintronics which includes mostly W and Mo such as WSe_2, $MoSe_2$ as well as functionalized MoS_2 with Gd, $BiIrO_3$ (111), or Fe-X_6 clustered structures where X can be F, O, N, C, S [44,45].

Within the idea of obtaining novel transition metal chalcogenide materials, Shang et al. [46] synthesized several of those with the general formula of M_6X_6 which consisted of atomic wires that were 66 atoms long. The analysis of these nanostructures showed satisfactory properties in terms of ferromagnetism along with chemical stability towards oxidation, moisture, and aggregation. Also, in terms of mechanical properties, high Young's modulus and a large fracture strain were obtained. One of the insights obtained through this study was the increase in metallic behavior for the structures of Cr_6X_6 where X was Te, Se, or S. It was observed that, as the electronegativity increases from Te to Se and S, there was a stronger interaction among the p orbitals from X atom with the d orbitals from the transition metal. Through that, after the Bader charge analysis, the authors observed an electron transfer that went from 0.40, 0.58, and 0.76 which were related to the Cr atom bonded with Te, Se, and S, respectively. In addition, it was observed that for both Cr_6Se_6 and Cr_6S_6 there was a downward movement of the chalcogen atoms from the bottom conduction band from p orbitals, which touched or crossed the Fermi level which led to considerable overlap with the valence bands on the d orbital of Cr. Because of that, the atomic wires composed of Cr, Mo, and W bonded with S or Se presented a metallic behavior. For the analysis of the magnetic properties, it was noticed that all the six magnetic materials (Re$_6$S$_6$, Fe$_6$Se$_6$, Fe$_6$S$_6$, Co$_6$Te$_6$, Co$_6$Se$_6$, and Co$_6$S$_6$) presented ferromagnetic order at the ground state. For the analysis of that, the electronic band structure for the Fe$_6$S$_6$ system is provided in Figure 18.8a which presents the metallic behavior based on the strong polarization in the two spin channels. The magnetism emerges from the unpaired d electrons in the transition metal. Based on that, the magnetic moment (μ_B) varied based on the transition metal as the values were 1.71–2.02, 0.97–1.18, and 0.62 μ_B for Fe$_6$X$_6$ with X = S or Se, Co$_6$X$_6$ with X = S, Se or Te, and Re$_6$S$_6$, respectively. Figure 18.8b shows the crystal field using Fe$_6$S$_6$ as an example with its d orbitals split into five components at which the pair d$_{xz}$ and d$_{yz}$ and the pair d$_{xy}$ and d$_{x^2-y^2}$ were degenerated. Since the Fe atom presents eight electrons in the valence band, the spin-up channels for the 5 d electrons are occupied, whereas the spin-down states of d$_{xy}$ and d$_{x^2-y^2}$ were vacant. That resulted in a magnetic moment of 2 μ_B on the Fe atom. Figure 18.8d showed both magnetic moment and exchange energy (E$_m$) in the top and bottom panels, respectively. Based on this analysis it was observed that the six magnetic 1D materials presented considerably large E$_m$ of 650 meV considering the M$_6$X$_6$ formula or 108 meV per transition metal. Through that, to calculate the Curie temperature Monte Carlo simulations were performed based on Ising

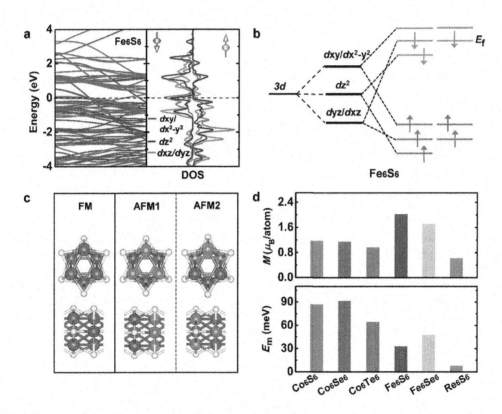

FIGURE 18.8 Analysis of Fe$_6$S$_6$ atomic wire based on (a) (left) structure of the spin-polarized band structure and (right) density of states. (b) Crystal field splitting for the Fe-3d orbital. (c) Spin density distribution within the ferromagnetic ground state along with the two antiferromagnetic states with the lowest energy presenting a surface value of 0.3 e/Å^3. (d) (top panel) μ_B for each transition metal along with (bottom panel) E$_m$ for the six M$_6$X$_6$ magnetic wires. Adapted with permission [46].

Source: Copyright (2020), The authors. Published by American Chemical Society. This is an open access article published under a Creative Commons Non-Commercial No Derivative Works (CC-BY-NC-ND) Attribution License. Further permissions related to the material excerpted should be directed to the ACS.

model as the Hamiltonian operator can be defined as in equation 6. At which S_i and S_j represent the magnetic moments at the region i and j, respectively. Also, J_{ij} represents the parameter of exchange coupling between the regions i and j. Through that, the anti-ferromagnetic configurations were determined at which as the ones with the lowest energy were determined for the six M_6X_6 nanowires as anti-ferromagnetic 1 for Co_6Se_6, Co_6Te_6, and Co_6S_6 and anti-ferromagnetic 2 for Re_6S_6, Fe_6S_6, and Fe_6Se_6 which are presented in Figure 18.8c. Such complexity in this study demonstrated a broad number of several magnetic materials that presented a structure of atomic wires which led to satisfactory magnetic properties.

$$H = -\sum_{i,j} J_{ij} S_i S_j \tag{6}$$

4 CONCLUSION

There is a considerable driving force within the scientific community to continue developing novel spintronic materials that can reach the nanoscale as well as be incorporated into flexible substrates to allow the manufacture of competitive components with high performance in terms of electronic and magnetic properties. The importance of this field also lies in its great necessity as the need to develop better and enhanced electronic devices is a requirement in every technological sector. The use of nanomaterials with 2D structures based on organometallic as well as transition metal systems are showing promising magnetic properties along with high Curie temperatures. In that sense, metal embedding is a viable approach to introduce magnetism in nonmagnetic materials which can provide a sustainable approach to decrease the load of valuable metallic resources. Also, graphene can play an important role in this manner as it can enable the fabrication of spintronics as a support material to improve the magnetic and electronic properties due to its weak spin-orbit coupling, long lifetime spin, high polarized-spin transport, and flexibility. Through that, there has been some progress in terms of inducing magnetism by e- and hole (h+) doping on metal-free 2D nanomaterials. Yet, the Curie temperature of these compounds is not within the range of those that are metal-based or metal embedded, thus requiring further progress to be able to compete. Within this factor, it is important to mention that obtaining materials with Curie temperatures that are at least higher than room temperature is one of the core factors for their large-scale applicability, yet it is also one of the current challenges to overcome. Alongside those studies, it has been found that graphene composited with metal dichalcogenides can also present valuable properties due to weak antilocalization, Hanle precession, and spin Hall effects which can enable their use as topological insulators, for instance. Despite that, some of the current issues remain on maintaining magnetic properties such as spin-orbit coupling, Hall effect, and transport at higher temperatures as it has been noticed that spin Hall conductivity tends to decrease in function of temperature. Thus, throughout the discussions in the chapter it is notable that spintronic systems present complex operational mechanisms which require further elucidation to effectively implement known materials such as transition metal oxides, dichalcogenides along with graphene into flexible substrates to fabricate flexible spintronics.

REFERENCES

[1] J. Kim, J. Lee, D. Son, M.K. Choi, D.H. Kim, Deformable devices with integrated functional nanomaterials for wearable electronics, *Nano Converg.* 3 (2016) 4.
[2] D.-H. Kim, R. Ghaffari, N. Lu, S. Wang, S.P. Lee, H. Keum, R. D'Angelo, L. Klinker, Y. Su, C. Lu, Y.-S. Kim, A. Ameen, Y. Li, Y. Zhang, B. de Graff, Y.-Y. Hsu, Z. Liu, J. Ruskin, L. Xu, C. Lu, F.G. Omenetto, Y. Huang, M. Mansour, M.J. Slepian, J.A. Rogers, Electronic sensor and actuator webs for large-area complex geometry cardiac mapping and therapy, *Proc. Natl. Acad. Sci.* 109 (2012) 19910 LP–19915.

[3] M. Alnassar, A. Alfadhel, Y.P. Ivanov, J. Kosel, Magnetoelectric polymer nanocomposite for flexible electronics, *J. Appl. Phys.* 117 (2015) 17D711.

[4] Y.-W. Liu, Q.-F. Zhan, R.-W. Li, Fabrication, properties, and applications of flexible magnetic films, *Chinese Phys. B.* 22 (2013) 127502.

[5] M.N. Baibich, J.M. Broto, A. Fert, F.N. Van Dau, F. Petroff, P. Etienne, G. Creuzet, A. Friederich, J. Chazelas, Giant magnetoresistance of (001)Fe/(001)Cr magnetic superlattices, *Phys. Rev. Lett.* 61 (1988) 2472–2475.

[6] S.A. Crooker, J.J. Baumberg, F. Flack, N. Samarth, D.D. Awschalom, Terahertz spin precession and coherent transfer of angular momenta in magnetic quantum wells, *Phys. Rev. Lett.* 77 (1996) 2814–2817.

[7] H. Munekata, H. Ohno, S. von Molnar, A. Segmüller, L.L. Chang, L. Esaki, Diluted magnetic III-V semiconductors, *Phys. Rev. Lett.* 63 (1989) 1849–1852.

[8] A. Hirohata, K. Takanashi, Future perspectives for spintronic devices, *J. Phys. D. Appl. Phys.* 47 (2014) 193001.

[9] A. Hirohata, K. Yamada, Y. Nakatani, I.-L. Prejbeanu, B. Diény, P. Pirro, B. Hillebrands, Review on spintronics: Principles and device applications, *J. Magn. Magn. Mater.* 509 (2020) 166711.

[10] J.M. Kikkawa, D.D. Awschalom, Lateral drag of spin coherence in gallium arsenide, *Nature.* 397 (1999) 139–141.

[11] A. Fujiwara, Y. Takahashi, Manipulation of elementary charge in a silicon charge-coupled device, *Nature.* 410 (2001) 560–562.

[12] N. Tombros, C. Jozsa, M. Popinciuc, H.T. Jonkman, B.J. van Wees, Electronic spin transport and spin precession in single graphene layers at room temperature, *Nature.* 448 (2007) 571–574.

[13] M. Julliere, Tunneling between ferromagnetic films, *Phys. Lett. A.* 54 (1975) 225–226.

[14] H.X. Wei, Q.H. Qin, M. Ma, R. Sharif, X.F. Han, 80% tunneling magnetoresistance at room temperature for thin Al – O barrier magnetic tunnel junction with CoFeB as free and reference layers, *J. Appl. Phys.* 101 (2007) 09B501.

[15] J. Mathon, A. Umerski, Theory of tunneling magnetoresistance of an epitaxial Fe/MgO/Fe(001) junction, *Phys. Rev. B.* 63 (2001) 220403.

[16] W.H. Butler, X.-G. Zhang, T.C. Schulthess, J.M. MacLaren, Spin-dependent tunneling conductance of $\mathrm{Fe}|\mathrm{MgO}|\mathrm{Fe}$ sandwiches, *Phys. Rev. B.* 63 (2001) 54416.

[17] D.-H. Kim, J.A. Rogers, Stretchable electronics: Materials strategies and devices, *Adv. Mater.* 20 (2008) 4887–4892.

[18] T. Someya, Y. Kato, T. Sekitani, S. Iba, Y. Noguchi, Y. Murase, H. Kawaguchi, T. Sakurai, Conformable, flexible, large-area networks of pressure and thermal sensors with organic transistor active matrixes, *Proc. Natl. Acad. Sci. U. S. A.* 102 (2005) 12321 LP – 12325.

[19] S.R. Forrest, The path to ubiquitous and low-cost organic electronic appliances on plastic, *Nature.* 428 (2004) 911–918.

[20] M.V. Kamalakar, C. Groenveld, A. Dankert, S.P. Dash, Long distance spin communication in chemical vapour deposited graphene, *Nat. Commun.* 6 (2015) 6766.

[21] P.J. Zomer, M.H.D. Guimarães, N. Tombros, B.J. van Wees, Long-distance spin transport in high-mobility graphene on hexagonal boron nitride, *Phys. Rev. B.* 86 (2012) 161416.

[22] M. Drögeler, C. Franzen, F. Volmer, T. Pohlmann, L. Banszerus, M. Wolter, K. Watanabe, T. Taniguchi, C. Stampfer, B. Beschoten, Spin lifetimes exceeding 12 ns in graphene nonlocal spin valve devices, *Nano Lett.* 16 (2016) 3533–3539.

[23] J. Ingla-Aynés, R.J. Meijerink, B.J. van Wees, Eighty-eight percent directional guiding of spin currents with 90 μm relaxation length in bilayer graphene using carrier drift, *Nano Lett.* 16 (2016) 4825–4830.

[24] J. Panda, M. Ramu, O. Karis, T. Sarkar, M.V. Kamalakar, Ultimate spin currents in commercial chemical vapor deposited graphene, *ACS Nano.* 14 (2020) 12771–12780.

[25] N.F.W. Thissen, R.H.J. Vervuurt, A.J.M. Mackus, J.J.L. Mulders, J.-W. Weber, W.M.M. Kessels, A.A. Bol, Graphene devices with bottom-up contacts by area-selective atomic layer deposition, *2D Mater.* 4 (2017) 25046.

[26] F.A. Chaves, D. Jiménez, A.W. Cummings, S. Roche, Physical model of the contact resistivity of metal-graphene junctions, *J. Appl. Phys.* 115 (2014) 164513.

[27] I.G. Serrano, J. Panda, F. Denoel, Ö. Vallin, D. Phuyal, O. Karis, M.V. Kamalakar, Two-dimensional flexible high diffusive spin circuits, *Nano Lett.* 19 (2019) 666–673.

[28] J. Zhou, Q. Sun, Magnetism of phthalocyanine-based organometallic single porous sheet, *J. Am. Chem. Soc.* 133 (2011) 15113–15119.

[29] J. Zhou, Q. Sun, Carrier induced magnetic coupling transitions in phthalocyanine-based organometallic sheet, *Nanoscale.* 6 (2014) 328–333.

[30] M. Zhao, A. Wang, X. Zhang, Half-metallicity of a kagome spin lattice: The case of a manganese bis-dithiolene monolayer, *Nanoscale.* 5 (2013) 10404–10408.

[31] J. Liu, Q. Sun, Enhanced ferromagnetism in a Mn3C12N12H12 sheet, *ChemPhysChem.* 16 (2015) 614–620.

[32] R.S. Keizer, S.T.B. Goennenwein, T.M. Klapwijk, G. Miao, G. Xiao, A. Gupta, A spin triplet supercurrent through the half-metallic ferromagnet CrO2, *Nature.* 439 (2006) 825–827.

[33] A. Urushibara, Y. Moritomo, T. Arima, A. Asamitsu, G. Kido, Y. Tokura, Insulator-metal transition and giant magnetoresistance in $\mathrm{La}_{1-x}\mathrm{Sr}_{x}\mathrm{MnO}_{3}$, *Phys. Rev. B.* 51 (1995) 14103–14109.

[34] H. Xiang, F. Shi, M.S. Rzchowski, P.M. Voyles, Y.A. Chang, Epitaxial growth and magnetic properties of Fe3O4 films on TiN buffered Si(001), Si(110), and Si(111) substrates, *Appl. Phys. Lett.* 97 (2010) 92508.

[35] C.-H. Ma, J.-C. Lin, H.-J. Liu, T.H. Do, Y.-M. Zhu, T.D. Ha, Q. Zhan, J.-Y. Juang, Q. He, E. Arenholz, P.-W. Chiu, Y.-H. Chu, Van der Waals epitaxy of functional MoO2 film on mica for flexible electronics, *Appl. Phys. Lett.* 108 (2016) 253104.

[36] P.-C. Wu, P.-F. Chen, T.H. Do, Y.-H. Hsieh, C.-H. Ma, T.D. Ha, K.-H. Wu, Y.-J. Wang, H.-B. Li, Y.-C. Chen, J.-Y. Juang, P. Yu, L.M. Eng, C.-F. Chang, P.-W. Chiu, L.H. Tjeng, Y.-H. Chu, Heteroepitaxy of Fe3O4/muscovite: A new perspective for flexible spintronics, *ACS Appl. Mater. Interfaces.* 8 (2016) 33794–33801.

[37] J. Li, J. Wu, L. Shen, J. Yang, L. Lu, C. Cao, C. Jiang, X. Lu, L. Zhang, H. Yu, M. Liu, Self-assembled La0.67Sr0.33MnO3 epitaxial nanocolumn films towards future flexible spintronic devices, *Mater. Today Phys.* 13 (2020) 100218.

[38] N. Lampis, P. Sciau, A.G. Lehmann, Rietveld refinements of the paraelectric and ferroelectric structures of PbFe0.5Nb0.5O3, *J. Phys. Condens. Matter.* 11 (1999) 3489–3500.

[39] S. Zhou, C.-C. Liu, J. Zhao, Y. Yao, Monolayer group-III monochalcogenides by oxygen functionalization: A promising class of two-dimensional topological insulators, *NPJ Quantum Mater.* 3 (2018) 16.

[40] J. Sławińska, F.T. Cerasoli, H. Wang, S. Postorino, A. Supka, S. Curtarolo, M. Fornari, M. Buongiorno Nardelli, Giant spin Hall effect in two-dimensional monochalcogenides, *2D Mater.* 6 (2019) 25012.

[41] B. Wen, Y. Zhu, D. Yudistira, A. Boes, L. Zhang, T. Yidirim, B. Liu, H. Yan, X. Sun, Y. Zhou, Y. Xue, Y. Zhang, L. Fu, A. Mitchell, H. Zhang, Y. Lu, Ferroelectric-driven exciton and trion modulation in monolayer molybdenum and tungsten diselenides, *ACS Nano.* 13 (2019) 5335–5343.

[42] J.-W. Chen, S.-T. Lo, S.-C. Ho, S.-S. Wong, T.-H.-Y. Vu, X.-Q. Zhang, Y.-D. Liu, Y.-Y. Chiou, Y.-X. Chen, J.-C. Yang, Y.-C. Chen, Y.-H. Chu, Y.-H. Lee, C.-J. Chung, T.-M. Chen, C.-H. Chen, C.-L. Wu, A gate-free monolayer WSe2 pn diode, *Nat. Commun.* 9 (2018) 3143.

[43] T. Wakamura, F. Reale, P. Palczynski, S. Guéron, C. Mattevi, H. Bouchiat, Strong anisotropic spin-orbit interaction induced in graphene by monolayer WS2, *Phys. Rev. Lett.* 120 (2018) 106802.

[44] N. Feng, W. Mi, Y. Cheng, Z. Guo, U. Schwingenschlögl, H. Bai, First principles prediction of the magnetic properties of Fe-X6 (X = S, C, N, O, F) doped monolayer MoS2, *Sci. Rep.* 4 (2014) 3987.

[45] Y. Ji, Y. Song, J. Zou, W. Mi, Spin splitting and p-/n-type doping of two-dimensional WSe2/BiIrO3(111) heterostructures, *Phys. Chem. Chem. Phys.* 20 (2018) 6100–6107.

[46] C. Shang, L. Fu, S. Zhou, J. Zhao, Atomic wires of transition metal chalcogenides: A family of 1D materials for flexible electronics and spintronics, *JACS Au.* 1 (2021) 147–155.

Index

1D, 117, 149, 152, 295–296, 304, 315
2D, 34–44, 50, 56, 59, 117, 149–150, 158, 295, 297, 307, 310–312, 314, 316
3D, 21, 117, 151, 233, 236, 305

A

antenna, 216–220, 223–227, 232, 246–257

B

batteries, 15, 264, 266–269, 271, 273, 274, 276–279, 283, 287, 288, 292, 314
biochar, 12, 23, 25, 26
biomedical, 6, 10, 77, 89, 118, 120, 129, 131, 137, 138, 142, 144, 303
bioseparation, 138, 139

C

catalysis, 2, 10–13, 17, 21–23, 26, 28, 138, 208, 230, 285
catalyst, 2, 3, 5, 10, 11, 13, 15–20, 26–28, 79, 124, 208, 210, 276, 277, 279
ceramics, 165, 216, 228, 230, 237, 286
chalcogenides, 150, 314, 318
chemical, 1–4, 10–15, 17, 19, 23, 24, 34–38, 41, 44, 78, 79, 84, 87, 116, 117, 120, 122, 132, 137, 144, 152, 154, 168, 172–174, 177, 183–185, 188–190, 198, 199, 201, 202, 204, 205, 209, 230, 264–270, 272–276, 278, 279, 285–288, 294, 307–309, 311–315
chemical vapor deposition, 41, 78, 79, 152, 168, 199, 286, 307
composites, 17, 19–24, 122, 142, 143, 199, 209, 216, 218, 219, 229–231, 233–237, 286, 290, 298, 299, 314
crystal, 20, 34, 37, 41, 80, 99, 104, 110, 137, 148, 156, 167, 171, 172, 174, 199, 201, 267, 271, 298, 315
crystallographic, 98, 99, 163, 164, 168, 188, 267, 276

D

deposition, 41, 78, 79, 84, 85, 99, 104, 106, 120, 124, 152, 167, 168, 199, 264, 277, 283, 285, 286, 307
devices, 40–44, 49–55, 57–59, 63, 65, 66, 71, 72, 83, 94–96, 98, 99, 101, 102, 104, 106, 108–111, 117–120, 137, 148, 150–153, 156–158, 163, 176, 177, 182–185, 193, 194, 216, 230, 245, 251, 252, 256, 257, 263, 267, 269, 271–273, 279, 283, 284, 290, 299, 303, 304, 307, 308, 312–314, 316
dielectric, 35, 56, 69, 94, 116, 171, 173, 175, 185, 186, 188, 189, 191, 216–221, 224, 225, 228–235, 246, 249, 253, 254, 287, 292
drug, 17, 28, 89, 128, 129, 138, 140–142, 207, 211, 230, 285
drug delivery, 17, 89, 128, 129, 138, 140–142, 230, 285
dye, 21, 23, 131, 197, 201, 203, 206, 208, 211

E

EDLC, 284, 287–289, 292, 294
electrochemical double-layer capacitors, 284, 287
electrodeposition, 78, 86, 87, 138, 199, 286, 287
electroless, 78, 79
electronic, 34, 42–44, 53, 57, 63, 78, 80, 84, 93–95, 104, 109, 127, 132, 133, 148, 149, 152–154, 157, 163, 164, 170, 171, 174–177, 194, 216, 225, 227, 231, 263, 264, 266, 267, 273, 277, 290, 296, 299, 303–305, 307, 308, 310, 312, 313, 315, 316
electrospinning, 22, 30, 77, 80, 84, 233, 259, 265, 273, 274, 276, 282
EMI, 183–185, 188, 189, 191–195, 216, 222
enzyme, 17, 23, 24, 26
exfoliations, 41

F

ferrimagnetic, 12, 72, 117, 168, 173, 174, 183, 184, 230, 297
ferrites, 3–5, 12, 13, 98, 122, 183–185, 187, 192, 195, 197, 201, 207, 216, 229–231, 233, 235, 237, 285, 286
ferromagnetic, 5, 11, 16, 39, 40, 50, 52, 58, 64, 69–72, 80, 83, 84, 117, 119, 122, 123, 125–127, 148, 151–153, 155, 156, 164, 172, 175, 177, 183, 200, 207, 223, 229, 233, 273, 274, 290, 297, 304–308, 310–313, 315
FET, 42, 52, 55
flexible, 15, 34, 42, 106, 120, 127, 134, 154, 185, 186, 188, 218, 228, 231, 257, 277, 286, 304, 307–310, 312, 314, 316
fuel cells, 16, 263, 268, 276, 283, 292

G

giant magnetoresistance, 63, 83, 94, 118, 119, 121, 129, 131, 152, 155, 304
graphene, 6, 16, 17, 34, 38, 41–43, 52, 53, 118, 148–158, 208, 234, 236, 237, 270, 277, 286, 292, 294, 307–311, 314, 316

H

Hall effect, 34, 44, 71, 72, 94, 110, 118, 120–124, 126, 127, 134, 149, 305, 311, 314, 316
hexaferrite, 184, 188, 191, 233, 234, 236, 237
hybrid, 17, 20, 21, 28, 71, 82, 124–126, 144, 216, 268, 276, 277, 284, 287, 289
hydrothermal, 16, 23, 27, 40, 165, 166, 233, 234, 264, 269–273, 276, 278, 285, 286, 293, 295, 297
hyperthermia, 89, 117, 128, 129, 138, 140, 141, 177, 257

I

indium phosphide, 56, 57, 118
ionic liquid, 11, 15

319

L

Li-air, 266

M

magnetic, 4, 6, 10–28, 35, 37–41, 43, 44, 50, 52, 53, 57–59, 64–67, 69–72, 78–80, 83–89, 94–101, 103–108, 110, 111, 116–120, 122–134, 138–143, 148, 149, 151–158, 164, 165, 168, 169, 171–177, 181, 183–195, 197, 199–202, 206–209, 211, 216, 222–224, 229, 230, 233, 234, 237, 246, 249, 264–266, 269–279, 285, 286, 290–299, 304–316
magnetic frameworks, 17
magnetic resonance imaging, 89, 91, 117, 128, 138, 143, 144
Magneto-Optic Kerr Effect, 120, 121, 123–125, 149
magnetoresistance, 63, 65, 94, 95, 97, 102, 109, 111, 118–121, 129, 131, 148, 154, 164, 165, 174, 175, 304, 307
magnetoresistive, 65, 95, 118, 131, 148
memory, 43, 52, 53, 64–69, 72, 73, 83, 88, 95, 96, 98–102, 104, 107, 108, 110, 111, 119, 148, 151, 157, 164, 176, 184, 284, 307, 312
metal-air, 263, 264, 276
metal carbide, 12, 15, 16, 311
metal-ion, 264, 268, 269
metallic, 1, 5, 6, 13, 14, 21, 34, 35, 50, 78, 80, 94–96, 110, 142, 149, 150, 153, 158, 183, 200, 222, 223, 249, 252, 264, 266–268, 272, 279, 296, 304, 310, 314–316
metal-sulfur, 264, 267, 273
meta-materials, 249, 250, 252
microbial, 24, 25, 207
microstrip, 216, 224, 246–249, 253–257
mixed metal oxide, 12
MOKE, 120, 121, 123–125, 149
MOSFET, 50, 98
MRAM, 64–67, 71, 73, 94, 95, 98, 99, 103, 104, 106–108, 110, 111
MRI, 89, 91, 117, 128, 138, 143, 144
MTJ, 50, 57, 63–67, 69–73, 94–111, 129, 306
MTMs, 252, 253
multiferroic, 164, 165, 169–172, 174, 175, 177, 314

N

nanocatalysts, 2, 3, 13, 21, 28
nanocomposites, 11, 12, 17, 19, 20, 23–25, 28, 185, 201, 207, 211, 231–234, 292, 295, 304
nanodevices, 34, 42, 152, 157
nanoparticles, 1–6, 10–15, 17–21, 23–26, 28, 78, 79, 88, 89, 117, 120, 122, 125, 126, 128–134, 138–144, 149, 152, 166, 174, 184, 185, 187–190, 196, 197, 199, 200, 207, 208, 211, 229, 230, 236, 237, 269–272, 277, 285, 286, 290, 295, 304, 310
nanosensors, 117, 120, 129, 130, 134
nanowire, 78–87, 89
neutron diffraction, 163, 168, 169
NMR, 120, 129, 133, 134
nonvolatile memory, 71, 77, 88
nuclear magnetic resonance, 120, 129, 133

O

one-dimensional, 83, 84, 152
optical, 14, 21, 34–36, 39, 40, 43, 44, 72, 116, 118, 120, 128, 138, 142, 167, 173, 201, 218, 230, 269, 285, 304, 314
oxides, 1, 5, 6, 12–14, 16, 19, 21, 129, 164, 165, 197, 200, 230, 270, 272, 278, 284–286, 294, 297, 298, 303, 312, 316

P

pathogens, 133, 207
permeability, 116, 183–185, 191, 216, 222, 223, 225, 233, 235, 246, 249, 252, 253
permittivity, 184, 191, 217, 220–222, 225, 233–235, 246, 249, 252, 253, 256
pesticides, 199, 201, 207, 208, 211
pharmaceutical, 20, 28, 206, 207, 211, 230
pollutants, 13, 19, 21, 23, 138, 199, 200, 203, 204, 206, 209–211
pseudocapacitor, 288, 289

Q

quantum dots, 17, 149, 152, 153, 278, 304

R

radome, 216–232, 235
Raman, 36–38, 131, 172, 201
refrigerator, 163, 177

S

sensor, 98, 116–118, 120, 122, 124–127, 131–133, 184, 225
silicene, 53, 55, 56
SOFC, 163, 177
sol-gel, 3, 5, 16, 22, 27, 78–80, 133, 165, 166, 199, 233, 234, 264, 265, 270, 285, 286
solid oxide fuel cells, 177
solvothermal, 16, 22, 26, 27, 199, 264, 270, 271, 274, 285, 286
spin, 39, 40, 43, 44, 50–55, 57–60, 63, 65, 67, 70–72, 80, 81, 84, 93–97, 99–110, 119, 121, 125–127, 132, 141, 143, 148–154, 156–158, 164, 165, 169, 171–177, 200, 304–116
spin-transistors, 53, 57
spintronics, 34, 40, 42, 43, 50, 53, 57, 58, 63, 66, 72, 93, 97, 109, 148–153, 155–158, 164, 304–308, 310–312, 314, 316
spin-tunnel, 52
STT-MRAM, 64, 66–71, 73, 94, 96, 98–102, 104–111
supercapacitors, 42, 230, 284–290, 314

T

telecommunication, 219
thin films, 39, 41, 79, 117, 119, 122, 125, 165, 167, 168, 177, 264, 286, 306, 307, 312, 313
three-dimensional, 108, 300
tunnel, 50, 52, 53, 55, 57, 64–66, 69–71, 94, 95, 97–99, 102, 103, 106, 107, 119, 129, 148, 305, 306
tunneling, 42, 64, 95, 97, 98, 101, 105, 106, 118, 119, 121, 148, 150, 304, 306, 310
two-dimensional, 42–44, 149, 155

V

vapor, 15, 41, 78, 79, 152, 167, 168, 177, 199, 264, 285, 286, 307

X

X-ray, 151, 168–170, 201

Z

Zn-air, 266, 277, 292